Lecture Notes in Computer Science 8054

Commenced Publication in 1973
Founding and Former Series Editors:
Gerhard Goos, Juris Hartmanis, and Jan van Leeuwen

Kaustubh Joshi   Markus Siegle
Mariëlle Stoelinga   Pedro R. D'Argenio (Eds.)

Quantitative Evaluation of Systems

10th International Conference, QEST 2013
Buenos Aires, Argentina, August 27-30, 2013
Proceedings

 Springer

Volume Editors

Kaustubh Joshi
AT&T Labs Research
180 Park Avenue, Building 103, Florham Park, NJ 07932, USA
E-mail: kaustubh@research.att.com

Markus Siegle
Universität der Bundeswehr München, Institut für Technische Informatik
Werner-Heisenberg-Weg 39, 85577 Neubiberg, Germany
E-mail: markus.siegle@unibw.de

Mariëlle Stoelinga
University of Twente
Faculty of Electrical Engineering, Mathematics and Computer Science
Drienerlolaan 5, 7522 NB Enschede, The Netherlands
E-mail: m.i.a.stoelinga@utwente.nl

Pedro R. D'Argenio
Universidad Nacional de Córdoba – CONICET
Facultad de Matemáticas, Astronomía y Física
Medina Allende s/n, X5000HUA Córdoba, Argentina
E-mail: dargenio@famaf.unc.edu.ar

ISSN 0302-9743   e-ISSN 1611-3349
ISBN 978-3-642-40195-4   e-ISBN 978-3-642-40196-1
DOI 10.1007/978-3-642-40196-1
Springer Heidelberg Dordrecht London New York

Library of Congress Control Number: 2013944564

CR Subject Classification (1998): F.1, F.3, C.4, D.2, G.3, C.2, D.3, F.4, J.7

LNCS Sublibrary: SL 1 – Theoretical Computer Science and General Issues

Typesetting: Camera-ready by author, data conversion by Scientific Publishing Services, Chennai, India

Printed on acid-free paper

Springer is part of Springer Science+Business Media (www.springer.com)

Preface

Welcome to QEST 2013, the International Conference on Quantitative Evaluation of SysTems. QEST is a leading forum on quantitative evaluation and verification of computer systems and networks, and celebrated its 10th anniversary this year. QEST was first held in Enschede, The Netherlands, in 2004, followed by meetings in Turin, Italy, in 2005, Riverside, USA, in 2006, Edinburgh, UK, in 2007, St. Malo, France, in 2008, Budapest, Hungary, in 2009, Williamsburg, USA, in 2010, Aachen, Germany, in 2011, and, most recently, London, UK, in 2012.

This year's QEST was held in Buenos Aires, Argentina, and collocated with the 24th International Conference on Concurrency Theory (CONCUR 2013), the 11th International Conference on Formal Modeling and Analysis of Timed Systems (FORMATS 2013), and the 8th International Symposium on Trustworthy Global Computing (TGC 2013).

As one of the premier fora for research on quantitative system evaluation and verification of computer systems and networks, QEST covers topics including classical measures involving performance, reliability, safety, correctness, and security. QEST welcomes measurement-based as well as analytic studies, and is also interested in case studies highlighting the role of quantitative evaluation in the design of systems. Tools supporting the practical application of research results in all of the above areas are of special interest, and tool papers are highly sought as well. In short, QEST aims to encourage all aspects of work centered around creating a sound methodological basis for assessing and designing systems using quantitative means.

The program for the 2013 edition of QEST was curated with the help of an international Program Committee (PC) of experts from 15 countries. We received a total of 52 submissions from 19 countries spanning five continents. Each submission was reviewed by at least four reviewers, either PC members or external reviewers. The reviews were submitted electronically followed by several rounds of discussions to reach consensus on acceptance decisions. In the end, 21 full papers and nine tool demonstration papers were selected.

The program was greatly enriched with the invited talks of Lorenzo Alvisi (joint invited speakers with CONCUR 2013), Gilles Barthe, and Edmundo de Souza e Silva. The program was completed with two tutorials preceding the main conference. They were presented by Diego Garbervetsky and Marco Vieira. We believe the outcome was a high-quality conference program of interest to QEST attendees and other researchers in the field.

We would like to thank a number of people. First of all, all authors who submitted their work to QEST: no papers, no conference! We are indebted to the PC members and additional reviewers for their thorough and valuable reviews. We also thank the Tools Chair Kai Lampka, the Tutorials Chair Lijun Zhang, the local Organization Chair Hernán Melgratti, the Publicity Chair Damián Barsotti, the Proceedings Chair Nicolás Wolovick, and the Steering Committee Chair Joost-Pieter Katoen, for their dedication and excellent work. In addition, we thank the Facultad de Ciencias Económicas of the University of Buenos Aires for providing the venue location. Furthermore, we gratefully acknowledge the financial support of the Consejo Nacional de Investigaciones Científicas y Técnicas (CONICET), the Agencia Nacional de Promoción Científica y Tecnológica (through the RC program of FONCYT and FONSOFT), and the EU FP7 grant agreement 295261 MEALS (Mobility between Europe and Argentina applying Logics to Systems).

Finally, we are also grateful to Andrei Voronkov for providing us with his conference software system EasyChair, which was extremely helpful for the PC discussions and the production of the proceedings.

August 2013

Kaustubh Joshi
Markus Siegle
Mariëlle Stoelinga
Pedro R. D'Argenio

Organization

General Chair

Pedro R. D'Argenio   Universidad Nacional de Córdoba, Argentina

Program Committee Co-chairs

Kaustubh Joshi   AT&T Florham Park, USA
Markus Siegle   Bundeswehr University of Munich, Germany
Mariëlle Stoelinga   University of Twente, The Netherlands

Steering Committee

Nathalie Bertrand   INRIA Rennes, France
Peter Buchholz   TU Dortmund, Germany
Susanna Donatelli   Università di Torino, Italy
Holger Hermanns   Saarland University, Germany
Joost-Pieter Katoen   RWTH Aachen University, Germany
Peter Kemper   College of William and Mary, USA
William Knottenbelt   Imperial College London, UK
Andrew S. Miner   Iowa State University, USA
Gethin Norman   University of Glasgow, UK
Gerardo Rubino   INRIA Rennes, France
Miklos Telek   Technical University of Budapest, Hungary

Program Committee

Jonatha Anselmi   Basque Center for Applied Mathematics, Spain
Christel Baier   Technical University of Dresden, Germany
Nathalie Bertrand   INRIA Rennes Bretagne Atlantique, France
Andrea Bobbio   Universitá del Piemonte Orientale, Italy
Peter Buchholz   TU Dortmund, Germany
Hector Cancela   Universidad de la República, Argentina
Gluliano Casale   Imperial College London, UK
Gianfranco Ciardo   University of California at Riverside, USA
Yuxin Deng   Shanghai Jiao Tong University, China
Derek Eager   University of Saskatchewan, Canada
Jane Hillston   University of Edinburgh, UK
Andras Horvath   University of Turin, Italy
David N. Jansen   Radboud Universiteit, The Netherlands

Kaustubh Joshi	AT&T Labs Research, USA
Krishna Kant	George Mason University, USA
Peter Kemper	College of William and Mary, USA
Marta Kwiatkowska	University of Oxford, UK
Boris Köpf	IMDEA Software Institute, Spain
Kai Lampka	Uppsala University, Sweden
Annabelle McIver	Macquarie University, Australia
Arif Merchant	Google, USA
Gethin Norman	University of Glasgow, UK
Anne Remke	University of Twente, The Netherlands
William Sanders	University of Illinois at Urbana-Champaign, USA
Roberto Segala	University of Verona, Italy
Markus Siegle	Universität der Bundeswehr München, Germany
Marielle Stoellinga	University of Twente, The Netherlands
Miklos Telek	Budapest University of Technology and Economics, Hungary
Bhuvan Urgaonkar	Pennsylvania State University, USA
Aad Van Moorsel	University of Newcastle, UK
Marco Vieira	University of Coimbra, Portugal
Verena Wolf	Saarland University, Germany

Additional Reviewers

Alhakami, Hind	Jin, Xiaoqing
Andreychenko, Alexander	Kolesnichenko, Anna
Angius, Alessio	Legay, Axel
Ballarini, Paolo	Mereacre, Alexandru
Bernardo, Marco	Mikeev, Linar
Bortolussi, Luca	Mumme, Malcolm
Crocce, Fabián	Parker, David
de Boer, Pieter-Tjerk	Perez, Juan F.
Delahaye, Benoit	Rabehaja, Tahiry
Diciolla, Marco	Randour, Mickael
Dräger, Klaus	Sandmann, Werner
Feng, Lu	Santinelli, Luca
Feng, Yuan	Song, Lei
Galpin, Vashti	Spieler, David
Ghasemieh, Hamed	Sproston, Jeremy
Gimbert, Hugo	Sundararaman, Akshay
Giusto, Álvaro	Ujma, Mateusz
Grampin, Eduardo	Waliji, Muhammad
Hahn, Ernst Moritz	Wang, Weikun
Hedin, Daniel	Zhao, Yang
Horvath, Illes	Zonouz, Saman

Table of Contents

Session 4: Tool Demos I

Session 5: Model Checking and Systems

Session 6: Systems

Session 7: Tools Demos II

Session 8: Control and Games

Session 9: Timed Automata and Simulation

Computer-Aided Security Proofs

Gilles Barthe

IMDEA Software Institute

Probabilistic programs provide a convenient formalism for defining probability distributions and have numerous applications in computer science. In particular, they are used pervasively in code-based provable security for modeling security properties of cryptographic constructions as well as cryptographic assumptions. Thanks to their well-defined semantics, probabilistic programming languages provide a natural framework to prove the correctness of probabilistic computations. Probabilistic program logics are program logics that allow to reason formally about executions of probabilistic programs, and can be used to verify complex probabilistic algorithms.

However, these program logics cannot be used to reason about the security of cryptographic constructions. Indeed, cryptographic proofs are reductionist, in the sense that they show that the probability that an adversary breaks the security of the cryptographic system in "reasonable time" is "small", provided the probability that a probabilistic algorithm solves a computationally intractable problem in "reasonable time" is also "small". Such reductionist arguments fall out of the scope of traditional program logics, that reason about properties of program executions. They can be captured by relational program logics that reason about properties of executions of two programs—or as a special case two executions of the same program.

CertiCrypt [5] and EasyCrypt [4] are computer-assisted frameworks for verifying relational properties of a core programming language with sequential composition, conditionals, loops, procedure calls, deterministic assignments and probabilistic assignments drawn from discrete distributions. Both tools implement mechanisms for deriving valid judgments in a relational program logic, and for carrying common (algebraic, arithmetic, information-theoretic, ...) forms of reasoning that arise in cryptographic proofs. Over the last few years, we have used CertiCrypt and EasyCrypt to prove security of many cryptographic constructions. Using extensions of the program logic to reason about approximate relational judgments, we have additionally verified differentially private computations [6].

EasyCrypt is accessible to working cryptographers. However, proofs must be built interactively, which is time consuming and requires some familiarity with formal verification. Although fully automated security analyses are not possible in general, we have been exploring two scenarios for which it is possible to generate proofs automatically. In [1], we have instrumented a zero-knowledge compiler with a proof generation mechanism: given a high-level proof goal, the compiler automatically generates a zero-knowledge protocol that realizes the goal and proofs in CertiCrypt and EasyCrypt that the protocol complies with some standard properties of soundness, completeness, and zero knowledge. In [3], we present the ZooCrypt framework, which supports automated analyses of

K. Joshi et al. (Eds.): QEST 2013, LNCS 8054, pp. 1–2, 2013.
© Springer-Verlag Berlin Heidelberg 2013

public-key encryption schemes built from one-way trapdoor permutations and random oracles. ZooCrypt implements automated procedures for proving the security of such constructions, or for discovering attacks. Using ZooCrypt, we have analyzed over a million (automatically generated) schemes, including many schemes from the literature.

Our most recent work intends to accomodate real-world descriptions of cryptographic constructions in machine-checked security proofs. Many practical attacks exploit implementation details, for instance error management or message formatting, that are typically not considered in pen-and-paper provable security proofs. Pleasingly, EasyCrypt provides effective mechanisms for managing the complexity of cryptographic proofs, and allows to build rigorous security proofs for realistic descriptions of cryptographic standards [2].

More information about the project is available from the project web page

$$\text{http://www.easycrypt.info}$$

References

1. Almeida, J.B., Barbosa, M., Bangerter, E., Barthe, G., Krenn, S., Zanella-Béguelin, S.: Full proof cryptography: verifiable compilation of efficient zero-knowledge protocols. In: ACM Conference on Computer and Communications Security, pp. 488–500. ACM (2012)
2. Almeida, J.B., Barbosa, M., Barthe, G., Dupressoir, F.: Certified computer-aided cryptography: efficient provably secure machine code from high-level implementations. Cryptology ePrint Archive, Report 2013/316 (2013)
3. Barthe, G., Crespo, J.M., Grégoire, B., Kunz, C., Lakhnech, Y., Schmidt, B., Zanella-Béguelin, S.: Automated analysis and synthesis of padding-based encryption schemes. Cryptology ePrint Archive, Report 2012/695 (2012)
4. Barthe, G., Grégoire, B., Heraud, S., Béguelin, S.Z.: Computer-aided security proofs for the working cryptographer. In: Rogaway, P. (ed.) CRYPTO 2011. LNCS, vol. 6841, pp. 71–90. Springer, Heidelberg (2011)
5. Barthe, G., Grégoire, B., Zanella-Béguelin, S.: Formal certification of code-based cryptographic proofs. In: 36th ACM SIGPLAN-SIGACT Symposium on Principles of Programming Languages, POPL 2009, pp. 90–101. ACM, New York (2009)
6. Barthe, G., Köpf, B., Olmedo, F., Zanella-Béguelin, S.: Probabilistic relational reasoning for differential privacy. In: 39th ACM SIGPLAN-SIGACT Symposium on Principles of Programming Languages, POPL 2012, pp. 97–110. ACM, New York (2012)

On the Interplay between Content Popularity and Performance in P2P Systems*

Edmundo de Souza e Silva[1], Rosa M.M. Leão[1],
Daniel Sadoc Menasché[1], and Antonio A. de A. Rocha[2]

[1] Federal University of Rio de Janeiro, Rio de Janeiro, Brazil
[2] Fluminense Federal University, Niteroi, Brazil
{edmundo,rosam,sadoc}@land.ufrj.br,
arocha@ic.uff.br

Abstract. Peer-to-peer swarming, as used by BitTorrent, is one of the
de facto solutions for content dissemination in today's Internet. By leveraging resources provided by users, peer-to-peer swarming is a simple
and efficient mechanism for content distribution. In this paper we survey recent work on peer-to-peer swarming, relating content popularity to
three performance metrics of such systems: fairness, content availability/
self-sustainability and scalability.

1 Introduction

Peer-to-peer (P2P) swarming system has been, in recent years, one of the most
successful architectures used to share information and has been adopted by numerous publishers [1, 21]. The success of the architecture can be evinced by its
utilization in real-world applications. For instance, updates of the game *World
of Warcraft* can be obtained via P2P at the site of the company responsible
for the game (Blizzard) [3] and Ubuntu Linux distributions are made available
via P2P [21]. In addition, the Wikipedia is considering the use of P2P to share
videos [22]. Another evidence of the success of P2P architectures is the large
volume of traffic generated by the applications that employ P2P. This traffic
grew approximately 12% between 2009 and 2010 [19].

P2P architectures have been studied for over a decade and there is a multitude
of performance models that address the efficiency of the approach. Most work
in the literature focuses on understanding the BitTorrent (BT) dynamics and
the impact of several aspects of its protocol on performance. For instance, the
impact of incentive mechanisms, free riding, heterogeneous and homogeneous
peer download/upload rates, file dissemination, etc. [24]. However, fundamental
issues are still open such as those related to the system's scalability. Examples
of questions that one may ask are: how does the system scale with the number
of users? What are the main parameters that influence the system's ability to
scale?

---

* This research is sponsored in part by grants from CAPES, CNPq and FAPERJ.

K. Joshi et al. (Eds.): QEST 2013, LNCS 8054, pp. 3–21, 2013.

It should be clear that content popularity severely impacts P2P performance and its ability to scale. Peers in a swarm act both as consumers as well as content servers. Intuitively, the more popular a certain content is, the more efficient should the P2P system be as there are more "resources" available to the user's swarm. Suppose a single publisher is available to serve a given content. Does the swarm depend on the publisher to retrieve the content? How large should the swarm be in order for the content to be available among peers?

Unpopular contents rarely find a hand full of interested users trying to access the same information at a given instant of time. Consequently, the publisher of the content is most likely the only resource from which peers can download the information at any time. If the publisher becomes unavailable, peers can't conclude their downloads. If there is a large population of requesters for unpopular contents, the publisher may be overloaded with requests for different files and the system would not scale with increasing user population. Quantifying the dependence of peers on publishers allows us to evaluate the performance impact of increasing the content catalog size and to understand the scalability constraints.

As content popularity plays a key role in the P2P system performance, an important problem is to characterize the empirical probability density function of the number of peers in a swarm. The large measurement study reported in [8] addresses this problem. That work indicates that, for the vast majority of the datasets studied, around 40% to 70% of the swarms have only 3 or less peers and more than 70% of the swarms are of size smaller than 10.

We have also conducted measurement studies. We developed a crawler to monitor one of the most popular Torrent Search Engines (Torlock.com) aimed at determining the daily size of all swarms announced in the website (around 150,000 swarms) for ten consecutive days. The swarm size was given by the number of connected peers (both seeders and leechers, i.e., peers that have all the content or just part of it, respectively).

Figure 1 shows the empirical complementary cumulative distribution of swarm sizes for the ten days monitored. We only consider swarms with, at least, one seed. These constitutes a total of approximately 130,000 swarms which corresponds to more than 85% of the collected measurements. Let $f(x, y)$ be the fraction of swarms with at least y peers at day x. Each color in the palette of

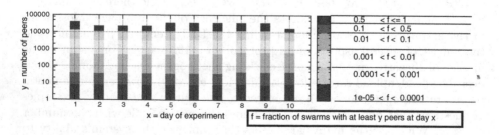

Fig. 1. Swarm sizes: Torlock.com, 10-day monitoring

Figure 1 characterizes a range of values for $f(x, y)$. Figure 1 indicates that a very small fraction of swarms are huge, with more than 30.000 peers, while most of the swarms are very small, that is, approximately 73% of the total are formed by less than 10 peers and 58% have less than 5 peers. From Figure 1 it is clear that one must study the performance of P2P systems for a wide range of swam sizes, from very large sizes (tens of thousands of peers) to those with around half a dozen of peers.

In this work, we focus on the the performance impact that content popularity has on the performance of P2P systems. We study three popularity ranges: low, medium and high. For contents that have low popularity, we survey recent results that show that users in a swarm may suffer from high variability concerning the time to retrieve the content. For contents that have low to medium popularity, we study the dependence that peers have on the publisher to retrieve the desired content. The less dependent the swarm is on the publisher, the more scalable the system is, since the load at the publisher should only marginally increase with the demand for content.

For highly popular contents one may expect that the system perfectly scales with increasing number of users. We elaborate on this issue and try to determine if the access rate for a given popular content can increase without bound or if there are limitations. In order to facilitate the organization of our presentation, we include Table 1 that presents an heuristic summary of the tradeoffs we study.

Table 1. Heuristic summary of tradeoffs in the peer-to-peer system metrics considered in this paper, as a function of content popularity

Content popularity (compared to service capacity of publisher)	Fairness Issues (see §2 and [17])	Content Availability Issues (see §3 and [13, 14, 16])	Scalability Issues (see §4 and [15])
small	✗	✗	
medium		✗	
very large			✗

We start by discussing, in Section 2 a fairness problem that may occur when content is unpopular. We show that peers might experience heterogeneous download times even in homogeneous BitTorrent swarms. It has been observed through measurement studies reported in the literature that a vast number of swarms suffers from unavailability. This motivates the models for content availability presented in Section 3. The first model indicates that, when publishers are intermittent, combining K contents in a single swarm file (bundling) increases content availability exponentially as function of K. We also estimate the dependence of peers on a stable publisher, which is useful for provisioning purposes as well as in deciding how to bundle. We present a metric which is closely related to availability, referred to as swarm self-sustainability, and present a model that yields swarm self-sustainability as a function of the file size, popularity and service capacity of peers. Finally, in Section 4, we consider very popular contents, and study the fundamental scalability limits of peer-to-peer swarming systems in this regime where the behavior of such systems might resemble that of client-server systems.

2   Fairness

In this section we study BitTorrent fairness. In the networking literature, fairness has mostly been studied in the context of bandwidth allocation and different definitions of this metric are considered in [25], such as *global proportional fairness* and *pairwise proportional fairness*. In the first case, for instance, the system is *fair* when the download capacity of a peer equalizes the overall contribution of that peer to the system. In the second case, fairness is achieved when a peer i allocates bandwidth to another peer j based on the service received from j. The BT chocking algorithm is a variation of this last notion of fairness.

In the literature of peer-to-peer systems, most works are devoted to study the behavior of large swarms and the effect of heterogeneous download rates on performance (e.g., [4]). In such cases, BT systems are reasonably fair and peers with the same characteristics behave similarly and get approximately the same amount of bandwidth [10]. Nonetheless, in smaller swarms the behavior might be radically different, and fairness issues are a concern as illustrated in the next experiment.

2.1   Unfairness Observed through an Experiment

Next, we illustrate the unfairness problem through a controlled experiment deployed at a private swarm at PlanetLab (Figure 2). The arrival rate of peers is chosen to maintain the swarm with a small number of peers. Let c_S and c_l be the upload capacities of the publisher and of the peers, respectively. The parameters for this experiment are: arrival rate equal to 8.0×10^{-3} peers per second, $c_S = c_l = 50$ KBps, and the file size is 20 MB. Each peer is represented in the figure by a line. Each line starts when a peer arrives, and ends when a peer concludes its download and departs. The first two peers arrive at approximately the same time, followed by another couple of peers. The fifth peer (marked in blue) arrived much later than the others, but experienced a much smaller download time compared to the first peers. Note that all the first five peers to arrive depart approximately at the same time (as compared to the file download time) which shows the high variability of the download times. This is explained by content synchronizations that occur among peers and are the root causes of unfairness. In what follows, we introduce a simple model that allow us to obtain the download rates of peers in a swarm as a function of time, and in turn explains why these synchronizations occur.

2.2   Why Unfairness?

We consider swarms in which peers have approximately the same upload and (relatively large) download capacities and are interested in the same content.

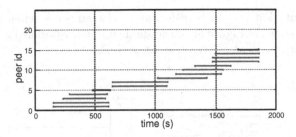

Fig. 2. Experimental results with PlanetLab. Each line in the figure starts when a peer arrives, and ends when a peer departs.

Such swarms are referred to as *homogeneous swarms*. As illustrated in Section 2.1, even in homogeneous swarms, BT protocol may lead to unfairness problems. Murai *et al* [17] observed that the peers download times can be highly variable and their arrival order influences the rate at which service is provided to each. In addition, peers in a swarm may be *content-synchronized*, that is, their current downloaded content may become identical during the download which results in serious performance degradation. This behavior is predominant in swarms with a small peer population, which in turn correspond to a large fraction of the BitTorrent swarms (see Figure 1). In what follows we discuss the model of [17] that sheds light into the unfairness issue.

Let us consider a swarm with a few *homogeneous peers*, a single seed and the BT protocol to exchange content. We first observe that, in a small swarm, it is usually the case that any peer can upload to all the others. In particular, the tit-for-tat and optimistic unchoke mechanisms of BT do not come into play if the swarm size is smaller than the maximum number of upload connections (typical of small swarms). The rarest-first policy is used for block selection.

Let $b_i(t)$ be the amount of content downloaded by peer i during interval $(0, t)$. Let $N(t)$ be the number of peers in a swarm at time t. Note that, if $b_j(t) > b_i(t)$, then peer j has at least $b_j(t) - b_i(t)$ pieces that i does not have. Assuming infinite upload capacities, let $u'_{ij}(t)$ be the rate at which peer i uploads content to peer j at time t. Accordingly, let $u_{ij}(t)$ be the corresponding rate when upload capacity constraints are considered.

The following assumptions are considered: (a) the content is divided into tiny pieces with respect to the total file size (fluid approximation); (b) due to the rarest-first policy, the (infinitesimal) pieces currently downloaded by a peer i from the (single) publisher are not present in the swarm and, as such, are of immediate interest to all other peers; (c) if $b_k(t) > b_j(t) > b_i(t)$ then peer k has pieces that both j and i do not have. Similarly to assumption (b), the pieces peer k sends to i at t are also of immediate interest to j due to the rarest-first policy; (d) the publisher capacity is equally divided among all the peers in the swarm; (e) peers have infinite download capacities.

In what follows, for notational convenience, we drop the dependence of the model variables on time. In addition, also for notational convenience, at any point in time, we index peers in an order such that if $i > j$ then $b_i < b_j$. Figure 3(a) illustrates the notation and the transfer rates between peers, when four peers are present in the system. In Figure 3(a), the single publisher (labeled S) sends pieces to each of the four peers at a given instant of time.

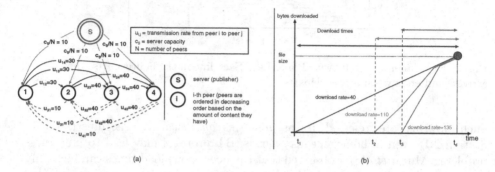

Fig. 3. (a) Upload rates between peers (4 peers); (b) download rates versus time (3 peers)

Next, we compute the upload rates from each peer to its neighbors. Recall that $u'_{ij}(t)$ is the potential transmission rate from i to j provided that there is no limit to the upload capacity of i. We compute u'_{ij} and u_{ij} in lexicographical order as follows.

From assumptions (b) and (c) above, if $b_j > b_i$, then j can potentially transmit at infinite rate to i since it has pieces that i does not have. In contrast, the only pieces from i that may interest j are those that i is receiving from the seed and from other peers that have more pieces than j (i.e., from peers in set $\{k : b_k > b_j\}$). Then,

$$
u'_{ij} = \begin{cases} \infty, & \text{if } b_i > b_j \quad\quad (1a) \\ \dfrac{c_S}{N} + \displaystyle\sum_{b_k > b_j} u_{ki}, & \text{if } b_i \leq b_j \quad (1b) \end{cases}
$$

We must now impose the restriction on the upload rates in order to obtain u_{ij}. We follow a max-min fair *progressive filling algorithm* to allocate bandwidth. Let $u'_{ik} = \min_l \{u'_{il}\}$. The transmission rate from i to every other peer is uniformly increased until either (a) it reaches u'_{ik} or (b) the upload capacity of peer i is exhausted. In the first case, the spare capacity is distributed by i in the same manner as described above, among all peers except k, and so on (refer to [17] for details).

In scenarios where the publisher offers a large catalog of files, the amount of bandwidth allocated to each swarm is small. In what follows we consider one such scenario, and assume $c_S < c_l$, letting $c_S = 40$ and $c_l = 90$. Figure 3(a) shows, for a swarm consisting of four peers, the upload rates obtained using the bandwidth allocation algorithm described above.

The download rate of peer i is $d_i = \sum_{j=1}^{N} u_{ji}$. For the example in Figure 3(a), $d_1 = 40$, $d_2 = d_3 = d_4 = 120$. Note that peers $2, 3$ and 4 download at a much higher rate than peer 1. This indicates that our model captures the unfair behavior referred to in the beginning of the section. We also observe the total download rate when more than one peer is present is larger than that of the publisher, which shows the benefits of the P2P architecture.

The download rates can be easily calculated as a function of time. To this aim, assume we are given the arrival times of peers to the swarm. Then, using equation (1) and the algorithm described above we can obtain $d_i(t)$ and $b_i(t)$. Figure 3(b) shows how $d_i(t)$ varies over time for a swarm consisting of three peers, letting $c_S = 40$ and $c_l = 90$. In this figure, peer 1 is the first to arrive at time t_1, and subsequently peers 2 and 3 arrive at t_2 and t_3, respectively. We note that peers 2 and 3 are able to download at a much faster rate than that of the first peer to arrive until they synchronize in the amount of content they downloaded. From this moment up to download completion, the rates equalize. At time t_d all content is downloaded and the peers may depart. Once again, the model supports the key observation made in Section 2.1: peers that have identical characteristics may expect different download times.

3   Content Availability and Self-sustainability

Despite the tremendous success of BitTorrent, it suffers from a fundamental problem: content unavailability [16]. Although peer-to-peer swarming in BitTorrent scales impressively to tolerate massive flash crowds for popular content, swarming does little to disseminate unpopular content as the availability of such unpopular content is limited by the presence of a seed or publisher. The extent of publisher unavailability can be severe. In addition to the statistics already presented at Section 1, according to measurements in Menasché et al. [16] 40% of swarms have no publishers available more than 50% of the time and according to Kaune et al. [9], in the absence of publishers, 86% of peers are unable to reconstruct their files.

To appreciate the availability problem, consider a swarm for an episode of a popular TV show. When a publisher first posts the episode, a flash crowd of peers joins the swarm to download the content. The original publisher goes offline at some point, but peers may continue to obtain the content from other peers while the swarm is active. If a peer arrives after the initial popularity wave, when the population of the swarm has dwindled to near-zero, it may likely find the content unavailable and must wait until a publisher reappears.

We review mathematical models to study content availability in swarming systems such as BitTorrent. First, we consider the single content case and assume that there is a publisher that may go offline. We present the simplest version of the availability model and, similarly to Section 2, we use a fluid approximation and assume that the file is divided into an infinite number of chunks. In addition, we assume that, whenever two peers meet, they have useful content to share (Section 3.1).

In the second half of Section 3.1 we assume that there is a stable publisher that is always online. Clearly the content is always available in this case, but we focus on the dependence of the swarm on the publisher and introduce a new metric to measure the availability among peers. We refer to the fraction of time that content is available among peers (without considering the publisher) as *self-sustainability*. In this case we relax the fluid approximation to account for the fact that a file is divided into a finite number of chunks. The model yields self-sustainability as a function of content popularity, the number of chunks in the file and peer capacities. Finally, we consider the multiple content case, and study the implications of bundling and playlists on content availability and self-sustainability (Section 3.2).

3.1   Single Content

Availability. We start by defining when we consider a content to be available.

Definition 1. *The system availability (A) is the fraction of time at which there is at least one publisher in the system or the peer population contiguously remains at least equal to a given threshold coverage t after all publishers depart.*

In what follows, to simplify presentation we consider a threshold coverage of 1, meaning that when all publishers depart, as far as the population of peers contiguously remains greater than or equal to 1, the content is available. This assumption is clearly a rough simplification, which we adopt to facilitate the explanation while preserving the main points we want to emphasize. The assumption is removed in [16].

We use an $M/G/\infty$ queue to model the self-scaling property of BitTorrent swarms, i.e., more peers bring in more capacity to the system. (In section 4 we discuss scalability limits, but for now we assume that the system scales without bound.)

The key insight is to model uninterrupted intervals during which the content is available as busy periods of that queue. The busy period increases exponentially with the arrival rate of peers and publishers and with the time spent by peers and publishers in the swarm (for details see [16]).

The scenario consists of a single publisher that distributes a file of size s and has service capacity μ. The publisher resides in the system for an interval exponentially distributed with mean s/μ and then departs. Assume peers take on average s/μ to complete their downloads and arrive according to a Poisson process with rate λ peers/s. A busy period is a contiguous period when at least one peer is online. Let B be the average available period length, $B = (e^{\lambda s/\mu} - 1)/\lambda$.

If publishers are intermittent (see Figure 4(a)), arriving according to a Poisson process with rate r, and have the same mean residence time as peers, the system passes through idle periods, during which content is unavailable, and busy periods, during which content is available. The average duration of each

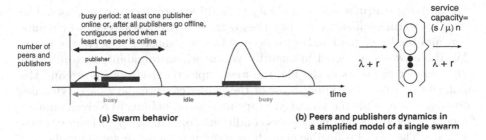

(a) Swarm behavior

(b) Peers and publishers dynamics in a simplified model of a single swarm

Fig. 4. Quantifying unavailability

busy period, B, corresponds to the busy period of the $M/G/\infty$ queue illustrated in Figure 4(b), and equals

$$B = \frac{e^{(\lambda+r)s/\mu} - 1}{(\lambda + r)} \tag{2}$$

Let A be the system availability. The unavailability is $1 - A$, and equals

$$1 - A = \frac{\text{length of idle period}}{\text{length of idle period} + \text{length of busy period}} = \frac{1/r}{1/r + B} = \frac{1}{1 + rB} \tag{3}$$

Finally, the mean download time of peers, D, can be approximated as the mean idle waiting time plus the active download time,

$$D \approx (1 - A)/r + s/\mu \tag{4}$$

Equations (3) and (4) show that both the unavailability and the expected download time decrease exponentially with content popularity. It is then expected that highly popular contents do not need the publisher to download. But what about unpopular contents? How much load the publisher is expect to receive from peers? In what follows we address these questions.

Self-sustainability. For unpopular content, peers will rely on publishers in order to complete their downloads. We investigate this dependence of peers on a publisher [14]. We consider a scenario in which each swarm includes one stable publisher that is always online and ready to serve content. The corresponding system is henceforth referred to as a hybrid peer-to-peer system, since peers can always rely on the publisher if they cannot find blocks of the content among themselves.

Definition 2. *The swarm self-sustainability, $\mathcal{A}$, is the fraction of time during which the swarm is self-sustaining, that is, it is the steady-state probability that the peers collectively have the entire file (each block in the file is present at one or more peers). If all blocks are available among peers, the swarm is referred to as self-sustaining.*

Quantifying swarm self-sustainability is useful for provisioning purposes. The larger the swarm's self-sustainability, the lower the dependency of peers on a publisher, and the lower the bandwidth needed by the publisher to serve the peers.

We use a two-layer model to quantify swarm self-sustainability as a function of the number of blocks in the file, the mean upload capacity of peers and the popularity of a file. The upper layer of the model captures how user dynamics evolve over time, while the lower layer captures the probability of a given number of blocks being available among peers conditioned on a fixed upper layer population state. The model is flexible enough to account for large or small numbers of blocks in the file, heterogeneous download times for different blocks, and peers residing in the system after completing their downloads.

In [14], closed-form expressions are derived for the distribution of the number of blocks available among the peers and an efficient algorithm is obtained to compute the swarm self-sustainability. We survey these results that show that self-sustainability increases as a function of the number of blocks in the file and investigate the minimum popularity needed to attain a given self-sustainability level.

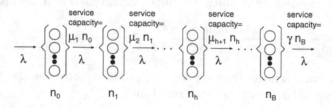

Fig. 5. User dynamics. In stage h, there are n_h users, each user owning h blocks, $0 \le h \le B$.

A file consists of B blocks. Requests for a file arrive according to a Poisson process with rate λ. We further assume that the time required for a user to download its jth block is a random variable with mean $1/\mu_j$, $1 \le j \le B$. After completing their downloads, peers remain in the system for mean time $1/\gamma$.

The user dynamics is modeled with $(B+1)$ M/G/∞ queues in series. Each of the first B M/G/∞ queues models the download of a single block, and capture the self-scaling property of BitTorrent swarms, i.e., each peer brings one unit of service capacity to the system. The last queue captures the residence time of seeds (see Figure 5).

The system population state is characterized by a $(B+1)$-tuple, $\mathbf{n} = (n_0, n_1, n_2, \ldots, n_B)$, where n_h represents the number of customers in queue h, i.e., the number of users that have downloaded h blocks of the file, $0 \le h \le B$. Let $\pi(n_0, \ldots, n_B)$ be the joint steady state population probability distribution, $\pi(n_0, \ldots, n_B) = P(\mathbf{N} = (n_0, \ldots, n_B))$, of finding n_h users in the hth queue, $0 \le h \le B$, and let $\pi_h(n_h) = P(N_h = n_h)$, $h = 0, \ldots, B$, be the corresponding marginal probability. The steady state distribution of the queueing system has the following product form,

$$\pi(n_0, \ldots, n_{B-1}, n_B) = \prod_{h=0}^{B} \pi_h(n_h) = \frac{(\lambda/\gamma)^{n_B}}{n_B!} e^{-(\lambda/\gamma)} \prod_{h=0}^{B-1} [\frac{\rho^{n_h}}{n_h!} e^{-\rho}] \qquad (5)$$

where $\rho = \lambda/\mu$ is the *load* of the system.

We now describe the lower layer of the model. Given the current population state, $\mathbf{n} = (n_0, \ldots, n_B)$, our goal is to determine the distribution of the number of blocks available among the peers. We state our key modeling assumption: in steady state, the set of blocks owned by a randomly selected user in stage h is chosen uniformly at random among the $\binom{B}{h}$ possibilities and independently among users (uniform and independent block allocation).

A user u in stage h, $0 \leq h \leq B$, has a signature $s_{h,u} \in \{0,1\}^B$, defined as a B bit vector where the ith bit is set to 1 if the user has block i and 0 otherwise. Each user in stage h owns h blocks and has one of $\binom{B}{h}$ possible signatures.

Under the uniform and independent block allocation assumption, signatures are chosen uniformly at random and independently among users; the latter is clearly a strong assumption since in any peer-to-peer swarming system the signatures of users are correlated. Nevertheless, in [14] it is shown that the effect of such correlations on swarm self-sustainability is negligible in many interesting scenarios.

Let V denote the steady state number of blocks available among the peers. Denote by $p(v)$ the steady state probability that v blocks are available among the peers,

$$p(v) = P(V = v) = \sum_{\mathbf{n} \in \mathbb{N}^{B+1}} P(V = v | \mathbf{N} = \mathbf{n}) \pi(\mathbf{n}) \qquad (6)$$

Then,

$$\mathcal{A} = p(B) \qquad (7)$$

To illustrate the applicability of the model presented above, we use it to approximate the minimum load, ρ^*, necessary to attain a given self-sustainability level, $\mathcal{A}^*$, when $\gamma = \infty$, i.e., nodes leave immediately after getting the file (typical values of $1/\gamma$ in BitTorrent are less than 1 hour [20]). It is shown in [14] that if $\gamma = \infty$ the probability that a tagged block is unavailable among the peers is $q = \exp(-\rho(B-1)/2)$. It is also shown in [14] that for values of q close to 0 ($q \lesssim 0.01$), $p(B) \approx 1 + E[V] - B = 1 - Bq$. This approximation, in turn, can be used to select the load ρ^* to attain self-sustainability level $\mathcal{A}^*$,

$$\rho^* \approx [2\log(B/(1-\mathcal{A}^*))]/(B-1), \qquad \gamma = \infty \qquad (8)$$

Figure 6 shows the minimum load, ρ^*, necessary to achieve a given self-sustainability level, $\mathcal{A}^*$, ($\mathcal{A}^* = 0.8, 0.9, 0.99, 0.999$), for file sizes varying 2.5 to 256MB (B=10,...,1000 blocks) using (8). The figure indicates that the required load to attain a given level of self-sustainability significantly decreases as the file size increases. This, in turn, indicates that an unpopular large file (e.g., a movie) will be more available among peers than a smaller file with the same popularity (e.g., the subtitle of that movie) in case they are distributed in two isolated swarms.

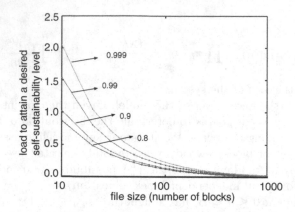

Fig. 6. Load necessary to attain self-sustainability as a function of file size

3.2 Multiple Contents

Bundling. A common strategy adopted in BitTorrent is bundling wherein, instead of disseminating individual files via isolated swarms, a publisher packages a number of related files and disseminates it via a single larger swarm [16]. Nowadays, more than a half of music content as well as nearly one quarter of TV shows are distributed in bundles. To appreciate why bundling improves content availability, consider a bundle of K files. Assume that the popularity of the bundle is roughly K times the popularity of an individual file, since a peer requesting any file requests the entire bundle. The size of the bundle is roughly K times the size of an individual file. Our model suggests that the busy period of the bundled swarm is a factor $\exp(K^2)$ larger than that of an individual swarm. Indeed, if busy periods supported by peers alone last until a publisher reappears, the content will be available throughout [7].

In some cases, improved availability can reduce the download time experienced by peers, i.e., peers can download more content in less time. The download time of peers in the system consists of the waiting time spent while content is unavailable and the service time spent in actively downloading content (see (4)). If the reduction in waiting time due to bundling is greater than the corresponding increase in service time, the download time decreases. This conclusion is validated in [16] through large-scale controlled experiments using the Mainline BitTorrent client over PlanetLab.

In what follows we quantify the implications of bundling on content availability. Bundling K files impacts the system parameters in two ways: (a) the arrival rate of peers to the bundle is K times the arrival rate of peers to the individual swarm, λ, since we assume that peers interested in any of the files download the entire bundle; (b) the size of bundle is K times the size of individual files, s, therefore the active download time of the bundle is K times longer than that of an individual file.

We use superscript (b) to denote bundled metrics. For instance, the mean duration of the bundled available periods is denoted by $B^{(b)}$. The bundled available period increases as $\Theta(\exp(K^2))$,

$$B^{(b)} = \frac{e^{(\bar{X}^{K\lambda}+r)\bar{s}^{\bar{X}Ks}/\mu} - 1}{(\bar{X}^{K\lambda}+r)} = \Theta(\exp(K^2)) \qquad (9)$$

Thus, bundled unavailability decreases as $\Theta(\exp(-K^2))$

$$1 - A^{(b)} = \Theta(\exp(-K^2)) \qquad (10)$$

Note that the asymptotic results concerning bundling (see (10)) can be linked to self-sustainability since the bundling parameter K can be related to the increase in the number of blocks in a file an the file size.

We present measurements on private swarms that corroborate the fact that bundling can decrease mean download times. Our experiments were conducted using approximately 200 PlanetLab hosts and two hosts at the University of Massachusetts at Amherst, one of which is designated as the controller of the experiment and another as a BitTorrent tracker. Figures 7(a)-(b) show peer arrivals and departures with time. Each horizontal line delineates the period a peer stays in the swarm. The publisher appears for a while and then it is turned off. Figure 7(a) shows that for $K = 1$, peers get *blocked* whenever the publisher goes offline and must wait until the publisher reappears in order to complete their downloads. On the other hand, when $K = 4$ (Figure 7(b)) nearly no blocking occurs. Figure 7(c) indicates that bundling also reduces the mean time to download content.

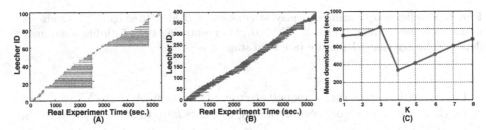

Fig. 7. Real experiment with bundling and an intermittent publisher: (a) Swarm dynamics for $K = 1$; (b) Swarm Dynamics for $K = 4$; (c) Download time with K

Implications of Playlists on Self-sustainability. Next, we consider streaming systems, such as Youtube or Last.fm. In this systems, users have the option to watch a channel or a playlist [5, 11, 23]. Channels or playlists are bundles of related content organized in a sequence. In order to simplify presentation, we assume that users always access the content of the playlists in the order at which they are setup in the playlist. Our goal is to compute the playlist self-sustainability, assuming that peers collaborate with each other while accessing a playlist, but that they also count on a stable publisher that is always online.

Consider a playlist consisting of N files (e.g., songs). The download and re-production of the i-th song lasts on average $1/\mu_i$ seconds. After listening to the i-th song, a user might depart from the system or start the streaming of the next song in the playlist. Requests to a playlist arrive to the system according to a Poisson process with rate λ peer/s. The users that listen to all songs can still remain in the system after listening to the last song, remaining online as seeds for an average of $1/\gamma$ seconds before departing. After listening to the i-th song, users might leave the system, which occurs with rate β_i, $1 \leq i \leq N$.

The model introduced in the previous subsection can be easily adapted to compute the playlist self-sustainability. We assume that users have enough storage capacity to store all the content that they watched in a playlist. In that way, users streaming the i-th content can collaborate with users streaming contents 0 up to $i-1$.

We model a playlist as a series of $(N+1)$ M/G/∞ queues, where queue j characterizes the number of users that are streaming the j-th content (see Figure 8). Users at the j-th queue store $j-1$ contents. As we assume that users will always access content in order, the identity of the contents stored by users at the j-th queue is determined.

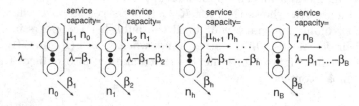

Fig. 8. Playlist user dynamics. A playlist consists of B files. In stage h, there are n_h users, each user owning h files, $0 \leq h \leq B$. After watching their h-th file, users may depart. The aggregate departure rate from stage h is β_h.

Let Λ_k be the arrival rate of requests to the k-th content, $\Lambda_k = \lambda - \sum_{i=1}^{k} \beta_i$. Let δ_k be the mean number of peers streaming contents k up to N, plus the seeders, $\delta_k = \left(\sum_{j=k}^{N} \Lambda_j / \mu_j \right) + \Lambda_{N+1} / \gamma$. Then, the probability that the k-th content is replicated at r_k users is $P(R_k = r_k) = (e^{-\delta_k} \delta_k^{r_k}) / r_k!$ and the probability that song k is available among users, $\mathcal{A}_k$, is the self-sustainability of the k-th content in the playlist,

$$\mathcal{A}_k = 1 - P(R_k = 0) = 1 - e^{-\delta_k} \tag{11}$$

Equation (11) gives the playlist self-sustainability as a function of its popularity, the song durations and the rate at which peers prematurely depart from the system. Publishers might also use (11) in order to decide how to construct their playlists, accounting for user preferences as well as content self-sustainability. The analysis above can be adapted to account for different configurations, such as: (1) playlists whose contents are dynamically ordered according to the system state; (2) users that decide to stream a single content and then depart and; (3) users that have limited caches.

4   Scalability

Swarming is a powerful, simple and scalable solution for content dissemination in today's Internet. However, there is a limit on the extent to which swarming can scale. In this section our goal is to answer the following question: what are the fundamental scalability limits of peer-to-peer swarming systems? To that aim, we consider peer-to-peer swarming systems in the limit when the arrival rate is very high in relation to the service capacity of a stable publisher assumed to be always online in the system.

We illustrate the fundamental limitations of the scalability of peer-to-peer swarming system through an analogy, involving Newton's third law. According to Newton's law, there is a linear relationship between the acceleration and the force, and the constant that relates the two is the mass, $F = ma$. Accordingly, in peer-to-peer swarming systems we have $T = N\eta$, where T is the system throughput (peer departure rate), and N is the number of peers in the system. The constant η characterizes the system efficiency, and is henceforth assumed to be equal to one. As the capacity of the system scales with the number of peers, peer-to-peer swarming system subject to a Poisson arrival process can be well approximated by an $M/G/\infty$ model, i.e., there is no queueing in the system. System capacity perfectly scales with the number of peers in the system, as illustrated in the top three curves of Figure 10(b), to be detailed in what follows.

Let us now consider a peer-to-peer swarming system in which peers depart as soon as they conclude their downloads. Let U be the service capacity of the single stable publisher. In physics, as the speed approaches the speed of light, the mass changes and $F = ma$ does not hold. Accordingly, in peer-to-peer swarming systems, when the arrival rate λ is larger than U, the relationship $T = N\eta$ does not hold anymore. In essence, this occurs because when $\lambda > U$ peers bring resources such as bandwidth and memory to the system, but such resources cannot be fully utilized. In this regime, the system behaves as a client-server system. The amount of resources increases linearly with the population size, but the throughput does not scale accordingly. Peers frequently have no content to exchange with their neighbors, and resources remain idle until a useful encounter takes place. In physics, if the speed approaches the speed of light, we need to rely on the theory of relativity. Accordingly, we need different models in order to understand the behavior of peer-to-peer swarming systems in face of extreme popularities.

4.1   Missing Piece Syndrome and the Most Deprived Peer Selection First

In swarming systems, files are divided into pieces. Eventually, one piece might become very rare. If peers immediately depart after obtaining a content, they can leave before helping other peers receive the rare piece, and the population of peers that does not have that piece will grow unboundedly. This is referred to as the *missing piece syndrome* [6], and the population of peers that does not have the rare (missing) piece is referred to as the *one club*. As mentioned above, the system behaves like a client-server system, because newcomers rapidly join the one club, and they can only leave the system after receiving the missing piece from the publisher. As soon as they receive the piece, they depart without contributing. If $\lambda > U$, the arrival rate of peers to the one club is larger than the departure rate, and the system is unstable.

Whenever a peer has spare service capacity, it needs to make the following two decisions: whom to contact? and; what piece to transmit? In the literature, it has been assumed that peers go through random encounters and that peers are paired uniformly at random [6,18,26]. Nonetheless, if peers can strategically select their neighbors it is possible to show that the capacity region of the system increases. To illustrate that claim, we consider a publisher that adopts the *most deprived peer selection first*.

Most Deprived First. According to the most deprived peer selection first, the publisher prioritizes the transmission to peers that have the fewest number of chunks of the file. The rationale consists of sending rare chunks to peers that will remain the longest time in the system, and who can potentially contribute most.

We now reason that if the file is divided into K chunks, the publisher adopts most deprived peer selection and rarest-first piece selection and if peers adopt random peer and random useful piece selection, the maximum achievable throughput is upper bounded by KU (see [12] for details). Let $\lambda > KU$ and let **n** be the system's state which represents the number of peers that have a given signature (a bit vector). First, we note that all states **n** are achievable. Eventually, the system reaches a state in which a large number of peers have all pieces except a tagged one. These peers are also referred to as *one-club peers* (see Figure 9).

As a consequence of the random peer selection adopted by peers, if the one-club is large enough then gifted peers (i.e., peers that have the missing piece) will transmit content only to one-club peers, with high probability. As shown next, if $\lambda > KU$ the one club grows unboundedly. Therefore, the effect of transmissions from gifted peers to members outside the one club reduces with time, and does not affect the maximum achievable throughput. For this reason, henceforth we neglect arrow (a) in Figure 9.

All uploads from the stable publisher are to newcomers, a fraction U/λ of which effectively receive pieces from the publisher. Each peer that receives a piece from the publisher has an additional expected lifetime of $(K-1)/\mu$. During this time, it will serve on average $K-1$ peers from the one-club, who will then leave the system. Therefore, the population of the one-club decreases at a rate of $U(K-1)$, and increases at a rate of $\lambda - U$. Hence, the total departure rate of peers is upper bounded by $U(K-1) + U = UK$ and if $\lambda < UK$ the system is stable.

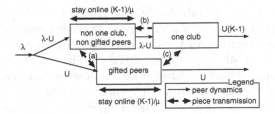

Fig. 9. Publisher adopts most deprived peer strategy

4.2 Closed System Analysis

Next, to analyze the throughput limits of swarming systems we consider a Markovian model of the peer-to-peer system such that every time a peer leaves a new one immediately arrives. This system has a fixed population size, N, and is referred to as a closed system.

Figure 10(a) plots the throughput as a function of the population size, for different publisher capacities U (varied between 0.5 and 1 blocks/s) and publisher strategies. Peers follow random peer, random useful piece selection. Figure 10 shows that the throughput obtained when publishers adopt rarest piece/most deprived peer selection is greater than that obtained with each of the other two strategies. The figure also shows that for large population sizes, the throughput of rarest first/random peer and random useful piece/random peer are roughly the same.

Figure 10(b) shows results for the case where peers reside in the system as seeds after completing their downloads. The parameters are the same as those used to generate 10(a). Let $1/\gamma$ be the mean time that peers reside in the system after completing their downloads. Note that if $\gamma = 1/U - 1$ the throughput increases with the population size and the system is scalable. As γ increases the throughput decreases, $\gamma = \infty$ corresponding to the scenario shown in Figure 10(a).

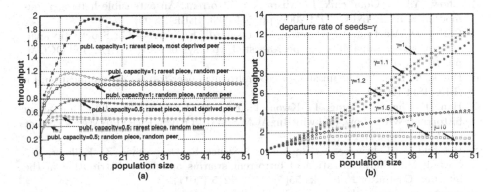

Fig. 10. System throughput (U=1) (a) immediate departures (γ=∞) (b) with lingering (γ<∞)

Since the work by Hajek and Zhu [6], different authors have considered the fundamental scalability limits of peer-to-peer swarming systems, under different assumptions. Accounting for the fact that the network might be the system bottleneck, Baccelli et al. [2] show conditions under which peer-to-peer networks might be super-scalable. Considering a non work-conserving system, i.e., that peers might prefer not to transmit content even in face of opportunities, Oguz and Anantharam [18] present a scheduling strategy according to which the system is always scalable even if peers leave the system immediately after completing their downloads. Finally, considering a collection of files, Zhu et al. [26] show conditions under which the system is stable if peers are willing to download content that they did not initially request, using a strategy that is similar in spirit to bundling.

5   Conclusion

Our contribution in this paper is to show how content popularity impacts different performance metrics of peer-to-peer swarming systems such as fairness, content availability, download time and scalability. Content popularity, in turn, is influenced by social networks and recommendations, among other factors. We believe that the understanding of how content popularity affects quality of service metrics is important not only to provision network capacity, but also to issue content recommendations and to decide how to organize content into bundles or playlists.

References

1. Amazon: Using BitTorrent with Amazon S3, http://aws.amazon.com/
2. Baccelli, F., Mathieu, F., Norros, I., Varloot, R.: Can p2p networks be super-scalable? In: IEEE INFOCOM (2013)
3. Blizzard: World of Warcraft, http://us.battle.net/wow
4. Chow, A.L.H., Golubchik, L., Misra, V.: Bittorrent: An extensible heterogeneous model. In: IEEE INFOCOM 2009, pp. 585–593. IEEE (2009)
5. Ciullo, D., Martina, V., Garetto, M., Leonardi, E., Torrisi, G.: Stochastic analysis of self-sustainability in peer-assisted vod systems. In: IEEE INFOCOM, pp. 1539–1547 (2012)
6. Hajek, B., Zhu, J.: The missing piece syndrome in peer-to-peer communication. In: IEEE ISIT (2010)
7. Han, J., Kim, S., Chung, T., Kwon, T., Kim, H., Choi, Y.: Bundling practice in bittorrent: What, how, and why. In: ACM SIGMETRICS/PERFORMANCE Joint Conference, pp. 77–88 (2012)
8. Hossfeld, T., Lehrieder, F., Hock, D., Oechsner, S., Despotovic, Z., Kellerer, W., Michel, M.: Characterization of bittorrent swarms and their distribution in the internet. Computer Networks 55(5), 1197–1215 (2011)
9. Kaune, S., Cuevas, R., Tyson, G., Mauthe, A., Guerrero, C., Steinmetz, R.: Unraveling bittorrent's file unavailability: Measurements, analysis and solution exploration. arXiv:0912.0625v1 (2009)

10. Liao, W., Papadopoulos, F., Psounis, K.: Performance analysis of bittorrent-like systems with heterogeneous users. Performance Evaluation 64(9-12), 876–891 (2007)
11. Melo, C., Oliveira, J., da Fonseca, N.: Promotion of content availability by playlist viewers in cdn-p2p systems. In: IEEE ICC (2013)
12. Menasché, D.S., de A. Rocha, A.A., de Souza e Silva, E., Leão, R.M.M., Towsley, D.: Stability of peer-to-peer swarming systems. In: SBRC, pp. 161–174 (2012)
13. Menasché, D.S., de A. Rocha, A.A., de Souza e Silva, E., Leão, R.M.M., Towsley, D., Venkataramani, A.: Modeling chunk availability in peer-to-peer swarming systems. ACM SIGMETRICS Performance Evaluation Review 37(2), 30–32 (2009)
14. Menasché, D.S., de A. Rocha, A.A., de Souza e Silva, E., Leão, R.M.M., Towsley, D., Venkataramani, A.: Estimating self sustainability in peer-to-peer systems. Performance Evaluation 67(11), 1243–1258 (2010)
15. Menasché, D.S., de A. Rocha, A.A., de Souza e Silva, E., Towsley, D., Leão, R.M.M.: Implications of peer selection strategies by publishers on the performance of p2p swarming systems. ACM SIGMETRICS Performance Evaluation Review 39(3), 55–57 (2011)
16. Menasché, D.S., de A. Rocha, A.A., Li, B., Towsley, D., Venkataramani, A.: Content availability and bundling in swarming systems. In: CONEXT (2009)
17. Murai, F., de A. Rocha, A.A., Figueiredo, D., de Souza e Silva, E.: Heterogeneous download times in a homogeneous bittorrent swarm. Computer Networks 56, 1983–2000 (2012)
18. Oguz, B., Anantharam, V., Norros, I.: Stable, distributed p2p protocols based on random peer sampling. In: 50th Allerton Conf. on Comm., Control and Comput., pp. 915–919. IEEE (2012)
19. Otto, J.S., Sanchez, M.A., Choffnes, D.R., Bustamante, F.E., Siganos, G.: On blind mice and the elephant – understanding the network impact of a large distributed system. In: SIGCOMM (2011)
20. Pouwelse, J., Garbacki, P., Epema, D.H.J., Sips, H.J.: The bittorrent P2P file-sharing system: Measurements and analysis. In: van Renesse, R. (ed.) IPTPS 2005. LNCS, vol. 3640, pp. 205–216. Springer, Heidelberg (2005)
21. Ubuntu: Download Ubuntu using BitTorrent, http://torrent.ubuntu.com:6969/
22. Wikipedia. Wikipedia Is Using BitTorrent P2P for HTML5 Video, http://gigaom.com/wikipedia-is-using-bittorrent-p2p-for-html5-video-2/
23. Wu, D., Liu, Y., Ross, K.: Queueing network models for multi-channel p2p live streaming systems. In: IEEE INFOCOM (2009)
24. Xia, R.L., Muppala, J.: A survey of bittorrent performance. IEEE Communications Surveys & Tutorials 12(2), 140–158 (2010)
25. Yang, X., de Veciana, G.: Performance of peer-to-peer networks: Service capacity and role of resource sharing policies. Performance Evaluation 63, 175–194 (2006)
26. Zhou, X., Ioannidis, S., Massoulié, L.: On the stability and optimality of universal swarms. ACM SIGMETRICS Performance Evaluation Review 39(1), 301–312 (2011)

Refinement and Difference for Probabilistic Automata

Benoît Delahaye[1], Uli Fahrenberg[1], Kim Guldstrand Larsen[2], and Axel Legay[1]

[1] INRIA/IRISA, France
{benoit.delahaye,ulrich.fahrenberg,axel.legay}@inria.fr
[2] Aalborg University, Denmark
kgl@cs.aau.dk

Abstract. This paper studies a difference operator for stochastic systems whose specifications are represented by Abstract Probabilistic Automata (APAs). In the case refinement fails between two specifications, the target of this operator is to produce a specification APA that represents all witness PAs of this failure. Our contribution is an algorithm that allows to approximate the difference of two deterministic APAs with arbitrary precision. Our technique relies on new quantitative notions of distances between APAs used to assess convergence of the approximations as well as on an in-depth inspection of the refinement relation for APAs. The procedure is effective and not more complex than refinement checking.

1 Introduction

Probabilistic automata as promoted by Segala and Lynch [37] are a widely-used formalism for modeling systems with probabilistic behavior. These include randomized security and communication protocols, distributed systems, biological processes and many other applications. Probabilistic model checking [23,5,41] is then used to analyze and verify the behavior of such systems. Given the prevalence of applications of such systems, probabilistic model checking is a field of great interest. However, and similarly to the situation for non-probabilistic model checking, probabilistic model checking suffers from *state space explosion*, which hinders its applicability considerably.

One generally successful technique for combating state space explosion is the use of *compositional* techniques, where a (probabilistic) system is model checked by verifying its components one by one. This compositionality can be obtained by *decomposition*, that is, to check whether a given system satisfies a property, the system is automatically decomposed into components which are then verified. Several attempts at such automatic decomposition techniques have been made [10,28], but in general, this approach has not been very successful [9].

As an alternative to the standard model checking approaches using logical specifications, such as e.g. LTL, MITL or PCTL [33,3,20], automata-based specification theories have been proposed, such as Input/Output Automata [31], Interface Automata [11], and Modal Specifications [29,34,6]. These support composition *at specification level*; hence a model which naturally consists of a composition of several components can be verified by model checking each component on its own, against its own specification. The overall model will then automatically satisfy the composition of the component specifications. Remark that this solves the decomposition problem mentioned above: instead

K. Joshi et al. (Eds.): QEST 2013, LNCS 8054, pp. 22–38, 2013.

of trying to automatically decompose a system for verification, specification theories make it possible to verify the system without constructing it in the first place.

Moreover, specification theories naturally support *stepwise refinement* of specifications, i.e. iterative implementation of specifications, and *quotient*, i.e. the synthesis of missing component specifications given an overall specification and a partial implementation. Hence they allow both logical and compositional reasoning at the same time, which makes them well-suited for compositional verification.

For probabilistic systems, such automata-based specification theories have been first introduced in [25], in the form of Interval Markov Chains. The focus there is only on refinement however; to be able to consider also composition and conjunction, we have in [7] proposed Constraint Markov Chains as a natural generalization which uses general constraints instead of intervals for next-state probabilities.

In [14], we have extended this specification theory to probabilistic automata, which combine stochastic and non-deterministic behaviors. These *Abstract Probabilistic Automata* (APA) combine modal specifications and constraint Markov chains. Our specification theory using APA should be viewed as an alternative to classical PCTL [20], probabilistic I/O automata [32] and stochastic extensions of CSP [21]. Like these, its purpose is model checking of probabilistic properties, but unlike the alternatives, APA support compositionality at specification level.

In the context of refinement of specifications, it is important that informative debugging information is given in case refinement fails. We hence need to be able to compare APA at the semantic level, i.e. to capture the *difference between their sets of implementations*. This is, then, what we attempt in this paper: given two APAs N_1 and N_2, to generate another APA N for which $[\![N]\!] = [\![N_1]\!] \setminus [\![N_2]\!]$ (where $[\![N]\!]$ denotes the set of implementations of N).

As a second contribution, we introduce a notion of *distance* between APAs which measures how far away one APA is from refining a second one. This distance, adapted from our work in [39,6], is *accumulating* and *discounted*, so that differences between APAs accumulate along executions, but in a way so that differences further in the future are discounted, i.e. have less influence on the result than had they occurred earlier.

Both difference and distances are important tools to compare APAs which are not in refinement. During an iterative development process, one usually wishes to successively replace specifications by more refined ones, but due to external circumstances such as e.g. cost of implementation, it may happen that a specification needs to be replaced by one which is not a refinement of the old one. This is especially important when models incorporate quantitative information, such as for APAs; the reason for the failed refinement might simply be some changes in probability constraints due to e.g. measurement updates. In this case, it is important to assess precisely *how much* the new specification differs from the old one. Both the distance between the new and old specifications, as well as their precise difference, can aid in this assessment.

Unfortunately, because APAs are finite-state structures, the difference between two APAs cannot always itself be represented by an APA. Instead of extending the formalism, we propose to *approximate* the difference for a subclass of APAs. We introduce both over- and under-approximations of the difference of two *deterministic* APAs. We construct a sequence of under-approximations which converges to the exact difference, hence eventually capturing all PAs in $[\![N_1]\!] \setminus [\![N_2]\!]$, and a fixed over-approximation which may capture also PAs which are not in the exact difference, but whose distance to the exact difference is zero: hence any superfluous PAs which are captured by the over-approximation are infinitesimally close to the real difference. Taken together, these approximations hence solve the problem of assessing the precise difference between deterministic APAs in case of failing refinement.

We restrict ourselves to the subclass of deterministic APAs, as it allows syntactic reasoning to decide and compute refinement. Indeed, for deterministic APAs, syntactic refinement coincides with semantic refinement, hence allowing for efficient procedures. Note that although the class of APAs we consider is called "deterministic", it still offers non-determinism in the sense that one can choose between different actions in a given state. For space reasons, detailed proofs and additional comments are given in [13].

Related Work.   This paper embeds into a series of articles on APA as a specification theory [14,15,16]. In [14] we introduce deterministic APA, generalizing earlier work on interval-based abstractions of probabilistic systems [18,25,26], and define notions of refinement, logical composition, and structural composition for them. We also introduce a notion of *compositional abstraction* for APA. In [15] we extend this setting to non-deterministic APA and give a notion of (lossy) determinization, and in [16] we introduce the tool APAC. The distance and difference we introduce in the present paper complement the refinement and abstraction from [14].

Compositional abstraction of APA is also considered in [38], but using a different refinement relation. Differences between specifications are developed in [35] for the formalism of modal transition systems, and distances between specifications, in the variant of weighted modal automata, have been considered in [6]. Distances between probabilistic systems have been introduced in [12,17,40].

The originality of our present work is, then, the ability to measure how far away one probabilistic specification is from being a refinement of another, using distances and our new difference operator. Both are important in assessing precisely how much one APA differs from another.

2   Background

Let $Dist(S)$ denote the set of all discrete probability distributions over a finite set S and $\mathbb{B}_2 = \{\top, \bot\}$.

Definition 1.   *A probabilistic automaton (PA) [37] is a tuple* (S, A, L, AP, V, s_0), *where* S *is a finite set of states with the initial state* $s_0 \in S$, A *is a finite set of actions,* L: $S \times A \times Dist(S) \to \mathbb{B}_2$ *is a (two-valued) transition function,* AP *is a finite set of atomic propositions and* $V: S \to 2^{AP}$ *is a state-labeling function.*

Consider a state s, an action a, and a probability distribution μ. The value of $L(s, a, \mu)$ is set to $\top$ in case there exists a transition from s under action a to a distribution μ on successor states. In other cases, we have $L(s, a, \mu) = \bot$. We now introduce Abstract Probabilistic Automata (APA) [14], that is a specification theory for PAs. For a finite set S, we let $C(S)$ denote the set of constraints over discrete probability distributions on S. Each element $\varphi \in C(S)$ describes a set of distributions: $Sat(\varphi) \subseteq Dist(S)$. Let $\mathbb{B}_3 = \{\top, ?, \bot\}$. APAs are formally defined as follows.

Definition 2. *An APA [14] is a tuple* (S, A, L, AP, V, S_0)*, where* S *is a finite set of states,* $S_0 \subseteq S$ *is a set of initial states,* A *is a finite set of actions, and* AP *is a finite set of atomic propositions.* $L : S \times A \times C(S) \to \mathbb{B}_3$ *is a three-valued distribution-constraint function, and* $V : S \to 2^{2^{AP}}$ *maps each state in* S *to a set of admissible labelings.*

APAs play the role of specifications in our framework. An APA transition abstracts transitions of a certain unknown PA, called its implementation. Given a state s, an action a, and a constraint φ, the value of $L(s, a, \varphi)$ gives the modality of the transition. More precisely, the value $\top$ means that transitions under a must exist in the PA to some distribution in $Sat(\varphi)$; ? means that these transitions are allowed to exist; $\bot$ means that such transitions must not exist. We will sometimes view L as a *partial* function, with the convention that a lack of value for a given argument is equivalent to the $\bot$ value. The function V labels each state with a subset of the powerset of AP, which models a disjunctive choice of possible combinations of atomic propositions. We say that an APA $N = (S, A, L, AP, V, S_0)$ is in *Single Valuation Normal Form* (SVNF) if the valuation function V assigns at most one valuation to all states, i.e. $\forall s \in S, |V(s)| \leq 1$. From [14], we know that every APA can be turned into an APA in SVNF with the same set of implementations. An APA is *deterministic* [14] if (1) there is at most one outgoing transition for each action in all states, (2) two states with overlapping atomic propositions can never be reached with the same transition, and (3) there is only one initial state.

Note that every PA is an APA in SVNF where all constraints represent single-point distributions. As a consequence, all the definitions we present for APAs in the following can be directly extended to PAs.

Let $N = (S, A, L, AP, V, \{s_0\})$ be an APA in SVNF and let $v \subseteq AP$. Given a state $s \in S$ and an action $a \in A$, we will use the notation $\mathsf{succ}_{s,a}(v)$ to represent the set of potential a-successors of s that have v as their valuation. Formally, $\mathsf{succ}_{s,a}(v) = \{s' \in S \mid V(s') = \{v\}, \exists \varphi \in C(S), \mu \in Sat(\varphi) : L(s, a, \varphi) \neq \bot, \mu(s') > 0\}$. When clear from the context, we may use $\mathsf{succ}_{s,a}(s')$ instead of $\mathsf{succ}_{s,a}(V(s'))$. Remark that when N is deterministic, we have $|\mathsf{succ}_{s,a}(v)| \leq 1$ for all s, a, v.

3 Refinement and Distances between APAs

We introduce the notion of refinement between APAs. Roughly speaking, refinement guarantees that if A_1 refines A_2, then the set of implementations of A_1 is included in the one of A_2. We first recall the notion of simulation $\Subset_{\mathcal{R}}$ between two given distributions.

Definition 3 ([14]). *Let* S *and* S' *be non-empty sets, and* μ*,* μ' *be distributions;* $\mu \in Dist(S)$ *and* $\mu' \in Dist(S')$*. We say that* μ *is simulated by* μ' *with respect to a relation* $\mathcal{R} \subseteq S \times S'$ *and a correspondence function* $\delta : S \to (S' \to [0, 1])$ *iff*
1. for all $s \in S$ *with* $\mu(s) > 0$*,* $\delta(s)$ *is a distribution on* S'*,*
2. for all $s' \in S'$*,* $\sum_{s \in S} \mu(s) \cdot \delta(s)(s') = \mu'(s')$*, and*

3. whenever $\delta(s)(s') > 0$, then $(s, s') \in \mathcal{R}$.

We write $\mu \in_{\mathcal{R}}^{\delta} \mu'$ if μ is simulated by μ' w.r.t $\mathcal{R}$ and δ, and $\mu \in_{\mathcal{R}} \mu'$ if there exists δ with $\mu \in_{\mathcal{R}}^{\delta} \mu'$.

We will also need distribution simulations without the requirement of a relation $\mathcal{R} \subseteq S \times S'$ (hence also without claim 3 above); these we denote by $\mu \in^{\delta} \mu'$.

Definition 4 ([14]). Let $N_1 = (S_1, A, L_1, AP, V_1, S_0^1)$ and $N_2 = (S_2, A, L_2, AP, V_2, S_0^2)$ be APAs. A relation $\mathcal{R} \subseteq S_1 \times S_2$ is a refinement relation if and only if, for all $(s_1, s_2) \in \mathcal{R}$, we have $V_1(s_1) \subseteq V_2(s_2)$ and
1. $\forall a \in A, \forall \varphi_2 \in C(S_2)$, if $L_2(s_2, a, \varphi_2) = \top$, then $\exists \varphi_1 \in C(S_1) : L_1(s_1, a, \varphi_1) = \top$ and $\forall \mu_1 \in Sat(\varphi_1), \exists \mu_2 \in Sat(\varphi_2)$ such that $\mu_1 \in_{\mathcal{R}} \mu_2$,
2. $\forall a \in A, \forall \varphi_1 \in C(S_1)$, if $L_1(s_1, a, \varphi_1) \neq \bot$, then $\exists \varphi_2 \in C(S_2)$ such that $L_2(s_2, a, \varphi_2) \neq \bot$ and $\forall \mu_1 \in Sat(\varphi_1), \exists \mu_2 \in Sat(\varphi_2)$ such that $\mu_1 \in_{\mathcal{R}} \mu_2$.

We say that N_1 refines N_2, denoted $N_1 \preceq N_2$, iff there exists a refinement relation such that $\forall s_0^1 \in S_0^1, \exists s_0^2 \in S_0^2 : (s_0^1, s_0^2) \in \mathcal{R}$. Since any PA P is also an APA, we say that P satisfies N (or equivalently P implements N), denoted $P \models N$, iff $P \preceq N$. In [14], it is shown that for deterministic APAs N_1, N_2, we have $N_1 \preceq N_2 \iff [\![N_1]\!] \subseteq [\![N_2]\!]$, where $[\![N_i]\!]$ denotes the set of implementations of APA N_i. Hence for deterministic APAs, the difference $[\![N_1]\!] \setminus [\![N_2]\!]$ is non-empty iff $N_1 \npreceq N_2$. This equivalence breaks for non-deterministic APAs [14], whence we develop our theory only for deterministic APAs.

To show a convergence theorem about our difference construction in Sect. 4.2 below, we need a relaxed notion of refinement which takes into account that APAs are a *quantitative* formalism. Indeed, refinement as of Def. 4 is a purely qualitative relation; if both $N_2 \npreceq N_1$ and $N_3 \npreceq N_1$, then there are no criteria to compare N_2 and N_3 with respect to N_1, saying which one is the closest to N_1. We provide such a relaxed notion by generalizing refinement to a *discounted distance* which provides precisely such criteria. In Sect. 4.2, we will show how those distances can be used to prove that increasingly precise difference approximations between APAs converge to the real difference. The next definition shows how a distance between states is lifted to a distance between constraints.

Definition 5. Let $d : S_1 \times S_2 \to \mathbb{R}^+$ and $\varphi_1 \in C(S_1)$, $\varphi_2 \in C(S_2)$ be constraints in N_1 and N_2. Define the distance D_{N_1,N_2} between φ_1 and φ_2 as follows:

$$D_{N_1,N_2}(\varphi_1, \varphi_2, d) =$$

$$\sup_{\mu_1 \in Sat(\varphi_1)} \left[\inf_{\mu_2 \in Sat(\varphi_2)} \left(\inf_{\delta : \mu_1 \in^{\delta} \mu_2} \sum_{(s_1, s_2) \in S_1 \times S_2} \mu_1(s_1) \delta(s_1)(s_2) d(s_1, s_2) \right) \right]$$

For the definition of d below, we say that states $s_1 \in S_1$, $s_2 \in S_2$ are *not compatible* if either (1) $V_1(s_1) \neq V_2(s_2)$, (2) there exists $a \in A$ and $\varphi_1 \in C(S_1)$ such that $L_1(s_1, a, \varphi_1) \neq \bot$ and for all $\varphi_2 \in C(S_2), L_2(s_2, a, \varphi_2) = \bot$, or (3) there exists $a \in A$ and $\varphi_2 \in C(S_2)$ such that $L_2(s_2, a, \varphi_2) = \top$ and for all $\varphi_1 \in C(S_1), L_1(s_1, a, \varphi_1) \neq \top$. For compatible states, their distance is similar to the accumulating branching distance on modal transition systems as introduced in [6,39], adapted to our formalism. In the rest of the paper, the real constant $0 < \lambda < 1$ represents a discount factor. Formally, $d : S_1 \times S_2 \to [0,1]$ is the least fixpoint to the following system of equations:

$$d(s_1, s_2) = \tag{1}$$

$$\begin{cases} 1 \text{ if } s_1 \text{ is not compatible with } s_2 \\ \max \begin{cases} \max\limits_{\{a,\varphi_1 : L_1(s_1,a,\varphi_1) \neq \perp\}} \min\limits_{\{\varphi_2 : L_2(s_2,a,\varphi_2) \neq \perp\}} \lambda D_{N_1,N_2}(\varphi_1, \varphi_2, d) \\ \max\limits_{\{a,\varphi_2 : L_2(s_2,a,\varphi_2) = \top\}} \min\limits_{\{\varphi_1 : L_1(s_1,a,\varphi_1) = \top\}} \lambda D_{N_1,N_2}(\varphi_1, \varphi_2, d) \end{cases} \text{ otherwise} \end{cases}$$

Since the above system of linear equations defines a *contraction*, the existence and uniqueness of its least fixpoint is ensured, cf. [30]. This definition intuitively extends to PAs, which allows us to propose the two following notions of distance:

Definition 6. *Let* $N_1 = (S_1, A, L_1, AP, V_1, S_0^1)$ *and* $N_2 = (S_2, A, L_2, AP, V_2, S_0^2)$ *be APAs in SVNF. The* syntactic *distance and* thorough *distances between* N_1 *and* N_2 *are defined as follows:*

- **Syntactic distance.** $d(N_1, N_2) = \max_{s_0^1 \in S_0^1} \left(\min_{s_0^2 \in S_0^2} d(s_0^1, s_0^2) \right)$.
- **Thorough distance.** $d_t(N_1, N_2) = \sup_{P_1 \in [\![N_1]\!]} \left(\inf_{P_2 \in [\![N_2]\!]} d(P_1, P_2) \right)$.

Note that the notion of thorough distance defined above intuitively extends to sets of PAs: given two sets of PAs $\mathbb{S}_1$, $\mathbb{S}_2$, we have $d_t(\mathbb{S}_1, \mathbb{S}_2) = \sup_{P_1 \in \mathbb{S}_1} \left(\inf_{P_2 \in \mathbb{S}_2} d(P_1, P_2) \right)$.

The intuition here is that $d(s_1, s_2)$ compares not only the probability distributions at s_1 and s_2, but also (recursively) the distributions at all states reachable from s_1 and s_2, weighted by their probability. Each step is discounted by λ, hence steps further in the future contribute less to the distance. We also remark that $N_1 \preceq N_2$ implies $d(N_1, N_2) = 0$. It can easily be shown, cf. [39], that both d and d_t are *asymmetric pseudometrics* (or *hemimetrics*), i.e. satisfying $d(N_1, N_1) = 0$ and $d(N_1, N_2) + d(N_2, N_3) \geq d(N_1, N_3)$ for all APAs N_1, N_2, N_3 (and similarly for d_t). The fact that they are only pseudometrics, i.e. that $d(N_1, N_2) = 0$ does not imply $N_1 = N_2$, will play a role in our convergence arguments later. The following proposition shows that the thorough distance is bounded above by the syntactic distance. Hence we can bound distances between (sets of) implementations by the syntactic distance between their specifications.

Proposition 1. *For all APAs* N_1 *and* N_2 *in SVNF, it holds that* $d_t(N_1, N_2) \leq d(N_1, N_2)$.

4   Difference Operators for Deterministic APAs

The difference $N_1 \setminus N_2$ of two APAs N_1, N_2 is meant to be a syntactic representation of *all counterexamples*, i.e. all PAs P for which $P \in [\![N_1]\!]$ but $P \notin [\![N_2]\!]$. We will see later that such difference cannot be an APA itself; instead we will *approximate* it using APAs.

Because N_1 and N_2 are deterministic, we know that the difference $[\![N_1]\!] \setminus [\![N_2]\!]$ is non-empty if and only if $N_1 \npreceq N_2$. So let us assume that $N_1 \npreceq N_2$, and let $\mathcal{R}$ be a maximal refinement relation between N_1 and N_2. Since $N_1 \not\preceq N_2$, we know that $(s_0^1, s_0^2) \notin \mathcal{R}$. Given $(s_1, s_2) \in S_1 \times S_2$, we can distinguish between the following cases:

1. $(s_1, s_2) \in \mathcal{R}$
2. $V_1(s_1) \neq V_2(s_2)$,
3. $(s_1, s_2) \notin \mathcal{R}$ and $V_1(s_1) = V_2(s_2)$, and

(a) there exists $e \in A$ and $\varphi_1 \in C(S_1)$ such that $L_1(s_1, e, \varphi_1) = \top$ and $\forall \varphi_2 \in C(S_2) : L_2(s_2, e, \varphi_2) = \bot$,

(b) there exists $e \in A$ and $\varphi_1 \in C(S_1)$ such that $L_1(s_1, e, \varphi_1) = ?$ and $\forall \varphi_2 \in C(S_2) : L_2(s_2, e, \varphi_2) = \bot$,

(c) there exists $e \in A$ and $\varphi_1 \in C(S_1)$ such that $L_1(s_1, e, \varphi_1) \geq ?$ and $\exists \varphi_2 \in C(S_2) : L_2(s_2, e, \varphi_2) = ?, \exists \mu \in Sat(\varphi_1)$ such that $\forall \mu' \in Sat(\varphi_2) : \mu \not\equiv_{\mathcal{R}} \mu'$,

(d) there exists $e \in A$ and $\varphi_2 \in C(S_2)$ such that $L_2(s_2, e, \varphi_2) = \top$ and $\forall \varphi_1 \in C(S_1) : L_1(s_1, e, \varphi_1) = \bot$,

(e) there exists $e \in A$ and $\varphi_2 \in C(S_2)$ such that $L_2(s_2, e, \varphi_2) = \top$ and $\exists \varphi_1 \in C(S_1) : L_1(s_1, e, \varphi_1) = ?$,

(f) there exists $e \in A$ and $\varphi_2 \in C(S_2)$ such that $L_2(s_2, e, \varphi_2) = \top, \exists \varphi_1 \in C(S_1) : L_1(s_1, e, \varphi_1) = \top$ and $\exists \mu \in Sat(\varphi_1)$ such that $\forall \mu' \in Sat(\varphi_2) : \mu \not\equiv_{\mathcal{R}} \mu'$.

Remark that because of the determinism and SVNF of APAs N_1 and N_2, cases 1, 2 and 3 cannot happen at the same time. Moreover, although the cases in 3 can happen simultaneously, they cannot be "triggered" by the same action. In order to keep track of these "concurrent" situations, we define the following sets.

Given a pair of states (s_1, s_2), let us define $B_a(s_1, s_2)$ to be the set of actions in A such that case $3.a$ above holds. If there is no such action, then $B_a(s_1, s_2) = \emptyset$. Similarly, we define $B_b(s_1, s_2), B_c(s_1, s_2), B_d(s_1, s_2), B_e(s_1, s_2)$ and $B_f(s_1, s_2)$ to be the sets of actions such that cases $3.b, c, d, e$ and $3.f$ holds respectively. Given a set $X \subseteq \{a, b, c, d, e, f\}$, let $B_X(s_1, s_2) = \cup_{x \in X} B_x(s_1, s_2)$. In addition, let $B(s_1, s_2) = B_{\{a,b,c,d,e,f\}}(s_1, s_2)$.

4.1   Over-Approximating Difference

We now try to compute an APA that represents the difference between the sets of implementations of two APAs. We first observe that such a set may not be representable by an APA, then we will propose over- and under-approximations. Consider the APAs N_1 and N_2 given in Figures 1a and 1b, where $\alpha \neq \beta \neq \gamma$. Consider the difference of their sets of implementations. It is easy to see that this set contains all the PAs that can finitely loop on valuation α and then move into a state with valuation β. Since there is no bound on the time spent in the loop, there is no finite-state APA that can represent this set of implementations.

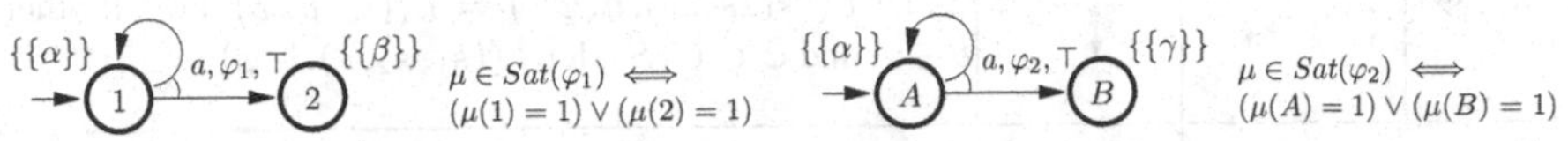

(a) APA N_1 $\qquad\qquad\qquad\qquad\qquad$ (b) APA N_2

Fig. 1. APAs N_1 and N_2 such that $[\![N_1]\!] \setminus [\![N_2]\!]$ cannot be represented using a finite-state APA

Now we propose a construction $\setminus^*$ that over-approximates the difference between APAs in the following sense: given two deterministic APAs $N_1 = (S_1, A, L_1, AP, V_1, \{s_0^1\})$ and $N_2 = (S_2, A, L_2, AP, V_2, \{s_0^2\})$ in SVNF, such that $N_1 \not\preceq N_2$, we have $[\![N_1]\!] \setminus [\![N_2]\!] \subseteq [\![N_1 \setminus^* N_2]\!]$. We first observe that if $V_1(s_0^1) \neq V_2(s_0^2)$, i.e. (s_0^1, s_0^2) in case 2, then $[\![N_1]\!] \cap [\![N_2]\!] = \emptyset$. In such case, we define $N_1 \setminus^* N_2$ as N_1. Otherwise, we build on the reasons for which refinement fails between N_1 and N_2. Note that the assumption $N_1 \not\preceq N_2$ implies that the pair (s_0^1, s_0^2) can never be in any refinement relation, hence in case 1. We first give an informal intuition of how the construction works and then define it formally.

In our construction, states in $N_1 \setminus^* N_2$ will be elements of $S_1 \times (S_2 \cup \{\bot\}) \times (A \cup \{\varepsilon\})$. Our objective is to ensure that any implementation of our constructed APA will satisfy N_1 and not N_2. In (s_1, s_2, e), states s_1 and s_2 keep track of executions of N_1 and N_2. Action e is the action of N_1 that will be used to break satisfaction with respect to N_2, i.e. the action that will be the cause for which any implementation of (s_1, s_2, e) cannot satisfy N_2. Since satisfaction is defined recursively, the breaking is not necessarily immediate and can be postponed to successors. $\bot$ is used to represent states that can only be reached after breaking the satisfaction relation to N_2. In these states, we do not need to keep track of the corresponding execution in N_2, thus only focus on satisfying N_1. States of the form (s_1, s_2, ε) with $s_2 \neq \bot$ are states where the satisfaction is broken by a distribution that does not match constraints in N_2 (cases 3.c and 3.f). In order to invalidate these constraints, we still need to keep track of the corresponding execution in N_2, hence the use of ε instead of $\bot$.

The transitions in our construction will match the different cases shown in the previous section, ensuring that in each state, either the relation is broken immediately or reported to at least one successor. Since there can be several ways of breaking the relation in state (s_0^1, s_0^2), each corresponding to an action $e \in B(s_0^1, s_0^2)$, the APA $N_1 \setminus^* N_2$

Table 1. Definition of the transition function L in $N_1 \setminus^* N_2$

$e \in$	N_1, N_2	$N_1 \setminus^* N_2$	Formal Definition of L
$B_a(s_1, s_2)$		(s_1, s_2, e)	For all $a \neq e \in A$ and $\varphi \in C(S_1)$ such that $L_1(s_1, a, \varphi) \neq \bot$, let $L((s_1, s_2, e), a, \varphi^{\bot}) = L_1(s_1, a, \varphi)$. In addition, let $L((s_1, s_2, e), e, \varphi_1^{\bot}) = \top$. For all other $b \in A$ and $\varphi \in C(S)$, let $L((s_1, s_2, e), b, \varphi) = \bot$.
$B_b(s_1, s_2)$		$\varphi_1^{\bot}$	
$B_d(s_1, s_2)$		(s_1, s_2, e)	For all $a \in A$ and $\varphi \in C(S_1)$ such that $L_1(s_1, a, \varphi) \neq \bot$, let $L((s_1, s_2, e), a, \varphi^{\bot}) = L_1(s_1, a, \varphi)$. For all other $b \in A$ and $\varphi \in C(S)$, let $L((s_1, s_2, e), b, \varphi) = \bot$.
$B_e(s_1, s_2)$		(s_1, s_2, e) φ_{12}^{B}	For all $a \neq e \in A$ and $\varphi \in C(S_1)$ such that $L_1(s_1, a, \varphi) \neq \bot$, let $L((s_1, s_2, e), a, \varphi^{\bot}) = L_1(s_1, a, \varphi)$. In addition, let $L((s_1, s_2, e), e, \varphi_{12}^{B}) = ?$. For all other $b \in A$ and $\varphi \in C(S)$, let $L((s_1, s_2, e), b, \varphi) = \bot$.
$B_c(s_1, s_2)$		(s_1, s_2, e)	For all $a \in A$ and $\varphi \in C(S_1)$ such that $L_1(s_1, a, \varphi) \neq \bot$ (including e and φ_1), let $L((s_1, s_2, e), a, \varphi^{\bot}) = L_1(s_1, a, \varphi)$. In addition, let $L((s_1, s_2, e), e, \varphi_{12}^{B}) = \top$. For all other $b \in A$ and $\varphi \in C(S)$, let $L((s_1, s_2, e), b, \varphi) = \bot$.
$B_f(s_1, s_2)$		$\varphi_{12}^{B} \neq \varphi_1^{\bot}$	

will have one initial state for each of them. Formally, if (s_0^1, s_0^2) is in case 3, we define the over-approximation of the difference of N_1 and N_2 as follows:

Definition 7. Let $N_1 \setminus^* N_2 = (S, A, L, AP, V, S_0)$, where $S = S_1 \times (S_2 \cup \{\bot\}) \times (A \cup \{\varepsilon\})$, $V(s_1, s_2, a) = V(s_1)$ for all s_2 and a, $S_0 = \{(s_0^1, s_0^2, f) \mid f \in B(s_0^1, s_0^2)\}$, and L is defined by:

- If $s_2 = \bot$ or $e = \varepsilon$ or (s_1, s_2) in case 1 or 2, then for all $a \in A$ and $\varphi \in C(S_1)$ such that $L_1(s_1, a, \varphi) \neq \bot$, let $L((s_1, s_2, e), a, \varphi^{\bot}) = L_1(s_1, a, \varphi)$, with $\varphi^{\bot}$ defined below. For all other $b \in A$ and $\varphi \in C(S)$, let $L((s_1, s_2, e), b, \varphi) = \bot$.
- Else, we have (s_1, s_2) in case 3 and $B(s_1, s_2) \neq \emptyset$ by construction. The definition of L is given in Table 1, with the constraints $\varphi^{\bot}$ and φ_{12}^{B} defined hereafter.

Given $\varphi \in C(S_1)$, $\varphi^{\bot} \in C(S)$ is defined as follows: $\mu \in Sat(\varphi^{\bot})$ iff $\forall s_1 \in S_1, \forall s_2 \neq \bot, \forall b \neq \varepsilon, \mu(s_1, s_2, b) = 0$ and the distribution $(\mu \downarrow_1 : s_1 \mapsto \mu(s_1, \bot, \varepsilon))$ is in $Sat(\varphi)$. Given a state $(s_1, s_2, e) \in S$ with $s_2 \neq \bot$ and $e \neq \varepsilon$ and two constraints $\varphi_1 \in C(S_1)$, $\varphi_2 \in C(S_2)$ such that $L_1(s_1, e, \varphi_1) \neq \bot$ and $L_2(s_2, e, \varphi_2) \neq \bot$, the constraint $\varphi_{12}^{B} \in C(S)$ is defined as follows: $\mu \in Sat(\varphi_{12}^{B})$ iff (1) for all $(s_1', s_2', c) \in S$, we have $\mu(s_1', s_2', c) > 0 \Rightarrow s_2' = \bot$ if $\mathsf{succ}_{s_2, e}(s_1') = \emptyset$ and

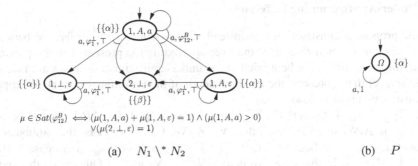

$$\mu \in Sat(\varphi_{12}^B) \iff (\mu(1,A,a) + \mu(1,A,\varepsilon) = 1) \wedge (\mu(1,A,a) > 0)$$
$$\vee (\mu(2,\bot,\varepsilon) = 1)$$

(a) $N_1 \setminus^* N_2$ (b) P

Fig. 2. Over-approximating difference $N_1 \setminus^* N_2$ of APAs N_1 and N_2 from Figure 1 and PA P such that $P \models N_1 \setminus^* N_2$ and $P \models N_2$

$\{s_2'\} = \text{succ}_{s_2,e}(s_1')$ otherwise, and $c \in B(s_1',s_2') \cup \{\varepsilon\}$, (2) the distribution $\mu_1 : s_1' \mapsto \sum_{c \in A \cup \{\varepsilon\}, s_2' \in S_2 \cup \{\bot\}} \mu(s_1',s_2',c)$ satisfies φ_1, and (3) either (a) there exists $(s_1',\bot,c)$ such that $\mu(s_1',\bot,c) > 0$ or (b) the distribution $\mu_2 : s_2' \mapsto \sum_{c \in A \cup \{\varepsilon\}, s_1' \in S_1} \mu(s_1',s_2',c)$ does not satisfy φ_2, or (c) there exists $s_1' \in S_1$, $s_2' \in S_2$ and $c \neq \varepsilon$ such that $\mu(s_1',s_2',c) > 0$. Informally, distributions in φ_{12}^B must (1) follow the corresponding execution is N_1 and N_2 if possible, (2) satisfy φ_1 and (3) either (a) reach a state in N_1 that cannot be matched in N_2 or (b) break the constraint φ_2, or (c) report breaking the relation to at least one successor state.

The following theorem shows that $N_1 \setminus^* N_2$ is an over-approximation of the difference of N_1 and N_2 in terms of sets of implementations.

Theorem 1. *For all deterministic APAs N_1 and N_2 in SVNF such that $N_1 \not\preceq N_2$, we have $[\![N_1]\!] \setminus [\![N_2]\!] \subseteq [\![N_1 \setminus^* N_2]\!]$.*

The reverse inclusion unfortunately does not hold. Intuitively, as explained in the construction of the constraint φ_{12}^B above, one can postpone the breaking of the satisfaction relation for N_2 to the next state (condition (3.c)). This assumption is necessary in order to produce an APA representing *all* counterexamples. However, when there are cycles in the execution of $N_1 \setminus^* N_2$, this assumption allows to postpone forever, thus allowing for implementations that will ultimately satisfy N_2. This is illustrated in the following example.

Example 1. Consider the APAs N_1 and N_2 given in Fig. 1. Their over-approximating difference $N_1 \setminus^* N_2$ is given in Fig. 2a. One can see that the PA P in Fig. 2b satisfies both $N_1 \setminus^* N_2$ and N_2.

We will later see in Corollary 1 that even though $N_1 \setminus^* N_2$ may be capturing too many counterexamples, the *distance* between $N_1 \setminus^* N_2$ and the real set of counterexamples $[\![N_1]\!] \setminus [\![N_2]\!]$ is zero. This means that the two sets are infinitesimally close to each other, so in this sense, $N_1 \setminus^* N_2$ is the *best possible* over-approximation.

4.2   Under-Approximating Difference

We now propose a construction that instead *under-estimates* the difference between APAs. This construction resembles the over-approximation presented in the previous section, the main difference being that in the under-approximation, states are indexed with an integer that represents the maximal depth of the unfolding of counterexamples. The construction is as follows.

Let $N_1 = (S_1, A, L_1, AP, V_1, \{s_0^1\})$ and $N_2 = (S_2, A, L_2, AP, V_2, \{s_0^2\})$ be two deterministic APAs in SVNF such that $N_1 \not\preceq N_2$. Let $K \in \mathbb{N}$ be the parameter of our construction. As in Section 4.1, if $V_1(s_0^1) \neq V_2(s_0^2)$, i.e. (s_0^1, s_0^2) in case 2, then $[\![N_1]\!] \cap [\![N_2]\!] = \emptyset$. In this case, we define $N_1 \setminus^K N_2$ as N_1. Otherwise, the under-approximation is defined as follows.

Definition 8. *Let* $N_1 \setminus^K N_2 = (S, A, L, AP, V, S_0^K)$, *where* $S = S_1 \times (S_2 \cup \{\perp\}) \times (A \cup \{\varepsilon\}) \times \{1, \ldots, K\}$, $V(s_1, s_2, a, k) = V(s_1)$ *for all* s_2, a, $k < K$, $S_0^K = \{(s_0^1, s_0^2, f, K) \mid f \in B(s_0^1, s_0^2)\}$, *and* L *is defined by:*
- *If* $s_2 = \perp$ *or* $e = \varepsilon$ *or* (s_1, s_2) *in case 1 or 2, then for all* $a \in A$ *and* $\varphi \in C(S_1)$ *such that* $L_1(s_1, a, \varphi) \neq \perp$, *let* $L((s_1, s_2, e, k), a, \varphi^\perp) = L_1(s_1, a, \varphi)$, *with* $\varphi^\perp$ *defined below. For all other* $b \in A$ *and* $\varphi \in C(S)$, *let* $L((s_1, s_2, e, k), b, \varphi) = \perp$.
- *Else we have* (s_1, s_2) *in case 3 and* $B(s_1, s_2) \neq \emptyset$ *by construction. The definition of* L *is given in Table 2. The constraints* $\varphi^\perp$ *and* $\varphi_{12}^{B,k}$ *are defined hereafter.*

Given a constraint $\varphi \in C(S_1)$, the constraint $\varphi^\perp \in C(S)$ is defined as follows: $\mu \in Sat(\varphi^\perp)$ iff $\forall s_1 \in S_1, \forall s_2 \neq \perp, \forall b \neq \varepsilon, \forall k \neq 1, \mu(s_1, s_2, b, k) = 0$ and the distribution $(\mu \downarrow_1 : s_1 \mapsto \mu(s_1, \perp, \varepsilon, 1))$ is in $Sat(\varphi)$. Given a state $(s_1, s_2, e, k) \in S$ with $s_2 \neq \perp$ and $e \neq \varepsilon$ and two constraints $\varphi_1 \in C(S_1)$ and $\varphi_2 \in C(S_2)$ such that $L_1(s_1, e, \varphi_1) \neq \perp$ and $L_2(s_2, e, \varphi_2) \neq \perp$, the constraint $\varphi_{12}^{B,k} \in C(S)$ is defined as follows: $\mu \in Sat(\varphi_{12}^{B,k})$ iff (1) for all $(s_1', s_2', c, k') \in S$, if $\mu(s_1', s_2', c, k') > 0$, then $c \in B(s_1', s_2') \cup \{\varepsilon\}$ and either $\text{succ}_{s_2, e}(s_1') = \emptyset$, $s_2' = \perp$ and $k' = 1$, or $\{s_2'\} = \text{succ}_{s_2, e}(s_1')$, (2) the distribution $\mu_1 : s_1' \mapsto \sum_{c \in A \cup \{\varepsilon\}, s_2' \in S_2 \cup \{\perp\}, k' \geq 1} \mu(s_1', s_2', c, k')$ satisfies φ_1, and (3) either (a) there exists $(s_1', \perp, c, 1)$ such that $\mu(s_1', \perp, c, 1) > 0$, or (b) the distribution $\mu_2 : s_2' \mapsto \sum_{c \in A \cup \{\varepsilon\}, s_1' \in S_1, k' \geq 1} \mu(s_1', s_2', c, k')$ does not satisfy φ_2, or (c) $k \neq 1$ and there exists $s_1' \in S_1$, $s_2' \in S_2$, $c \neq \varepsilon$ and $k' < k$ such that $\mu(s_1', s_2', c, k') > 0$. The construction is illustrated in Figure 3.

4.3   Properties

We already saw in Theorem 1 that $N_1 \setminus^* N_2$ is a correct over-approximation of the difference of N_1 by N_2 in terms of sets of implementations. The next theorem shows that, similarly, all $N_1 \setminus^K N_2$ are correct under-approximations. Moreover, for increasing K the approximation is improving, and eventually all PAs in $[\![N_1]\!] \setminus [\![N_2]\!]$ are getting caught. (Hence in a set-theoretic sense, $\lim_{K \to \infty} [\![N_1 \setminus^K N_2]\!] = [\![N_1]\!] \setminus [\![N_2]\!]$.)

Theorem 2. *For all deterministic APAs* N_1 *and* N_2 *in SVNF such that* $N_1 \not\preceq N_2$:
1. *for all* $K \in \mathbb{N}$, *we have* $N_1 \setminus^K N_2 \preceq N_1 \setminus^{K+1} N_2$,
2. *for all* $K \in \mathbb{N}$, $[\![N_1 \setminus^K N_2]\!] \subseteq [\![N_1]\!] \setminus [\![N_2]\!]$, *and*

Table 2. Definition of the transition function L in $N_1 \setminus^K N_2$

$e \in$	N_1, N_2	$N_1 \setminus^K N_2$	Formal Definition of L
$B_a(s_1, s_2)$		(s_1, s_2, e, k) $e, \top$	For all $a \neq e \in A$ and $\varphi \in C(S_1)$ such that $L_1(s_1, a, \varphi) \neq \bot$, let $L((s_1, s_2, e, k), a, \varphi^\bot) = L_1(s_1, a, \varphi)$. In addition, let $L((s_1, s_2, e, k), e, \varphi_1^\bot) = \top$. For all other $b \in A$ and $\varphi \in C(S)$, let $L((s_1, s_2, e, k), b, \varphi) = \bot$.
$B_b(s_1, s_2)$		$\varphi_1^\bot$	
$B_d(s_1, s_2)$		(s_1, s_2, e, k) e	For all $a \in A$ and $\varphi \in C(S_1)$ such that $L_1(s_1, a, \varphi) \neq \bot$, let $L((s_1, s_2, e, k), a, \varphi^\bot) = L_1(s_1, a, \varphi)$. For all other $b \in A$ and $\varphi \in C(S)$, let $L((s_1, s_2, e, k), b, \varphi) = \bot$.
$B_e(s_1, s_2)$		(s_1, s_2, e, k) $e, ?$ $\varphi_{12}^{B,k}$	For all $a \neq e \in A$ and $\varphi \in C(S_1)$ such that $L_1(s_1, a, \varphi) \neq \bot$, let $L((s_1, s_2, e, k), a, \varphi^\bot) = L_1(s_1, a, \varphi)$. In addition, let $L((s_1, s_2, e, k), e, \varphi_{12}^{B,k}) = ?$. For all other $b \in A$ and $\varphi \in C(S)$, let $L((s_1, s_2, e, k), b, \varphi) = \bot$.
$B_c(s_1, s_2)$		(s_1, s_2, e, k) $e, \top$ $e, \{?, \top\}$ $\varphi_{12}^{B,k}$ $\varphi_1^\bot$	For all $a \in A$ and $\varphi \in C(S_1)$ such that $L_1(s_1, a, \varphi) \neq \bot$ (including e and φ_1), let $L((s_1, s_2, e, k), a, \varphi^\bot) = L_1(s_1, a, \varphi)$. In addition, let $L((s_1, s_2, e, k), e, \varphi_{12}^{B,k}) = \top$. For all other $b \in A$ and $\varphi \in C(S)$, let $L((s_1, s_2, e, k), b, \varphi) = \bot$.
$B_f(s_1, s_2)$			

3. *for all PA $P \in [\![N_1]\!] \setminus [\![N_2]\!]$, there exists $K \in \mathbb{N}$ such that $P \in [\![N_1 \setminus^K N_2]\!]$.*

Note that item 3 implies that for all PA $P \in [\![N_1]\!] \setminus [\![N_2]\!]$, there is a finite specification capturing $[\![N_1]\!] \setminus [\![N_2]\!]$ "up to" P.

Using our distance defined in Section 3, we can make the above convergence result more precise. The next proposition shows that the speed of convergence is exponential in K; hence in practice, K will typically not need to be very large.

Proposition 2. *Let N_1 and N_2 be two deterministic APAs in SVNF such that $N_1 \not\preceq N_2$, and let $K \in \mathbb{N}$. Then $d_t([\![N_1]\!] \setminus [\![N_2]\!], [\![N_1 \setminus^K N_2]\!]) \leq \lambda^K (1 - \lambda)^{-1}$.*

For the actual application at hand however, the particular accumulating distance d we have introduced in Section 3 may have limited interest, especially considering that one has to choose a discounting factor for actually calculating it.

What is more interesting are results of a *topological* nature which abstract away from the particular distance used and apply to all distances which are *topologically equivalent* to d. The results we present below are of this nature.

It can be shown, c.f. [39], that accumulating distances for different choices of λ are topologically equivalent (indeed, even Lipschitz equivalent), hence the particular choice

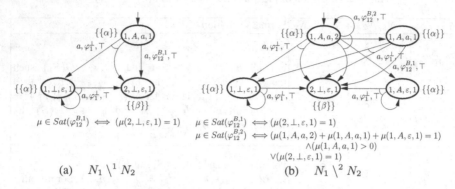

Fig. 3. Under-approximations at level 1 and 2 of the difference of APAs N_1 and N_2 from Figure 1

of discounting factor is not important. Also some other system distances are Lipschitz equivalent to the accumulating one, in particular the so-called *point-wise* and *maximum-lead* ones, see again [39].

Theorem 3. *Let N_1 and N_2 be two deterministic APAs in SVNF such that $N_1 \not\leq N_2$.*
1. *The sequence $(N_1 \setminus^K N_2)_{K \in \mathbb{N}}$ converges in the distance d, and $\lim_{K \to \infty} d(N_1 \setminus^* N_2, N_1 \setminus^K N_2) = 0$.*
2. *The sequence $(\llbracket N_1 \setminus^K N_2 \rrbracket)_{K \in \mathbb{N}}$ converges in the distance d_t, and $\lim_{K \to \infty} d_t(\llbracket N_1 \rrbracket \setminus \llbracket N_2 \rrbracket, \llbracket N_1 \setminus^K N_2 \rrbracket) = 0$.*

Recall that as d and d_t are not metrics, but only (asymmetric) pseudometrics (i.e. hemi-metrics), the above sequences may have more than one limit; hence the particular formulation. The theorem's statements are topological as they only allure to convergence of sequences and distance 0; topologically equivalent distances obey precisely the property of having the same convergence behaviour and the same kernel, c.f. [1].

The next corollary, which is easily proven from the above theorem by noticing that its first part implies that also $\lim_{K \to \infty} d_t(\llbracket N_1 \setminus^* N_2 \rrbracket, \llbracket N_1 \setminus^K N_2 \rrbracket) = 0$, shows what we mentioned already at the end of Section 4.1: $N_1 \setminus^* N_2$ is the best possible over-approximation of $\llbracket N_1 \rrbracket \setminus \llbracket N_2 \rrbracket$.

Corollary 1. *Let N_1 and N_2 be two deterministic APAs in SVNF such that $N_1 \not\leq N_2$. Then $d_t(\llbracket N_1 \setminus^* N_2 \rrbracket, \llbracket N_1 \rrbracket \setminus \llbracket N_2 \rrbracket) = 0$.*

Again, as d_t is not a metric, the distance being zero does not imply that the sets $\llbracket N_1 \setminus^* N_2 \rrbracket$ and $\llbracket N_1 \rrbracket \setminus \llbracket N_2 \rrbracket$ are equal; it merely means that they are *indistinguishable* by the distance d_t, or infinitesimally close to each other.

5   Conclusion

We have in this paper added an important aspect to the specification theory of Abstract Probabilistic Automata, in that we have shown how to exhaustively characterize the *difference* between two deterministic specifications. In a stepwise refinement methodology, difference is an important tool to gauge refinement failures.

We have also introduced a notion of *discounted distance* between specifications which can be used as another measure for how far one specification is from being a refinement of another. Using this distance, we were able to show that our sequence of under-approximations converges, semantically, to the real difference of sets of implementations, and that our over-approximation is infinitesimally close to the real difference.

There are many different ways to measure distances between implementations and specifications, allowing to put the focus on either transient or steady-state behavior. In this paper we have chosen one specific discounted distance, placing the focus on transient behavior. Apart from the fact that this can indeed be a useful distance in practice, we remark that the convergence results about our under- and over-approximations are topological in nature and hence apply with respect to all distances which are topologically equivalent to the specific one used here, typically discounted distances. Although the results presented in the paper do not hold in general for the accumulating (undiscounted) distance, there are other notions of distances that are more relevant for steady-state behavior, e.g. limit-average. Whether our results hold in this setting remains future work.

We also remark that we have shown that it is not more difficult to compute the difference of two APAs than to check for their refinement. Hence if a refinement failure is detected (using e.g. the methods presented in our APAC tool), it is not difficult to also compute the difference for information about the reason for refinement failure.

One limitation of our approach is the use of *deterministic* APAs. Even though deterministic specifications are generally considered to suffice from a modeling point of view [29], non-determinism may be introduced e.g. when composing specifications. Indeed, our constructions themselves introduce non-determinism: for deterministic APAs N_1, N_2, both $N_1 \setminus^* N_2$ and $N_1 \setminus^K N_2$ may be non-deterministic. Hence it is of interest to extend our approach to non-deterministic specifications. The problem here is, however, that for non-deterministic specifications, the relation between refinement and inclusion of sets of implementations $N_1 \preceq N_2 \iff [\![N_1]\!] \subseteq [\![N_2]\!]$ breaks: we may well have $N_1 \npreceq N_2$ but $[\![N_1]\!] \subseteq [\![N_2]\!]$, cf. [14]. So the technique we have used in this paper to compute differences will not work for non-deterministic APAs, and techniques based on *thorough refinement* will have to be used.

As a last note, we wish to compare our approach of difference between APA specifications with the use of *counterexamples* in probabilistic model checking. Counterexample generation is studied in a number of papers [2,19,42,4,24,36,22,43,8,27], typically with the purpose of embedding it into a procedure of counterexample guided abstraction refinement (CEGAR). The focus typically is on generation of *one* particular counterexample to refinement, which can then be used to adapt the abstraction accordingly.

In contrast, our approach at computing APA difference generates a representation of *all counterexamples*. Our focus is not on refinement of abstractions at *system* level, using counterexamples, but on assessment of *specifications*. This is, then, the reason why we want to compute all counterexamples instead of only one. We remark, however, that our approach also can be used, in a quite simplified version, to generate only one counterexample; details of this are given in [13]. Our work is hence supplementary and orthogonal to the CEGAR-type use of counterexamples: CEGAR procedures can be used also to refine APA specifications, but only our difference can assess the precise distinction between specifications.

Acknowledgement. The authors wish to thank Joost-Pieter Katoen for interesting discussions and insightful comments on the subject of this work.

References

1. Aliprantis, C.D., Border, K.C.: Infinite Dimensional Analysis: A Hitchhiker's Guide, 3rd edn. Springer (2007)
2. Aljazzar, H., Leue, S.: Directed explicit state-space search in the generation of counterexamples for stochastic model checking. IEEE Trans. Software Eng. (2010)
3. Alur, R., Feder, T., Henzinger, T.A.: The benefits of relaxing punctuality. J. ACM 43(1), 116–146 (1996)
4. Andrés, M.E., D'Argenio, P., van Rossum, P.: Significant diagnostic counterexamples in probabilistic model checking. In: Chockler, H., Hu, A.J. (eds.) HVC 2008. LNCS, vol. 5394, pp. 129–148. Springer, Heidelberg (2009)
5. Baier, C., Katoen, J.-P.: Principles of Model Checking. MIT Press (2008)
6. Bauer, S.S., Fahrenberg, U., Legay, A., Thrane, C.: General quantitative specification theories with modalities. In: Hirsch, E.A., Karhumäki, J., Lepistö, A., Prilutskii, M. (eds.) CSR 2012. LNCS, vol. 7353, pp. 18–30. Springer, Heidelberg (2012)
7. Caillaud, B., Delahaye, B., Larsen, K.G., Legay, A., Pedersen, M.L., Wasowski, A.: Constraint Markov chains. TCS 412(34), 4373–4404 (2011)
8. Chadha, R., Viswanathan, M.: A counterexample-guided abstraction-refinement framework for Markov decision processes. ACM Trans. Comput. Log. 12(1), 1 (2010)
9. Cobleigh, J.M., Avrunin, G.S., Clarke, L.A.: Breaking up is hard to do: An evaluation of automated assume-guarantee reasoning. ACM Trans. Softw. Eng. Methodol. 17(2) (2008)
10. Cobleigh, J.M., Giannakopoulou, D., Pǎsǎreanu, C.S.: Learning assumptions for compositional verification. In: Garavel, H., Hatcliff, J. (eds.) TACAS 2003. LNCS, vol. 2619, pp. 331–346. Springer, Heidelberg (2003)
11. de Alfaro, L., Henzinger, T.A.: Interface automata. In: FSE, pp. 109–120. ACM (2001)
12. de Alfaro, L., Majumdar, R., Raman, V., Stoelinga, M.: Game relations and metrics. In: LICS, pp. 99–108. IEEE Computer Society (2007)
13. Delahaye, B., Fahrenberg, U., Larsen, K.G., Legay, A.: Refinement and difference for probabilistic automata - long version (2013),
    http://delahaye.benoit.free.fr/rapports/QEST13-long.pdf
14. Delahaye, B., Katoen, J.-P., Larsen, K.G., Legay, A., Pedersen, M.L., Sher, F., Wąsowski, A.: Abstract probabilistic automata. In: Jhala, R., Schmidt, D. (eds.) VMCAI 2011. LNCS, vol. 6538, pp. 324–339. Springer, Heidelberg (2011)
15. Delahaye, B., Katoen, J.-P., Larsen, K.G., Legay, A., Pedersen, M.L., Sher, F., Wąsowski, A.: New results on abstract probabilistic automata. In: ACSD, pp. 118–127. IEEE (2011)
16. Delahaye, B., Larsen, K.G., Legay, A., Pedersen, M.L., Wąsowski, A.: APAC: A tool for reasoning about abstract probabilistic automata. In: QEST, pp. 151–152. IEEE (2011)
17. Desharnais, J., Gupta, V., Jagadeesan, R., Panangaden, P.: Metrics for labelled Markov processes. TCS 318(3), 323–354 (2004)

18. Fecher, H., Leucker, M., Wolf, V.: *Don't know* in probabilistic systems. In: Valmari, A. (ed.) SPIN 2006. LNCS, vol. 3925, pp. 71–88. Springer, Heidelberg (2006)
19. Han, T., Katoen, J.-P., Damman, B.: Counterexample generation in probabilistic model checking. IEEE Trans. Software Eng. 35(2), 241–257 (2009)
20. Hansson, H., Jonsson, B.: A logic for reasoning about time and reliability. Formal Asp. Comput. 6(5), 512–535 (1994)
21. Hermanns, H., Herzog, U., Katoen, J.: Process algebra for performance evaluation. TCS 274(1-2), 43–87 (2002)
22. Hermanns, H., Wachter, B., Zhang, L.: Probabilistic CEGAR. In: Gupta, A., Malik, S. (eds.) CAV 2008. LNCS, vol. 5123, pp. 162–175. Springer, Heidelberg (2008)
23. Hinton, A., Kwiatkowska, M., Norman, G., Parker, D.: PRISM: A tool for automatic verification of probabilistic systems. In: Hermanns, H., Palsberg, J. (eds.) TACAS 2006. LNCS, vol. 3920, pp. 441–444. Springer, Heidelberg (2006)
24. Jansen, N., Ábrahám, E., Katelaan, J., Wimmer, R., Katoen, J.-P., Becker, B.: Hierarchical counterexamples for discrete-time Markov chains. In: Bultan, T., Hsiung, P.-A. (eds.) ATVA 2011. LNCS, vol. 6996, pp. 443–452. Springer, Heidelberg (2011)
25. Jonsson, B., Larsen, K.G.: Specification and refinement of probabilistic processes. In: LICS, pp. 266–277. IEEE (1991)
26. Katoen, J.-P., Klink, D., Leucker, M., Wolf, V.: Three-valued abstraction for continuous-time Markov chains. In: Damm, W., Hermanns, H. (eds.) CAV 2007. LNCS, vol. 4590, pp. 311–324. Springer, Heidelberg (2007)
27. Komuravelli, A., Păsăreanu, C.S., Clarke, E.M.: Assume-guarantee abstraction refinement for probabilistic systems. In: Madhusudan, P., Seshia, S.A. (eds.) CAV 2012. LNCS, vol. 7358, pp. 310–326. Springer, Heidelberg (2012)
28. Kwiatkowska, M.Z., Norman, G., Parker, D., Qu, H.: Assume-guarantee verification for probabilistic systems. In: Esparza, J., Majumdar, R. (eds.) TACAS 2010. LNCS, vol. 6015, pp. 23–37. Springer, Heidelberg (2010)
29. Larsen, K.G.: Modal specifications. In: Sifakis, J. (ed.) CAV 1989. LNCS, vol. 407, pp. 232–246. Springer, Heidelberg (1990)
30. Larsen, K.G., Fahrenberg, U., Thrane, C.: Metrics for weighted transition systems: Axiomatization and complexity. TCS 412(28), 3358–3369 (2011)
31. Lynch, N., Tuttle, M.R.: An introduction to Input/Output automata. CWI 2(3) (1989)
32. Lynch, N.A.: Distributed Algorithms. Morgan Kaufmann (1996)
33. Manna, Z., Pnueli, A.: The Temporal Logic of Reactive and Concurrent Systems. Springer (1992)
34. Raclet, J.-B.: Quotient de spécifications pour la réutilisation de composants. PhD thesis, Université de Rennes I (December 2007) (in French)
35. Sassolas, M., Chechik, M., Uchitel, S.: Exploring inconsistencies between modal transition systems. Software and System Modeling 10(1), 117–142 (2011)
36. Schmalz, M., Varacca, D., Völzer, H.: Counterexamples in probabilistic LTL model checking for Markov chains. In: Bravetti, M., Zavattaro, G. (eds.) CONCUR 2009. LNCS, vol. 5710, pp. 587–602. Springer, Heidelberg (2009)
37. Segala, R., Lynch, N.A.: Probabilistic simulations for probabilistic processes. In: Jonsson, B., Parrow, J. (eds.) CONCUR 1994. LNCS, vol. 836, pp. 481–496. Springer, Heidelberg (1994)
38. Sher, F., Katoen, J.-P.: Compositional abstraction techniques for probabilistic automata. In: Baeten, J.C.M., Ball, T., de Boer, F.S. (eds.) TCS 2012. LNCS, vol. 7604, pp. 325–341. Springer, Heidelberg (2012)

39. Thrane, C., Fahrenberg, U., Larsen, K.G.: Quantitative analysis of weighted transition systems. JLAP 79(7), 689–703 (2010)
40. van Breugel, F., Mislove, M.W., Ouaknine, J., Worrell, J.: An intrinsic characterization of approximate probabilistic bisimilarity. In: Gordon, A.D. (ed.) FOSSACS 2003. LNCS, vol. 2620, pp. 200–215. Springer, Heidelberg (2003)
41. Vardi, M.Y.: Automatic verification of probabilistic concurrent finite-state programs. In: FOCS, pp. 327–338. IEEE (1985)
42. Wimmer, R., Braitling, B., Becker, B.: Counterexample generation for discrete-time Markov chains using bounded model checking. In: Jones, N.D., Müller-Olm, M. (eds.) VMCAI 2009. LNCS, vol. 5403, pp. 366–380. Springer, Heidelberg (2009)
43. Wimmer, R., Jansen, N., Ábrahám, E., Becker, B., Katoen, J.-P.: Minimal critical subsystems for discrete-time Markov models. In: Flanagan, C., König, B. (eds.) TACAS 2012. LNCS, vol. 7214, pp. 299–314. Springer, Heidelberg (2012)

High-Level Counterexamples
for Probabilistic Automata

Ralf Wimmer[1], Nils Jansen[2], Andreas Vorpahl[2], Erika Ábrahám[2],
Joost-Pieter Katoen[2], and Bernd Becker[1]

[1] Albert-Ludwigs-University Freiburg, Germany
{wimmer,becker}@informatik.uni-freiburg.de
[2] RWTH Aachen University, Germany
{nils.jansen,abraham,katoen}@cs.rwth-aachen.de,
andreas.vorpahl@rwth-aachen.de[*]

Abstract. Providing compact and understandable counterexamples for violated system properties is an essential task in model checking. Existing works on counterexamples for probabilistic systems so far computed either a large set of system runs or a subset of the system's states, both of which are of limited use in manual debugging. Many probabilistic systems are described in a guarded command language like the one used by the popular model checker PRISM. In this paper we describe how a minimal subset of the commands can be identified which together already make the system erroneous. We additionally show how the selected commands can be further simplified to obtain a well-understandable counterexample.

1 Introduction

The ability to provide *counterexamples* for violated properties is one of the most essential features of model checking [1]. Counterexamples make errors reproducible and are used to guide the designer of an erroneous system during the debugging process. Furthermore, they play an important role in counterexample-guided abstraction refinement (CEGAR) [2,3,4,5]. For linear-time properties of digital or hybrid systems, a single violating run suffices to refute the property. Thereby, this run—acquired during model checking—directly forms a counterexample.

Probabilistic formalisms like discrete-time Markov chains (DTMCs), Markov decision processes (MDPs) and probabilistic automata (PAs) are well-suited to model systems with uncertainties. Violating behavior in the probabilistic setting means that the probability that a certain property holds is outside of some required bounds. For probabilistic reachability properties, this can be reduced to the case where an upper bound on the probability is exceeded [6]. Thereby, a *probabilistic counterexample* is formed by a set of runs that all satisfy a given property while their probability mass is larger than the allowed upper bound.

[*] This work was partly supported by the German Research Council (DFG) as part of the Transregional Collaborative Research Center AVACS (SFB/TR 14), the DFG project CEBug (AB 461/1-1), and the EU-FP7 IRSES project MEALS.

K. Joshi et al. (Eds.): QEST 2013, LNCS 8054, pp. 39–54, 2013.

Tools like PRISM [7] verify probabilistic systems by computing the solution of a linear equation system. While this technique is very efficient, the simultaneous generation of counterexamples is not supported.

During the last years, a number of approaches have been proposed to compute probabilistic counterexamples by enumerating certain paths of a system [8,6,9]. In general, such a set may be extremely large; for some systems it is at least double exponential in the number of system states [6]. Also different compact representations of counterexamples have been devised, e. g., counterexamples are described symbolically by regular expressions in [6], while in [10] and [11] the abstraction of strongly connected components yields loop-free systems.

A different representation is obtained by taking a preferably small subset of the state space, forming a *critical subsystem*. Inside this part of the original system the property is already violated, see [8] and [11]. Both approaches use heuristic path search algorithms to incrementally build such critical subsystems for probabilistic reachability properties. In [12,13,14], a different approach was suggested: not only a small subsystem, but a minimal one is computed for a large class of properties, namely probabilistic reachability and ω-regular properties for both DTMCs and MDPs. This is achieved using solver techniques such as mixed integer linear programming (MILP) [15].

An unanswered question for all these approaches is how they can actually be used for debugging. Most practical examples are built by the parallel composition of modules forming a flat state-space with millions of states. Although critical subsystems are often smaller by orders of magnitude than the original system, they may still be very large, rendering manual debugging practically impossible.

In this paper, we focus on the non-deterministic and fully compositional model of probabilistic automata (PA) [16,17]. The specification of such models is generally done in a high-level language allowing the parallel composition of modules. The modules of the system are not specified by enumerating states and transitions but can be described using a *guarded command language* [18,19] like the one used by PRISM. The communication between different modules takes place using synchronization on common actions and via shared variables. Having this human-readable specification language, it seems natural that a user should be pointed to the part of the system description which causes the error, instead of referring to the probabilistic automaton defined by the composition. To the best of our knowledge, no work on probabilistic counterexamples has considered this sort of *high-level counterexamples* yet.

We show how to identify a *smallest set of guarded commands* which induces a critical subsystem. In order to correct the system, at least one of the returned commands has to be changed. We additionally simplify the commands by removing branching choices which are not necessary to obtain a counterexample. We present this as a special case of a method where the number of different transition labels for a PA is minimized. This offers great flexibility in terms of human-readable counterexamples. The NP-hard computation of such a smallest critical label set is done by the established approach of mixed integer linear programming.

Structure of the paper. In Section 2 we review some foundations. Our approach to obtain smallest command sets is presented in Section 3. How the essential commands can be simplified is described in Section 4. After some experimental results in Section 5 we conclude the paper in Section 6.

2   Foundations

Let S be a countable set. A *sub-distribution* on S is a function $\mu : S \to [0,1]$ such that $0 \le \sum_{s \in S} \mu(s) \le 1$. We use the notation $\mu(S') = \sum_{s \in S'} \mu(s)$ for a subset $S' \subseteq S$. A sub-distribution with $\mu(S) = 1$ is called a *probability distribution*. We denote the set of all probability distributions on S by $\mathrm{Distr}(S)$ and analogously by $\mathrm{SubDistr}(S)$ for sub-distributions.

Probabilistic Automata

Definition 1 (Probabilistic automaton). *A probabilistic automaton (PA) is a tuple $\mathcal{M} = (S, s_{init}, \mathrm{Act}, P)$ such that S is a finite set of states, $s_{init} \in S$ is an initial state, Act is a finite set of actions, and $P : S \to 2^{\mathrm{Act} \times \mathrm{Distr}(S)}$ is a probabilistic transition relation such that $P(s)$ is finite for all $s \in S$.*

In the following we also use η to denote an action-distribution pair (α, μ). We further define $\mathrm{succ}(s, \alpha, \mu) = \{s' \in S \mid \mu(s') > 0\}$ for $(\alpha, \mu) \in P(s)$, $\mathrm{succ}(s) = \bigcup_{(\alpha,\mu) \in P(s)} \mathrm{succ}(s, \alpha, \mu)$, and $\mathrm{pred}(s) = \{s' \in S \mid \exists (\alpha, \mu) \in P(s') : \mu(s) > 0\}$.

The evolution of a probabilistic automaton is as follows: Starting in the initial state $s = s_{\mathrm{init}}$, first a transition $(\alpha, \mu) \in P(s)$ is chosen non-deterministically. Then the successor state $s' \in \mathrm{succ}(s, \alpha, \mu)$ is determined probabilistically according to the distribution μ. This process is repeated for the successor state s'. To prevent deadlocks we assume $P(s) \ne \emptyset$ for all $s \in S$.

An *infinite path* of a PA $\mathcal{M}$ is an infinite sequence $s_0(\alpha_0, \mu_0)s_1(\alpha_1, \mu_1)\ldots$ with $s_i \in S$, $(\alpha_i, \mu_i) \in P(s_i)$ and $s_{i+1} \in \mathrm{succ}(s_i, \alpha_i, \mu_i)$ for all $i \ge 0$. A *finite path* π of $\mathcal{M}$ is a finite prefix $s_0(\alpha_0, \mu_0)s_1(\alpha_1, \mu_1)\ldots s_n$ of an infinite path of $\mathcal{M}$ with last state $\mathrm{last}(\pi) = s_n$. We denote the set of all finite paths of $\mathcal{M}$ by $\mathrm{Paths}_{\mathcal{M}}^{\mathrm{fin}}$.

A *sub-PA* is like a PA, but it allows sub-distributions instead of probability distributions in the definition of P.

Definition 2 (Subsystem). *A sub-PA $\mathcal{M}' = (S', s'_{init}, \mathrm{Act}', P')$ is a subsystem of a sub-PA $\mathcal{M} = (S, s_{init}, \mathrm{Act}, P)$, written $\mathcal{M}' \sqsubseteq \mathcal{M}$, iff $S' \subseteq S$, $s'_{init} = s_{init}$, $\mathrm{Act}' \subseteq \mathrm{Act}$ and for all $s \in S'$ there is an injective function $f : P'(s) \to P(s)$ such that for all $(\alpha', \mu') \in P'(s)$ with $f((\alpha', \mu')) = (\alpha, \mu)$ we have that $\alpha' = \alpha$ and for all $s' \in S'$ either $\mu'(s') = 0$ or $\mu'(s') = \mu(s')$.*

A sub-PA $\mathcal{M} = (S, s_{\mathrm{init}}, \mathrm{Act}, P)$ can be transformed into a PA as follows: We add a new state $s_\perp \notin S$, turn all sub-distributions into probability distributions by defining $\mu(s_\perp) := 1 - \mu(S)$ for each $s \in S$ and $(\alpha, \mu) \in P(s)$, and make $s_\perp$ absorbing by setting $P(s_\perp) := \{(\tau, \mu) \in \mathrm{Act} \times \mathrm{Distr}(S \cup \{s_\perp\}) \mid \mu(s_\perp) = 1\}$. This way all methods we formalize for PAs can also be applied to sub-PAs.

Before a probability measure on PAs can be defined, the nondeterminism has to be resolved. This is done by an entity called *scheduler*.

Definition 3 (Scheduler). *A scheduler for a PA* $\mathcal{M} = (S, s_{init}, \text{Act}, P)$ *is a function* $\sigma : \text{Paths}_{\mathcal{M}}^{\text{fin}} \to \text{SubDistr}(\text{Act} \times \text{Distr}(S))$ *such that* $\sigma(\pi)(\alpha, \mu) > 0$ *implies* $(\alpha, \mu) \in P(\text{last}(\pi))$ *for all* $\pi \in \text{Paths}_{\mathcal{M}}^{\text{fin}}$ *and* $(\alpha, \mu) \in \text{Act} \times \text{Distr}(S)$. *We use* $\text{Sched}_{\mathcal{M}}$ *to denote the set of all schedulers of* $\mathcal{M}$.

By resolving the nondeterminism, a scheduler turns a PA into a fully probabilistic model, for which a standard probability measure can be defined [20, Chapter 10.1]. In this paper we are interested in *probabilistic reachability properties*: Is the probability to reach a set $T \subseteq S$ of target states from s_{init} at most equal to a given bound $\lambda \in [0, 1] \subseteq \mathbb{R}$? Such a reachability property will be denoted with $\mathcal{P}_{\leq\lambda}(\lozenge T)$. Note that checking ω-regular properties can be reduced to checking reachability properties. For a fixed scheduler σ, this probability $\text{Pr}_{\mathcal{M}}^{\sigma}(s_{init}, \lozenge T)$ can be computed by solving a linear equation system. However, for a PA without a scheduler, this question is not well-posed. Instead we ask: Is the probability to reach a set $T \subseteq S$ of target states from s_{init} at most λ for all schedulers? That means, $\mathcal{P}_{\leq\lambda}(\lozenge T)$ has to hold for all schedulers. To check this, it suffices to compute the maximal probability over all schedulers that T is reached from s_{init}, which we denote with $\text{Pr}_{\mathcal{M}}^{+}(s_{init}, \lozenge T)$. One can show that for this kind of properties maximizing over a certain subclass of all schedulers suffices, namely the so-called memoryless deterministic schedulers, which can be seen as functions $\sigma : S \to \text{Act} \times \text{SubDistr}(S)$.

Definition 4 (Memoryless deterministic scheduler). *A scheduler* σ *of* $\mathcal{M} = (S, s_{init}, \text{Act}, P)$ *is memoryless if* $\text{last}(\pi) = \text{last}(\pi')$ *implies* $\sigma(\pi) = \sigma(\pi')$ *for all* $\pi, \pi' \in \text{Paths}_{\mathcal{M}}^{\text{fin}}$. *The scheduler* σ *is deterministic if* $\sigma(\pi)(\eta) \in \{0, 1\}$ *for all* $\pi \in \text{Paths}_{\mathcal{M}}^{\text{fin}}$ *and* $\eta \in \text{Act} \times \text{Distr}(S)$.

The maximal probability $\text{Pr}_{\mathcal{M}}^{+}(s, \lozenge T)$ to reach T from s is obtained as the unique solution of the following equation system: $\text{Pr}_{\mathcal{M}}^{+}(s, \lozenge T) = 1$, if $s \in T$; $\text{Pr}_{\mathcal{M}}^{+}(s, \lozenge T) = 0$, if T is unreachable from s under all schedulers, and $\text{Pr}_{\mathcal{M}}^{+}(s, \lozenge T) = \max_{(\alpha,\mu)\in P(s)} \sum_{s'\in S} \mu(s') \cdot \text{Pr}_{\mathcal{M}}^{+}(s', \lozenge T)$ otherwise. It can be solved by either rewriting it into a linear program, by applying a technique called value iteration, or by iterating over the possible schedulers (policy iteration) (see, e. g., [20, Chapter 10.6]). A memoryless deterministic scheduler is obtained from the solution by taking an arbitrary element of $P(s)$ in the first two cases and an element of $P(s)$ for which the maximum is obtained in the third case.

PRISM's Guarded Command Language. For a set Var of Boolean variables, let $\mathcal{A}_{\text{Var}}$ denote the set of variable assignments, i. e., of functions $\nu : \text{Var} \to \{0, 1\}$.

Definition 5 (Model, module, command). *A model is a tuple* $(\text{Var}, s_{init}, M)$ *where* Var *is a finite set of Boolean variables,* $s_{init} : \text{Var} \to \{0, 1\}$ *the initial state, and* $M = \{M_1, \ldots, M_k\}$ *a finite set of modules.*

A module is a tuple $M_i = (\mathrm{Var}_i, \mathrm{Act}_i, C_i)$ *with* $\mathrm{Var}_i \subseteq \mathrm{Var}$ *a set of variables such that* $\mathrm{Var}_i \cap \mathrm{Var}_j = \emptyset$ *for* $i \neq j$, Act_i *a finite set of synchronizing actions, and* C_i *a finite set of commands. The action* τ *with* $\tau \notin \bigcup_{i=1}^{k} \mathrm{Act}_i$ *denotes the internal non-synchronizing action. A command* $c \in C_i$ *has the form*

$$c = [\alpha]\, g \to\ p_1 : f_1 + \ldots + p_n : f_n$$

with $\alpha \in \mathrm{Act}_i \,\dot{\cup}\, \{\tau\}$, g *a Boolean predicate ("guard") over the variables in* Var, $p_i \in [0,1]$ *a rational number with* $\sum_{i=1}^{n} p_i = 1$, *and* $f_i : \mathcal{A}_{\mathrm{Var}} \to \mathcal{A}_{\mathrm{Var}_i}$ *being a variable update function. We refer to the action* α *of* c *by* $act(c)$.

Note that each variable may be written by only one module, but the update may depend on variables of other modules. Each model with several modules is equivalent to a model with a single module which is obtained by computing the parallel composition of these modules. We give a short intuition on how this composition is built. For more details we refer to the documentation of PRISM. Assume two modules $M_1 = (\mathrm{Var}_1, \mathrm{Act}_1, C_1)$ and $M_2 = (\mathrm{Var}_2, \mathrm{Act}_2, C_2)$ with $\mathrm{Var}_1 \cap \mathrm{Var}_2 = \emptyset$. The *parallel composition* $M = M_1 \| M_2 = (\mathrm{Var}, \mathrm{Act}, C)$ is given by $\mathrm{Var} = \mathrm{Var}_1 \cup \mathrm{Var}_2$, $\mathrm{Act} = \mathrm{Act}_1 \cup \mathrm{Act}_2$ and

$$C = \{\, c \quad |\, c \in C_1 \cup C_2 \wedge act(c) \in \{\tau\} \cup (\mathrm{Act}_1 \setminus \mathrm{Act}_2) \cup (\mathrm{Act}_2 \setminus \mathrm{Act}_1) \,\} \cup$$
$$\{\, c \otimes c' \,|\, c \in C_1 \wedge c' \in C_2 \wedge act(c) = act(c') \in \mathrm{Act}_1 \cap \mathrm{Act}_2 \qquad\quad \},$$

where $c \otimes c'$ for $c = [\alpha]\, g \to\ p_1 : f_1 + \ldots + p_n : f_n \in C_1$ and $c' = [\alpha]\, g' \to\ p_1' : f_1' + \ldots + p_m' : f_m' \in C_2$ is defined as

$$c \otimes c' = [\alpha]\, g \wedge g' \to\ p_1 \cdot p_1' : f_1 \otimes f_1' + \ldots + p_n \cdot p_1' : f_n \otimes f_1'$$
$$\ldots$$
$$+ p_1 \cdot p_m' : f_1 \otimes f_n' + \ldots + p_n \cdot p_n' : f_n \otimes f_m'.$$

Here, for $f_i : \mathcal{A}_{\mathrm{Var}} \to \mathcal{A}_{\mathrm{Var}_1}$ and $f_j' : \mathcal{A}_{\mathrm{Var}} \to \mathcal{A}_{\mathrm{Var}_2}$ we define $f_i \otimes f_j' : \mathcal{A}_{\mathrm{Var}} \to \mathcal{A}_{\mathrm{Var}_1 \cup \mathrm{Var}_2}$ such that for all $\nu \in \mathcal{A}_{\mathrm{Var}}$ we have that $(f_i \otimes f_j')(\nu)(x)$ equals $f_i(\nu)(x)$ for each $x \in \mathrm{Var}_1$ and $f_j'(\nu)(x)$ for each $x \in \mathrm{Var}_2$.

Intuitively, commands labeled with non-synchronizing actions are executed on their own, while for synchronizing actions a command from each synchronizing module is executed simultaneously. Note that if a module has an action in its synchronizing action set but no commands labeled with this action, this module will block the execution of commands with this action in the composition. This is considered to be a modeling error and the corresponding commands are ignored.

The PA-semantics of a model is as follows. Assume a model $(\mathrm{Var}, s_{\mathrm{init}}, M)$ with a single module $M = (\mathrm{Var}, \mathrm{Act}, C)$ which will not be subject to parallel composition any more. The *state space* S of the corresponding PA $\mathcal{M} = (S, s_{\mathrm{init}}, \mathrm{Act}, P)$ is given by the set of all possible variable assignments $\mathcal{A}_{\mathrm{Var}}$, i.e., a state s is a vector $(x_1, \ldots, x_m)$ with x_i being a value of the variable $v_i \in \mathrm{Var} = \{v_1, \ldots, v_m\}$. To construct the transitions, we observe that the guard g of each command

$$c = [\alpha]\, g \to\ p_1 : f_1 + \ldots + p_n : f_n \ \in C$$

defines a subset of the state space $S_c \subseteq \mathcal{A}_{\mathrm{Var}}$ with $s \in S_c$ iff s satisfies g. Each update $f_i : \mathcal{A}_{\mathrm{Var}} \to \mathcal{A}_{\mathrm{Var}}$ maps a state $s' \in S$ to each $s \in S_c$. Together with the associated values p_i, we define a probability distribution $\mu_{c,s} : S \to [0, 1]$ with

$$\mu_{c,s}(s') = \sum_{\{i \,|\, 1 \leq i \leq n \wedge f_i(s) = s'\}} p_i$$

for each $s' \in \mathcal{A}_{\mathrm{Var}}$. The probabilistic transition relation $P : \mathcal{A}_{\mathrm{Var}} \to 2^{\mathrm{Act} \times \mathrm{Distr}(\mathcal{A}_{\mathrm{Var}})}$ is given by $P(s) = \{(\alpha, \mu_{c,s}) \mid c \in C \wedge act(c) = \alpha \wedge s \in S_c\}$ for all $s \in \mathcal{A}_{\mathrm{Var}}$.

Mixed Integer Programming. A *mixed integer linear program* optimizes a linear objective function under a condition specified by a conjunction of linear inequalities. A subset of the variables in the inequalities is restricted to take only integer values, which makes solving MILPs NP-hard [21, Problem MP1].

Definition 6 (Mixed integer linear program). *Let* $A \in \mathbb{Q}^{m \times n}$, $B \in \mathbb{Q}^{m \times k}$, $b \in \mathbb{Q}^m$, $c \in \mathbb{Q}^n$, *and* $d \in \mathbb{Q}^k$. *A mixed integer linear program (MILP) consists in computing* $\min c^T x + d^T y$ *such that* $Ax + By \leq b$ *and* $x \in \mathbb{R}^n$, $y \in \mathbb{Z}^k$.

MILPs are typically solved by a combination of a branch-and-bound algorithm and the generation of so-called cutting planes. These algorithms heavily rely on the fact that relaxations of MILPs which result by removing the integrality constraints can be efficiently solved. MILPs are widely used in operations research, hardware-software co-design, and numerous other applications. Efficient open source as well as commercial implementations are available like `Scip` [22], `Cplex` [23], or `Gurobi` [24]. We refer to, e. g., [15] for more information on solving MILPs.

3   Computing Counterexamples

In this section we show how to compute smallest critical command sets. For this, we introduce a generalization of this problem, namely smallest critical labelings, state the complexity of the problem, and specify an MILP formulation which yields a smallest critical labeling.

Let $\mathcal{M} = (S, s_{\mathrm{init}}, \mathrm{Act}, P)$ be a PA, $T \subseteq S$, Lab a finite set of labels, and $L : S \times \mathrm{Act} \times \mathrm{Distr}(S) \nrightarrow 2^{\mathrm{Lab}}$ a partial labeling function such that $L(s, \eta)$ is defined iff $\eta \in P(s)$. Let $\mathrm{Lab}' \subseteq \mathrm{Lab}$ be a subset of the labels. The *PA induced by* Lab' is $\mathcal{M}_{|\mathrm{Lab}'} = (S, s_{\mathrm{init}}, \mathrm{Act}, P')$ such that for all $s \in S$ we have $P'(s) = \{\eta \in P(s) \mid L(s, \eta) \subseteq \mathrm{Lab}'\}$.

Definition 7 (Smallest critical labeling problem). *Let* $\mathcal{M}$, T, Lab *and* L *be defined as above and* $\mathcal{P}_{\leq \lambda}(\lozenge T)$ *be a reachability property that is violated by* s_{init} *in* $\mathcal{M}$. *A subset* $\mathrm{Lab}' \subseteq \mathrm{Lab}$ *is critical if* $\mathrm{Pr}^{+}_{\mathcal{M}_{|\mathrm{Lab}'}}(s_{\mathrm{init}}, \lozenge T) > \lambda$.

Given a weight function $w : \mathrm{Lab} \to \mathbb{R}^{\geq 0}$, *the smallest critical labeling problem is to determine a critical subset* $\mathrm{Lab}' \subseteq \mathrm{Lab}$ *such that* $w(\mathrm{Lab}') := \sum_{\ell \in \mathrm{Lab}'} w(\ell)$ *is minimal among all critical subsets of* Lab.

Theorem 1. *To decide whether there is a critical labeling* $\mathrm{Lab}' \subseteq \mathrm{Lab}$ *with* $w(\mathrm{Lab}') \leq k$ *for a given integer* $k \geq 0$ *is NP-complete.*

A proof of this theorem, which is based on the reduction of exact 3-cover [21, Problem SP2] is given in the extended version [25] of this paper.

The concept of smallest critical labelings gives us a flexible description of counterexamples being minimal with respect to different quantities.

Commands. In order to minimize the number of commands that together induce an erroneous system, let $\mathcal{M} = (S, s_{\mathrm{init}}, \mathrm{Act}, P)$ be a PA generated by modules $M_i = (\mathrm{Var}_i, \mathrm{Act}_i, C_i)$, $i = 1, \ldots, k$. For each module M_i and each command $c \in C_i$ we introduce a unique label[1] $\ell_{c,i}$ with weight 1 and define the labeling function $L : S \times \mathrm{Act} \times \mathrm{Distr}(S) \to 2^{\mathrm{Lab}}$ such that each transition is labeled with the set of commands which together generate this transition[2]. Note that in case of synchronization several commands together create a certain transition. A smallest critical labeling corresponds to a *smallest critical command set*, being a smallest set of commands which together generate an erroneous system.

Modules. We can also minimize the number of modules involved in a counterexample by using the same label for all commands in a module. Often systems consist of a number of copies of the same module, containing the same commands, only with the variables renamed, plus a few extra modules. Consider for example a wireless network: n nodes want to transmit messages using a protocol for medium access control [26]. All nodes run the same protocol. Additionally there may be a module describing the channel. When fixing an erroneous system, one wants to preserve the identical structure of the nodes. Therefore the selected commands should contain the same subset of commands from all identical modules. This can be obtained by assigning the same label to all corresponding commands from the symmetric modules and using the number of symmetric modules as its weight.

States. The state-minimal subsystems as introduced in [12] can be obtained as special case of smallest critical labelings: For each state $s \in S$ introduce a label ℓ_s and set $L(s, \eta) = \{\ell_s\}$ for all $\eta \in P(s)$. $\mathrm{Lab}' \subseteq \mathrm{Lab} = \{\ell_s \mid s \in S\}$ is a smallest critical labeling iff $S' = \{s \in S \mid \ell_s \in \mathrm{Lab}'\}$ induces a minimal critical subsystem.

We will now explain how these smallest critical labelings are computed. First, the notions of *relevant* and *problematic* states are considered. Intuitively, a state s is relevant, if there exists a scheduler such that a target state is reachable from s. A state s is problematic, if there additionally exists a deadlock-free scheduler under that no target state is reachable from s.

Definition 8 (Relevant and problematic states). *Let* $\mathcal{M}$, T, *and* L *be as above. The relevant states of* $\mathcal{M}$ *for* T *are given by* $S_T^{\mathrm{rel}} = \{s \in S \mid \exists \sigma \in \mathrm{Sched}_{\mathcal{M}} : \mathrm{Pr}_{\mathcal{M}}^{\sigma}(s, \Diamond T) > 0\}$. *A label* ℓ *is relevant for* T *if there is* $s \in S_T^{\mathrm{rel}}$ *and* $\eta \in P(s)$ *such that* $S_T^{\mathrm{rel}} \cap \mathrm{succ}(s, \eta) \neq \emptyset$ *and* $\ell \in L(s, \eta)$.

---

[1] In the following we write short ℓ_c instead of $\ell_{c,i}$ if the index i is clear from the context.
[2] If several command sets generate the same transition, we make copies of the transition.

Let $\mathrm{Sched}^+_{\mathcal{M}}$ be the set of all schedulers σ with $\{\eta \mid \sigma(\pi)(\eta) > 0\} \neq \emptyset$ for all π. The states in $S_T^{\mathrm{prob}} = \{s \in S_T^{\mathrm{rel}} \mid \exists \sigma \in \mathrm{Sched}^+_{\mathcal{M}} : \mathrm{Pr}^\sigma_{\mathcal{M}}(s, \Diamond T) = 0\}$ are problematic states and the set $P_T^{\mathrm{prob}} = \{(s, \eta) \in S_T^{\mathrm{prob}} \times \mathrm{Act} \times \mathrm{Distr}(S) \mid \eta \in P(s) \wedge \mathrm{succ}(s, \eta) \subseteq S_T^{\mathrm{prob}}\}$ are problematic transitions regarding T.

Both relevant states and problematic states and actions can be computed in linear time using graph algorithms [27].

States that are not relevant can be removed from the PA together with all their incident edges without changing the probability of reaching T from s_{init}. Additionally, all labels that do not occur in the relevant part of the PA can be deleted. We therefore assume that the (sub-)PA under consideration contains only states and labels that are relevant for T.

In our computation, we need to ensure that from each problematic state an unproblematic state is reachable under the selected scheduler, otherwise the probability of the problematic states is not well defined by the constraints [14]. We solve this problem by attaching a value r_s to each problematic state $s \in S_T^{\mathrm{prob}}$ and encoding that a distribution of s is selected only if it has at least one successor state s' with a value $r_{s'} > r_s$ attached to it. This requirement assures by induction that there is an *increasing path* from s to an unproblematic state, along which the values attached to the states are strictly increasing.

To encode the selection of smallest critical command sets as an MILP, we need the following variables:

- for each $\ell \in \mathrm{Lab}$ a variable $x_\ell \in \{0, 1\}$ which is 1 iff ℓ is part of the critical labeling,
- for each state $s \in S \setminus T$ and each transition $\eta \in P(s)$ a variable $\sigma_{s,\eta} \in \{0, 1\}$ which is 1 iff η is chosen in s by the scheduler; the scheduler is free not to choose any transition,
- for each state $s \in S$ a variable $p_s \in [0, 1]$ which stores the probability to reach a target state from s under the selected scheduler within the subsystem defined by the selected labeling,
- for each state $s \in S$ being either a problematic state or a successor of a problematic state a variable $r_s \in [0, 1] \subseteq \mathbb{R}$ for the encoding of increasing paths, and
- for each problematic state $s \in S_T^{\mathrm{prob}}$ and each successor state $s' \in \mathrm{succ}(s)$ a variable $t_{s,s'} \in \{0, 1\}$, where $t_{s,s'} = 1$ implies that the values attached to the states increase along the edge (s, s'), i.e., $r_s < r_{s'}$.

Let $w_{\min} := \min\{w(\ell) \mid \ell \in \mathrm{Lab} \wedge w(\ell) > 0\}$ be the smallest positive weight that is assigned to any label. The MILP for the smallest critical labeling problem is then as follows:

$$\text{minimize} \quad -\frac{1}{2} w_{\min} \cdot p_{s_{\mathrm{init}}} + \sum_{\ell \in \mathrm{Lab}} w(\ell) \cdot x_\ell \qquad (1a)$$

such that

$$p_{s_{\mathrm{init}}} > \lambda \qquad (1b)$$

$$\forall s \in S \setminus T : \qquad \sum_{\eta \in P(s)} \sigma_{s,\eta} \leq 1 \tag{1c}$$

$$\forall s \in S \; \forall \eta \in P(s) \; \forall \ell \in L(s,\eta) : \quad x_\ell \geq \sigma_{s,\eta} \tag{1d}$$

$$\forall s \in T : \quad p_s = 1 \tag{1e}$$

$$\forall s \in S \setminus T : \quad p_s \leq \sum_{\eta \in P(s)} \sigma_{s,\eta} \tag{1f}$$

$$\forall s \in S \setminus T \; \forall \eta \in P(s) : \quad p_s \leq \sum_{s' \in \mathrm{succ}(s,\eta)} \mu(s') \cdot p_{s'} + (1 - \sigma_{s,\eta}) \tag{1g}$$

$$\forall (s,\eta) \in P_T^{\mathrm{prob}} : \quad \sigma_{s,\eta} \leq \sum_{s' \in \mathrm{succ}(s,\eta)} t_{ss'} \tag{1h}$$

$$\forall s \in S_T^{\mathrm{prob}} \; \forall s' \in \mathrm{succ}(s) : \quad r_s < r_{s'} + (1 - t_{ss'}) \; . \tag{1i}$$

The number of variables in this MILP is in $O(l + n + m)$ and the number of constraints in $O(n + l \cdot m)$ where l is the number of labels, n the number of states, and m the number of transitions of $\mathcal{M}$, i. e., $m = \big| \{(s, \eta, s') \mid s' \in \mathrm{succ}(s,\eta)\} \big|$.

We first explain the constraints in lines (1b)–(1i) of the MILP, which describe a critical labeling. First, we ensure that the probability of the initial state is greater than the probability bound λ (1b). For reachability properties, we can restrict ourselves to memoryless deterministic schedulers. So for each state $s \in S \setminus T$ at most one scheduler variable $\sigma_{s,\eta} \in P(s)$ can be set to 1 (1c). Note, that there may be states where no transition is chosen. For target states we do not need any restriction. If the scheduler selects a transition $\eta \in P(s)$, all labels $\ell \in L(s,\eta)$ have to be chosen (1d). For all target states $s \in T$ the probability p_s is set to 1 (1e), while for all non-target states without chosen transition ($\sigma_{s,\eta} = 0$ for all $\eta \in P(s)$), the probability is set to zero (1f); if $\sigma_{s,\eta} = 1$ for some $\eta \in P(s)$, this constraint is no restriction to probability p_s. However, in this case constraint (1g) is responsible for assigning a valid probability to p_s. The constraint is trivially satisfied if $\sigma_{s,\eta} = 0$. If transition η is selected, the probability p_s is bounded from above by the probability to go to one of the successor states of η and to reach the target states from there.

The reachability of at least one unproblematic state is ensured by (1h) and (1i). First, for every state s with transition η that is problematic regarding T, at least one transition variable must be activated. Second, for a path according to these transition variables, an increasing order is enforced for the problematic states. Because of this order, no problematic states can be revisited on an increasing path which enforces the final reachability of a non-problematic state.

These constraints enforce that each satisfying assignment of the label variables x_ℓ corresponds to a critical labeling. By minimizing the weight of the selected labels we obtain a smallest critical labeling. By the additional term $-\frac{1}{2} w_{\min} \cdot p_{s_{\mathrm{init}}}$ we obtain not only a smallest critical labeling but one with maximal probability. The coefficient $-\frac{1}{2} w_{\min}$ is needed to ensure that the benefit from maximizing the probability is smaller than the loss by adding an additional label. Please note, that any coefficient c with $0 < c < w_{\min}$ could be used.

Theorem 2. *The MILP given in* (1a)–(1i) *yields a smallest critical labeling.*

A proof of this theorem can be found in the extended version [25] of this paper.

Optimizations. The constraints of the MILP describe *critical labelings*, whereas *minimality* is enforced by the objective function. In this section we describe how some additional constraints can be imposed, which explicitly exclude variable assignments that are either not optimal or encode labelings that are also encoded by other assignments. Adding such redundant constraints to the MILP often speeds up the search.

Scheduler Cuts. We want to exclude solutions of the constraint set for which a state $s \in S$ has a selected action-distribution pair $\eta \in P(s)$ with $\sigma_{s,\eta} = 1$ but all successors of s under η are non-target states without any selected action-distribution pairs. Note that such solutions would define $p_s = 0$. We add for all $s \in S \setminus T$ and all $\eta \in P(s)$ with $\mathrm{succ}(s, \eta) \cap T = \emptyset$ the constraint

$$\sigma_{s,\eta} \leq \sum_{s' \in \mathrm{succ}(s,\eta) \setminus \{s\}} \sum_{\eta' \in P(s')} \sigma_{s',\eta'} . \tag{2}$$

Analogously, we require for each non-initial state s with a selected action-distribution pair $\eta \in P(s)$ that there is a selected action-distribution pair leading to s. Thus, we add for all states $s \in S \setminus \{s_{\mathrm{init}}\}$ the constraint

$$\sum_{\eta \in P(s)} \sigma_{s,\eta} \leq \sum_{s' \in \mathrm{pred}(s) \setminus \{s\}} \sum_{\{\eta' \in P(s') \,|\, s' \in \mathrm{succ}(s,\eta)\}} \sigma_{s',\eta'} . \tag{3}$$

As special cases of these cuts, we can encode that the initial state has at least one activated outgoing transition and that at least one of the target states has an selected incoming transition. These special cuts come with very few additional constraints and often have a great impact on the solving times.

Label cuts In order to guide the solver to select the correct combinations of labels and scheduler variables, we want to enforce that for every selected label ℓ there is at least one scheduler variable $\sigma_{s,\eta}$ activated such that $\ell \in L(s, \eta)$:

$$x_\ell \leq \sum_{s \in S} \sum_{\{\eta \in P(s) \,|\, \ell \in L(s,\eta)\}} \sigma_{s,\eta} . \tag{4}$$

Synchronization cuts While scheduler and label cuts are applicable to the general smallest critical labeling problem, synchronization cuts take the proper synchronization of commands into account. They are therefore only applicable for the computation of smallest critical command sets.

Let M_i, M_j $(i \neq j)$ be two modules which synchronize on action α, c a command of M_i with action α, and $C_{j,\alpha}$ the set of commands with action α in module M_j. The following constraint ensures that if command c is selected by activating the variable x_{l_c}, then at least one command $d \in C_{j,\alpha}$ is selected, too.

$$x_{\ell_c} \leq \sum_{d \in C_{j,\alpha}} x_{\ell_d} . \tag{5}$$

4   Simplification of Counterexamples

Even though we can obtain a smallest set of commands which together induce an erroneous system by the techniques described in the previous section, further simplifications may be possible. For this we identify branching choices of each command in the counterexample which can be removed, still yielding an erroneous system. To accomplish this, we specify an MILP formulation which identifies a smallest set of branching choices that need to be preserved for the critical command set, such that the induced sub-PA still violates the property under consideration.

For this we need a more detailed labeling of the commands. Given a command c_i of the form $[\alpha]\ g \to p_1 : f_1 + p_2 : f_2 + \cdots + p_n : f_n$, we assign to each branching choice $p_j : f_j$ a unique label $b_{i,j}$. Let Lab_b be the set of all such labels.

When composing the modules, we compute the union of the labeling of the branching choices being executed together. When computing the corresponding PA $\mathcal{M}$, we transfer this labeling to the branching choices of the transition relation of $\mathcal{M}$. We define the partial function $L_b : S \times \mathrm{Act} \times \mathrm{Distr}(S) \times S \nrightarrow 2^{\mathrm{Lab}_b}$ such that $L_b(s, \nu, s')$ is defined iff $\nu \in P(s)$ and $s' \in \mathrm{succ}(s, \nu)$. In this case, $L_b(s, \nu, s')$ contains the labels of the branching choices of all commands that are involved in generating the transition from s to s' via the transition ν.

The following MILP identifies a largest number of branching choices which can be removed. The program is similar to the MILP for command selection, but instead of selecting commands it uses decision variables x_b to select branching choices in the commands. Additionally to the probability p_s of the composed states $s \in S$, we use variables $p_{s,\nu,s'} \in [0,1] \subseteq \mathbb{R}$ for $s \in S$, $\nu \in P(s)$ and $s' \in \mathrm{succ}(s, \nu)$, which are forced to be zero if not all branching choices which are needed to generate the transition from s to s' in ν are available (6g). For the definition of p_s in (6h), the expression $\mu(s') \cdot p_{s'}$ of (1g) is replaced by $p_{s,\eta,s'}$. The remaining constraints are unchanged.

$$\text{minimize} \quad -\frac{1}{2}p_{s_{\mathrm{init}}} + \sum_{b \in \mathrm{Lab}_b} x_b \tag{6a}$$

such that

$$p_{s_{\mathrm{init}}} > \lambda \tag{6b}$$

$$\forall s \in S \setminus T : \quad \sum_{\eta \in P(s)} \sigma_{s,\eta} \leq 1 \tag{6c}$$

$$\forall s \in T : \quad p_s = 1 \tag{6d}$$

$$\forall s \in S \setminus T \ \forall \eta \in P(s) \ \forall s' \in \mathrm{succ}(s, \eta) :$$

$$p_{s,\eta,s'} \leq \mu(s') \cdot p_{s'} \tag{6e}$$

$$p_{s,\eta,s'} \leq \sigma_{s,\eta} \tag{6f}$$

$$\forall b \in L_b(s, \eta, s') : p_{s,\eta,s'} \leq x_b \tag{6g}$$

$$\forall s \in S \setminus T \ \forall \eta \in P(s) : \quad p_s \leq \sum_{s' \in \mathrm{succ}(s, \eta)} p_{s,\eta,s'} + (1 - \sigma_{s,\eta}) \tag{6h}$$

$$\forall s \in S \setminus T: \quad p_s \leq \sum_{\eta \in P(s)} \sigma_{s,\eta} \tag{6i}$$

$$\forall (s,\eta) \in P_T^{\text{prob}}: \quad \sigma_{s,\eta} \leq \sum_{s' \in \text{succ}(s,\eta)} t_{s,s'} \tag{6j}$$

$$\forall s \in S_T^{\text{prob}} \forall s' \in \text{succ}(s): \quad r_s < r_{s'} + (1 - t_{s,s'}) \tag{6k}$$

5 Experiments

We have implemented the described techniques in C++ using the MILP solver Gurobi [24]. The experiments were performed on an Intel® Xeon® CPU E5-2450 with 2.10 GHz clock frequency and 32 GB of main memory, running Ubuntu 12.04 Linux in 64 bit mode. We focus on the minimization of the number of commands needed to obtain a counterexample and simplify them by deleting a maximum number of branchings. We do not consider symmetries in the models. We ran our tool with two threads in parallel and aborted any experiment which did not finish within 10 min (1200 CPU seconds). We conducted a number of experiments that are publicly available on the web page of PRISM [28].

▶ coin-N-K models the shared coin protocol of a randomized consensus algorithm [29]. The protocol returns a preference between two choices with a certain probability, whenever requested by a process at some point in the execution of the consensus algorithm. The shared coin protocol is parameterized by the number N of involved processes and a constant $K > 1$. Internally, the protocol is based on flipping a coin to come to a decision. We consider the property $\mathcal{P}_{\leq \lambda}(\Diamond\,(\text{finished} \wedge \text{all\_coins\_equal}))$, which is satisfied if the probability to finish the protocol with all coins equal is at most λ.

▶ wlan-B-C models the two-way handshake mechanism of the IEEE 802.11 Wireless LAN protocol. Two stations try to send data, but run into a collision. Therefore they enter the randomized exponential backoff scheme. The parameter B denotes the maximal allowed value of the backoff counter. We check the property $\mathcal{P}_{\leq \lambda}(\Diamond\,(\text{num\_collisions} = C))$ putting an upper bound on the probability that a maximal allowed number C of collisions occur.

▶ csma-N-C concerns the IEEE 802.3 CSMA/CD network protocol. N is the number of processes that want to access a common channel, C is the maximal value of the backoff counter. We check $\mathcal{P}_{\leq \lambda}(\neg\text{collision\_max\_backoff}\,U\,\text{delivered})$ expressing that the probability that all stations successfully send their messages before a collision with maximal backoff occurs is at most λ.

▶ fw-N models the Tree Identify Protocol of the IEEE 1394 High Performance Serial Bus (called "FireWire") [30]. It is a leader election protocol which is executed each time a node enters or leaves the network. The parameter N denotes the delay of the wire as multiples of 10 ns. We check $\mathcal{P}_{\leq \lambda}(\Diamond\,\text{leader\_elected})$, i.e., that the probability of finally electing a leader is at most λ.

Some statistics of the models for different parameter values are shown in Table 1. The columns contain the name of the model, its number of states,

Table 1. Model statistics

Model	#states	#trans.	#mod.	#comm.	$Pr^+(s_{init}, \Diamond T)$	λ	MCS
coin-2-1	144	252	2	14 (12)	0.6	0.4	13
coin-2-2	272	492	2	14 (12)	0.5556	0.4	25*
coin-2-4	528	972	2	14 (12)	0.529 40	0.4	55*
coin-2-5	656	1212	2	14 (12)	0.523 79	0.4	67*
coin-2-6	784	1452	2	14 (12)	0.519 98	0.4	83*
coin-4-1	12416	40672	4	28 (20)	0.636 26	0.4	171*
coin-4-2	22656	75232	4	28 (20)	0.578 94	0.4	244*
csma-2-2	1038	1282	3	34 (34)	0.875	0.5	540
csma-2-4	7958	10594	3	38 (38)	0.999 02	0.5	1769*
fw-1	1743	2197	4	68 (64)	1.0	0.5	412
fw-4	5452	7724	4	68 (64)	1.0	0.5	412*
fw-10	17190	29364	4	68 (64)	1.0	0.5	412*
fw-15	33425	63379	4	68 (64)	1.0	0.5	412*
wlan-0-2	6063	10619	3	70 (42)	0.183 59	0.1	121
wlan-0-5	14883	26138	3	70 (42)	0.001 14	0.001	952*
wlan-2-1	28597	57331	3	76 (14)	1.0	0.5	7
wlan-2-2	28598	57332	3	76 (42)	0.182 60	0.1	121*
wlan-2-3	35197	70216	3	76 (42)	0.017 93	0.01	514*
wlan-3-1	96419	204743	3	78 (14)	1.0	0.5	7
wlan-3-2	96420	204744	3	78 (42)	0.183 59	0.1	121*

transitions, modules, and commands. The value in braces is the number of relevant commands. Column 6 contains the reachability probability and column 7 the bound λ. The last column shows the number of states in the minimal critical subsystem, i.e., the smallest subsystem of the PA such that the probability to reach a target state inside the subsystem is still above the bound. Entries which are marked with a star, correspond to the smallest critical subsystem we could find within the time bound of 10 min using our tool LTLSubsys [12], but they are not necessarily optimal.

The results of our experiments are displayed in Table 2. The first column contains the name of the model. The following three blocks contain the results of runs without any cuts, with all cuts, and with the best combination of cuts: If there were cut combinations with which the MILP could be solved within the time limit, we report the one with the shortest solving time. If all combinations timed out, we report the one that yielded the largest lower bound.

For each block we give the computation time in seconds ("Time"), the memory consumption in MB ("Mem."), the number of commands in the critical command set ("n") and, in case the time limit was exceeded, a lower bound on the size of the smallest critical command set ("lb"), which the solver obtains by solving a linear programming relaxation of the MILP. An entry "??" for the number of commands means that the solver was not able to find a non-trivial critical command set within the time limit. For the run without cuts we additionally give the number of variables ("Var.") and constraints ("Constr.") of the MILP.

In the last block we give information about the number of branching choices which could be removed from the critical command set ("simp."). In case the

Table 2. Experimental results (time limit = 600 seconds)

Model	no cuts						all cuts				best cut combination				branches	
	Var.	Constr.	Time	Mem.	n	lb	Time	Mem.	n	lb	Time	Mem.	n	lb	simp.	$|S'|$
coin-2-1	277	491	TO	773	9	8	298.56	146	9	opt	145.76	95	9	opt	1/12	28
coin-2-2	533	1004	TO	864	9	6	TO	676	9	7	TO	562	9	7	1/12	72
coin-2-4	1045	2028	TO	511	9	6	TO	162	9	6	TO	426	9	7	1/12	105
coin-2-5	1301	2540	TO	485	9	5	TO	121	9	6	TO	408	9	6	1/12	165
coin-2-6	1557	3052	TO	550	9	5	TO	159	9	6	TO	495	9	6	1/12	103
coin-4-1	26767	50079	TO	642	??	3	TO	627	20	3	TO	703	20	5	2/24	391
coin-4-2	47759	92063	TO	947	??	3	TO	993	??	3	TO	961	??	4	??	??
csma-2-2	2123	5990	2.49	24	32	opt	17.88	50	32	opt	2.11	24	32	opt	3/42	879
csma-2-4	15977	46882	195.39	208	36	opt	263.89	397	36	opt	184.05	208	36	opt	20/90	4522
fw-1	3974	13121	TO	205	28	27	184.49	119	28	opt	44.21	135	28	opt	38/68	419
fw-4	13144	43836	TO	268	28	21	TO	367	28	21	107.71	328	28	opt	38/68	424
fw-10	46282	153764	TO	790	28	13	TO	1141	28	18	545.68	993	28	opt	38/68	428
fw-15	96222	318579	TO	1496	28	9	TO	958	31	14	TO	1789	28	18	33/68	416
wlan-0-2	7072	6602	TO	324	33	15	TO	209	33	30	TO	174	33	32	23/72	3178
wlan-0-5	19012	25808	TO	570	??	10	TO	351	??	30	TO	357	??	30	??	??
wlan-2-1	28538	192	0.04	43	8	opt	0.07	44	8	opt	0.04	43	8	opt	6/14	7
wlan-2-2	29607	15768	TO	413	33	14	TO	188	33	30	TO	180	33	30	23/72	25708
wlan-2-3	36351	18922	TO	600	38	14	TO	315	37	32	TO	275	38	32	31/72	25173
wlan-3-1	96360	192	0.09	137	8	opt	0.13	137	8	opt	0.08	137	8	opt	6/14	7
wlan-3-2	97429	6602	TO	450	33	15	TO	292	33	30	TO	260	33	31	23/72	93639

different runs did not compute the same set, we used the one obtained with all cuts. An entry k/m means that we could remove k out of m relevant branching choices. We omit the running times of the simplification since in all cases it was faster than the command selection due to the reduced state space. The last column ("$|S'|$") contains the number of states in the PA that is induced by the minimized command set.

Although we ran into timeouts for many instances, in particular without any cuts, in almost all cases (with the exception of `coin4-2` and `wlan0-5`) a solution could be found within the time limit. We suppose that also the solutions of the aborted instances are optimal or close to optimal. It seems that the MILP solver is able to quickly find good (or even optimal) solutions due to sophisticated heuristics, but proving their optimality is hard. A solution is proven optimal as soon as the objective value of the best solution and the lower bound coincide. The additional cuts strengthen this lower bound considerably. Further experiments have shown that the scheduler cuts of Eq.(2) have the strongest effect on the lower bound. Choosing good cuts consequently enables the solver to obtain optimal solutions for more benchmarks.

Our method provides the user not only with a smallest set of simplified commands which induce an erroneous system, but also with a critical subsystem of the state space. Comparing its size with the size of the minimal critical subsystem (cf. Table 1) we can observe that for some models it is is close to optimal (e. g., the `coin`-instances), for others it is much larger (e. g., the `wlan`-instances). In all cases, however, we are able to reduce the number of commands and to simplify the commands, in some cases considerably.

6   Conclusion

We have presented a new type of counterexamples for probabilistic automata which are described using a guarded command language: We computed a smallest subset of the commands which alone induces an erroneous system. This requires the solution of a mixed integer linear program whose size is linear in the size of the state space of the PA. State-of-the-art MILP solvers apply sophisticated techniques to find small command sets quickly, but they are often unable to prove the optimality of their solution.

For the MILP formulation of the smallest critical labeling problem we both need decision variables for the labels and for the scheduler inducing the maximal reachability probabilities of the subsystem. On the other hand, model checking can be executed without any decision variables. Therefore we plan to develop a dedicated branch & bound algorithm which only branches on the decision variables for the labels. We expect a considerable speedup by using this method. Furthermore, we will investigate heuristic methods based on graph algorithms.

References

1. Clarke, E.M.: The birth of model checking. In: Grumberg, O., Veith, H. (eds.) 25 Years of Model Checking. LNCS, vol. 5000, pp. 1–26. Springer, Heidelberg (2008)
2. Clarke, E.M., Grumberg, O., Jha, S., Lu, Y., Veith, H.: Counterexample-guided abstraction refinement. In: Emerson, E.A., Sistla, A.P. (eds.) CAV 2000. LNCS, vol. 1855, pp. 154–169. Springer, Heidelberg (2000)
3. Hermanns, H., Wachter, B., Zhang, L.: Probabilistic CEGAR. In: Gupta, A., Malik, S. (eds.) CAV 2008. LNCS, vol. 5123, pp. 162–175. Springer, Heidelberg (2008)
4. Bobaru, M.G., Păsăreanu, C.S., Giannakopoulou, D.: Automated assume-guarantee reasoning by abstraction refinement. In: Gupta, A., Malik, S. (eds.) CAV 2008. LNCS, vol. 5123, pp. 135–148. Springer, Heidelberg (2008)
5. Komuravelli, A., Păsăreanu, C.S., Clarke, E.M.: Assume-guarantee abstraction refinement for probabilistic systems. In: Madhusudan, P., Seshia, S.A. (eds.) CAV 2012. LNCS, vol. 7358, pp. 310–326. Springer, Heidelberg (2012)
6. Han, T., Katoen, J.P., Damman, B.: Counterexample generation in probabilistic model checking. IEEE Trans. on Software Engineering 35(2), 241–257 (2009)
7. Kwiatkowska, M., Norman, G., Parker, D.: PRISM 4.0: Verification of probabilistic real-time systems. In: Gopalakrishnan, G., Qadeer, S. (eds.) CAV 2011. LNCS, vol. 6806, pp. 585–591. Springer, Heidelberg (2011)
8. Aljazzar, H., Leue, S.: Directed explicit state-space search in the generation of counterexamples for stochastic model checking. IEEE Trans. on Software Engineering 36(1), 37–60 (2010)
9. Wimmer, R., Braitling, B., Becker, B.: Counterexample generation for discrete-time Markov chains using bounded model checking. In: Jones, N.D., Müller-Olm, M. (eds.) VMCAI 2009. LNCS, vol. 5403, pp. 366–380. Springer, Heidelberg (2009)
10. Andrés, M.E., D'Argenio, P., van Rossum, P.: Significant diagnostic counterexamples in probabilistic model checking. In: Chockler, H., Hu, A.J. (eds.) HVC 2008. LNCS, vol. 5394, pp. 129–148. Springer, Heidelberg (2009)

11. Jansen, N., Ábrahám, E., Katelaan, J., Wimmer, R., Katoen, J.-P., Becker, B.: Hierarchical counterexamples for discrete-time markov chains. In: Bultan, T., Hsiung, P.-A. (eds.) ATVA 2011. LNCS, vol. 6996, pp. 443–452. Springer, Heidelberg (2011)
12. Wimmer, R., Jansen, N., Ábrahám, E., Becker, B., Katoen, J.-P.: Minimal critical subsystems for discrete-time markov models. In: Flanagan, C., König, B. (eds.) TACAS 2012. LNCS, vol. 7214, pp. 299–314. Springer, Heidelberg (2012)
13. Wimmer, R., Becker, B., Jansen, N., Ábrahám, E., Katoen, J.P.: Minimal critical subsystems as counterexamples for ω-regular DTMC properties. In: Proc. of MBMV, 169–180. Verlag Dr. Kovač (2012)
14. Wimmer, R., Jansen, N., Ábrahám, E., Katoen, J.P., Becker, B.: Minimal counterexamples for refuting ω-regular properties of Markov decision processes. Reports of SFB/TR 14 AVACS 88 (2012) ISSN: 1860-9821, http://www.avacs.org
15. Schrijver, A.: Theory of Linear and Integer Programming. Wiley (1986)
16. Segala, R.: Modeling and Verification of Randomized Distributed Real-Time Systems. PhD thesis, Massachusetts Institute of Technology (1995), available as Technical Report MIT/LCS/TR-676
17. Segala, R.: A compositional trace-based semantics for probabilistic automata. In: Lee, I., Smolka, S.A. (eds.) CONCUR 1995. LNCS, vol. 962, pp. 234–248. Springer, Heidelberg (1995)
18. Dijkstra, E.W.: Guarded commands, non-determinacy and formal derivation of programs. Communications of the ACM 18(8), 453–457 (1975)
19. He, J., Seidel, K., McIver, A.: Probabilistic models for the guarded command language. Science of Computer Programming 28(2-3), 171–192 (1997)
20. Baier, C., Katoen, J.P.: Principles of Model Checking. The MIT Press (2008)
21. Garey, M.R., Johnson, D.S.: Computers and Intractability: A Guide to the Theory of NP-Completeness. W.H. Freeman & Co. Ltd. (1979)
22. Achterberg, T.: SCIP: Solving constraint integer programs. Mathematical Programming Computation 1(1), 1–41 (2009)
23. IBM: CPLEX optimization studio, version 12.5 (2012), http://www-01.ibm.com/software/integration/optimization/cplex-optimization-studio/
24. Gurobi Optimization, Inc.: Gurobi optimizer reference manual (2012), http://www.gurobi.com
25. Wimmer, R., Jansen, N., Vorpahl, A., Ábrahám, E., Katoen, J.P., Becker, B.: High-level counterexamples for probabilistic automata (extended version). Technical Report arXiv:1305.5055 (2013), http://arxiv.org/abs/1305.5055
26. Kwiatkowska, M., Norman, G., Sproston, J.: Probabilistic model checking of the IEEE 802.11 wireless local area network protocol. In: Hermanns, H., Segala, R. (eds.) PAPM-PROBMIV 2002. LNCS, vol. 2399, pp. 169–187. Springer, Heidelberg (2002)
27. Bianco, A., de Alfaro, L.: Model checking of probabilistic and nondeterministic systems. In: Thiagarajan, P.S. (ed.) FSTTCS 1995. LNCS, vol. 1026, pp. 499–513. Springer, Heidelberg (1995)
28. Kwiatkowska, M., Norman, G., Parker, D.: The PRISM benchmark suite. In: Proc. of QEST, pp. 203–204. IEEE CS Press (2012)
29. Aspnes, J., Herlihy, M.: Fast randomized consensus using shared memory. Journal of Algorithms 15(1), 441–460 (1990)
30. Stoelinga, M.: Fun with FireWire: A comparative study of formal verification methods applied to the IEEE 1394 Root Contention Protocol. Formal Aspects of Computing 14(3), 328–337 (2003)

Modelling, Reduction and Analysis of Markov Automata*

Dennis Guck[1,3], Hassan Hatefi[2], Holger Hermanns[2],
Joost-Pieter Katoen[1,3], and Mark Timmer[3]

[1] Software Modelling and Verification, RWTH Aachen University, Germany
[2] Dependable Systems and Software, Saarland University, Germany
[3] Formal Methods and Tools, University of Twente, The Netherlands

Abstract. Markov automata (MA) constitute an expressive continuous-time compositional modelling formalism. They appear as semantic backbones for engineering frameworks including dynamic fault trees, Generalised Stochastic Petri Nets, and AADL. Their expressive power has thus far precluded them from effective analysis by probabilistic (and statistical) model checkers, stochastic game solvers, or analysis tools for Petri net-like formalisms. This paper presents the foundations and underlying algorithms for efficient MA modelling, reduction using static analysis, and most importantly, quantitative analysis. We also discuss implementation pragmatics of supporting tools and present several case studies demonstrating feasibility and usability of MA in practice.

1 Introduction

Markov automata (MA, for short) have been introduced in [13] as a continuous-time version of Segala's (simple) probabilistic automata [26]. They are closed under parallel composition and hiding. An MA-transition is either labelled with an action, or with a positive real number representing the rate of a negative exponential distribution. An action transition leads to a discrete probability distribution over states. MA can thus model action transitions as in labelled transition systems, probabilistic branching, as well as delays that are governed by exponential distributions.

The semantics of MA has been recently investigated in quite some detail. Weak and strong (bi)simulation semantics have been presented in [13,12], whereas it is shown in [10] that weak bisimulation provides a sound and complete proof methodology for reduction barbed congruence. A process algebra with data for the efficient modelling of MA, accompanied with some reduction techniques using static analysis, has been presented in [29]. Although the MA model raises several challenging theoretical issues, both from a semantical and from an analysis

* This work is funded by the EU FP7-projects MoVeS, SENSATION and MEALS, the DFG-NWO bilateral project ROCKS, the NWO projects SYRUP (grant 612.063.817), the STW project ArRangeer (grant 12238), and the DFG Sonderforschungsbereich AVACS.

K. Joshi et al. (Eds.): QEST 2013, LNCS 8054, pp. 55–71, 2013.
© Springer-Verlag Berlin Heidelberg 2013

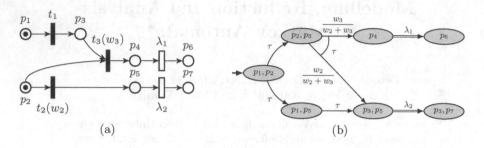

Fig. 1. (a) Confused GSPN, see [22, Fig. 21] with partial weights and (b) its MA semantics

point of view, our main interest is in their practical applicability. As MA extend Hermanns' interactive Markov chains (IMCs) [18], they inherit IMC application domains, ranging from GALS hardware designs [6] and dynamic fault trees [3] to the standardised modeling language AADL [4,17]. The added feature of probabilistic branching yields a natural operational model for generalised stochastic Petri nets (GSPNs) [23] and stochastic activity networks (SANs) [24], both popular modelling formalisms for performance and dependability analysis. Let us briefly motivate this by considering GSPNs. Whereas in SPNs all transitions are subject to a random delay, GSPNs also incorporate immediate transitions, transitions that happen instantaneously. The traditional GSPN semantics yields a continuous-time Markov chain (CTMC), i.e., an MA without action transitions, but is restricted to GSPNs that do not exhibit non-determinism. Such "well-defined" GSPNs occur if the net is free of confusion. It has recently been detailed in [19,11] that MA are a natural semantic model for *every* GSPN. Without going into the technical details, consider the confused GSPN in Fig. 1(a). This net is confused, as the transitions t_1 and t_2 are not in conflict, but firing transition t_1 leads to a conflict between t_2 and t_3, which does not occur if t_2 fires before t_1. Transitions t_2 and t_3 are weighted so that in a marking $\{p_2, p_3\}$ in which both transitions are enabled, t_2 fires with probability $\frac{w_2}{w_2+w_3}$ and t_3 with its complement probability. Classical GSPN semantics and analysis algorithms cannot cope with this net due to the presence of confusion (i.e., non-determinism). Figure 1(b) depicts the MA semantics of this net. Here, states correspond to sets of net places that contain a token. In the initial state, there is a non-deterministic choice between the transitions t_1 and t_2. Note that the presence of weights is naturally represented by discrete probabilistic branching. One can show that for confusion-free GSPNs, the classical semantics and the MA semantics are weakly bisimilar [11].

This paper focuses on the quantitative analysis of MA—and thus (possibly confused) GSPNs and probabilistic AADL error models. We present analysis algorithms for three objectives: expected time, long-run average, and timed (interval) reachability. As the model exhibits non-determinism, we focus on maximal and minimal values for all three objectives. We show that expected time and long-run average objectives can be efficiently reduced to well-known problems on

MDPs such as stochastic shortest path, maximal end-component decomposition, and long-run ratio objectives. This generalizes (and slightly improves) the results reported in [14] for IMCs to MA. Secondly, we present a discretisation algorithm for timed interval reachability objectives which extends [33]. Finally, we present the MaMa tool-chain, an easily accessible publicly available tool chain [1] for the specification, mechanised simplification—such as confluence reduction [31], a form of on-the-fly partial-order reduction—and quantitative evaluation of MA. We describe the overall architectural design, as well as the tool components, and report on empirical results obtained with MaMa on a selection of case studies taken from different domains. The experiments give insight into the effectiveness of our reduction techniques and demonstrate that MA provide the basis of a very expressive stochastic timed modelling approach without sacrificing the ability of time and memory efficient numerical evaluation.

Organisation of the Paper. After introducing Markov Automata in Section 2, we discuss a fully compositional modelling formalism in Section 3. Section 4 considers the evaluation of expected time properties. Section 5 discusses the analysis of long run properties, and Section 6 focusses on reachability properties with time interval bounds. Implementation details of our tool as well as experimental results are discussed in detail in Section 7. Section 8 concludes the paper. Due to space constraints, we refer to [15] for the proofs of our main results.

2   Preliminaries

Markov Automata. An MA is a transition system with two types of transitions: probabilistic (as in PAs) and Markovian transitions (as in CTMCs). Let Act be a universe of actions with internal action $\tau \in Act$, and $\mathsf{Distr}(S)$ denote the set of distribution functions over the countable set S.

Definition 1 (Markov automaton). *A* Markov automaton (MA) *is a tuple* $\mathcal{M} = (S, A, \rightarrow, \Longrightarrow, s_0)$ *where* S *is a nonempty, finite set of states with* initial *state* $s_0 \in S$, $A \subseteq Act$ *is a finite set of actions, and*

- $\rightarrow \subseteq S \times A \times \mathsf{Distr}(S)$ *is the* probabilistic *transition relation, and*
- $\Longrightarrow \subseteq S \times \mathbb{R}_{>0} \times S$ *is the* Markovian *transition relation.*

We abbreviate $(s, \alpha, \mu) \in \rightarrow$ by $s \xrightarrow{\alpha} \mu$ and $(s, \lambda, s') \in \Longrightarrow$ by $s \xrightarrow{\lambda} s'$. An MA can move between states via its probabilistic and Markovian transitions. If $s \xrightarrow{a} \mu$, it can leave state s by executing the action a, after which the probability to go to some state $s' \in S$ is given by $\mu(s')$. If $s \xrightarrow{\lambda} s'$, it moves from s to s' with rate λ, except if s enables a τ-labelled transition. In that case, the MA will always take such a transition and never delays. This is the *maximal progress* assumption [13]. The rationale behind this assumption is that internal transitions are not subject to interaction and thus can happen immediately, whereas

---

[1] Stand-alone download as well as web-based interface available from http://fmt.cs. utwente.nl/~timmer/mama.

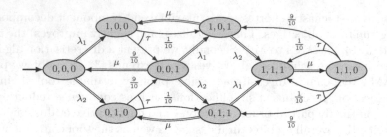

Fig. 2. A queueing system, consisting of a server and two stations. The two stations have incoming requests with rates λ_1, λ_2, which are stored until fetched by the server. If both stations contain a job, the server chooses nondeterministically (in state $(1,1,0)$). Jobs are processed with rate μ, and when polling a station, there is a $\frac{1}{10}$ probability that the job is erroneously kept in the station after being fetched. Each state is represented as a tuple (s_1, s_2, j), with s_i the number of jobs in station i, and j the number of jobs in the server. For simplicity we assume that each component can hold at most one job.

the probability for a Markovian transition to happen immediately is zero. As an example of an MA, consider Fig. 2.

We briefly explain the semantics of Markovian transitions. For a state with Markovian transitions, let $\mathbf{R}(s, s') = \sum\{\lambda \mid s \overset{\lambda}{\Longrightarrow} s'\}$ be the total rate to move from state s to state s', and let $E(s) = \sum_{s' \in S} \mathbf{R}(s, s')$ be the total outgoing rate of s. If $E(s) > 0$, a competition between the transitions of s exists. Then, the probability to move from s to state s' within d time units is

$$\frac{\mathbf{R}(s, s')}{E(s)} \cdot \left(1 - e^{-E(s)d}\right).$$

This asserts that after a delay of at most d time units (second factor), the MA moves to a direct successor state s' with probability $\mathbf{P}(s, s') = \frac{\mathbf{R}(s,s')}{E(s)}$.

Paths. A path in an MA is an infinite sequence $\pi = s_0 \xrightarrow{\sigma_0, \mu_0, t_0} s_1 \xrightarrow{\sigma_1, \mu_1, t_1} \dots$ with $s_i \in S$, $\sigma_i \in Act \cup \{\bot\}$, and $t_i \in \mathbb{R}_{\geq 0}$. For $\sigma_i \in Act$, $s_i \xrightarrow{\sigma_i, \mu_i, t_i} s_{i+1}$ denotes that after residing t_i time units in s_i, the MA has moved via action σ_i to s_{i+1} with probability $\mu_i(s_{i+1})$. Instead, $s_i \xrightarrow{\bot, \mu_i, t_i} s_{i+1}$ denotes that after residing t_i time units in s, a Markovian transition led to s_{i+1} with probability $\mu_i(s_{i+1}) = \mathbf{P}(s_i, s_{i+1})$. For $t \in \mathbb{R}_{\geq 0}$, let $\pi@t$ denote the sequence of states that π occupies at time t. Due to instantaneous action transitions, $\pi@t$ need not be a single state, as an MA may occupy various states at the same time instant. Let *Paths* denote the set of infinite paths. The time elapsed along the path π is $\sum_{i=0}^{\infty} t_i$. Path π is Zeno whenever this sum converges. As the probability of a Zeno path in an MA that only contains Markovian transitions is zero [1], an MA is non-Zeno if and only if no SCC with only probabilistic states is reachable with positive probability. In the rest of this paper, we assume MAs to be non-Zeno.

Policies. Nondeterminism occurs when there is more than one action transition emanating from a state. To define a probability space, the choice is resolved using *policies*. A policy (ranged over by D) is a measurable function which yields for each finite path ending in state s a probability distribution over the set of enabled actions in s. The information on basis of which a policy may decide yields different classes of policies. Let GM denote the class of the general measurable policies. A stationary deterministic policy is a mapping $D\colon PS \to Act$ where PS is the set of states with outgoing probabilistic transitions; such policies always take the same decision in a state s. A time-abstract policy may decide on basis of the states visited so far, but not on their timings; we use TA denote this class. For more details on different classes of policies (and their relation) on models such as MA, we refer to [25]. Using a cylinder set construction we obtain a σ-algebra of subsets of *Paths*; given a policy D and an initial state s, a measurable set of paths is equipped with probability measure $\mathrm{Pr}_{s,D}$.

Stochastic Shortest Path (SSP) Problems. As some objectives on MA are reduced to SSP problems, we briefly introduce them. A non-negative SSP problem is an MDP $(S, Act, \mathbf{P}, s_0)$ with set $G \subseteq S$ of goal states, cost function $c\colon S \setminus G \times Act \to \mathbb{R}_{\geq 0}$ and terminal cost function $g\colon G \to \mathbb{R}_{\geq 0}$. The accumulated cost along a path π through the MDP before reaching G, denoted $C_G(\pi)$, is $\sum_{j=0}^{k-1} c(s_j, \alpha_j) + g(s_k)$ where k is the state index of reaching G. Let $cR^{\min}(s, \Diamond G)$ denote the minimum expected cost reachability of G in the SSP when starting from s. This expected cost can be obtained by solving an LP problem [2].

3 Efficient Modeling of Markov Automata

As argued in the introduction, MA can be used as semantical model for various modeling formalisms. We show this for the process-algebraic specification language MAPA (MA Process Algebra) [29]. This language is rather expressive and supports several reductions techniques for MA specifications. In fact, it turns out to be beneficial to map a language (like GSPNs) to MAPA so as to profit from these reductions. We present the syntax and a brief informal overview of the reduction techniques.

The Markov Automata Process Algebra. MAPA relies on external mechanisms for evaluating expressions, able to handle boolean and real-valued expressions. We assume that any variable-free expression in this language can be evaluated. Our tool uses a simple and intuitive fixed data language that includes basic arithmetic and boolean operators, conditionals, and dynamic lists. For expression t in our data language and vectors $\boldsymbol{x} = (x_1, \ldots, x_n)$ and $\boldsymbol{d} = (d_1, \ldots, d_n)$, let $t[\boldsymbol{x} := \boldsymbol{d}]$ denote the result of substituting every x_i in t by d_i.

A *MAPA specification* consists of a set of uniquely-named *processes* X_i, each defined by a *process equation* $X_i(\boldsymbol{x_i} : \boldsymbol{D_i}) = p_i$. In such an equation, $\boldsymbol{x_i}$ is a vector of process variables with type $\boldsymbol{D_i}$, and p_i is a *process term* specifying the behaviour of X_i. Additionally, each specification has an *initial process* $X_j(\boldsymbol{t})$. We abbreviate $X((x_1, \ldots, x_n) : (D_1 \times \cdots \times D_n))$ by $X(x_1 : D_1, \ldots, x_n : D_n)$. A MAPA *process term* adheres to the grammar:

$$p ::= Y(t) \mid c \Rightarrow p \mid p + p \mid \textstyle\sum_{x:D} p \mid a(t)\textstyle\sum_{x:D} f : p \mid (\lambda) \cdot p$$

---

constant $queueSize = 10, nrOfJobTypes = 3$

type $Stations = \{1, 2\}, \ Jobs = \{1, \ldots, nrOfJobTypes\}$

$Station(i : Stations, q : Queue, size : \{0..queueSize\})$

$\quad = size < queueSize \Rightarrow (2i + 1) \cdot \sum_{j:Jobs} arrive(j) \cdot Station(i, \text{enqueue}(q, j), size + 1)$

$\quad + size > 0 \qquad \Rightarrow deliver(i, \text{head}(q)) \sum_{k \in \{1,9\}} \frac{k}{10} : k = 1 \Rightarrow Station(i, q, size)$

$\qquad\qquad\qquad\qquad\qquad\qquad + k = 9 \Rightarrow Station(i, \text{tail}(q), size - 1)$

$Server = \sum_{n:Stations} \sum_{j:Jobs} poll(n, j) \cdot (2 * j) \cdot finish(j) \cdot Server$

$\gamma(poll, deliver) = copy \qquad$ // actions $poll$ and $deliver$ synchronise and yield action $copy$

$System = \tau_{\{copy, arrive, finish\}}(\partial_{\{poll, deliver\}}(Station(1, \text{empty}, 0) \parallel Station(2, \text{empty}, 0) \parallel Server))$

Fig. 3. MAPA specification of a polling system

Here, Y is a process name, t a vector of expressions, c a boolean expression, x a vector of variables ranging over a finite type D, $a \in Act$ a (parameterised) atomic action, f a real-valued expression yielding a value in $[0, 1]$, and λ an expression yielding a positive real number. Note that, if $|x| > 1$, D is a Cartesian product, as for instance in $\sum_{(m,i):\{m_1, m_2\} \times \{1,2,3\}} send(m, i) \ldots$. In a process term, $Y(t)$ denotes *process instantiation*, where t instantiates Y's process variables (allowing recursion). The term $c \Rightarrow p$ behaves as p if the *condition* c holds, and cannot do anything otherwise. The $+$ operator denotes *nondeterministic choice*, and $\sum_{x:D} p$ a *nondeterministic choice over data type* D. The term $a(t) \sum_{x:D} f : p$ performs the action $a(t)$ and then does a *probabilistic choice* over D. It uses the value $f[x := d]$ as the probability of choosing each $d \in D$. We write $a(t) \cdot p$ for the action $a(t)$ that goes to p with probability 1. Finally, $(\lambda) \cdot p$ can behave as p after a delay, determined by an exponential distribution with rate λ. Using MAPA processes as basic building blocks, the language also supports the modular construction of large systems via top-level parallelism (denoted $\parallel$), encapsulation (denoted ∂), hiding (denoted τ), and renaming (denoted γ), cf. [30, App. B]. The operational semantics of a MAPA specification yields an MA; for details we refer to [29].

Example 1. Fig. 3 depicts the MAPA specification [29] of a polling system—inspired by [27]—which generalised the system of Fig. 2. Now, there are incoming requests of 3 possible types, each of which has a different service rate. Additionally, the stations store these in a queue of size 10. □

Reduction Techniques. To simplify state space generation and reduction, we use a linearised format referred to as MLPPE (Markovian linear probabilistic process equation). In this format, there is precisely one process consisting of a nondeterministic choice between a set of summands. Each summand can contain a nondeterministic choice, followed by a condition, and either an interactive action with a probabilistic choice (determining the next state) or a rate and a next state. Every MAPA specification can be translated efficiently into an MLPPE [29] while

preserving strong bisimulation. On MLPPEs two types of reduction techniques
have been defined: simplifications and state space reductions:

- *Maximal progress reduction* removes Markovian transitions from states also
  having τ-transitions. It is more efficient to perform this on MLPPEs than
  on the initial MAPA specification. We use heuristics (as in [32]) to omit all
  Markovian summands in presence of internal non-Markovian ones.
- *Constant elimination* [20] replaces MLPPE parameters that remain constants by their initial value.
- *Expression simplification* [20] evaluates functions for which all parameters
  are constants and applies basic laws from logic.
- *Summation elimination* [20] removes unnecessary summations, transforming
  e.g., $\sum_{d:\mathbb{N}} d = 5 \Rightarrow send(d) \cdot X$ to $send(5) \cdot X$, $\sum_{d:\{1,2\}} a \cdot X$ to $a \cdot X$, and
  $\sum_{d:D}(\lambda) \cdot X$ to $(|D| \times \lambda) \cdot X$, to preserve the total rate to X.
- *Dead-variable reduction* [32] detects states in which the value of some data
  variable d is irrelevant. This is the case if d will be overwritten before being
  used for all possible futures. Then, d is reset to its initial value.
- *Confluence reduction* [31] detects spurious nondeterminism, resulting from
  parallel composition. It denotes a subset of the probabilistic transitions of
  a MAPA specification as confluent, meaning that they can safely be given
  priority if enabled together with other transitions.

4 Expected Time Objectives

The actions of an MA are only used for composing models from smaller ones.
For the analysis of MA, they are not relevant and we may safely assume that
all actions are internal[2]. Due to the maximal progress assumption, the outgoing transitions of a state s are all either probabilistic transitions or Markovian
transitions. Such states are called probabilistic and Markovian, respectively; let
$PS \subseteq S$ and $MS \subseteq S$ denote these sets.

Let $\mathcal{M}$ be an MA with state space S and $G \subseteq S$ a set of goal states. Define the
(extended) random variable $V_G \colon Paths \to \mathbb{R}^\infty_{\geq 0}$ as the elapsed time before first
visiting some state in G. That is, for an infinite path $\pi = s_0 \xrightarrow{\sigma_0,\mu_0,t_0} s_1 \xrightarrow{\sigma_1,\mu_1,t_1} \cdots$,
let $V_G(\pi) = \min\{t \in \mathbb{R}_{\geq 0} \mid G \cap \pi@t \neq \emptyset\}$ where $\min(\emptyset) = +\infty$. (With slight
abuse of notation we use $\pi@t$ as the set of states occurring in the sequence
$\pi@t$.) The minimal expected time to reach G from $s \in S$ is defined by

$$eT^{\min}(s, \Diamond G) = \inf_D \mathbb{E}_{s,D}(V_G) = \inf_D \int_{Paths} V_G(\pi) \, \mathrm{Pr}_{s,D}(d\pi)$$

where D is a policy on $\mathcal{M}$. Note that by definition of V_G, only the amount of time
before entering the first G-state is relevant. Hence, we may turn all G-states into
absorbing Markovian states without affecting the expected time reachability. In
the remainder we assume all goal states to be absorbing.

---

[2] Like in the MAPA specification of the queueing system in Fig. 3, the actions used
in parallel composition are explicitly turned into internal actions by hiding.

Theorem 1. *The function $eT^{\min}$ is a fixpoint of the Bellman operator*

$$[L(v)](s) = \begin{cases} \dfrac{1}{E(s)} + \displaystyle\sum_{s' \in S} \mathbf{P}(s,s') \cdot v(s') & \text{if } s \in MS \setminus G \\[2ex] \displaystyle\min_{\alpha \in Act(s)} \sum_{s' \in S} \mu_\alpha^s(s') \cdot v(s') & \text{if } s \in PS \setminus G \\[2ex] 0 & \text{if } s \in G. \end{cases}$$

For a goal state, the expected time obviously is zero. For a Markovian state $s \notin G$, the minimal expected time to G is the expected sojourn time in s plus the expected time to reach G via its successor states. For a probabilistic state, an action is selected that minimises the expected reachability time according to the distribution μ_α^s corresponding to α. The characterization of $eT^{\min}(s, \Diamond G)$ in Thm. 1 allows us to reduce the problem of computing the minimum expected time reachability in an MA to a non-negative SSP problem [2,9].

Definition 2 (SSP for minimum expected time reachability). *The SSP of MA $\mathcal{M} = (S, Act, \rightarrow, \Longrightarrow, s_0)$ for the expected time reachability of $G \subseteq S$ is $\mathsf{ssp}_{et}(\mathcal{M}) = (S, Act \cup \{\bot\}, \mathbf{P}, s_0, G, c, g)$ where $g(s) = 0$ for all $s \in G$ and*

$$\mathbf{P}(s, \sigma, s') = \begin{cases} \frac{\mathbf{R}(s,s')}{E(s)} & \text{if } s \in MS, \sigma = \bot \\ \mu_\sigma^s(s') & \text{if } s \in PS, s \xrightarrow{\sigma} \mu_\sigma^s \\ 0 & \text{otherwise, and} \end{cases} \qquad c(s, \sigma) = \begin{cases} \frac{1}{E(s)} & \text{if } s \in MS \setminus G, \sigma = \bot \\ 0 & \text{otherwise.} \end{cases}$$

Terminal costs are zero. Transition probabilities are defined in the standard way. The reward of a Markovian state is its expected sojourn time, and zero otherwise.

Theorem 2. *For MA $\mathcal{M}$, $eT^{\min}(s, \Diamond G)$ equals $cR^{\min}(s, \Diamond G)$ in $\mathsf{ssp}_{et}(\mathcal{M})$.*

Thus here is a stationary deterministic policy on $\mathcal{M}$ yielding $eT^{\min}(s, \Diamond G)$. Moreover, the uniqueness of the minimum expected cost of an SSP [2,9] now yields that $eT^{\min}(s, \Diamond G)$ is the unique fixpoint of L (see Thm. 1). The uniqueness result enables the usage of standard solution techniques such as value iteration and linear programming to compute $eT^{\min}(s, \Diamond G)$. For maximal expected time objectives, a similar fixpoint theorem is obtained, and it can be proven that those objectives correspond to the maximal expected reward in the SSP problem defined above. In the above, we have assumed MA to not contain any Zeno cycle, i.e., a cycle solely consisting of probabilistic transitions. The above notions can all be extended to deal with such Zeno cycles, by, e.g., setting the minimal expected time of states in Zeno BSCCs that do not contain G-states to be infinite (as such states cannot reach G). Similarly, the maximal expected time of states in Zeno end components (that do not containg G-states) can be defined as ∞, as in the worst case these states will never reach G.

5   Long Run Objectives

Let $\mathcal{M}$ be an MA with state space S and $G \subseteq S$ a set of goal states. Let $\mathbf{1}_G$ be the characteristic function of G, i.e., $\mathbf{1}_G(s) = 1$ if and only if $s \in G$.

Following the ideas of [8,21], the fraction of time spent in G on an infinite path π in $\mathcal{M}$ up to time bound $t \in \mathbb{R}_{\geq 0}$ is given by the random variable (r. v.) $A_{G,t}(\pi) = \frac{1}{t} \int_0^t 1_G(\pi@u)\, du$. Taking the limit $t \to \infty$, we obtain the r. v.

$$A_G(\pi) = \lim_{t \to \infty} A_{G,t}(\pi) = \lim_{t \to \infty} \frac{1}{t} \int_0^t 1_G(\pi@u)\, du.$$

The expectation of A_G for policy D and initial state s yields the corresponding long-run average time spent in G:

$$LRA^D(s,G) = \mathbb{E}_{s,D}(A_G) = \int_{Paths} A_G(\pi) \Pr_{s,D}(d\pi).$$

The minimum long-run average time spent in G starting from state s is then:

$$LRA^{\min}(s,G) = \inf_D LRA^D(s,G) = \inf_D \mathbb{E}_{s,D}(A_G).$$

For the long-run average analysis, we may assume w.l.o.g. that $G \subseteq MS$, as the long-run average time spent in any probabilistic state is always 0. This claim follows directly from the fact that probabilistic states are instantaneous, i.e. their sojourn time is 0 by definition. Note that in contrast to the expected time analysis, G-states cannot be made absorbing in the long-run average analysis. It turns out that stationary deterministic policies are sufficient for yielding minimal or maximal long-run average objectives.

In the remainder of this section, we discuss in detail how to compute the minimum long-run average fraction of time to be in G in an MA $\mathcal{M}$ with initial state s_0. The general idea is the following three-step procedure:

1. Determine the maximal end components[3] $\{\mathcal{M}_1, \ldots, \mathcal{M}_k\}$ of MA $\mathcal{M}$.
2. Determine $LRA^{\min}(G)$ in maximal end component $\mathcal{M}_j$ for all $j \in \{1, \ldots, k\}$.
3. Reduce the computation of $LRA^{\min}(s_0, G)$ in MA $\mathcal{M}$ to an SSP problem.

The first phase can be performed by a graph-based algorithm [7,5], whereas the last two phases boil down to solving LP problems.

Unichain MA. We first show that for unichain MA, i.e., MA that under any stationary deterministic policy yield a strongly connected graph structure, computing $LRA^{\min}(s,G)$ can be reduced to determining long-ratio objectives in MDPs. Let us first explain such objectives. Let $M = (S, Act, \mathbf{P}, s_0)$ be an MDP. Assume w.l.o.g. that for each state s in M there exists $\alpha \in Act$ such that $\mathbf{P}(s, \alpha, s') > 0$. Let $c_1, c_2: S \times (Act \cup \{\perp\}) \to \mathbb{R}_{\geq 0}$ be cost functions. The operational interpretation is that a cost $c_1(s, \alpha)$ is incurred when selecting action α in state s, and similar for c_2. Our interest is the *ratio* between c_1 and c_2 along a path. The

---

[3] A sub-MA of MA $\mathcal{M}$ is a pair (S', K) where $S' \subseteq S$ and K is a function that assigns to each $s \in S'$ a non-empty set of actions such that for all $\alpha \in K(s)$, $s \xrightarrow{\alpha} \mu$ with $\mu(s') > 0$ or $s \xRightarrow{\lambda} s'$ imply $s' \in S'$. An end component is a sub-MA whose underlying graph is strongly connected; it is maximal w.r.t. K if it is not contained in any other end component (S'', K).

long-run ratio $\mathcal{R}$ between the accumulated costs c_1 and c_2 along the infinite path $\pi = s_0 \xrightarrow{\alpha_0} s_1 \xrightarrow{\alpha_1} \ldots$ in the MDP M is defined by[4]:

$$\mathcal{R}(\pi) = \lim_{n \to \infty} \frac{\sum_{i=0}^{n-1} c_1(s_i, \alpha_i)}{\sum_{j=0}^{n-1} c_2(s_j, \alpha_j)}.$$

The minimum long-run ratio objective for state s of MDP M is defined by:

$$R^{\min}(s) = \inf_D \mathbb{E}_{s,D}(\mathcal{R}) = \inf_D \sum_{\pi \in Paths} \mathcal{R}(\pi) \cdot \mathrm{Pr}_{s,D}(\pi).$$

Here, *Paths* is the set of paths in the MDP, D an MDP-policy, and Pr the probability mass on MDP-paths. From [7], it follows that $R^{\min}(s)$ can be obtained by solving the following LP problem with real variables k and x_s for each $s \in S$: Maximize k subject to:

$$x_s \leq c_1(s, \alpha) - k \cdot c_2(s, \alpha) + \sum_{s' \in S} \mathbf{P}(s, \alpha, s') \cdot x_{s'} \quad \text{for each } s \in S, \alpha \in Act.$$

We now transform an MA into an MDP with 2 cost functions as follows.

Definition 3 (From MA to two-cost MDPs). *Let* $\mathcal{M} = (S, Act, \to, \Longrightarrow, s_0)$ *be an MA and* $G \subseteq S$ *a set of goal states. The MDP* $\mathsf{mdp}(\mathcal{M}) = (S, Act \cup \{\bot\}, \mathbf{P}, s_0)$ *with cost functions* c_1 *and* c_2, *where* $\mathbf{P}$ *is defined as in Def. 2, and*

$$c_1(s, \sigma) = \begin{cases} \frac{1}{E(s)} & \text{if } s \in MS \cap G \wedge \sigma = \bot \\ 0 & \text{otherwise,} \end{cases} \qquad c_2(s, \sigma) = \begin{cases} \frac{1}{E(s)} & \text{if } s \in MS \wedge \sigma = \bot \\ 0 & \text{otherwise.} \end{cases}$$

Observe that cost function c_2 keeps track of the average residence time in state s whereas c_1 only does so for states in G.

Theorem 3. *For unichain MA* $\mathcal{M}$, $LRA^{\min}(s, G)$ *equals* $R^{\min}(s)$ *in* $\mathsf{mdp}(\mathcal{M})$.

To summarise, computing the minimum long-run average fraction of time that is spent in some goal state in $G \subseteq S$ in an unichain MA $\mathcal{M}$ equals the minimum long-run ratio objective in an MDP with two cost functions. The latter can be obtained by solving an LP problem. Observe that for any two states s, s' in a unichain MA, $LRA^{\min}(s, G)$ and $LRA^{\min}(s', G)$ coincide. We therefore omit the state and simply write $LRA^{\min}(G)$ when considering unichain MA.

Arbitrary MA. Let $\mathcal{M}$ be an MA with initial state s_0 and maximal end components $\{\mathcal{M}_1, \ldots, \mathcal{M}_k\}$ for $k > 0$ where MA $\mathcal{M}_j$ has state space S_j. Note that each $\mathcal{M}_j$ is a unichain MA. Using this decomposition of $\mathcal{M}$ into maximal end components, we obtain the following result:

---

[4] In our setting, $\mathcal{R}(\pi)$ is well-defined as the cost functions c_1 and c_2 are obtained from non-Zeno MA. Thus for any infinite path π, $c_2(s_j, \alpha_j) > 0$ for some index j.

Theorem 4. [5] *For MA* $\mathcal{M} = (S, Act, \rightarrow, \Longrightarrow, s_0)$ *with MECs* $\{\mathcal{M}_1, \ldots, \mathcal{M}_k\}$ *with state spaces* $S_1, \ldots, S_k \subseteq S$, *and set of goal states* $G \subseteq S$:

$$LRA^{\min}(s_0, G) = \inf_D \sum_{j=1}^{k} LRA_j^{\min}(G) \cdot \mathrm{Pr}^D(s_0 \models \Diamond \Box S_j),$$

where $\mathrm{Pr}^D(s_0 \models \Diamond \Box S_j)$ *is the probability to eventually reach and continuously stay in some state in* S_j *from* s_0 *under policy* D *and* $LRA_j^{\min}(G)$ *is the LRA of* $G \cap S_j$ *in unichain MA* $\mathcal{M}_j$.

Computing minimal LRA for arbitrary MA is now reducible to a non-negative SSP problem. This proceeds as follows. In MA $\mathcal{M}$, we replace each maximal end component $\mathcal{M}_j$ by two fresh states q_j and u_j. Intuitively, q_j represents $\mathcal{M}_j$ whereas u_j represents a decision state. State u_j has a transition to q_j and contains all probabilistic transitions leaving S_j. Let U denote the set of u_j states and Q the set of q_j states.

Definition 4 (SSP for long run average). *The SSP of MA* $\mathcal{M}$ *for the LRA in* $G \subseteq S$ *is* $\mathsf{ssp}_{lra}(\mathcal{M}) = (S \setminus \bigcup_{i=1}^{k} S_i \cup U \cup Q, Act \cup \{\bot\}, \mathbf{P}', s_0, Q, c, g)$, *where* $g(q_i) = LRA_i^{\min}(G)$ *for* $q_i \in Q$ *and* $c(s, \sigma) = 0$ *for all* s *and* $\sigma \in Act \cup \{\bot\}$. $\mathbf{P}'$ *is defined as follows. Let* $S' = S \setminus \bigcup_{i=1}^{k} S_i$. $\mathbf{P}'$ *equals* $\mathbf{P}$ *for all* $s, s' \in S'$. *For the new states* u_j:

$\mathbf{P}'(u_j, \tau, s') = \mathbf{P}(S_j, \tau, s')$ *if* $s' \in S' \setminus S_j$ *and* $\mathbf{P}'(u_i, \tau, u_j) = \mathbf{P}(S_i, \tau, S_j)$ *for* $i \neq j$.

Finally, we have: $\mathbf{P}'(q_j, \bot, q_j) = 1 = \mathbf{P}'(u_j, \bot, q_j)$ *and* $\mathbf{P}'(s, \sigma, u_j) = \mathbf{P}(s, \sigma, S_j)$.

Here, $\mathbf{P}(s, \alpha, S')$ is a shorthand for $\sum_{s' \in S'} \mathbf{P}(s, \alpha, s')$; similarly, $\mathbf{P}(S', \alpha, s') = \sum_{s \in S'} \mathbf{P}(s, \alpha, s')$. The terminal costs of the new q_i-states are set to $LRA_i^{\min}(G)$.

Theorem 5. *For MA* $\mathcal{M}$, $LRA^{\min}(s, G)$ *equals* $cR^{\min}(s, \Diamond U)$ *in SSP* $\mathsf{ssp}_{lra}(\mathcal{M})$.

6   Timed Reachability Objectives

This section presents an algorithm that approximates time-bounded reachability probabilities in MA. We start with a fixed point characterisation, and then explain how these probabilities can be approximated using digitisation.

Fixed Point Characterisation. Our goal is to come up with a fixed point characterisation for the maximum (minimum) probability to reach a set of goal states in a time interval. Let $\mathcal{I}$ and $\mathcal{Q}$ be the set of all nonempty nonnegative real intervals with real and rational bounds, respectively. For interval $I \in \mathcal{I}$ and $t \in \mathbb{R}_{\geq 0}$, let $I \ominus t = \{x - t \mid x \in I \wedge x \geq t\}$. Given MA $\mathcal{M}$, $I \in \mathcal{I}$ and a set $G \subset S$ of goal states, the set of all paths that reach some goal states within interval I is denoted by $\Diamond^I G$. Let $p_{\max}^{\mathcal{M}}(s, \Diamond^I G)$ be the maximum probability of reaching G within interval I if starting in state s at time 0. Here, the maximum is taken over all possible general measurable policies. The next result provides a characterisation of $p_{\max}^{\mathcal{M}}(s, \Diamond^I G)$ as a fixed point.

---

[5] This theorem corrects a small flaw in the corresponding theorem for IMCs in [14].

Lemma 1. *Let $\mathcal{M}$ be an MA, $G \subseteq S$ and $I \in \mathcal{I}$ with $\inf I = a$ and $\sup I = b$. Then, $p_{\max}^{\mathcal{M}}(s, \diamondsuit^I G)$ is the least fixed point of the higher-order operator $\Omega \colon (S \times \mathcal{I} \rightarrowtail [0,1]) \rightarrowtail (S \times \mathcal{I} \rightarrowtail [0,1])$, which for $s \in MS$ is given by:*

$$\Omega(F)(s, I) = \begin{cases} \int_0^b E(s)e^{-E(s)t}\sum_{s' \in S} \mathbf{P}(s, \bot, s')F(s', I \ominus t)\,\mathrm{d}t & s \notin G \\ e^{-E(s)a} + \int_0^a E(s)e^{-E(s)t}\sum_{s' \in S}\mathbf{P}(s,\bot,s')F(s', I \ominus t)\,\mathrm{d}t & s \in G \end{cases}$$

and for $s \in PS$ is defined by:

$$\Omega(F)(s, I) = \begin{cases} 1 & s \in G \wedge a = 0 \\ \max_{\alpha \in Act_{\backslash\bot}(s)} \sum_{s' \in S} \mathbf{P}(s, \alpha, s')F(s', I) & otherwise. \end{cases}$$

This characterisation is a simple generalisation of that for IMCs [33], reflecting the fact that taking an action from an probabilistic state leads to a distribution over the states (rather than a single state). The above characterisation yields an integral equation system which is in general not directly tractable [1]. To tackle this problem, we approximate the fixed point characterisation using digitisation, extending ideas developed in [33]. We split the time interval into equally-sized digitisation steps, assuming a digitisation constant δ, small enough such that with high probability at most one Markovian transition firing occurs in any digitisation step. This allows us to construct a digitised MA (dMA), a variant of a semi-MDP, obtained by summarising the behaviour of the MA at equidistant time points. Paths in a dMA can be seen as time-abstract paths in the corresponding MA, implicitly still counting digitisation steps, and thus discrete time. Digitisation of MA $\mathcal{M} = (S, Act, \rightarrow, \Longrightarrow, s_0)$ and digitisation constant δ, proceeds by replacing $\Longrightarrow$ by $\Longrightarrow_\delta = \{(s, \mu^s) \mid s \in MS\}$, where

$$\mu^s(s') = \begin{cases} (1 - e^{-E(s)\delta})\mathbf{P}(s, \bot, s') & \text{if } s' \neq s \\ (1 - e^{-E(s)\delta})\mathbf{P}(s, \bot, s') + e^{-E(s)\delta} & \text{otherwise.} \end{cases}$$

Using the above fixed point characterisation, it is now possible to relate reachability probabilities in an MA $\mathcal{M}$ to reachability probabilities in its dMA $\mathcal{M}_\delta$.

Theorem 6. *Given MA $\mathcal{M} = (S, Act, \rightarrow, \Longrightarrow, s_0)$, $G \subseteq S$, interval $I = [0, b] \in \mathcal{Q}$ with $b \geq 0$ and $\lambda = \max_{s \in MS} E(s)$. Let $\delta > 0$ be such that $b = k_b\delta$ for some $k_b \in \mathbb{N}$. Then, for all $s \in S$ it holds that*

$$p_{\max}^{\mathcal{M}_\delta}(s, \diamondsuit^{[0,k_b]} G) \ \leq\ p_{\max}^{\mathcal{M}}(s, \diamondsuit^{[0,b]} G) \ \leq\ p_{\max}^{\mathcal{M}_\delta}(s, \diamondsuit^{[0,k_b]} G) + 1 - e^{-\lambda b}(1 + \lambda\delta)^{k_b}.$$

This theorem can be extended to intervals with non-zero lower bounds; for the sake of brevity, the details are omitted here. The remaining problem is to compute the maximum (or minimum) probability to reach G in a dMA within a step bound $k \in \mathbb{N}$. Let $\diamondsuit^{[0,k]} G$ be the set of infinite paths in a dMA that reach a G state within k steps, and $p_{\max}^{\mathcal{D}}(s, \diamondsuit^{[0,k]} G)$ denote the maximum probability of this set. Then we have $p_{\max}^{\mathcal{D}}(s, \diamondsuit^{[0,k]} G) = \sup_{D \in TA} \Pr_{s,D}(\diamondsuit^{[0,k]} G)$. Our algorithm is now an adaptation (to dMA) of the well-known value iteration scheme for MDPs.

The algorithm proceeds by backward unfolding of the dMA in an iterative manner, starting from the goal states. Each iteration intertwines the analysis of Markov states and of probabilistic states. The key issue is that a path from probabilistic states to G is split into two parts: reaching Markov states from probabilistic states in zero time and reaching goal states from Markov states in interval $[0, j]$, where j is the step count of the iteration. The former computation can be reduced to an unbounded reachability problem in the MDP induced by probabilistic states with rewards on Markov states. For the latter, the algorithm operates on the previously computed reachability probabilities from all Markov states up to step count j. We can generalize this recipe from step-bounded reachability to step interval-bounded reachability, details are described in [16].

7   Tool-Chain and Case Studies

This section describes the implementation of the algorithms discussed, together with the modelling features resulting in our MAMA tool-chain. Furthermore, we present two case studies that provide empirical evidence of the strengths and weaknesses of the MAMA tool chain.

7.1   MAMA Tool Chain

Our tool chain consists of several tool components: SCOOP [28,29], IMCA [14], and GEMMA (realized in Haskell), see Figure 4. The tool-chain comprises about 8,000 LOC (without comments). SCOOP (in Haskell) supports the generation from MA from MAPA specifications by a translation into the MLPPE format. It implements all the reduction techniques described in Section 3, in particular confluence reduction. The capabilities of the IMCA tool-component (written in C++) have been lifted to expected time and long-run objectives for MA, and extended with timed reachability objectives. It also supports (untimed) reachability objectives which are not further treated here. A prototypical translator from GSPNs to MA, in fact MAPA specifications, has been realized (the GEMMA component). We connected the three components into a single tool chain, by making SCOOP export the (reduced) state space of an MLPPE in the IMCA input language. Additionally, SCOOP has been extended to translate properties, based on the actions and parameters of a MAPA specification, to a set of goal states in the underlying MA. That way, in one easy process systems and their properties can be modelled in MAPA, translated to an optimised MLPPE by SCOOP, exported to the IMCA tool and then analysed.

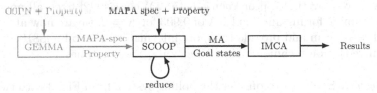

Fig. 4. Analysing Markov Automata using the MAMA tool chain

Table 1. Interval reachability probabilities for the grid. (Time in seconds.)

K	unreduced			reduced			ϵ	I	$p^{min}(s_0, \square^I G)$	time(unred)	time(red)	$p^{max}(s_0, \square^I G)$	time(unred)	time(red)
	$|S|$	$|G|$	time	$|S|$	$|G|$	time								
2	2,508	1,398	0.6	1,789	1,122	0.8	10^{-2}	[0,3]	0.91	58.5	31.0	0.95	54.9	21.7
							10^{-2}	[0,4]	0.96	103.0	54.7	0.98	97.3	38.8
							10^{-2}	[1,4]	0.91	117.3	64.4	0.96	109.9	49.0
							10^{-3}	[0,3]	0.910	580.1	309.4	0.950	544.3	218.4
3	10,852	4,504	3.1	7,201	3,613	3.5	10^{-2}	[0,3]	0.18	361.5	202.8	0.23	382.8	161.1
							10^{-2}	[0,4]	0.23	643.1	360.0	0.30	681.4	286.0
							10^{-2}	[1,4]	0.18	666.6	377.3	0.25	696.4	317.7
							10^{-3}	[0,3]	0.176	3,619.5	2,032.1	0.231	3,837.3	1,611.9
4	31,832	10,424	9.8	20,021	8,357	10.5	10^{-2}	[0,3]	0.01	1,156.8	614.9	0.03	1,196.5	486.4

7.2   Case Studies

This section reports on experiments with MaMa. All experiments were conducted on a 2.5 GHz Intel Core i5 processor with 4GB RAM, running on Mac OS X 10.8.3.

Processor Grid. First, we consider a model of a 2×2 concurrent processor architecture. Using GEMMA, we automatically derived the MA model from the GSPN model in [22, Fig. 11.7]. Previous analysis of this model required weights for all immediate transitions, requiring complete knowledge of the mutual behaviour of all these transitions. We allow a weight assignment to just a (possibly empty) subset of the immediate transitions—reflecting the practical scenario of only knowing the mutual behaviour for a selection of the transitions. For this case study we indeed kept weights for only a few of the transitions, obtaining probabilistic behaviour for them and nondeterministic behaviour for the others.

Table 1 reports on the time-bounded and time-interval bounded probabilities for reaching a state such that the first processor has an empty task queue. We vary the degree of multitasking K, the error bound ϵ and the interval I. For each setting, we report the number of states $|S|$ and goal states $|G|$, and the generation time with SCOOP (both with and without the reductions from Section 3).

The runtime demands grow with both the upper and lower time bound, as well as with the required accuracy. The model size also affects the per-iteration cost and thus the overall complexity of reachability computation. Note that our reductions speed-up the analysis times by a factor between 1.7 and 3.5: even more than the reduction in state space size. This is due to our techniques significantly reducing the degree of nondeterminism.

Table 2 displays results for expected time until an empty task queue, as well as the long-run average that a processor is active. Whereas [22] fixed all nondeterminism, obtaining for instance an LRA of 0.903 for $K = 2$, we are now able to retain nondeterminism and provide the more informative interval $[0.8810, 0.9953]$. Again, our reduction techniques significantly improve runtimes.

Polling System. Second, we consider the polling system from Fig. 3 with two stations and one server. We varied the queue sizes Q and the number of job types N,

Table 2. Expected times and long-run averages for the grid. (Time in seconds.)

K	$eT^{min}(s_0, \Diamond G)$	time(unred)	time(red)	$eT^{max}(s_0, \Diamond G)$	time(unred)	time(red)	$LRA^{min}(s_0, G)$	time(unred)	time(red)	$LRA^{max}(s_0, G)$	time(unred)	time(red)
2	1.0000	0.3	0.1	1.2330	0.7	0.3	0.8110	1.3	0.7	0.9953	0.5	0.2
3	11.1168	18.3	7.7	15.2768	135.4	40.6	0.8173	36.1	16.1	0.9998	4.7	2.6
4	102.1921	527.1	209.9	287.8616	6,695.2	1,869.7	0.8181	505.1	222.3	1.0000	57.0	34.5

Table 3. Interval reachability probabilities for the polling system. (Time in seconds.)

Q	N	unreduced $\|S\|$	$\|G\|$	time	reduced $\|S\|$	$\|G\|$	time	ϵ	I	$p^{min}(s_0, \Diamond^I G)$	time(unred)	time(red)	$p^{max}(s_0, \Diamond^I G)$	time(unred)	time(red)
2	3	1,497	567	0.4	990	324	0.2	10^{-3}	[0,1]	0.277	4.7	2.9	0.558	4.6	2.5
								10^{-3}	[1,2]	0.486	22.1	14.9	0.917	22.7	12.5
2	4	4,811	2,304	1.0	3,047	1,280	0.6	10^{-3}	[0,1]	0.201	25.1	14.4	0.558	24.0	13.5
								10^{-3}	[1,2]	0.344	106.1	65.8	0.917	102.5	60.5
3	3	14,322	5,103	3.0	9,522	2,916	1.7	10^{-3}	[0,1]	0.090	66.2	40.4	0.291	60.0	38.5
								10^{-3}	[1,2]	0.249	248.1	180.9	0.811	241.9	158.8
3	4	79,307	36,864	51.6	50,407	20,480	19.1	10^{-3}	[0,1]	0.054	541.6	303.6	0.291	578.2	311.0
								10^{-3}	[1,2]	0.141	2,289.3	1,305.0	0.811	2,201.5	1,225.9
4	2	6,667	1,280	1.1	4,745	768	0.8	10^{-3}	[0,1]	0.049	19.6	14.0	0.118	19.7	12.8
								10^{-3}	[1,2]	0.240	83.2	58.7	0.651	80.9	53.1
4	3	131,529	45,927	85.2	87,606	26,244	30.8	10^{-3}	[0,1]	0.025	835.3	479.0	0.118	800.7	466.1
								10^{-3}	[1,2]	0.114	3,535.5	2,062.3	0.651	3,358.9	2,099.5

Table 4. Expected times and long-run averages for the polling system. (Time in seconds.)

Q	N	$eT^{min}(s_0, \Diamond G)$	time(unred)	time(red)	$eT^{max}(s_0, \Diamond G)$	time(unred)	time(red)	$LRA^{min}(s_0, G)$	time(unred)	time(red)	$LRA^{max}(s_0, G)$	time(unred)	time(red)
2	3	1.0478	0.2	0.1	2.2489	0.3	0.2	0.1230	0.8	0.5	0.6596	0.2	0.1
2	4	1.0478	0.2	0.1	3.2053	2.0	1.0	0.0635	9.0	5.2	0.6596	1.3	0.6
3	3	1.4425	1.0	0.6	4.6685	8.4	5.0	0.0689	177.9	123.6	0.6600	26.2	13.0
3	4	1.4425	9.7	4.6	8.0294	117.4	67.2	0.0277	7,696.7	5,959.5	0.6600	1,537.2	862.4
4	2	1.8226	0.4	0.3	4.6032	2.4	1.6	0.1312	45.6	32.5	0.6601	5.6	3.9
4	3	1.8226	29.8	14.2	9.0300	232.8	130.8	– timeout (18 hours) –			0.6601	5,339.8	3,099.0

analysing a total of six different settings. Since—as for the previous case—analysis scales proportionally with the error bound, we keep this constant here.

Table 3 reports results for time-bounded and time-interval bounded properties, and Table 4 displays probabilities and runtime results for expected times and long-run averages. For all analyses, the goal set consists of all states for which both station queues are full.

8 Conclusion

This paper presented new algorithms for the quantitative analysis of Markov automata (MA) and proved their correctness. Three objectives have been considered: expected time, long-run average, and timed reachability. The MaMa tool-chain supports the modelling and reduction of MA, and can analyse these three objectives. It is also equipped with a prototypical tool to map GSPNs onto MA. The MaMa is accessible via its easy-to-use web interface that can be found at http://wwwhome.cs.utwente.nl/~timmer/mama. Experimental results on a processor grid and a polling system give insight into the accuracy and scalability of the presented algorithms. Future work will focus on efficiency improvements and reward extensions.

References

1. Baier, C., Haverkort, B.R., Hermanns, H., Katoen, J.-P.: Model-checking algorithms for continuous-time Markov chains. IEEE TSE 29(6), 524–541 (2003)
2. Bertsekas, D.P., Tsitsiklis, J.N.: An analysis of stochastic shortest path problems. Mathematics of Operations Research 16(3), 580–595 (1991)
3. Boudali, H., Crouzen, P., Stoelinga, M.I.A.: A rigorous, compositional, and extensible framework for dynamic fault tree analysis. IEEE Trans. Dependable Sec. Comput. 7(2), 128–143 (2010)
4. Bozzano, M., Cimatti, A., Katoen, J.-P., Nguyen, V.Y., Noll, T., Roveri, M.: Safety, dependability and performance analysis of extended AADL models. The Computer Journal 54(5), 754–775 (2011)
5. Chatterjee, K., Henzinger, M.: Faster and dynamic algorithms for maximal end-component decomposition and related graph problems in probabilistic verification. In: SODA, pp. 1318–1336. SIAM (2011)
6. Coste, N., Hermanns, H., Lantreibecq, E., Serwe, W.: Towards performance prediction of compositional models in industrial GALS designs. In: Bouajjani, A., Maler, O. (eds.) CAV 2009. LNCS, vol. 5643, pp. 204–218. Springer, Heidelberg (2009)
7. de Alfaro, L.: Formal Verification of Probabilistic Systems. PhD thesis, Stanford University (1997)
8. de Alfaro, L.: How to specify and verify the long-run average behavior of probabilistic systems. In: LICS, pp. 454–465. IEEE (1998)
9. de Alfaro, L.: Computing minimum and maximum reachability times in probabilistic systems. In: Baeten, J.C.M., Mauw, S. (eds.) CONCUR 1999. LNCS, vol. 1664, pp. 66–81. Springer, Heidelberg (1999)
10. Deng, Y., Hennessy, M.: On the semantics of Markov automata. Inf. Comput. 222, 139–168 (2013)
11. Eisentraut, C., Hermanns, H., Katoen, J.-P., Zhang, L.: A semantics for every GSPN. In: Colom, J.-M., Desel, J. (eds.) ICATPN 2013. LNCS, vol. 7927, pp. 90–109. Springer, Heidelberg (2013)
12. Eisentraut, C., Hermanns, H., Zhang, L.: Concurrency and composition in a stochastic world. In: Gastin, P., Laroussinie, F. (eds.) CONCUR 2010. LNCS, vol. 6269, pp. 21–39. Springer, Heidelberg (2010)
13. Eisentraut, C., Hermanns, H., Zhang, L.: On probabilistic automata in continuous time. In: LICS, pp. 342–351. IEEE (2010)
14. Guck, D., Han, T., Katoen, J.-P., Neuhäußer, M.R.: Quantitative timed analysis of interactive Markov chains. In: Goodloe, A.E., Person, S. (eds.) NFM 2012. LNCS, vol. 7226, pp. 8–23. Springer, Heidelberg (2012)
15. Guck, D., Hatefi, H., Hermanns, H., Katoen, J.-P., Timmer, M.: Modelling, reduction and analysis of Markov automata (extended version). Technical Report 1305.7050, ArXiv e-prints (2013)

16. Hatefi, H., Hermanns, H.: Model checking algorithms for Markov automata. In: ECEASST (AVoCS proceedings), vol. 53 (2012) (to appear)
17. Haverkort, B.R., Kuntz, M., Remke, A., Roolvink, S., Stoelinga, M.I.A.: Evaluating repair strategies for a water-treatment facility using Arcade. In: DSN, pp. 419–424. IEEE (2010)
18. Hermanns, H.: Interactive Markov Chains: The Quest for Quantified Quality. LNCS, vol. 2428. Springer, Heidelberg (2002)
19. Katoen, J.-P.: GSPNs revisited: Simple semantics and new analysis algorithms. In: ACSD, pp. 6–11. IEEE (2012)
20. Katoen, J.-P., van de Pol, J.C., Stoelinga, M.I.A., Timmer, M.: A linear process-algebraic format with data for probabilistic automata. TCS 413(1), 36–57 (2012)
21. López, G.G.I., Hermanns, H., Katoen, J.-P.: Beyond memoryless distributions: Model checking semi-markov chains. In: de Luca, L., Gilmore, S. (eds.) PAPM-PROBMIV 2001. LNCS, vol. 2165, pp. 57–70. Springer, Heidelberg (2001)
22. Ajmone Marsan, M., Balbo, G., Conte, G., Donatelli, S., Franceschinis, G.: Modelling with Generalized Stochastic Petri Nets. John Wiley & Sons (1995)
23. Ajmone Marsan, M., Conte, G., Balbo, G.: A class of generalized stochastic Petri nets for the performance evaluation of multiprocessor systems. ACM Transactions on Computer Systems 2(2), 93–122 (1984)
24. Meyer, J.F., Movaghar, A., Sanders, W.H.: Stochastic activity networks: Structure, behavior, and application. In: PNPM, pp. 106–115. IEEE (1985)
25. Neuhäußer, M.R., Stoelinga, M.I.A., Katoen, J.-P.: Delayed nondeterminism in continuous-time Markov decision processes. In: de Alfaro, L. (ed.) FOSSACS 2009. LNCS, vol. 5504, pp. 364–379. Springer, Heidelberg (2009)
26. Segala, R.: Modeling and Verification of Randomized Distributed Real-Time Systems. PhD thesis, MIT (1995)
27. Srinivasan, M.M.: Nondeterministic polling systems. Management Science 37(6), 667–681 (1991)
28. Timmer, M.: SCOOP: A tool for symbolic optimisations of probabilistic processes. In: QEST, pp. 149–150. IEEE (2011)
29. Timmer, M., Katoen, J.-P., van de Pol, J.C., Stoelinga, M.I.A.: Efficient modelling and generation of Markov automata. In: Koutny, M., Ulidowski, I. (eds.) CONCUR 2012. LNCS, vol. 7454, pp. 364–379. Springer, Heidelberg (2012)
30. Timmer, M., Katoen, J.-P., van de Pol, J.C., Stoelinga, M.I.A.: Efficient modelling and generation of Markov automata (extended version). Technical Report TR-CTIT-12-16, CTIT, University of Twente (2012)
31. Timmer, M., van de Pol, J.C., Stoelinga, M.I.A.: Confluence reduction for markov automata. In: Braberman, V., Fribourg, L. (eds.) FORMATS 2013. LNCS, vol. 8053, pp. 240–254. Springer, Heidelberg (2013)
32. van de Pol, J.C., Timmer, M.: State space reduction of linear processes using control flow reconstruction. In: Liu, Z., Ravn, A.P. (eds.) ATVA 2009. LNCS, vol. 5799, pp. 54–68. Springer, Heidelberg (2009)
33. Zhang, L., Neuhäußer, M.R.: Model checking interactive markov chains. In: Esparza, J., Majumdar, R. (eds.) TACAS 2010. LNCS, vol. 6015, pp. 53–68. Springer, Heidelberg (2010)

Deciding Bisimilarities on Distributions

Christian Eisentraut[1], Holger Hermanns[1], Julia Krämer[1],
Andrea Turrini[1], and Lijun Zhang[2,3,1]

[1] Saarland University – Computer Science, Saarbrücken, Germany
[2] State Key Laboratory of Computer Science, Institute of Software,
Chinese Academy of Sciences, Beijing, China
[3] DTU Informatics, Technical University of Denmark, Denmark

Abstract. Probabilistic automata (*PA*) are a prominent compositional concurrency model. As a way to justify property-preserving abstractions, in the last years, bisimulation relations over *probability distributions* have been proposed both in the strong and the weak setting. Different to the usual bisimulation relations, which are defined over *states*, an algorithmic treatment of these relations is inherently hard, as their carrier set is uncountable, even for finite *PA*s. The coarsest of these relation, weak distribution bisimulation, stands out from the others in that no equivalent state-based characterisation is known so far. This paper presents an equivalent state-based reformulation for weak distribution bisimulation, rendering it amenable for algorithmic treatment. Then, decision procedures for the probability distribution-based bisimulation relations are presented.

1 Introduction

Weak probabilistic bisimilarity is a well-established behavioural equivalence on probabilistic automata (*PA*) [20]. However, it is arguably too fine [6, 9]. As an example, consider the two automata in Fig. 1, where a single visible step, embedding a probabilistic decision is depicted on the left, while on the right this is split into a visible step followed by an internal, thus invisible probabilistic decision of the very same kind (indicated by τ). Intuitively, an observer should not be able to distinguish between the two automata. However, they are not weak probabilistic bisimilar.

Markov Automata are a compositional behavioural model for continuous time stochastic and nondeterministic systems [5, 8, 9] subsuming Interactive Markov Chains (*IMC*) [12] and Probabilistic Automata. Markov automata weak bisimilarity has been

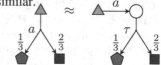

Fig. 1. Distribution bisimilarity

introduced as an elegant and powerful way of abstracting from internal computation cascades, yielding the coarsest reasonable bisimilarity [5]. It is a conservative extension of *IMC* weak bisimilarity, and also extends weak probabilistic bisimilarity on *PA*. But different from standard bisimulation notions, Markov automata weak bisimulations are defined as relations on subprobability distributions instead of states. Translated back to the *PA* setting, this *weak distribution bisimilarity* enables to equate automata such as the ones in Fig. 1. The equivalence of these two systems rests on the ability to relate distributions. If we are only allowed to relate states, we must fail to prove bisimilarity since

K. Joshi et al. (Eds.): QEST 2013, LNCS 8054, pp. 72–88, 2013.
© Springer-Verlag Berlin Heidelberg 2013

we would need to require the presence of a state bisimilar to state ○ on the left. This indicates that weak distribution bisimilarity is coarser than weak probabilistic bisimilarity on *PA*. It can be regarded as the symmetric version [8] of weak probabilistic forward similarity [20], the coarsest precongruence preserving trace distributions [16, 17]. The idea of distribution bisimilarity can also be instantiated in the strong setting [11], where internal computations are not abstracted from.

In this paper, we present decision algorithms for distribution bisimilarities in the strong and weak sense. Strong distribution bisimilarity requires only a minor adaptation of the polynomial time decision algorithm for strong probabilistic bisimilarity [1]. However, a decision algorithm for weak distribution bisimilarity cannot follow the traditional partition refinement approach directly. This is caused by the uncountability of the underlying carrier set, which here is the set of all distributions over the automaton's state space. The key contribution of this paper is an equivalent reformulation of weak distribution bisimulation in a state-based manner. This makes it eventually amenable to an algorithmic treatment. To arrive there, we have to tweak the usual approach to state-based characterisations of bisimulations: instead of all, *only specific* transitions of one state must be matched by its bisimilar counterpart. To identify those transitions, we introduce the concept of behaviour *preserving* transitions.

Based on this state-based characterisation, we then adapt the standard partition refinement algorithm [2, 14, 18] to decide weak bisimilarity. The algorithm successively refines the current equivalence relation by checking the conditions of the state-based characterisations. identifying the set of preserving transitions, the overall complexity of the algorithm is exponential.

The main contribution of this paper is a state-based characterisation of weak distribution bisimilarity, and a decision algorithm based on it. We develop our findings in the setting of probabilistic automata, they however carry over to Markov automata weak bisimilarity, where only the notion of maximal progress, inherited from *IMC*, requires technical care.

Organisation of the paper. After the preliminaries in Sec. 2, we introduce in Sec. 3 the state-based characterisation of the weak bisimilarity in the context of probabilistic automata. We devote Sec. 4 to prove the equivalence between state-based and distribution-based weak bisimilarities. We describe in Sec. 5 the decision procedure and we conclude the paper by Sec. 6 with a discussion on related and future work and by Sec. 7 with some general remarks.

2   Preliminaries

For a set X, we denote by $\mathrm{SubDisc}(X)$ the set of discrete sub-probability distributions over X. Given $\rho \in \mathrm{SubDisc}(X)$, we denote by $|\rho|$ the probability mass $\rho(X)$ of a subdistribution, by $\mathrm{Supp}(\rho)$ the set $\{\, x \in X \mid \rho(x) > 0 \,\}$, by $\rho(\perp)$ the value $1 - \rho(X)$ where $\perp \notin X$, and by δ_x, where $x \in X \cup \{\perp\}$, the *Dirac* distribution such that $\rho(y) = 1$ for $y = x$, 0 otherwise; $\delta_\perp$ represents the empty distribution such that $\rho(X) = 0$. We call a distribution ρ *full*, or simply a *probability* distribution, if $|\rho| = 1$. The set of all discrete probability distributions over X is denoted by $\mathrm{Disc}(X)$.

The lifting $\mathcal{L}(\mathcal{B}) \subseteq \text{SubDisc}(X) \times \text{SubDisc}(X)$ [15] of an equivalence relation $\mathcal{B}$ on X is defined as: for $\rho_1, \rho_2 \in \text{SubDisc}(X)$, $\rho_1 \mathcal{L}(\mathcal{B}) \rho_2$ if and only if for each $\mathcal{C} \in X/\mathcal{B}$, $\rho_1(\mathcal{C}) = \rho_2(\mathcal{C})$. We define the distribution $\rho := \rho_1 \oplus \rho_2$ by $\rho(s) = \rho_1(s) + \rho_2(s)$ provided $|\rho| \leq 1$, and conversely we say ρ can be split into ρ_1 and ρ_2. Since $\oplus$ is associative and commutative, we may use the notation $\bigoplus$ for arbitrary finite sums. Similarly, we define $\rho := \rho_1 \ominus \rho_2$ by $\rho(s) = \max\{\rho_1(s) - \rho_2(s), 0\}$. For notation convenience, for a state s, we denote by $\rho \ominus s$ the distribution $\rho \ominus \delta_s$.

It is often convenient to consider distributions as relations over $X \times \mathbb{R}^{\geq 0}$ and thus explicitly denote the distribution μ by the relation $\{ (s : p_s) \mid s \in X, p_s = \mu(s) \}$.

A *Probabilistic Automaton* (PA) [20] $\mathcal{A}$ is a quadruple $(S, \bar{s}, \Sigma, D)$, where S is a finite set of *states*, $\bar{s} \in S$ is the *start* state, Σ is the set of *actions*, and $D \subseteq S \times \Sigma \times \text{Disc}(S)$ is a *probabilistic transition relation*. The set Σ is partitioned into two sets $H = \{\tau\}$ and E of internal (hidden) and external actions, respectively; we refer to $\bar{s}$ also as the *initial* state and we let s,t,u,v, and their variants with indexes range over S and a, b over actions. In this work we consider only finite PAs, i.e., automata such that S and D are finite.

A transition $tr = (s, a, \mu) \in D$, also denoted by $s \xrightarrow{a} \mu$, is said to *leave* from state s, to be *labelled* by a, and to *lead* to μ, also denoted by μ_{tr}. We denote by $src(tr)$ the *source* state s, by $act(tr)$ the *action* a, and by $trg(tr)$ the *target* distribution μ. We also say that s enables action a, that action a is enabled from s, and that (s, a, μ) is enabled from s. Finally, we denote by $D(s)$ the set of transitions enabled from s, i.e., $D(s) = \{ tr \in D \mid src(tr) = s \}$, and similarly by $D(a)$ the set of transitions with action a, i.e., $D(a) = \{ tr \in D \mid act(tr) = a \}$.

Weak Transitions. An *execution fragment* of a PA $\mathcal{A}$ is a finite or infinite sequence of alternating states and actions $\alpha = s_0 a_1 s_1 a_2 s_2 \ldots$ starting from a state s_0, also denoted by $first(\alpha)$, and, if the sequence is finite, ending with a state denoted by $last(\alpha)$, such that for each $i > 0$ there exists a transition $(s_{i-1}, a_i, \mu_i) \in D$ such that $\mu_i(s_i) > 0$. The *length* of α, denoted by $|\alpha|$, is the number of occurrences of actions in α. If α is infinite, then $|\alpha| = \infty$. Denote by $frags(\mathcal{A})$ the set of execution fragments of $\mathcal{A}$ and by $frags^*(\mathcal{A})$ the set of finite execution fragments of $\mathcal{A}$. An execution fragment α is a *prefix* of an execution fragment α', denoted by $\alpha \leqslant \alpha'$, if the sequence α is a prefix of the sequence α'. The *trace* $trace(\alpha)$ of α is the sub-sequence of external actions of α; we denote by ε the empty trace. Similarly, we define $trace(a) = a$ for $a \in E$ and $trace(\tau) = \varepsilon$.

A *scheduler* for a PA $\mathcal{A}$ is a function $\sigma\colon frags^*(\mathcal{A}) \to \text{SubDisc}(D)$ such that for each finite execution fragment α, $\sigma(\alpha) \in \text{SubDisc}(D(last(\alpha)))$. Note that by using sub-probability distributions, it is possible that with some non-zero probability no transition is chosen after α, that is, the computation stops after α. A scheduler is *determinate* [2] if for each pair of execution fragments α, α', if $trace(\alpha) = trace(\alpha')$ and $last(\alpha) = last(\alpha')$, then $\sigma(\alpha) = \sigma(\alpha')$. A scheduler is *Dirac* if for each α, $\sigma(\alpha)$ is a Dirac distribution. Given a scheduler σ and a finite execution fragment α, the distribution $\sigma(\alpha)$ describes how transitions are chosen to move on from $last(\alpha)$. A scheduler σ and a state s induce a probability distribution $\mu_{\sigma,s}$ over execution fragments as follows. The basic measurable events are the cones of finite execution fragments, where the cone of α, denoted by C_α, is the set $\{\alpha' \in frags(\mathcal{A}) \mid \alpha \leqslant \alpha'\}$. The probability $\mu_{\sigma,s}$ of a cone C_α is recursively defined as:

$$\mu_{\sigma,s}(C_\alpha) = \begin{cases} 0 & \text{if } \alpha = t \text{ for a state } t \neq s, \\ 1 & \text{if } \alpha = s, \\ \mu_{\sigma,s}(C_{\alpha'}) \cdot \sum_{tr \in D(a)} \sigma(\alpha')(tr) \cdot \mu_{tr}(t) & \text{if } \alpha = \alpha' at. \end{cases}$$

Standard measure theoretical arguments ensure that $\mu_{\sigma,s}$ extends uniquely to the σ-field generated by cones. We call the resulting measure $\mu_{\sigma,s}$ a *probabilistic execution fragment* of $\mathcal{A}$ and we say that it is generated by σ from s. Given a finite execution fragment α, we define $\mu_{\sigma,s}(\alpha)$ as $\mu_{\sigma,s}(\alpha) = \mu_{\sigma,s}(C_\alpha) \cdot \sigma(\alpha)(\bot)$, where $\sigma(\alpha)(\bot)$ is the probability of terminating the computation after α has occurred.

We say that there is a *weak combined transition* from $s \in S$ to $\mu \in Disc(S)$ labelled by $a \in \Sigma$, denoted by $s \overset{a}{\Longrightarrow}_c \mu$, if there exists a scheduler σ such that the following holds for the induced probabilistic execution fragment $\mu_{\sigma,s}$: (1) $\mu_{\sigma,s}(frags^*(\mathcal{A})) = 1$; (2) for each $\alpha \in frags^*(\mathcal{A})$, if $\mu_{\sigma,s}(\alpha) > 0$ then $trace(\alpha) = trace(a)$ (3) for each state t, $\mu_{\sigma,s}(\{\alpha \in frags^*(\mathcal{A}) \mid last(\alpha) = t\}) = \mu(t)$. In this case, we say that the weak combined transition $s \overset{a}{\Longrightarrow}_c \mu$ is induced by σ.

We remark that $trace(\alpha) = trace(a)$ is equivalent to $trace(\alpha) = \varepsilon$ for $a = \tau$ and $trace(\alpha) = a$ for $a \in E$. Moreover, the first two conditions can be equivalently replaced by $\mu_{\sigma,s}(\{\alpha \in frags^*(\mathcal{A}) \mid trace(\alpha) = trace(a)\}) = 1$.

Given a set of *allowed* transitions $\check{A} \subseteq D$, we say that there is an *allowed weak combined transition* [13] from s to μ with label a respecting $\check{A}$, denoted by $s \overset{a\downarrow\check{A}}{\Longrightarrow}_c \mu$, if there exists a scheduler σ inducing $s \overset{a}{\Longrightarrow}_c \mu$ such that for each $\alpha \in frags^*(\mathcal{A})$, $Supp(\sigma(\alpha)) \subseteq \check{A}$.

Albeit the definition of weak combined transitions is somewhat intricate, this definition is just the obvious extension of weak transitions on labelled transition systems to the setting with probabilities. See [21] for more details on weak combined transitions.

Example 1. As an example of weak combined transition, consider the probabilistic automaton depicted in Fig. 2 and the probability distribution $\mu = \{(\blacktriangle : \frac{3}{4}), (\text{⑤} : \frac{1}{4})\}$. It is immediate to verify that the weak combined transition $\text{①} \overset{\tau}{\Longrightarrow}_c \mu$ is induced by the Dirac determinate scheduler σ defined as follows: $\sigma(\text{①}) = \delta_{tr_1}$, $\sigma(\text{①}\tau\text{②}) = \delta_{tr_2}$, $\sigma(\text{①}\tau\text{③}) = \delta_{tr_3}$, $\sigma(\text{①}\tau\text{②}\tau\text{④}) = \sigma(\text{①}\tau\text{③}\tau\text{④}) = \delta_{tr_4}$, and $\sigma(\alpha) = \delta_\bot$ for each other finite execution fragment α. If we consider all transitions but tr_2 as allowed transitions $\check{A}$, then there is no scheduler inducing $\text{①} \overset{\tau\downarrow\check{A}}{\Longrightarrow}_c \mu$. In fact, using this set of allowed transitions, the maximal probability of reaching $\blacktriangle$ from ① is $\frac{1}{4}$ by the execution fragment $\text{①}\tau\text{③}\tau\text{④}\tau\blacktriangle$. $\square$

We say that there is a *weak (allowed) hyper transition* from $\rho \in SubDisc(S)$ to $\mu \in SubDisc(S)$ labelled by $a \in \Sigma$, denoted by $\rho \overset{a}{\Longrightarrow}_c \mu$ ($\rho \overset{a\downarrow\check{A}}{\Longrightarrow}_c \mu$), if there exists a family of (allowed) weak combined transitions $\{s \overset{a}{\Longrightarrow}_c \mu_s\}_{s \in Supp(\rho)}$ ($\{s \overset{a\downarrow\check{A}}{\Longrightarrow}_c \mu_s\}_{s \in Supp(\rho)}$) such that $\mu = \bigoplus_{s \in Supp(\rho)} \rho(s) \cdot \mu_s$.

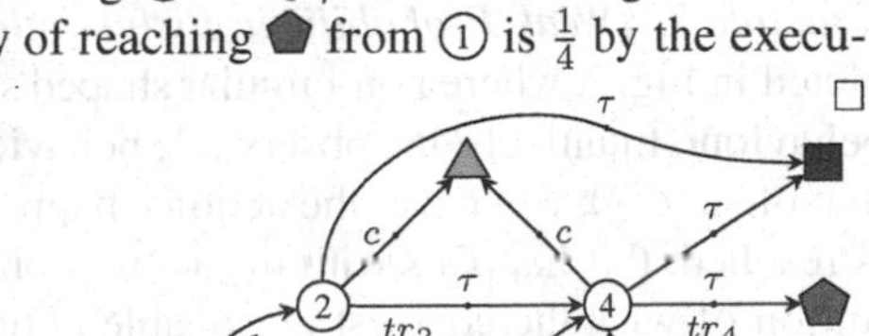

Fig. 2. A probabilistic automaton

3 Probabilistic Bisimulations

For non-stochastic systems, the idea of bisimulation can be formalised as a binary symmetric relation $\mathcal{B}$ over states where each pair of states $(s, t) \in \mathcal{B}$ satisfies that whenever $s \xrightarrow{a} s'$ for some state s', then there exists a state t' such that $t \xrightarrow{a} t'$ and $s \mathcal{B} t$. Strong bisimilarity is the union of all such strong bisimulations. Bisimulation can be seen as a game [7, 22, 23], and therefore one often calls s the *challenger* proposing a transition and t the *defender*. Phrased differently, in a bisimulation, *every* transition of a challenger must be matched by *some* transition of its corresponding defender. Weak bisimulation and bisimilarity is defined analogously, but with the strong transition arrow $\xrightarrow{a}$ replaced by its weak variant $\xRightarrow{a}$ that in addition allows to perform arbitrary sequences of τ actions before and after the action a is performed.

When translating the idea of bisimulation to probabilistic systems, it is generalised in order to account for the probabilistic setting: Transitions $\longrightarrow$ and $\Longrightarrow$ are replaced by their combined variants $\longrightarrow_c$ and $\Longrightarrow_c$, and target states s' and t' become target distributions μ and γ over states, respectively. Finally, target distributions must match up to the lifting of $\mathcal{B}$ to distributions $(\mathcal{L}(\mathcal{B}))$. For a detailed motivation of these adaptions we refer the interested reader to [20]. Strong and weak probabilistic bisimulation can then be defined as follows.

Definition 1 (Strong and Weak Probabilistic Bisimulations). *For a probabilistic automaton $\mathcal{A} = (S, \bar{s}, \Sigma, D)$, a symmetric relation $\mathcal{B}$ over S is a probabilistic bisimulation, if each pair of states $(s, t) \in \mathcal{B}$ satisfies for every $a \in \Sigma$: $s \overset{a}{\leadsto} \mu$ implies $t \overset{a}{\leadsto} \gamma$ for some $\gamma \in \mathrm{Disc}(S)$ and $\mu \, \mathcal{L}(\mathcal{B}) \, \gamma$.*

We call $\mathcal{B}$ strong, if $\leadsto \; = \; \longrightarrow_c$ and weak if $\leadsto \; = \; \Longrightarrow_c$. The union of all strong (weak) bisimulation relations is called strong (weak) bisimilarity. For a uniform presentation, our definitions differ from the standard in the challenger's transition, which usually chooses a strong and not combined transition. The resulting bisimilarities can, however, be shown to be identical.

It is worthwhile to observe that weak probabilistic bisimulation is often considered too fine when it comes to intuitively unobservable behavioural differences [6, 9]. This has been already illustrated in Fig. 1, where weak probabilistic bisimulation fails to equate the automata on the left and the right hand side. We are going to shed some more light on this.

Example 2. (Weak Probabilistic Bisimulation is Too Fine) Consider again the PA depicted in Fig. 2, where non-circular shaped states are supposed to have pairwise distinct behaviour. Intuitively, the observable behaviour of state ① cannot be distinguished from that of state ⑥: whenever the action c happens, or likewise, any of the non-round states is reached, this happens with the same probability for both ① and ⑥. In [20], this intuition of what the coarsest reasonable notion of observability is, has been formalised as *trace distribution precongruence*, that has been proven equivalent [16] to the notion of *weak probabilistic forward similarity*. The latter relates states to probability distributions over states. However, weak probabilistic bisimilarity distinguishes between the two states, as already the first transition of ① to the distribution $\gamma = \left(\frac{1}{2}\delta_{②}\right) \oplus \left(\frac{1}{2}\delta_{③}\right)$ cannot be matched by ⑥. The reason is that the only distribution reachable from ⑥ is

$\mu = \left(\frac{3}{4}\delta_{④}\right) \oplus \left(\frac{1}{4}\delta_{⑤}\right)$. Clearly, for $\mu\ \mathcal{L}(\mathcal{B})\ \gamma$ to hold, all states ②, ③, ④ and ⑤ must be equivalent. However, this cannot be the case, as for example ⑤ cannot perform any transition, while state ④ can perform a transition labelled with c. This means that although, ① and ⑥ show the same *observable* behaviour with the same probability, they are distinguished by weak probabilistic bisimilarity.

Notably, all the distributions $\delta_{①}$, $\left(\frac{3}{4}\delta_{④}\right) \oplus \left(\frac{1}{4}\delta_{⑤}\right)$, and γ are pairwise trace distribution precongruent (and weak probabilistic forward similar). □

With these motivations in mind, several probabilistic bisimulation variants defined on *probability (sub)distributions* over states have been introduced for the strong setting [11] and for the weak setting [5, 8, 9]. For the weak setting, there currently exist three different variations, however, two of them essentially coincide [6]. We will recall these notions in the following. Again, our definitions differ from the original definitions for the sake of a uniform presentation, which allows to highlight differences and similarities clearly.

Definition 2 (Strong and Weak Probabilistic Distribution Bisimulations). *For a PA* $\mathcal{A} = (S, \bar{s}, \Sigma, D)$, *a symmetric relation* $\mathcal{B}$ *over* $\mathrm{SubDisc}(S)$ *is a probabilistic distribution bisimulation, if each pair of subdistributions* $(\mu, \gamma) \in \mathcal{B}$ *satisfies* $|\mu| = |\gamma|$ *and for every* $a \in \Sigma$

(a) $\mu \xrightarrow{a} \mu'$ *implies* $\gamma \xrightarrow{a} \gamma'$ *for some* $\gamma \in \mathrm{SubDisc}(S)$ *and* $\mu'\ \mathcal{B}\ \gamma'$.
(b) $\mu = \mu_1 \oplus \mu_2$ *implies* $\gamma = \gamma_1 \oplus \gamma_2$ *for some* $\gamma_1, \gamma_2 \in \mathrm{SubDisc}(S)$ *such that* $\mu_i\ \mathcal{B}\ \gamma_i$ *for* $i \in \{1, 2\}$.

As before, we obtain the strong and weak variants by replacing $\rightsquigarrow$ by $\longrightarrow_c$ and $\Longrightarrow_c$ respectively; the corresponding bisimilarities are defined as the union of all respective bisimulations.

As shown in [11] (for strong) and [8] (for weak), these distribution-based bisimilarities are indeed reformulations of their state-based counterparts, in so far that for two states s and t, the distributions δ_s and δ_t are bisimilar in the distribution-based bisimulations, if and only if s and t are bisimilar in the respective state-based counterparts.

The weak bisimilarities defined in [9] and [5] (for Markov automata) coincide [6], if restricted to probabilistic automata, but do not correspond to any known state-based bisimilarity. We can define them as follows.

Definition 3 (Weak Distribution Bisimulation). *For a PA* $\mathcal{A} = (S, \bar{s}, \Sigma, D)$, *a symmetric relation* $\mathcal{B}$ *over* $\mathrm{SubDisc}(S)$ *is a weak distribution bisimulation, if each pair of subdistributions* $(\mu, \gamma) \in \mathcal{B}$ *satisfies* $|\mu| = |\gamma|$ *and for every* $a \in \Sigma$

(a) $\mu \xrightarrow{a}_c \mu'$ *implies* $\gamma \xrightarrow{a}_c \gamma'$ *for some* $\gamma' \in \mathrm{SubDisc}(S)$ *and* $\mu'\ \mathcal{B}\ \gamma'$.
(b) $\mu = \mu_1 \oplus \mu_2$ *implies* $\gamma \xrightarrow{\tau}_c \gamma_1 \oplus \gamma_2$ *for some* $\gamma_1, \gamma_2 \in \mathrm{SubDisc}(S)$ *such that* $\mu_i\ \mathcal{B}\ \gamma_i$ *for* $i \in \{1, 2\}$.

The union of all weak distribution bisimulation relations is called weak distribution bisimilarity, denoted by $\approx$. It is an equivalence relation and the coarsest weak distribution bisimulation relation. Two PAs are weak distribution bisimilar if the Dirac distributions of their initial states are weak distribution bisimilar in the direct sum of the two

PAs, i.e., in the automaton whose components are the disjoint union of the components of the two automata. We project the relation $\approx$ to states (denoted $\approx_\delta$) as follows. We say that two states s, t are related by $\approx_\delta \subseteq S \times S$, if and only if $\delta_s \approx \delta_t$.

The strength of this definition is the introduction of a weak transition in Condition (b). As already noted in [8], this is in fact the only difference to weak *probabilistic* distribution bisimulation (Def. 2).

While in Ex. 2 we have argued that the distributions $\gamma = \left(\frac{1}{2}\delta_{\textcircled{2}}\right) \oplus \left(\frac{1}{2}\delta_{\textcircled{3}}\right)$ and $\mu = \left(\frac{3}{4}\delta_{\textcircled{4}}\right) \oplus \left(\frac{1}{4}\delta_{\textcircled{5}}\right)$ are not weak probabilistic bisimilar in the PA of Fig. 2, they satisfy $\mu \approx \gamma$, because $\gamma \overset{\tau}{\Longrightarrow}_c \mu$, which is effectively the only transition of γ, and it thus directly satisfies Condition (a) and (b) of Def. 3.

So while distribution-based bisimulations give rise to coarser and more natural notions of equality, they also have severe drawbacks. A distribution-based bisimulation relation that is to be constructed in order to prove two systems bisimilar is much harder to define than for a state-based relation: For state-based bisimulations only the set of reachable states must be considered and suitably related pairwise. In contrast for distribution-based systems the potentially uncountable set of all reachable distributions needs to be considered. This gets problematic when it comes to algorithmic checks for bisimilarity, for example, in the context of verification of systems and state-space minimisation by bisimulation quotienting. Standard partition refinement approaches usually applied in this context seem infeasible here, as even for finite state space, the problem space (i.e., the reachable distributions) is uncountable.

For the strong and weak distribution-based bisimilarities according to Def. 2 the above issue is not a problem, since they can be reduced to the state-based setting. For weak distribution bisimilarity according to Def. 3, the situation is more complicated as no state-based characterisation is known, and it is by far not obvious how to arrive at such a characterisation. To approach this, we will now give an intuitive explanation why the fact that weak probabilistic bisimilarity is too distinctive seems rooted in the fact that it is a naturally state-based relation, and then explain how to overcome the problem while maintaining the state-based bisimulation approach as far as possible.

For the discussion that follows, we assume a generic underlying notion of observation equivalence such as a trace distribution-based equivalence. We call a state s *behaviourally pivotal*, if $s \overset{\tau}{\rightarrow} \mu$ implies that s and μ are not observation equivalent, i.e., μ is not able to perform $\mu \overset{\tau}{\Longrightarrow}_c \rho$ such that s and ρ are observation equivalent.

Ex. 2. (cont'd) (Behaviourally Pivotal States) Assume again that all non-round states of the PA in Fig. 2 induce pairwise distinct behaviour (for example each state can only perform a different external action). Then state $\textcircled{4}$ is behaviourally pivotal, since none of its internal successor distributions $\delta_\blacksquare$ and $\delta_\blacktriangle$ can behaviourally match the other, and thus cannot preserve the behaviour of s. Trivially, also $\textcircled{5}$ is behaviourally pivotal, since it has no successors. In contrast, all other states are *not* behaviourally pivotal, as for each of them the behaviour is fully preserved by one of its respective τ-successor distributions. In particular, state $\textcircled{2}$ is not behaviourally pivotal since its behaviour is fully preserved by $\delta_{\textcircled{4}}$ via transition $\textcircled{2} \overset{\tau}{\rightarrow} \delta_{\textcircled{4}}$. □

Consider the probability distribution $\mu = \left(\frac{3}{4}\delta_{\textcircled{4}}\right) \oplus \left(\frac{1}{4}\delta_{\textcircled{5}}\right)$ over behaviourally pivotal states. From the perspective of the individual behaviour of the single states in its support, this distribution is different from the distribution $\gamma = \left(\frac{1}{2}\delta_{\textcircled{2}}\right) \oplus \left(\frac{1}{2}\delta_{\textcircled{3}}\right)$ over non-pivotal states. For example, from the perspective of an observer, ③ $\in$ Supp(γ) can perform the transition to ⬤ with *at most* probability $\frac{1}{2}$. In comparison, state ④ $\in$ Supp(μ) can perform this transition with probability 1, while ⑤ $\in$ Supp(μ) cannot perform this transition at all.

However, as we have discussed in Ex. 2, both distributions as such can be regarded as observation equivalent. Weak probabilistic bisimilarity, however, focusing on state-wise behaviour, needs to distinguish between μ and γ regardless of the fact that distribution γ, consisting only of non-pivotal states, can by no means be noticed by an observer, as it is merely skipped over on the way from ① to μ.

From the discussion so far, we will now derive necessary steps to recast Def. 3 in a state-based setting. As we have seen, the fact that weak probabilistic bisimilarity is arguably too fine is mainly due to the fact that it is too much focused on single state behaviour. More precisely, the problem is that it treats behaviourally non-pivotal states (e.g., ② and ③) in the same way as pivotal states (e.g., ④ and ⑤). To overcome this, a state-based characterisation of weak distribution bisimilarity will first of all identify pivotal states, and then, speaking from the game perspective on bisimulation, allow the bisimulation challenger only to propose a challenging transition to a distribution over pivotal states.

Example 3. When we want to show that ① and ⑥ are weak distribution bisimilar, then the challenger should not be allowed to propose the transition to $\gamma = \left(\frac{1}{2}\delta_{②}\right) \oplus \left(\frac{1}{2}\delta_{③}\right)$, which has non-pivotal states in its support (actually both states are non-pivotal). Instead, it may only propose $\left(\frac{3}{4}\delta_{④}\right) \oplus \left(\frac{1}{4}\delta_{⑤}\right)$.

In fact, our approach will not characterise pivotal states explicitly, but rather use a set of distinguished internal transitions $(s, \tau, \mu) \in D(\tau)$ with the property that δ_s and μ are behaviourally equivalent. We call such transitions *preserving*. As a state is pivotal if it has no internal successor distribution that can fully mimic its behaviour, the set of pivotal state then is exactly the set of all states that *do not* enable a preserving transition.

The technically crucial idea of our approach is to define the bisimulation relation $\mathcal{B}$ over states and the set P of distinguished transitions simultaneously. The definitions of $\mathcal{B}$ and P will be mutually dependent. This allows us to use the information from set P to identify pivotal states when defining the bisimulation $\mathcal{B}$. Vice versa, the information provided from the bisimulation $\mathcal{B}$ allows us to determine when a state has a τ-successor distribution, that is behaviourally equivalent. As it is technically more convenient, we will not formally define the notion of pivotal states in the sequel, but directly work with the notion of preserving transitions instead.

Definition 4 (Preserving Transitions). *Let $\mathcal{B}$ be an equivalence relation on S. We call an internal transition $(s, \tau, \gamma) \in D(\tau)$ preserving with respect to $\mathcal{B}$ if whenever $s \stackrel{a}{\Rightarrow}_c \mu$ then there exist μ', γ' such that $\mu \stackrel{\mid\!\Gamma}{\Rightarrow}_c \mu', \gamma \stackrel{a}{\Rightarrow}_c \gamma'$, and $\mu' \, \mathcal{L}(\mathcal{B}) \, \gamma'$. We call a set $P \subseteq D(\tau)$ preserving with respect to $\mathcal{B}$ if it only consists of preserving transitions.*

Example 4. In Fig 2, transitions tr_1, tr_2, tr_3 and tr_5 are preserving, while all other transitions are not. It is especially interesting to note that tr_2 is preserving while the other internal transition leaving ② is not, as ▪ is not behaviourally equivalent to ②.

Given a set P of preserving transitions, we from now on call weak (hyper) transitions of the form $\overset{\tau \downarrow P}{\Longrightarrow}$ *preserving* weak (hyper) transitions, and $\overset{\tau \downarrow P}{\Longrightarrow}_c$ *preserving* weak combined (hyper) transitions.

Definition 5 (State-Based Characterisation of Weak Distribution Bisimulation).
An equivalence relation $\mathcal{B}$ on S is called a state-based weak distribution bisimulation, *if there is a set $P \subseteq D(\tau)$ that is preserving with respect to $\mathcal{B}$ and whenever $s \, \mathcal{B} \, t$,*

1. *if $s \overset{a}{\Longrightarrow}_c \mu$ for some μ, then $t \overset{a}{\Longrightarrow}_c \gamma$ for some γ, such that there exists μ' such that $\mu \overset{\tau \downarrow P}{\Longrightarrow}_c \mu'$ and $\mu' \, \mathcal{L}(\mathcal{B}) \, \gamma$;*
2. *if $s \overset{\tau \downarrow P}{\Longrightarrow}_c \mu$ for some μ, then $t \overset{\tau \downarrow P}{\Longrightarrow}_c \gamma$ for some γ, such that there exists μ' such that $\mu \overset{\tau \downarrow P}{\Longrightarrow}_c \mu'$ and $\mu' \, \mathcal{L}(\mathcal{B}) \, \gamma$.*

We write $s \approx_s t$ if there exists a state-based weak distribution bisimulation relating s and t.

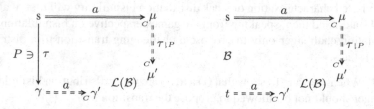

Fig. 3. Preserving transitions (left) and Condition 1 of state-based weak distribution bisimulation

In Fig. 3, preserving transitions and state-based weak distribution bisimulation are explained graphically. Solid lines denote challenger transitions, dashed lines defender transitions. Different to weak probabilistic bisimulation, the role of the defender in the bisimulation game is no longer linked exclusively to transitions of t. Although the weak allowed hyper transition from μ to μ' originates from a successor distribution of s rather than t, the defender can choose this transition. As a consequence, the defender does not need to match the challenging distribution μ directly, but it is allowed to choose an arbitrary distribution μ', which it is able to match, as long as $\mu \overset{\tau \downarrow P}{\Longrightarrow}_c \mu'$. In intuitive terms, a transition (t, τ, ξ) is in P if t is non-pivotal; the existence of a transition $\mu \overset{\tau \downarrow P}{\Longrightarrow}_c \mu'$ (with $\mu \neq \mu'$) means that μ must contain non-pivotal states, and thus, we liberate the defender from its obligation to match μ by allowing it to match μ' instead. At the same time, if μ was a distribution exclusively over pivotal states, then no $\mu' \neq \mu$ would exist such that $\mu \overset{\tau \downarrow P}{\Longrightarrow}_c \mu'$. Thus, the defender is forced to match exactly distributions over pivotal states. Intuitively, we want a transition $s \overset{\tau}{\longrightarrow} \gamma$ to be contained in P, exactly if s and γ allow the same observations, which in turn means that s is non-pivotal. Formally, this is achieved by defining P completely analogous to the state-based characterisation of weak distribution bisimulation. The only difference is that the role of the defender is played by a distribution γ, instead of a state.

So far, we have left Condition 2 of Def. 5 unmentioned, which expresses that if one of two related states can perform transitions within P, then also the other state must be able to match these transitions within P. This condition might come unexpected, and we claim, that the condition can be dropped without affecting the resulting notion of bisimilarity. Yet, currently it is needed for the proof of Thm. 1, which establishes that the distribution-based and the state-based characterisation of weak distribution bisimilarity are indeed equivalent.

Example 5. (State-Based Weak Distribution Bisimulation) Consider again the PA depicted in Fig. 2 and suppose that states ●, ▲, and ■ are not weak bisimilar. The equivalence relation $\mathcal{B}$ whose non-singleton classes are $\{①, ⑥\}$ and $\{②, ④\}$ is a state-based weak distribution bisimulation, given $P = \{tr_1, tr_2, tr_3, tr_5\}$.

Checking the step condition for the pair $(②, ④)$ is trivial, so let us focus on the pair $(①, ⑥)$. Each weak combined transition $⑥ \xrightarrow{\tau}_c \mu$ enabled by $⑥$, can be matched by $①$ by reaching μ_{tr_5} (via preserving transitions tr_1, tr_2, and tr_3 chosen with probability 1) and then behaving as in $⑥ \Longrightarrow_c \mu$. If we stay in $⑥$ with non-zero probability, then we remain in $①$ with the same probability and the lifting condition is satisfied.

Now, consider the weak transition $① \Longrightarrow_c \mu$ enabled by $①$, where $\mu = \{(② : \frac{1}{2}),$ $(③ : \frac{1}{2})\}$ (this is actually the ordinary transition tr_1). $⑥$ has no way to reach μ so it needs help of $①$ to match such a transition: $⑥$ performs the transition $⑥ \xrightarrow{\tau}_c \gamma$ where $\gamma = \{(④ : \frac{3}{4}), (⑤ : \frac{1}{4})\}$, i.e., it performs tr_5, while μ reaches γ by the preserving weak hyper transition $\mu \xrightarrow{\tau \downarrow P}_c \gamma$ by choosing with probability 1 preserving transitions tr_2 from $②$ and tr_3 from $③$ and then stopping.

The transition $① \Longrightarrow_c \mu$ is not the only weak combined transition enabled by $①$. It enables, for instance, the weak combined transition $① \Longrightarrow_c \rho$ where $\rho = \{(■ : \frac{1}{2}), (③ : \frac{1}{2})\}$. $⑥$ matches this transition by enabling $⑥ \Longrightarrow_c \phi$ where $\phi = \{(■ : \frac{1}{2}), (④ : \frac{1}{4}), (⑤ : \frac{1}{4})\}$ that can be reached from ρ by the preserving weak hyper transition $\rho \xrightarrow{\tau \downarrow P}_c \phi$ obtained by performing no transitions from ■ and choosing tr_3 (that is preserving) with probability 1 and then stopping. There are several other transitions enabled by $①$ that can be matched in a similar way.

Finally, we want to remark that weak probabilistic distribution bisimulation given in Def. 2 is obtained from Def. 5 by requiring $P = \emptyset$, since, when $P = \emptyset$, we have that $s \xrightarrow{\tau \downarrow \emptyset}_c \mu$ implies $\mu = \delta_s$ as well as $\mu \xrightarrow{\tau \downarrow \emptyset}_c \mu'$ implies $\mu' = \mu$.

4   Correctness of the Characterisation

The correctness of the state-based characterisation of weak distribution bisimilarity will be formalised by Thm. 1. We obtain this equality in a slightly restricted setting where we collapse probability 1 cycles, or *maximal end components* (mecs) [4], i.e., τ-cycles where it is possible to return to each state of the cycle with probability 1. This restriction, that is due to technical reasons, does not affect the general applicability of Thm. 1 since collapsing *mecs* preserves $\approx_\delta$, as stated by Lemma 1.

We will now define the restricted setting in which we will then establish the correctness proof. Along the way, we will present insightful examples where the unrestricted setting has caused unexpected difficulties. In the restricted setting, we will only consider PAs, where no cyclic structure in the following sense exists.

Definition 6 (Maximal End Components). *Given a PA $\mathcal{A}$ with set of states S, a maximal end component (mec) is a maximal set $C \subseteq S$ such that for each $s, t \in C$: $s \Longrightarrow_c \delta_t$ and $t \Longrightarrow_c \delta_s$.*

The definition stems from [4]. The set of all *mecs* is a disjoint partitioning of S. Thus, the relation $=_{mec}$, where $s =_{mec} t$ if and only if s and t lie in the same *mec*, is an equivalence relation on states. All states that lie in the same *mec* can mutually reach each other with τ transitions with probability 1^1. It is thus straightforward to show that such states are weak distribution bisimilar.

Lemma 1. $s =_{mec} t$ *implies* $s \approx_\delta t$.

Surprisingly, the presence of *mecs* in a *PA* leads to unexpected results. In general, it is folklore knowledge that replacing the weak challenger transition $s \overset{a}{\Rightarrow}_c \mu$ of weak bisimulation by a strong challenger transition $s \overset{a}{\longrightarrow} \mu$ leads to equipotent characterisations of the induced bisimilarities. For state-based weak distribution bisimilarity, this is not the case. We will refer to this variation of Definition 5 as the *strong challenger characterisation* in the sequel.

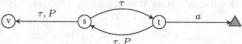

Example 6. (Strong Challenger Characterisation is Broken in the Presence of Mecs.) Consider the automaton above. All transitions in this example are Dirac transitions. We label two transitions with τ, P in order to express that they are elements of P, the set of supposedly preserving transitions considering the strong challenger characterisation. Note that, however, the transition from $\text{\textcircled{s}}$ to $\text{\textcircled{v}}$ is not a preserving transition in the sense of the original definition with respect to any bisimulation relation $\mathcal{B}$, since $\text{\textcircled{s}}$ can reach $\blacktriangle$ with a weak a transition, whereas $\text{\textcircled{v}}$ cannot perform an a transition at all. However, all conditions of the strong challenger characterisation are satisfied. The only non-preserving strong transition $\text{\textcircled{s}}$ can perform is the one to $\text{\textcircled{t}}$. Now it is enough that $\text{\textcircled{t}}$ can reach $\text{\textcircled{v}}$ via preserving transitions, by using $t \overset{\tau}{\longrightarrow} \delta_s$ and $s \overset{\tau}{\longrightarrow} \delta_v$. For completeness, it is easy to check that the transition from $\text{\textcircled{t}}$ to $\text{\textcircled{s}}$ satisfies the conditions to be a preserving transition. With this result, it is straightforward to construct two bisimulations $\mathcal{B}_1$ and $\mathcal{B}_2$ (satisfying the strong challenger characterisation), where $\mathcal{B}_1$ is the reflexive, transitive and symmetric closure of the relation containing only the pair $(\text{\textcircled{v}}, \text{\textcircled{s}})$ and $\mathcal{B}_2$ accordingly containing $(\text{\textcircled{s}}, \text{\textcircled{t}})$. It is easy to check that for both $\mathcal{B}_1$ and $\mathcal{B}_2$ our choice of preserving transitions satisfies the strong challenger characterisation. If this characterisation now indeed was equivalent to $\approx_\delta$, the restriction of $\approx$ to states, then also $\text{\textcircled{v}} \approx_\delta \text{\textcircled{s}}$ and $\text{\textcircled{s}} \approx_\delta \text{\textcircled{t}}$ would hold and thus, by transitivity, also $\text{\textcircled{v}} \approx_\delta \text{\textcircled{t}}$. But clearly, this cannot hold, as $\text{\textcircled{t}}$ can perform an a-transition while $\text{\textcircled{v}}$ cannot.

These considerations led us to only consider *mec*-contracted *PAs* in the following.

Definition 7 (Mec-Contracted PA). *A PA* $\mathcal{A}$ *is called* mec-contracted, *if for each pair of states* $s, t \in S$, $s \overset{\tau}{\Rightarrow}_c \delta_t$ *and* $t \overset{\tau}{\Rightarrow}_c \delta_s$ *implies* $s = t$.

Obviously, the quotient under $=_{mec}$ is a *mec*-contracted automaton, where the quotient under $=_{mec}$ of a *PA* $\mathcal{A}$ is defined as follows:

---

[1] Note that *mecs* are not necessarily bottom strongly connected components, as a *mec* may well be escaped by τ transitions.

Definition 8 (Quotient under $=_{mec}$). *Given a PA $\mathcal{A} = (S, \bar{s}, \Sigma, D)$ and the equivalence relation $=_{mec}$ on S, the* quotient under $=_{mec}$ *of $\mathcal{A}$ is the automaton $\mathcal{A}' = (S/=_{mec}, [\bar{s}]_{=_{mec}}, \Sigma, D/=_{mec})$ where $D/=_{mec} = \{\, ([s]_{=_{mec}}, a, [\mu]_{=_{mec}}) \mid (s, a, \mu) \in D \,\}$ and $[\mu]_{=_{mec}} \in \mathrm{Disc}(S/=_{mec})$ is the probability distribution defined for each $\mathcal{C} \in S/=_{mec}$ as $[\mu]_{=_{mec}}(\mathcal{C}) = \mu(\mathcal{C})$.*

In the restricted setting we have introduced, the following theorem states that the state-based characterisation (Def. 5) of weak distribution bisimulation is indeed equivalent to the original distribution-based definition (Def. 3).

Theorem 1 (Equivalence of Characterisations). *If $\mathcal{A}$ is a mec-contracted PA, then for every t and t' in S, $t \approx_\delta t'$ if and only if $t \approx_s t'$.*

5   Decision Procedure

In this section we investigate algorithms for distribution bisimilarity that decide whether two states of an automaton are equivalent. For strong and weak probabilistic distribution bisimilarity, we can rely on existing decision algorithms for the corresponding state-based characterisations [1, 2]. If we want to relate a pair of distributions (μ, ν), we can introduce two fresh states from which a transition with a fresh label goes to distribution μ, respectively ν, and then check the bisimilarity of these two states with the above mentioned algorithms tailored to the state-based setting. For weak distribution bisimilarity, we can proceed accordingly, provided we have a decision algorithm for state-based weak distribution bisimilarity. In the rest of this section, we will devise such an algorithm.

More precisely, the algorithm constructs $S/\approx_s$, the set of equivalence classes of states under $\approx_s$. In contrast to all known bisimulation variants, we cannot blindly apply the standard partition refinement approach [2, 14, 18], since we potentially split equivalence classes that should not be split as the result of a negative interference between the set of preserving transitions and the current partitioning, as Ex. 7 will show. We shortly repeat the general idea of partition refinement to illustrate the problems we face. Partition refinement starts with an initial partition $\mathbf{W}$, which only consists of a single set (called a block) containing all states. Thus, all states are assumed to be pairwise state-based weak distribution bisimilar. This assumption is then checked, and usually there is a reason to split the block. Refining the partition then means we successively split a block in two (or more) blocks, whenever it contains states still assumed state-based weak distribution bisimilar in the previous iteration of the refinement loop, while in the current iteration they violate any of the state-based weak distribution bisimulation conditions. When no more splitting is possible, the algorithm has found the largest state based weak distribution bisimulation and returns that.

In our setting, we have to manipulate also the set of preserving transitions P, since it depends on the equivalence relation induced by the partition. The obvious way is to start initially with the set $D(\tau)$ of all internal transitions. Transitions are then eliminated from this set, as soon as they violate Def. 4. However, as it turns out, the two procedures, partition refinement and transition elimination, interfere negatively. Focusing on Condition 1 of Def. 5, the challenging transition $s \xRightarrow{a}_c \mu$ is only dependent on the transition relation underlying the given *PA*, but not on the current partition or P. In contrast, Condition 2 demands $s \xrightarrow{\tau \downarrow P}_c \mu$. However, the existence of such transition depends on P, which itself varies over the refinement process. As a consequence, we can obtain false negatives, if P still contains a transition starting from s that will not be contained in the final P, while the corresponding transition from t has already been eliminated from P during an earlier refinement step.

Example 7. Let $s \xrightarrow{\tau} \triangle$ and $s \xrightarrow{\tau} \pentagon$ and also $t \xrightarrow{\tau} \triangle$ and $t \xrightarrow{\tau} \pentagon$. Assume that states $\triangle$ and $\pentagon$ are not weak distribution bisimilar. Then, clearly, none of the transitions is preserving. However, s and t are obviously weak distribution bisimilar. Assume the transition $t \xrightarrow{\tau} \triangle$ has been eliminated from the candidate set P, but $s \xrightarrow{\tau} \triangle$ has not. Then, when we check whether s and t satisfy the second condition of Def. 5, $s \xrightarrow{\tau \downarrow P}_c \delta_\triangle$ holds, but $t \xrightarrow{\tau \downarrow P}_c \delta_\triangle$ does not. Thus, s and t will be erroneously split.

DECIDE($\mathcal{A}$)
1: $\mathcal{A}' = $ QUOTIENT-UNDER-MEC($\mathcal{A}$)
2: $\mathbf{W} = \emptyset$
3: **for all** $P \subseteq D(\tau)$ **do**
4:      $\mathbf{W}' = $ QUOTIENT-WRT-PRES($\mathcal{A}', P$)
5:      $\mathbf{W} = $ JOIN($\mathbf{W}, \mathbf{W}'$)
6: **return** $\mathbf{W}$

QUOTIENT-WRT-PRES($\mathcal{A}, P$)
1: $\mathbf{W} = \{S\}$;
2: **repeat**
3:      $\mathbf{W}' = \mathbf{W}$;
4:      **if not** CONSISTENCY($P, \mathbf{W}$) **then**
5:          **return** $\emptyset$
6:      $(\mathcal{C}, a, \rho) = $ FINDSPLIT($\mathbf{W}, P$);
7:      $\mathbf{W} = $ REFINE($\mathbf{W}, (\mathcal{C}, a, \rho)$);
8: **until** $\mathbf{W} = \mathbf{W}'$
9: **return** $\mathbf{W}$

If we remove Condition 2 from Def. 5, then we can show that the set P can be correctly refined with respect to $\mathbf{W}$. Since currently we have to maintain such condition, we adopt a brute force approach, where we first fix P, and refine $\mathbf{W}$ according to the standard partition refinement approach with respect to the set P.[2]

We repeat the refinement described for every possible set of preserving transitions. This is done inside the **for** loop of the main procedure, DECIDE, of the algorithm. This means we consider all subsets of $D(\tau)$, which, unfortunately, is of size in $\mathcal{O}(2^{|D|})$.

The partition refinement happens in procedure QUOTIENT-WRT-PRES, which is parameterised by P. This procedure is entirely unsurprising except for a consistency check performed in procedure CONSISTENCY: During each refinement iteration of $\mathbf{W}$ in QUOTIENT-WRT-PRES, we check whether the currently assumed set P actually still satisfies Def. 4. If it does not, we stop refining and immediately return $\mathbf{W} = \emptyset$.

After each call of QUOTIENT-WRT-PRES, in procedure DECIDE the returned partitioning $\mathbf{W}$ is joined with the previously computed partitioning $\mathbf{W}'$. Procedure JOIN computes the partitioning that results from the union of the two partitionings. Treating $\mathbf{W}$ and $\mathbf{W}'$ as equivalence relations over S, it computes the reflexive, transitive and symmetric closure of $\mathbf{W} \cup \mathbf{W}'$. Thus, when QUOTIENT-WRT-PRES returns $\emptyset$ in order to indicate that no weak distribution bisimulation exists for the current candidate P, this result will not change $\mathbf{W}'$.

---

[2] In the following, we will treat $\mathbf{W}$ both as a set of partitions and as an equivalence relation, wherever convenient, without further mentioning.

As the algorithm is based on the state-based characterisation of weak distribution bisimulation, we cannot apply the algorithm on arbitrary PAs directly, but only on *mec*-contracted. Therefore, we have to transform every input PA into a *mec*-contracted PA before further processing. This is done in Line 1 of procedure DECIDE, where procedure QUOTIENT-UNDER-MEC is applied. This procedure computes the quotient PA with respect to $=_{mec}$. Clearly, this quotient is *mec*-contracted by definition. Deciding $=_{mec}$ is very efficient [3]. Lemma 1 guarantees the soundness of this approach with respect to deciding $\approx_\delta$.

5.1 Matching Weak Transitions, Consistency Checking, and Splitting

Before we provide explanations of the procedures FINDSPLIT and REFINE, we first discuss how to construct matching weak transitions. The following enables us to effectively compute the existence of two matching weak transitions.

Proposition 1 (cf. [13, Prop. 3]). *Given a PA $\mathcal{A}$, two sub-probability distributions $\rho_1, \rho_2 \in \mathrm{SubDisc}(S)$ such that $|\rho_1| = |\rho_2| > 0$, two actions $a_1, a_2 \in \Sigma$, two sets $\check{A}_1, \check{A}_2 \subseteq D$ of transitions, and an equivalence relation W on S, the existence of $\mu_1, \mu_2 \in \mathrm{SubDisc}(S)$ such that*

$$\rho_1 \stackrel{a_1 \downharpoonright \check{A}_1}{\Longrightarrow}_c \mu_1, \rho_2 \stackrel{a_2 \downharpoonright \check{A}_2}{\Longrightarrow}_c \mu_2, \text{ and } \mu_1 \mathcal{L}(W) \mu_2$$

can be checked in polynomial time.

The proof that this check, that we denote by $P(W, \rho_1, a_1, \check{A}_1, \rho_2, a_2, \check{A}_2)$, can be performed in polynomial time relies on the construction of a generalised flow problem, that in turn can be encoded into an LP-problem of polynomial size spanned by the parameters $\rho_1, \rho_2, a_1, a_2, \check{A}_1, \check{A}_2$, and W. Details are given in [13] whose Prop. 3 considers $\rho_1', \rho_2' \in \mathrm{Disc}(S)$; the above proposition follows by choosing the normalised distributions $\rho_i' = \rho_i/|\rho_i|$ for $i = 1, 2$. An exponential algorithm solving this task has been given in [2].

CONSISTENCY(P, W)	FINDSPLIT(W, P)
1: **for all** $(s, \tau, \rho) \in P$ **do**	1: **for all** $(s, a, \rho) \in \mathcal{T}$ **do**
2: **for all** $(s, a, \mu) \in \mathcal{T}$ **do**	2: **for all** $t \in [s]_W$ **do**
3: **if** $P(W, \mu, \tau, P, \rho, a, D)$	3: **if** $(s, a, \rho) \in \mathcal{T}_P$ **then**
has no solution **then**	4: **if** $P(W, \rho, \tau, P, \delta_t, a, P)$ has no
4: **return false**	solution **then**
5: **return true**	5: **return** $([s]_W, a, \rho)$
	6: **else**
	7: **if** $P(W, \rho, \tau, P, \delta_t, a, D)$ has no
	solution **then**
	8: **return** $([s]_W, a, \rho)$
	9: **return** $(\emptyset, \tau, \delta_\perp)$

Now we are ready to explain the remaining procedures. Following the same line as for instance [2], QUOTIENT-WRT-PRES makes use of a sub-procedure REFINE, which actually creates a finer partitioning, as long as there is a partition containing two states that violate the bisimulation condition, which is checked for in procedure FINDSPLIT.

More precisely, as in [2], procedure REFINE divides partition $\mathcal{C}$ into two new partitions according to the discriminating behaviour $\xrightarrow{a} \mu$, which has been identified by FINDSPLIT before. We do not provide REFINE explicitly.

In FINDSPLIT, the sets $\mathcal{T}$ and $\mathcal{T}_X \subseteq \mathcal{T}$ contain all transitions and all candidate preserving transitions, respectively, we have to match: $\mathcal{T}$ is the set of combined weak transitions and $\mathcal{T}_X$ is the set of combined candidate preserving weak transitions (defined by a scheduler using only candidate transitions in X) for state-based weak distribution bisimulation. Note that it is sufficient to use for $\mathcal{T}$ ($\mathcal{T}_X$) the set of (preserving) weak transitions defined by Dirac determinate schedulers (on preserving transitions X), which is a finite set (cf. [2, Prop. 3, 4]). Unfortunately, this set may be exponential, which also gives rise to an overall exponential run-time complexity of the algorithm.

Both procedures FINDSPLIT and CONSISTENCY rely on Prop. 1. By verifying $P(\mathbf{W}, \mu, \tau, P, \delta_t, a, D)$ in their conditional statement, they check the corresponding conditions from Def. 4 (preserving transitions) and Def. 5 (state-based characterisation of weak distribution bisimulation), respectively.

6   Related Work

Recently, the problem of a decision algorithm for *MA* weak bisimilarity has been addressed by Schuster and Siegle [19]. The treatment uses the concept of tangible states, which seems dual to our preserving transitions in the sense that a state is tangible if and only if it has no outgoing preserving transitions. The algorithm presented is a nested fixed-point computation with exponential time complexity. It iteratively refines a candidate state partition while iteratively enlarging the set of candidate tangible states. No correctness proof is provided. A particular obstacle we see is that some of the crucial correctness arguments need to be applied to candidate partitions which by construction do not represent weak bisimulation relations (except for the last one, provided the algorithm is correct). But these arguments are established to hold only in case the partitions do indeed represent weak bisimulation relations.

7   Concluding Remarks

This paper has developed a decision algorithm for weak distribution bisimulation on probabilistic automata. It can be extended straightforwardly to Markov automata. This algorithm can be considered as the nucleus for extending the compositional specification and reasoning means in use for *IMC* to the more expressive *MA* setting. Albeit being a distribution-based relation, we managed to circumvent uncountability in the carrier set by a state-based characterisation. The main obstacle has not been the issue of finding an alternative characterisation of $\approx_\delta$ and deriving a decision algorithm from there. Rather, the formal proof that the characterisation is indeed equivalent to the one of [9] has been very challenging. As Ex. 6 and Ex. 7 show, the pitfalls are hidden in seemingly obvious places. The presented algorithm uses worst-case exponential time and polynomial space, and we are investigating its theoretical and practical runtime characteristics further.

Acknowledgements. This work is supported by the DFG/NWO bilateral research programme ROCKS, by the DFG as part of the SFB/TR 14 AVACS, by the EU FP7 Programme under grant agreement no. 295261 (MEALS), 318490 (SENSATION), and 318003 (TREsPASS), by IDEA4CPS and MT-LAB, a VKR Centre of Excellence. Andrea Turrini is supported by the Cluster of Excellence "Multimodal Computing and Interaction" (MMCI), part of the German Excellence Initiative.

References

1. Baier, C., Engelen, B., Majster-Cederbaum, M.E.: Deciding bisimilarity and similarity for probabilistic processes. J. Comput. Syst. Sci. 60(1), 187–231 (2000)
2. Cattani, S., Segala, R.: Decision algorithms for probabilistic bisimulation. In: Brim, L., Jančar, P., Křetínský, M., Kučera, A. (eds.) CONCUR 2002. LNCS, vol. 2421, pp. 371–385. Springer, Heidelberg (2002)
3. Chatterjee, K., Henzinger, M.R.: Faster and dynamic algorithms for maximal end-component decomposition and related graph problems in probabilistic verification. In: SODA, pp. 1318–1336 (2011)
4. de Alfaro, L.: Formal Verification of Probabilistic Systems. PhD thesis, Stanford University (1997)
5. Deng, Y., Hennessy, M.: On the semantics of Markov automata. In: Aceto, L., Henzinger, M., Sgall, J. (eds.) ICALP 2011, Part II. LNCS, vol. 6756, pp. 307–318. Springer, Heidelberg (2011)
6. Deng, Y., Hennessy, M.: On the semantics of Markov automata. I&C 222, 139–168 (2012)
7. Ehrenfeucht, A.: An application of games to the completeness problem for formalized theories. Fundamenta Mathematicae 49, 129–144 (1961)
8. Eisentraut, C., Hermanns, H., Zhang, L.: Concurrency and composition in a stochastic world. In: Gastin, P., Laroussinie, F. (eds.) CONCUR 2010. LNCS, vol. 6269, pp. 21–39. Springer, Heidelberg (2010)
9. Eisentraut, C., Hermanns, H., Zhang, L.: On probabilistic automata in continuous time. In: LICS, pp. 342–351 (2010)
10. Eisentraut, C., Hermanns, H., Zhang, L.: On probabilistic automata in continuous time. Reports of SFB/TR 14 AVACS 62, SFB/TR 14 AVACS (2010)
11. Hennessy, M.: Exploring probabilistic bisimulations, part I. Formal Aspects of Computing 24(4-6), 749–768 (2012)
12. Hermanns, H.: Interactive Markov Chains: The Quest for Quantified Quality. LNCS, vol. 2428. Springer, Heidelberg (2002)
13. Hermanns, H., Turrini, A.: Deciding probabilistic automata weak bisimulation in polynomial time. In: FSTTCS, pp. 435–447 (2012)
14. Kanellakis, P.C., Smolka, S.A.: CCS expressions, finite state processes, and three problems of equivalence. I&C 86(1), 43–68 (1990)
15. Larsen, K.G., Skou, A.: Bisimulation through probabilistic testing (preliminary report). In: POPL, pp. 344–352 (1989)
16. Lynch, N.A., Segala, R., Vaandrager, F.W.: Compositionality for probabilistic automata. In: Amadio, R.M., Lugiez, D. (eds.) CONCUR 2003. LNCS, vol. 2761, pp. 208–221. Springer, Heidelberg (2003)
17. Lynch, N.A., Segala, R., Vaandrager, F.W.: Observing branching structure through probabilistic contexts. SIAM J. on Computing 37(4), 977–1013 (2007)
18. Philippou, A., Lee, I., Sokolsky, O.: Weak bisimulation for probabilistic systems. In: Palamidessi, C. (ed.) CONCUR 2000. LNCS, vol. 1877, pp. 334–349. Springer, Heidelberg (2000)

19. Schuster, J., Siegle, M.: Markov automata: Deciding weak bisimulation by means of "non-naïvely" vanishing states, http://arxiv.org/abs/1205.6192
20. Segala, R.: Modeling and Verification of Randomized Distributed Real-Time Systems. PhD thesis, MIT (1995)
21. Segala, R.: Probability and nondeterminism in operational models of concurrency. In: Baier, C., Hermanns, H. (eds.) CONCUR 2006. LNCS, vol. 4137, pp. 64–78. Springer, Heidelberg (2006)
22. Stirling, C.: Local model checking games (extended abstract). In: Lee, I., Smolka, S.A. (eds.) CONCUR 1995. LNCS, vol. 962, pp. 1–11. Springer, Heidelberg (1995)
23. Thomas, W.: On the Ehrenfeucht-Fraïssé game in theoretical computer science. In: Gaudel, M.-C., Jouannaud, J.-P. (eds.) TAPSOFT 1993. LNCS, vol. 668, pp. 559–568. Springer, Heidelberg (1993)

Learning and Designing Stochastic Processes from Logical Constraints

Luca Bortolussi[1,2,*] and Guido Sanguinetti[3,4]

[1] Department of Mathematics and Geosciences, University of Trieste
[2] CNR/ISTI, Pisa, Italy
[3] School of Informatics, University of Edinburgh
[4] SynthSys, Centre for Synthetic and Systems Biology, University of Edinburgh

Abstract. Continuous time Markov Chains (CTMCs) are a convenient mathematical model for a broad range of natural and computer systems. As a result, they have received considerable attention in the theoretical computer science community, with many important techniques such as model checking being now mainstream. However, most methodologies start with an assumption of complete specification of the CTMC, in terms of both initial conditions and parameters. While this may be plausible in some cases (e.g. small scale engineered systems) it is certainly not valid nor desirable in many cases (e.g. biological systems), and it does not lead to a constructive approach to rational design of systems based on specific requirements. Here we consider the problems of learning and designing CTMCs from observations/ requirements formulated in terms of satisfaction of temporal logic formulae. We recast the problem in terms of learning and maximising an unknown function (the likelihood of the parameters) which can be numerically estimated at any value of the parameter space (at a non-negligible computational cost). We adapt a recently proposed, provably convergent global optimisation algorithm developed in the machine learning community, and demonstrate its efficacy on a number of non-trivial test cases.

1 Introduction

Stochastic processes are convenient mathematical models of a number of real world problems, ranging from computer systems to biochemical reactions within single cells. Typically, such models are formulated intensionally by specifying the transition kernel of a *continuous time Markov chain* (CTMC, [10]). A classical question in formal modelling is to calculate the probability that a certain temporal logic formula is true, given a certain process (with specified parameters); this is the question addressed by *stochastic model checking*, one of the major success stories of formal modelling in the last thirty years [4,15].

---

* Work partially supported by EU-FET project QUANTICOL (nr. 600708) and by FRA-UniTS.

K. Joshi et al. (Eds.): QEST 2013, LNCS 8054, pp. 89–105, 2013.

While probabilistic model checking is indubitably a success story, it is not an unqualified one. Computationally, model checking suffers from limitations, either due to state space explosion or to the difficulty (impossibility) in checking analytically formulae in specific logics [4,8]. Simulation-based approaches, such as statistical model checking, can be used to circumvent these problems: these methods are usually asymptotically exact, in the limit when the number of simulations used is large; nevertheless, establishing what is a sufficiently large number of simulations to achieve a certain accuracy is a nontrivial problem. Conceptually, both model checking and statistical model checking start from the premise that a CTMC model of the system is entirely specified, i.e. the underlying parameters of the CTMC are known exactly. This is generally not true: it is certainly never true when employing CTMCs as models of physical systems (such as systems biology models, where parameters are at best known with considerable uncertainty), but it is often not appropriate even when modelling large-scale computer systems, when a coarse grained abstraction may be useful. In these cases, one would wish to use observations of the system or of its properties to determine (approximately) its parameters: this is the *system identification* problem. Moreover, the assumption of complete specification is not productive in an engineering context: rather than checking properties of systems with specific parameters, one is often interested in specifying *a priori* the properties of the system (requirements), and then adjust (few) control parameters in order to best match the requirements (the *system design* problem).

The identification of parameters of CTMCs from observations has recently received considerable interest in both the statistical machine learning and formal modelling communities, where a number of approximate methods have been proposed [3,17]. All of these methods assume that the *state* of the system, e.g. the counts of particles of each molecular species, is observed at discrete time points. Here we consider the more general case where the observations are represented by truth values of linear time temporal logic formulae representing qualitative properties of the system. This may be more appropriate in a computer systems scenario, as it may represent an easier type of data to store/ observe, or in a systems biology scenario, when one observes a qualitative phenotype in multiple cells as opposed to directly measuring protein counts. It is also a more natural framework in which to address the design problem, as it is easier to formulate requirements in terms of logical constraints than in terms of particle counts. The restriction to linear time properties is justified because we can only observe single realisations (trajectories) of a system. Naturally, the amount of information contained in these qualitative observations is lower, making the problem more challenging.

For both the design and identification problems the outstanding difficulty is the lack of an objective function that can be used in an optimisation routine: the fit of a CTMC with specific parameters to observations (or the match to requirements) cannot in general be estimated analytically. We therefore need to optimise an unknown function with the smallest number of function evaluations. The key observation in this paper is that a similar problem also occurs in the classical AI problem of *reinforcement learning*: there, the goal is to devise a *strategy* (i.e. an update rule) which will lead to the optimisation of an unknown reward function with the smallest number of trials (function evaluations). This observation allows us to leverage powerful, provably convergent algorithms from the statistical machine learning community: in particular, we adapt to our situation the Gaussian Process Upper Confidence Bound (GP-UCB) algorithm [20], and show on a number of examples that this provides a practical and reliable approach to both the identification and design problem. We further extend the algorithm in order to provide confidence estimates for the obtained parameters, and to detect possible non-identifiability issues. The paper is organised as follows: the problems we tackle and the formal methods tools we use are introduced in Section 2. We then present the machine learning tools we use in Section 3, while in Section 4 we present some computational experiments that give a proof of concept demonstration of our approach. We then briefly discuss our results, highlighting how the coupling of advanced machine learning and formal modelling can open innovative directions in both fields.

2 Problem Definition

Let the probability distribution on trajectories of stochastic process of interest be denoted as $P(x_{0:T}|\theta)$, where $x_{0:T}$ denotes a trajectory of the system up to time T, θ is a set of parameters, and $P(\cdot)$ denotes the probability distribution/density. Let $\varphi_1, \ldots, \varphi_d$ be d (temporal) logic formulae whose truth depends on the specific trajectory of the process which is observed. We are interested in the following two problems:

Identification Problem: Given evaluations of each of the d formulae over N independent runs of the process, arranged into a $d \times N$ binary design matrix D, determine the value(s) of the parameters θ that make these observations most probable.

Design Problem: Given a probability table P for the joint occurrence of a number of formulae, determine the parameters of the stochastic process which optimally match these probabilities.

We will see that a very similar approach can be adopted to solve both problems. We introduce now the main logical and algorithmic ingredients of our approach.

2.1 Metric Interval Temporal Logic

We will consider properties of stochastic trajectories specified by Metric interval Temporal Logic (MiTL), see [1,16]. This logic belongs to the family of linear temporal logics, whose truth can be assessed over single trajectories of the system. MiTL, in particular, is used to reason on real time systems, like those specified by CTMC, and its temporal operators are all time-bounded. We decided to focus on MiTL because when we observe a system, e.g. a biological one, we always observe *single time-bounded realisations* (essentially, time-bounded samples from its trajectory space). Hence, MiTL is the natural choice to formalise the qualitative outcome of experiments.

The syntax of MiTL is given by the following grammar:

$$\varphi ::= \mathtt{tt} \mid \mu \mid \neg\varphi \mid \varphi_1 \wedge \varphi_2 \mid \varphi_1 \mathbf{U}^{[T_1,T_2]}\varphi_2,$$

where $\mathtt{tt}$ is the true formula, conjunction and negation are the standard boolean connectives, and there is only one temporal modality, the time-bounded until $\mathbf{U}^{[T_1,T_2]}$. Atomic propositions μ are defined like in Signal Temporal Logic (STL [16]) as boolean predicate transformers: they take a real valued function $\boldsymbol{x}(t)$, $\boldsymbol{x} : [0,T] \to \mathbb{R}^n$, as input, and produce a boolean signal $s(t) = \mu(\boldsymbol{x}(t))$ as output, where $S : [0,T] \to \{\mathtt{tt}, \mathtt{ff}\}$. As customary, boolean predicates μ are (non-linear) inequalities on vectors of n variables, that are extended point-wise to the time domain. Temporal modalities like time-bounded eventually and always can be defined in the usual way from the until operator: $\mathbf{F}^{[T_1,T_2]}\varphi \equiv \mathtt{tt}\mathbf{U}^{[T_1,T_2]}\varphi$ and $\mathbf{G}^{[T_1,T_2]}\varphi \equiv \neg\mathbf{F}^{[T_1,T_2]}\neg\varphi$.

A MiTL formula is interpreted over a real valued function of time $\boldsymbol{x}$, and its satisfaction relation is given in a standard way, see e.g. [1,16]. We report here only the rules for atomic propositions and the temporal operator, as those for boolean connectives are standard:

- $\boldsymbol{x}, t \models \mu$ if and only if $\mu(\boldsymbol{x}(t)) = \mathtt{tt}$;
- $\boldsymbol{x}, t \models \varphi_1 \mathbf{U}^{[T_1,T_2]}\varphi_2$ if and only if $\exists t_1 \in [t + T_1, t + T_2]$ such that $\boldsymbol{x}, t_1 \models \varphi_2$ and $\forall t_0 \in [t, t_1]$, $\boldsymbol{x}, t_0 \models \varphi_1$ (here we follow the treatment of STL [16]).

The temporal logic MiTL can be easily extended to the probabilistic setting, and interpreted over CTMC [12,8]. Essentially, one is interested in the path probability of a formula φ, defined as $P(\varphi|\theta) = P(\{\boldsymbol{x}_{0:T}|\boldsymbol{x}_{0:T}, 0 \models \varphi\}|\theta)$, i.e. as the probability of the set of time-bounded CTMC trajectories that satisfy the formula[1].

2.2 Likelihood Function

Consider now a CTMC depending on a set of parameters θ, and a set of d MiTL formulae $\varphi_1, \ldots, \varphi_d$ whose truth values have been observed over N independent runs of the process. Let D be the $d \times N$ *design matrix*, whose column vectors correspond to joint observations of the properties. Given a specific value of the parameters θ, the probability of a particular joint truth value for the set of formulae of interest is uniquely determined. Let $P(D_i|\theta)$ be the probability of the joint truth value of formulae of the i_{th} column of the matrix D given the parameters θ. Under the assumption of independent runs, the likelihood of the observations D is then simply

---

[1] We assume implicitly that T is sufficiently large so that the truth of φ at time 0 can always be established from $\boldsymbol{x}$. The minimum of such times can be easily deduced from the formula φ, see [12,16].

$$\mathcal{L}(D, \theta) = \prod_{i=1}^{N} P(D_i | \theta). \tag{1}$$

Alternatively, if prior knowledge over the parameters is available as a prior distribution $P(\theta)$, we may want to consider the un-normalised posterior distribution $P(\theta, D) = P(\theta) \prod_{i=1}^{N} P(D_i | \theta)$. The identification problem can be carried out by maximising the likelihood (1) (maximum likelihood, ML) or the un-normalised posterior (maximum a posteriori, MAP).

Numerical evaluation of $P(D_i | \theta)$ is a major challenge: computing the path probability of a MiTL formula is an extremely difficult problem, with current algorithms [8] suffering severely from the state space explosion. Furthermore, numerical methods for stochastic model checking have always been developed to compute the path probability of a *single* formula, while computing $P(D_i | \theta)$ requires to know the *joint* probability distribution of the formulae $\varphi_1, \ldots, \varphi_d$. We therefore resort to statistical model checking to approximately evaluate the likelihood $\mathcal{L}(D, \theta)$.

2.3 Statistical Model Checking

We now briefly review Statistical Model Checking (SMC [12,22]), a class of methods that try to estimate the probability of a path formula or the truth of a state formula relying on simulation and statistical means. In the context of MiTL, SMC works as follows. Given a CTMC with fixed parameters θ, a simulation algorithm, like SSA [11], is used to sample trajectories of the process. For each sampled trajectory, we run a model checking algorithm for MiTL and establish if φ is true or false. The process therefore generates samples from a Bernoulli random variable Z_φ, equal to 1 if and only if φ is true. SMC uses a statistical treatment of those samples, like Wald sequential testing [22] or Bayesian alternatives [12], to establish if the query $P(\varphi | \theta) > q$ is true, with a chosen confidence level α, given the evidence seen so far. Bayesian SMC, in particular, uses a Beta prior distribution $Beta(q | a, b)$ for the probability of $q = P(\varphi = 1)$; by exploiting the conjugacy of the Beta and Bernoulli distributions [6], applying Bayes' theorem we get

$$P(q | D_\varphi) = \frac{1}{P(D_\varphi)} P(D_\varphi | q) P(q) = Beta(q, a + k_1, b + k_0).$$

The parameters a and b of the Beta prior distribution (usually set to 1) can be seen as pseudo-counts that regularise the estimate when a truth value is rarely observed. Given the simulated data D_φ, our best guess about the true probability $P(Z_\varphi = \mathtt{tt})$ is then given by the predictive distribution [6]:

$$P(Z_\varphi = \mathtt{tt} | D_\varphi) = \int_0^1 P(Z_\varphi = \mathtt{tt} | q) P(q | D_\varphi) dq = \mathbb{E}[q | D_\varphi] = \frac{k_1 + a}{k_1 + a + k_0 + b}$$

The Bayesian approach to SMC, especially the use of prior distributions as a form of regularization of sampled truth values of formulae, is particularly relevant for our setting, since we need to estimate probabilities over the much larger set of joint truth values of several formulae.

To extend Bayesian SMC to checking the joint truth probabilities of multiple formulae, we choose a Dirichlet prior distribution with parameters $\alpha_1, \ldots, \alpha_{2^d}$ equal to 1 (corresponding to adding one pseudo-count to every possible joint truth value). Given observations $D_{\varphi_1, \ldots, \varphi_d}$ of the truth values of $Z_{\varphi_1, \ldots, \varphi_d}{}^2$, analogous calculations yield the predictive distribution

$$P(Z_{\varphi_1, \ldots, \varphi_d} = d_j | D_{\varphi_1, \ldots, \varphi_d}) = (\alpha_j + k_j)/(\alpha_0 + k)$$

where k_j is the number of times we observed the jth truth combination, corresponding to a point $d_j \in \mathcal{D}$ and $\alpha_0 = \sum_j \alpha_j$. This probability is then used to estimate the likelihood $\mathcal{L}(D, \theta)$, as $\mathcal{L}(D, \theta) = \prod_{i=1}^{N} P(D_i | \theta)$. By the law of large numbers, with probability one, this quantity will converge to the true likelihood when the number of samples in the SMC procedure becomes large, and the deviation from the true likelihood will become approximately Gaussian.

3   Global Optimisation

As we have seen, the identification problem entails the maximisation of an unknown function which can be only estimated (with approximately Gaussian noise) at isolated points at considerable computational cost. One can approach this problem also from a Bayesian angle by treating the unknown function as a random function (arising from a suitable prior stochastic process) and then use the numerical evaluations as (noisy) observations of the function value, which in turn enable a *posterior* prediction of the function values at new input points. This is the idea underlying *statistical emulation* [14]. This leads to a very elegant algorithm for optimisation; we now briefly review the main concepts and the algorithm we use.

3.1   Gaussian Processes

Gaussian Processes (GPs) are a natural extension of the multivariate normal distribution to infinite dimensional spaces of functions. A GP is a probability measure over the space of continuous functions (over a suitable input space) such that the random vector obtained by evaluating a sample function at a finite set of points $x_1, \ldots, x_N$ follows a multivariate normal distribution. A GP is uniquely defined by its *mean* and *covariance* functions, denoted by $\mu(x)$ and $k(x, x')$. By definition, we have that for every finite set of points

$$f \sim \mathcal{GP}(\mu, k) \leftrightarrow \mathbf{f} = (f(x_1), \ldots, f(x_N)) \sim \mathcal{N}(\boldsymbol{\mu}, K) \qquad (2)$$

---

[2] Note that $D_{\varphi_1, \ldots, \varphi_d}$ is a matrix, similarly the design matrix discussed in Section 2, but we treat each column/ observation as a single point of $\mathcal{D}$.

where $\boldsymbol{\mu}$ is the vector obtained evaluating the mean function μ at every point, and K is the matrix obtained by evaluating the covariance function k at every pair of points. In the following, we will assume for simplicity that the prior mean function is identically zero (a non-zero mean can be added post-hoc to the predictions w.l.o.g.).

The choice of covariance function is an important modelling decision, as it essentially determines the type of functions which can be sampled from a GP (more precisely, it can assign prior probability zero to large subsets of the space of continuous functions). A popular choice of covariance function is the *radial basis function* (RBF) covariance

$$k(x, x') = \gamma \exp\left[-\frac{\|x - x'\|^2}{\lambda^2}\right] \tag{3}$$

which depends on two hyper-parameters, the *amplitude* γ and the *lengthscale* λ. Sample functions from a GP with RBF covariance are with probability one infinitely differentiable functions. For more details, we refer the interested reader to the excellent review book of Rasmussen and Williams [18].

3.2   GP Regression and Prediction

Suppose now that we are given a set of noisy observations $\mathbf{y}$ of the function value at input values $\mathbf{x} = x_1, \ldots, x_N$, distributed around an unknown true value $f(\mathbf{x})$ with spherical Gaussian noise of variance σ^2. We are interested in determining how these observations influence our belief over the function value at a further input value x^* where the function value is unobserved.

By using the basic rules of probability and matrix algebra, we have that the predictive distribution at x^* is again Gaussian with mean

$$\mu^* = (k(x^*, x_1), \ldots, k(x^*, x_N)) \, \hat{K}_N^{-1} \mathbf{y} \tag{4}$$

and variance

$$k^* = k(x^*, x^*) - (k(x^*, x_1), \ldots, k(x^*, x_N)) \, \hat{K}_N^{-1} \, (k(x^*, x_1), \ldots, k(x^*, x_N))^T. \tag{5}$$

where $\hat{K}_N$ is obtained by evaluating the covariance function at each pair of training points and adding σ^2 times the identity. Notice that the first term on the r.h.s of equation (5) is the prior variance at the new input point; therefore, we see that the observations lead to a *reduction* of the uncertainty over the function value at the new point. The variance however returns to the prior variance when the new point becomes very far from the observation points.

Equation (4) warrants two important observations. first, as a function of the new point x^*, μ^* is a linear combination of a finite number of *basis* functions $k(x^*, x)$ centred at the observation points. Secondly, the posterior mean at a fixed x^* is a linear combination of the observed values, with weights determined by the specific covariance function used. For the RBF covariance, input points further from the new point x^* are penalised exponentially, hence contribute less to the predicted value.

3.3 Upper Confidence Bound Optimisation

We now return to the problem of finding the maximum of an unknown function with the minimum possible number of function evaluations. This is related to the problem of performing sensitivity analysis w.r.t. the parameters of complex computer models, e.g. climate models, where a quantification of uncertainty on the model outputs is essential. An elegant approach to solving this problem has been proposed by Kennedy and O'Hagan [14] by recasting the problem in a Bayesian formalism: the true function linking the parameters to the model outputs is assumed unknown and is assigned a GP prior. A (limited) number of function evaluation are then used as (noiseless) observations to obtain a GP posterior mean function which *emulates* the true unknown function, and is used for subsequent analyses.

In the optimisation case, the situation is slightly different: given an initial set of function evaluations, we are interested in determining a sequence of input values that converges to the optimal value of the function. A naive approach would be to use GP regression to emulate the unknown function, and to explore the region near the maximum of the posterior mean. It is easy to see, though, that this approach is vulnerable to remaining trapped in local optima. On the other hand, one could sample uniformly across the input domain of interest; this is guaranteed to eventually find the global optimum but is unlikely to do so in a reasonable time. It is therefore clear that one needs to trade off the *exploitation* of promising regions (high posterior mean) with the *exploration* of new regions (high posterior variance).

The GP Upper Confidence Bound (GP-UCB) algorithm [20] prescribes an exploration-exploitation trade-off which provably converges to the global optimum of the function. The idea is intuitively very simple: rather than maximising the posterior mean function, one maximises an upper quantile of the distribution, obtained as mean value plus a constant times the standard deviation (e.g., the 95% quantile, approximately given as $\mu + 2\sigma$). The GP-UCB rule is therefore defined as follows: let $\mu_t(x)$ and $var_t(x)$ be the GP posterior mean and variance at x after t iterations of the algorithm. The next input point is then selected as

$$x_{t+1} = \text{argmax}_x \left[\mu_t(x) + \beta_t \sqrt{var_t(x)} \right] \qquad (6)$$

where β_t is a constant that depends on the iteration of the algorithm.

To specify in which sense the algorithm converges, we need a definition.

Definition 1. *Let x^* be the value at which a function f attains its maximum. The instantaneous regret of selecting a point x_t is defined as $r_t = f(x^*) - f(x_t)$ and the cumulative regret at time T is defined as $\sum_{t=1}^{T} r_t$. An iterative optimisation algorithm is no-regret if*

$$\lim_{T \to \infty} \frac{1}{T} \sum_{t=1}^{T} r_t = 0.$$

Srinivas et al [20] then proved the following theorem

Theorem 1. *Let $\beta_t = k + \alpha \log t$, where k and α are positive constants. Then the GP-UCB algorithm in equation (6) is no-regret. More specifically, with high probability, the cumulative regret is bounded by $O(\sqrt{T})$.*

This theorem indicates that, as the algorithm proceeds, exploration needs to become gradually more important than exploitation (β_t is monotonically increasing), as one would intuitively expect. The algorithm has been successfully employed in a number of difficult optimisation problems, from determining optimal structure of synthetic proteins [19] to computer vision [21].

3.4   Estimating Uncertainty

The GP-UCB algorithm enables us to find the maximum of a function (in our case, the likelihood function or the un-normalised posterior); in many cases, however, it is very desirable to be able to provide uncertainty estimates over the parameter values returned. Given the intractable nature of the likelihood, which requires a computationally expensive statistical model checking procedure at every parameter value, a fully Bayesian treatment (e.g. based on Markov chain Monte Carlo simulations [6]) is ruled out.

We therefore resort to a simple deterministic approximation which estimates the variance/ covariance in the parameter estimates by inverting the Hessian of the likelihood at its maximum. This approach, known as *Laplace approximation* in statistics/ machine learning, is equivalent to approximate the posterior around the maximum with a Gaussian which locally optimally matches the posterior. In order to estimate the Hessian, there are at least two strategies available: one can estimate the likelihood (numerically) on a fine (small) grid around the maximum and then use standard numerical estimation methods, or one can use the GP emulation as a surrogate of the function and directly differentiate the GP mean. This second option has the advantages of handling the noise in the estimation of the likelihood arising from statistical model checking (which is smoothed out in GP regression), and of being analytically tractable. Recalling that the GP mean at a point is a linear combination of basis functions, one can just differentiate twice equation (4) to obtain the result.

3.5   Model Design

The problem of model design is intimately linked to the inference problem: in fact, one could characterise model design as *inference with the data one would like to have* [5]. In our case, we are given a probability table for the joint occurrence of a number of formulae $\varphi_1, \ldots, \varphi_N$.[3] As explained earlier, the probability of a specific truth configuration of a number of formulae is an intractable function of the parameters, which in many cases can only be approximately computed by statistical model checking. However, in the design case, we do not aim to

---

[3] This problem formulation is different from a recent approach on parameter synthesis for CTMC using SMC, [13], in which the authors look for a subset of parameters in which a single formula φ is satisfied with probability greater than q.

use this function to estimate the likelihood of observations, rather to match (or be as near as possible to) some predefined values. We therefore need to define a different objective function that measures the distance between two probability distributions; we choose to use the Jensen-Shannon divergence due to its information theoretic properties and computational good behaviour (being always finite) [9]. This is defined as

$$JSD(p\|q) = \frac{1}{2} \sum_i \left[p_i \log \frac{2p_i}{p_i + q_i} + q_i \log \frac{2q_i}{p_i + q_i} \right]$$

where p and q are two probability distributions over a finite set. The Jensen-Shannon divergence is symmetric and always non negative, being zero if and only if $q = p$. The GP-UCB minimisation of the Jensen-Shannon divergence between an empirical q and the prescribed p can then be carried out as described above.

4 Experiments

We now illustrate our approach on a number of test cases. We benchmark the approach on a simple example where the calculations can be performed analytically: a Poisson process where the truth values of a single logical formula are observed. We then show how our approach can solve both the identification and the design problems on a non-trivial computer infection model.

4.1 Poisson Process

Poisson processes are random processes with values in $\mathbb{N} \cup \{0\}$; they play a fundamental role in physics (e.g. as models of radioactive decay), biology (e.g. as models of RNA production) and computer science (e.g. as models of arrivals of packets at servers). They can be defined equivalently in several different ways; here, we take the operational definition that a Poisson process with rate μ is an increasing, integer valued process such that

$$P(k = n | \mu, t) = \frac{(\mu t)^n}{n!} \exp[-\mu t]. \tag{7}$$

We consider a very simple scenario where we have observed five times independently the truth value of the formula $\varphi(k) = \mathbf{F}^{[0,1]}\{k > 3\}$, i.e. the formula expressing the fact that k has become bigger than 3 within 1 time units, evaluated on individual trajectories sampled from a process with $\mu = 2$. The probability of φ being true for a trajectory given the value of μ can be calculated analytically as

$$p = P(\varphi = true) = 1 - P(\varphi = false) = 1 - \sum_{n=0}^{3} \frac{(\mu)^n}{n!} \exp[-\mu]. \tag{8}$$

and hence we have an analytical expression for the log-likelihood (or unnormalised posterior given a prior).

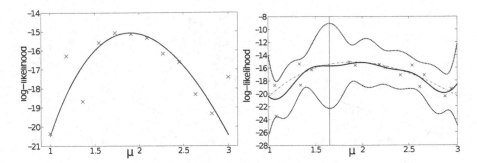

Fig. 1. Simulation on a Poisson process: *left* exact log likelihood and SMC estimation (red crosses) for $\mu \in [1,3]$; *right* Illustration of the GP-UCB algorithm: GP likelihood estimation (black line), true likelihood (green dashed-dotted line), and GP-UCB upper bound (dotted line).

Figure 1 left panel shows the log-likelihood for 40 independent observations of process trajectories, overlayed with the estimation obtained by SMC over a grid using 12 samples. As we can see, SMC provides a noisy (but overall accurate) measurement of the log-likelihood function. Figure 1 right panel instead shows the working of the GP-UCB algorithm (with constant $\beta_t \equiv 2$): here, we have observed only 15 SMC evaluations of the log-likelihood (red crosses); the GP mean is given by the solid black line, and the mean $\pm$ 2 standard deviations by the dashed line. The vertical line represents the next point chosen by the GP-UCB algorithm. The dashed-dotted line is the analytical log-likelihood.

4.2 Network Epidemics

We consider now a more structured example of the spread of a worm epidemics in a computer network with a fixed number of nodes [7]. We consider a simple variation of the classical SIR infection model [2], in which an initial population of susceptible nodes can be infected either from outside the network (e.g. by receiving an infected email message) or by the active spread of the virus by infected computers in the network. Infected nodes can be patched, and become immune to the worm for some time, after which they are susceptible again (for instance, to a new version of the worm).

This system is modelled as a population CTMC, in which the state space is described by a vector $\boldsymbol{X}$ of three variables, counting how many nodes are in the susceptible (X_S), infected (X_I), and patched state (X_R). The dynamics of the CTMC is described by a list of transitions, or reactions, together with their rate functions. We represent them in the biochemical notation style (see e.g. [11]). All rates of this model follow the law of mass action.

External infection: $S \xrightarrow{k_e} I$, with rate function $k_e X_S$;
Internal infection: $S + I \xrightarrow{k_i} I + I$, with rate function $k_i X_S X_I$;
Patching: $I \xrightarrow{k_r} R$, with rate function $k_r X_I$;
Immunity loss: $R \xrightarrow{k_s} S$, with rate function $k_s X_R$;

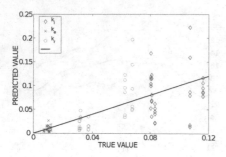 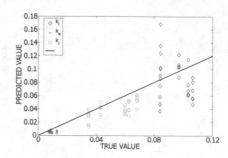

Fig. 2. True versus predicted value of epidemics model parameters k_i, k_e, k_r. The black line represents the identity true=predicted. Left: ML; right: MAP.

For this system, we considered three temporal logical properties, expressed as MiTL formulae, all concerned with the number of infected nodes (total number of nodes is 100). The properties are:

1. $\mathbf{G}^{[0,100]}(X_I < 40)$: the fraction of infected nodes never exceeds 40% in the first 100 time units;
2. $\mathbf{F}^{[0,60]}\mathbf{G}^{[0,40]}(5 \le X_I \le 20)$: within time 60, the fraction of infected nodes is between 5 and 20 and remains so for 40 time units.
3. $\mathbf{G}^{[30,50]}(X_I > 30)$: the fraction of infected nodes is above 30% between time 30 and time 50.

The first property puts a bound on the peak of infection, while the third constrains it to happen around time 40. The second property, instead, is intended to control the number of infected nodes after the infection peak.

Given the model and the properties, we set up the experiment as follows. We fixed the rate of immunity loss to 0.01; the remaining parameters are those we explored. First we fixed these parameters to a value sampled uniformly in $k_i \in [0.08, 0.12]$, $k_e \in [0.007, 0.013]$, $k_r \in [0.03, 0.07]$, and use the sampled configuration to generate 40 observations D of the value of the logical formulae. Then, we ran the GP-UCB optimisation algorithm with the following search space: $k_i \in [0.01, 1]$, $k_e \in [0.001, 0.1]$, $k_r \in [0.005, 0.5]$, so that each parameter domain spans over two orders of magnitude. To treat equally each order of magnitude, as customary we transformed logarithmically the search space, and rescaled each coordinate into $[-1, 1]$ (log-normalisation). The algorithm first computes the likelihood, using statistical model checking, for 60 points sampled randomly and uniformly from the log-normalized space, and then uses the GP-UCB algorithm to estimate the position of a potential maximum of the upper bound function in a grid of 800 points, sampled uniformly at each iteration. If in this grid a point is found with a larger value than those of the observation points, we compute the likelihood also for this point, and add it to the observations (thus changing the GP approximation). Termination happens when no improvement can be made after three grid resamplings. The algorithm terminated after doing only 12 additional likelihood evaluations on average.

Table 1. True parameter and three predictions randomly chosen, both for ML and MAP, after running a gradient ascent optimisation on the GP mean. We show the predicted value and the uncertainty estimate, obtained from the estimated hessian of the likelihood function.

Max Likelihood

Param	true value	pred1	sd	pred2	sd	pred3	sd
k_i	0.0811	0.0803	0.0142	0.0670	0.0084	0.1100	0.0132
k_e	0.0118	0.0114	0.0029	0.0106	0.0014	0.0065	0.0011
k_r	0.0319	0.0304	0.0032	0.0330	0.0020	0.0293	0.0034

MAP

Param	true value	pred1	sd	pred2	sd	pred3	sd
k_i	0.1034	0.0927	0.0048	0.0946	0.0079	0.0744	0.0062
k_e	0.0084	0.0081	0.0005	0.0106	0.0004	0.0076	0.0011
k_r	0.0683	0.0719	0.0044	0.0683	0.0039	0.0643	0.0113

We consider both the maximum likelihood (ML) and maximum a posteriori (MAP) identification problems; in the MAP case, we use independent, vaguely informative Gamma priors, with mean 0.1 for k_i, 0.01 for k_e and 0.05 for k_r, and shape equal to 10. To assess statistically our results, we repeated the experiments (both ML and MAP) on 5 different parameter configurations, doing 6 runs per configuration. In the test, we fixed the length-scale hyperparameter of the Gaussian kernel to 0.1, and the amplitude to 60% of the difference between the maximum and the mean value of the likelihood for the 60 initial observations. Results are reported in Figure 2, where we plot the values of the true parameters that generated the samples against the estimated values. As can be seen, the predicted values are quite close to the original ones, both in the ML and in the MAP cases. Indeed, the average observed error (euclidean distance from the true configuration) is 0.0492 for ML and 0.0346 for MAP. These results show that the use of prior information can improve the performances of the algorithm. Furthermore, it tends to reduce the sensitivity of the algorithm to the hyperparameters, especially the length-scale. We report also the relative errors, obtained dividing the absolute ones by the diameter of the search space: 4.43% for ML and 3.11% for MAP. In Table 1, we report for a random subset of runs both the true/ inferred parameter values, and the uncertainty estimates obtained using the Laplace approximation described in Section 3.4. Empirically, we observed that in some instances the Hessian became not negative definite: this may indicate identifiability problems given a specific set of observations. To circumvent this problem, we ran a gradient ascent optimisation on the GP mean before computing the Hessian, to ensure that the point is a local maximum. Also empirically, we observed that the estimation of the uncertainty is affected by the values of the hyperparameters of the GP; we discuss further this issue in the conclusions.

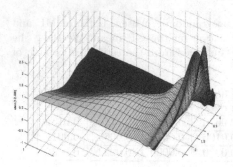
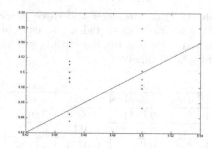

Fig. 3. Left: atanh($1 - 2JSD$) GP-estimated landscape for the network epidemics model and target probability $p(1,1) = 0.5$, $p(1,0) = 0.45$, $p(0,1) = 0.04$, $p(0,0) = 0.01$. Right: true versus predicted value of target probability for the network epidemics design problem. The black line represents the identity true=predicted.

4.3 System Design

We consider now an example of system design, in which we try to minimise the Jensen-Shannon divergence (JSD) between the estimated joint probability distribution and a target one (for numerical stability, we consider atanh($1 - 2JSD$)). We consider again the network epidemics model, and look at the following two properties:

1. $\mathbf{G}^{[0,100]}(X_I < 20)$: the fraction of infected nodes never exceeds 20% in the first 100 time units;
2. $\mathbf{F}^{[0,60]}\mathbf{G}^{[0,40]}X_I \leq 5$: within time 60, the fraction of infected nodes is less than or equal to 5 and remains so for 40 time units.

In our experimental setting, we fixed the internal infection rate to $k_i = 1$ and the immunity loss rate to $k_s = 0.01$. Hence, the search space is made of two parameters, k_e, the external infection rate, and k_r, the patch rate. Intuitively, those two parameters are somehow controllable, by filtering suspected messages from outside the network or by active patching. In the first experiment, we set the target probability p to the following: $p(1,1) = 0.5$, $p(1,0) = 0.45$, $p(0,1) = 0.04$, $p(0,0) = 0.01$. The idea is that we really want the first property to hold, but we are somehow less restrictive on the second one (conditional on the first being true). Having a 2 dimension parameter space to explore, allows us to visualise the GP estimate of the the JSD function, by sampling a 12x12 grid of equispaced points in $[0.01, 1] \times [0.01, 5]$ (after log-normalisation, we use length-scale 0.5 and amplitude 1 as hyperparameters). The result can be seen in Figure 3 left. As we can see, there is a relatively small region in the parameter space that seems to collect the larger score. Running ten experiments, we obtained an average JSD of 0.0155, while the probability values estimated for $p(1,1)$ and $p(1,0)$ are visually reported in Figure 3 right.

We also run an experiment varying the target probability distribution, sampling it from a Dirichelet distribution with parameters $10, 0.8, 9, 0.2$, thus giving

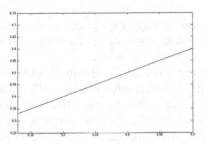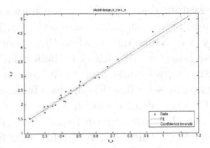

Fig. 4. Left: true versus predicted value of target probability for the network epidemics design problem. The black line represents the identity true=predicted. Right: k_r versus k_e of the predicted parameters for the 5x5 grid.

higher probability to $(1,1)$ and $(0,1)$, differently from the previous test. We sampled 5 different target distributions, and run 5 experiments for each combination, obtaining an average JSD of 0.0168. Probabilities obtained, plotted against target probabilities, are reported in Figure 4 left, while in Figure 4 right, we plot k_r versus k_e for parameter combinations found by the algorithm. While the overall results are good, there is a strong linear dependency between the two parameters, raising an issue of identifiability for this input specification of model design.

5   Conclusions

In this paper, we considered the problem of identifying and designing stochastic processes from logical constraints, given by the (probability of) satisfaction of specific formulae in MiTL. This complements approaches to learning parameters of stochastic processes from observations of the state of the system [17,3], and can be arguably more appropriate in a number of situations; however, the information loss resulting from having only access to qualitative observations makes the problem considerably more challenging. Another benefit of our approach is that it provides a conceptually unified framework that can also be used to address the system design problem, where logical constraints are a much more natural form of specifying requirements. A significant strength of our approach is its computational efficiency and scalability: in our experiments, the global maximum was usually found with few tens of function evaluations, hence the bottleneck is essentially the SMC step. Moreover, the GP regression approach naturally allows to incorporate and smooth the inaccuracies resulting from SMC (modelled as Gaussian observation noise), which means that relatively short runs of SMC are needed at each function evaluation. While we believe the results we have shown are a promising first step in addressing these challenging problems, there is considerable scope for further extension and improvements. Setting the hyperparameters of the GP covariance (3) is currently done heuristically; they could also be optimised, but at a non-negligible computational cost [18]. For the system design problem, one may incur in identifiability problems when the requirements cannot be satisfied (e.g. because of logical contradictions), when they are redundant, or when they under-constrain the system. Tools to address these issues would clearly be beneficial.

Finally, we would want to remark on how these results could only be obtained by the cross-fertilisation of ideas from advanced formal modelling (e.g. Bayesian SMC) and advanced machine learning (Gaussian processes, the GP-UCB algorithm). It is our opinion that increased interaction between these two areas of computer science will be highly beneficial to both, in particular towards the practical deployment of advanced algorithmic tools.

References

1. Alur, R., Feder, T., Henzinger, T.A.: The benefits of relaxing punctuality. J. ACM 43(1), 116–146 (1996)
2. Andersson, H., Britton, T.: Stochastic Epidemic Models and Their Statistical Analysis. Springer (2000)
3. Andreychenko, A., Mikeev, L., Spieler, D., Wolf, V.: Approximate maximum likelihood estimation for stochastic chemical kinetics. EURASIP Journal on Bioinf. and Sys. Bio. 9 (2012)
4. Baier, C., Haverkort, B., Hermanns, H., Katoen, J.P.: Model checking continuous-time Markov chains by transient analysis. IEEE TSE 29(6), 524–541 (2003)
5. Barnes, C.P., Silk, D., Sheng, X., Stumpf, M.P.: Bayesian design of synthetic biological systems. PNAS USA 108(37), 15190–15195 (2011)
6. Bishop, C.M.: Pattern Recognition and Machine Learning. Springer (2006)
7. Bradley, J.T., Gilmore, S.T., Hillston, J.: Analysing distributed internet worm attacks using continuous state-space approximation of process algebra models. J. Comput. Syst. Sci. 74(6), 1013–1032 (2008)
8. Chen, T., Diciolla, M., Kwiatkowska, M.Z., Mereacre, A.: Time-bounded verification of CTMCs against real-time specifications. In: Fahrenberg, U., Tripakis, S. (eds.) FORMATS 2011. LNCS, vol. 6919, pp. 26–42. Springer, Heidelberg (2011)
9. Cover, T., Thomas, J.: Elements of Information Theory, 2nd edn. Wiley (2006)
10. Durrett, R.: Essentials of stochastic processes. Springer (2012)
11. Gillespie, D.T.: Exact stochastic simulation of coupled chemical reactions. J. of Physical Chemistry 81(25) (1977)
12. Jha, S.K., Clarke, E.M., Langmead, C.J., Legay, A., Platzer, A., Zuliani, P.: A bayesian approach to model checking biological systems. In: Degano, P., Gorrieri, R. (eds.) CMSB 2009. LNCS, vol. 5688, pp. 218–234. Springer, Heidelberg (2009)
13. Jha, S.K., Langmead, C.J.: Synthesis and infeasibility analysis for stochastic models of biochemical systems using statistical model checking and abstraction refinement. Theor. Comp. Sc. 412(21), 2162–2187 (2011)
14. Kennedy, M., O'Hagan, A.: Bayesian calibration of computer models. Journal of the Royal Stat. Soc. Ser. B 63(3), 425–464 (2001)
15. Kwiatkowska, M., Norman, G., Parker, D.: Probabilistic symbolic model checking with PRISM: A hybrid approach. Int. Jour. on Softw. Tools for Tech. Transf. 6(2), 128–142 (2004)
16. Maler, O., Nickovic, D.: Monitoring temporal properties of continuous signals. In: Lakhnech, Y., Yovine, S. (eds.) FORMATS/FTRTFT 2004. LNCS, vol. 3253, pp. 152–166. Springer, Heidelberg (2004)

17. Opper, M., Sanguinetti, G.: Variational inference for Markov jump processes. In: Proc. of NIPS (2007)
18. Rasmussen, C.E., Williams, C.K.I.: Gaussian Processes for Machine Learning. MIT Press (2006)
19. Romero, P.A., Krause, A., Arnold, F.H.: Navigating the protein fitness landscape with Gaussian processes. PNAS USA 110(3), E193–E201 (2013)
20. Srinivas, N., Krause, A., Kakade, S., Seeger, M.: Information-theoretic regret bounds for Gaussian process optimisation in the bandit setting. IEEE Trans. Inf. Th. 58(5), 3250–3265 (2012)
21. Vezhnevets, A., Ferrari, V., Buhmann, J.: Weakly supervised structured output learning for semantic segmentation. In: Comp. Vision and Pattern Recog. (2012)
22. Younes, H.L.S., Simmons, R.G.: Statistical probabilistic model checking with a focus on time-bounded properties. Inf. Comput. 204(9), 1368–1409 (2006)

Characterizing Oscillatory and Noisy Periodic Behavior in Markov Population Models

David Spieler

Saarland University[*]
spieler@cs.uni-saarland.de

Abstract. In systems biology, an interesting problem is to analyze and characterize the oscillatory and periodic behavior of a chemical reaction system. Traditionally, those systems have been treated deterministically and continuously via ordinary differential equations. In case of high molecule counts with respect to the volume this treatment is justified. But otherwise, stochastic fluctuations can have a high influence on the characteristics of a system as has been shown in recent publications.

In this paper we develop an efficient numerical approach for analyzing the oscillatory and periodic character of user-defined observations on Markov population models (MPMs). MPMs are a special kind of continuous-time Markov chains that allow for a discrete representation of unbounded population counts for several population types and transformations between populations. Examples are chemical species and the reactions between them.

1 Introduction

Oscillation is a prevalent phenomenon that can be observed within biological systems at all kinds of granularity, e.g. on a microscopic level within individual cells [17] as well as on a macroscopic level within the growth of Savanna patches [18]. An example for oscillatory behavior is the day/night rhythm of many living organisms, with a *period length*, i.e. the time needed for one cycle, of approximately 24 hours. Oscillatory biological networks are the underlying structure of rhythmic behavior on a cellular level. In the case of the day/night cycle, the primary mechanism is based on *circadian clocks* [6].

It is the task of *systems biology*, an inter-disciplinary field of biology, to study those complex biological systems, in order to understand their underlying basic mechanisms. In the systems biology research area, the system under consideration is reduced to a formal model, which is then analyzed for emergent behavior. Traditionally, those models have been modelled by sets of ordinary differential equations (ODEs) used to describe the kinetics of the system's chemical reaction network. Those ODEs have been used to retrieve continuous deterministic

---

[*] This research has been partially funded by the German Research Council (DFG) as part of the Cluster of Excellence on Multimodal Computing and Interaction at Saarland University and the Transregional Collaborative Research Center "Automatic Verification and Analysis of Complex Systems" (SFB/TR 14 AVACS).

K. Joshi et al. (Eds.): QEST 2013, LNCS 8054, pp. 106–122, 2013.

solutions, i.e. the concentrations of the involved chemical species over time. Analyzing the oscillatory character for those deterministic systems by examination of its *limit-cycles* is a well-known problem [19]. An automated approach to retrieve information about the period length of the limit cycle in case of a single equilibrium point is described in [14].

But it has been shown recently, that for some classes of systems, a deterministic formalization is not appropriate. One example is the λ-phage decision circuit, where λ-phages, when infecting E.coli bacteria, either enter the *lytic* cycle (i.e. they directly force the host cell to produce replicas of the phage and then destroy the bacteria's cell membrane, which is called *lysis*) or they enter the *lysogenic* cycle (i.e. they inject their genetic code into the bacteria's DNA, such that phage replicas are produced in later generations of the host). The decision between lytic and lysogenic cycle is assumed to be *probabilistic* [2], whereas a deterministic model would result in the phage always choosing one of the two pathways or a mixture in-between. Likewise, a *stochastic* model is needed in various cases, with another example being the circadian clocks as argued in [6]. More precisely, in the case of a low number of particles, molecules should be modeled as such, i.e. via population counters. Hence, the state space of possible configurations, a certain system is in at each point in time, is *discrete*. In [7] for example, Bortolussi and Policriti point out that for some systems, to show oscillatory behavior on the model level, at least a certain subset of the involved species should be represented discretely.

This paper therefore uses a stochastic and discrete-state modeling approach via Continuous Time Markov Chains (CTMCs) as proposed by Gillespie [12]. The usual way of analyzing those models in systems biology as described in Gillespie's work is via *simulation* [11], i.e. the generation of large numbers of sample trajectories. Those trajectories are evaluated using statistical methods in order to retrieve estimations of stochastic quantities (like the probability of certain events happening within certain time bounds) together with a confidence interval, expressing the quality of that estimation.

Related Work: A novel idea of analyzing a system with respect to oscillatory behavior is the use of *model checking*. In model checking, the property to check is first translated into a (modal) logical formula with a clearly defined semantics. In the case of CTMCs, a prominent logic is Continuous Stochastic Logic (CSL). The underlying principle of CSL model checking, as proposed in [3], involves iterative transient and steady-state analysis methods in order to check the validity of real-time probabilistic properties. The seminal work of Ballarini et al. [4,5] features several logical characterizations of different aspects of oscillation as well as the work of Oana et al. [1]. Examples are the presence or absence of permanent fluctuations as well as whether the model everlastingly shows deviations by certain amplitude levels. Where classic model checking provides an intuitive way of describing a property it is always limited by the expressiveness of the underlying logic. It is therefore well suited for queries like bounded reachability but if the properties are getting more complex, logical formulas have to be nested, which is a highly error-prone task. Also, the chosen logic might turn out not to be expressive enough, or the model checking procedure might involve an unnecessary computational overhead since for some properties or parts thereof, the logic is too expressive. These are lessons learned during the conception of my Master's thesis [21].

Contribution: This paper is based on key ideas of the author's Master's thesis [21] as well as on transient and steady state analysis methods developed by the author and colleagues [13,9,8] and (i) provides a compact way to define oscillatory and periodic behavior in the stochastic setting inspired by [5], (ii) shows how to efficiently analyze such behavior in Markov population models by combining these existing methods, (iii) simplifies and optimizes the used methods as well as (iv) extends the method to handle infinite state spaces.

2   Preliminaries

Before we define the main concepts, we will clarify the notation and introduce needed definitions and algorithms. Matrices like $\mathbf{Q}$ are symbolized by capital boldface letters with their components denoted by the indexed lower case version (like q_{ij}). Vectors are represented by boldface lower case letters (except distributions which are non-boldface). Vector $\mathbf{0}$ is the zero vector, $\mathbf{e}$ represents a vector of ones, and $\mathbf{e}_i$ is the vector of zeros with a one at position i. Given vector $\mathbf{x} = [\mathbf{x}_1 \ldots \mathbf{x}_d]$ and scalar p we define their concatenation as $[\mathbf{x}\ p] = [\mathbf{x}_1 \ldots \mathbf{x}_d\ p]$. Sets are denoted by capital letters (sometimes calligraphic).

2.1   Markov Population Models

We will use *continuous-time stochastic processes* $\{X(t) \mid t \in \mathbb{R}_{\geq 0}\}$, i.e., families of random variables $X(t)$, where index t denotes time. At each moment in time $t \in \mathbb{R}_{\geq 0}$, the system is in a state $X(t) \in S$. We further restrict to *homogeneous continuous-time Markov chains (CTMCs)*, i.e., we demand that the possible future behavior only depends on the current state and does not change over time, as formalized by Equation (1).

$$\mathbf{Pr}[X(t_n) = s_n \mid X(t_{n-1}) = s_{n-1}, \ldots, X(t_0) = s_0]$$
$$= \mathbf{Pr}[X(t_n) = s_n \mid X(t_{n-1}) = s_{n-1}]$$
$$= \mathbf{Pr}[X(t_n - t_{n-1}) = s_n \mid X(0) = s_{n-1}]. \tag{1}$$

We define the *transient probability distribution* at time point t as a row vector $\pi(t)$ such that $\pi_s(t) = \mathbf{Pr}[X(t) = s] \in [0, 1]$ and $\pi(t) \cdot \mathbf{e} = 1$, where we assume a suitable enumeration scheme for its components $s \in S$. Due to the above constraints, the behavior of a CTMC is fully described by an *infinitesimal generator matrix* $\mathbf{Q} = (q_{ij})_{ij} \in \mathbb{R}^{S \times S}$ with $q_{ii} = -\sum_{j \neq i} q_{ij}$ and *initial distribution* $\pi(0)$. More precisely, the transient distribution satisfies the Kolmogorov differential equations

$$\frac{d}{dt}\pi(t) = \pi(t) \cdot \mathbf{Q}. \tag{2}$$

Furthermore, we assume the processes to be *ergodic* such that the *steady state distribution* $\pi = \lim_{t \to \infty} \pi(t)$ which satisfies

$$\pi \cdot \mathbf{Q} = 0 \text{ and } \pi \cdot \mathbf{e} = 1, \tag{3}$$

exists and is unique. We will elaborate that constraint in Section 2.3 and refer to [8] for details. In our case, we distinguish between $N \in \mathbb{N}$ population types such that a state $\mathbf{x} = [\mathbf{x}_1 \ldots \mathbf{x}_N] \in S$ represents the number of individuals $\mathbf{x}_i$ of each type $1 \leq i \leq N$. Consequently, we define the *state space* S as $S \subseteq \mathbb{N}^N$ and can summarize the model class in the following definition.

Definition 1 (Markov population model).
A Markov population model (MPM) with N population types is a continuous-time Markov chain represented by a tuple $(S, \mathbf{Q}, \pi(0))$ where $S \subseteq \mathbb{N}^N$ is the state space, $\mathbf{Q}$ is the infinitesimal generator matrix on S and $\pi(0)$ is the initial distribution.

Since S is not bounded a priori in any dimension and is therefore potentially an infinite set, we can not directly specify all entries of $\mathbf{Q}$ individually. Consequently, we will use a compact symbolic representation in the form of *transition classes*.

Definition 2 (Transition class).
A transition class on state space $S \subseteq \mathbb{N}^N$ is a tuple $(\alpha, \mathbf{v})$ where $\alpha : S \to \mathbb{R}_{\geq 0}$ is the propensity function and $\mathbf{v} \in \mathbb{Z}^N \setminus \{\mathbf{0}\}$ is the change vector.

A set of transition classes $\{(\alpha_r, \mathbf{v}_r)\}_{1 \leq r \leq R}$ on S induces the infinitesimal generator matrix $\mathbf{Q}$ of an MPM $(S, \mathbf{Q}, \pi(0))$ via

$$q_{\mathbf{xy}} = \begin{cases} \sum_{\{r \ | \ \mathbf{x}+\mathbf{v}_r=\mathbf{y}\}} \alpha_r(\mathbf{x}) & \text{if } \mathbf{x} \neq \mathbf{y}, \\ -\sum_{\mathbf{z} \neq \mathbf{x}} q_{\mathbf{xz}} & \text{if } \mathbf{x} = \mathbf{y}. \end{cases}$$

2.2 Markov Population Models of Chemical Reaction Networks

For many biological models based on chemical reaction networks, a treatment on the granularity of single molecules is needed to account for the inherent stochastic effects that govern its key mechanics [7]. An example for that will be shown later in Section 3. It has been shown by the seminal work of Gillespie et al. that under specific conditions [12], the underlying stochastic processes of those networks are continuous-time Markov chains. The use of Markov population models allows an intuitive modeling of those systems. More precisely, assuming we are given a chemical reaction network involving N different chemical species $C_1, \ldots, C_N$ and R different reactions, each reaction is of the form

$$\mathbf{u}_{\mathbf{r}1} \cdot C_1 + \cdots + \mathbf{u}_{\mathbf{r}N} \cdot C_N \xrightarrow{c_r} \mathbf{w}_{\mathbf{r}1} \cdot C_1 + \cdots + \mathbf{w}_{\mathbf{r}N} \cdot C_N,$$

where $1 \leq r \leq R$, $\mathbf{u}_{\mathbf{r}} = (\mathbf{u}_{\mathbf{r}1}, \ldots, \mathbf{u}_{\mathbf{r}N})^T \in \mathbb{N}^N$ and $\mathbf{w}_{\mathbf{r}} = (\mathbf{w}_{\mathbf{r}1}, \ldots, \mathbf{w}_{\mathbf{r}N})^T \in \mathbb{N}^N$ are the *stoichiometric coefficients* and $c_r \in \mathbb{R}_{\geq 0}$ is the *reaction rate*. Every reaction r induces a transition class $(\alpha_r, \mathbf{v_r})$ for the MPM $(S, \mathbf{Q}, \pi(0))$ with $S \subseteq \mathbb{N}^N$, where $\mathbf{v_r} = \mathbf{w_r} - \mathbf{u_r}$ and

$$\alpha_r(\mathbf{x}_1, \ldots, \mathbf{x}_N) = c_r \cdot \prod_{i=1}^{N} \frac{\mathbf{x}_i!}{\mathbf{u}_{r_i}! \cdot (\mathbf{x}_i - \mathbf{u}_{r_i})!}.$$

Example 1 (Repressilator).

We consider a self-regulating gene network inspired by the *repressilator*, a model from synthetic biology that was *designed* on paper first and afterwards implemented *in-vivo* [10]. Our system is based on two genes G_A and G_B that express proteins A and B, respectively. The behavior of that system is described by the chemical reactions stated in Table 1a. The state space of the underlying MPM is $S = \mathbb{N}^2 \times \{0, 1\}^4$ and the infinitesimal generator is induced by eight transition classes $(\alpha_r, \mathbf{v_r})_r$ with $1 \leq r \leq 8$ (cf. Table 1b). We assume that in a state $\mathbf{x} = [\mathbf{x}_1, \ldots, \mathbf{x}_6]^T \in S$, component $\mathbf{x}_1$ ($\mathbf{x}_2$) encodes the number of A (B) molecules, component $\mathbf{x}_3$ ($\mathbf{x}_4$) represents the number of G_A (G_B) molecules and component $\mathbf{x}_5$ ($\mathbf{x}_6$) the number of $\overline{G_A}$ ($\overline{G_B}$) inactive G_A (G_B) molecules. With the initial condition $\pi_{\mathbf{x}_0}(0) = 1$ for $\mathbf{x}_0 = [0, 0, 1, 1, 0, 0]$, we have the invariants $\mathbf{x}_3, \mathbf{x}_4, \mathbf{x}_5, \mathbf{x}_6 \in \{0, 1\}$, $\mathbf{x}_5 = 1 - \mathbf{x}_3$, and $\mathbf{x}_6 = 1 - \mathbf{x}_4$ that are valid for any point in time.

Table 1. The repressilator model

(a) Chemical reactions.

$R_1 : G_A$	$\xrightarrow{\rho_A}$	$G_A + A$
$R_2 : G_B$	$\xrightarrow{\rho_B}$	$G_B + B$
$R_3 : A$	$\xrightarrow{\delta_A}$	$\emptyset$
$R_4 : B$	$\xrightarrow{\delta_B}$	$\emptyset$
$R_5 : A + G_B$	$\xrightarrow{\beta_A}$	$A + \overline{G_B}$
$R_6 : B + G_A$	$\xrightarrow{\beta_B}$	$B + \overline{G_A}$
$R_7 : \overline{G_A}$	$\xrightarrow{\nu_A}$	G_A
$R_8 : \overline{G_B}$	$\xrightarrow{\nu_B}$	G_B

(b) Transition classes.

$\alpha_1(\mathbf{x}) = \rho_A,$	$\mathbf{v_1} = \mathbf{e}_1,$
$\alpha_2(\mathbf{x}) = \rho_B,$	$\mathbf{v_2} = \mathbf{e}_2,$
$\alpha_3(\mathbf{x}) = \delta_A \cdot \mathbf{x}_1,$	$\mathbf{v_3} = -\mathbf{e}_1,$
$\alpha_4(\mathbf{x}) = \delta_B \cdot \mathbf{x}_2,$	$\mathbf{v_4} = -\mathbf{e}_2,$
$\alpha_5(\mathbf{x}) = \beta_A \cdot \mathbf{x}_1 \cdot \mathbf{x}_4,$	$\mathbf{v_5} = -\mathbf{e}_4 + \mathbf{e}_6,$
$\alpha_6(\mathbf{x}) = \beta_B \cdot \mathbf{x}_2 \cdot \mathbf{x}_3,$	$\mathbf{v_6} = -\mathbf{e}_3 + \mathbf{e}_5,$
$\alpha_7(\mathbf{x}) = \nu_A \cdot \mathbf{x}_5,$	$\mathbf{v_7} = -\mathbf{e}_5 + \mathbf{e}_3,$
$\alpha_8(\mathbf{x}) = \nu_B \cdot \mathbf{x}_6,$	$\mathbf{v_8} = -\mathbf{e}_6 + \mathbf{e}_4.$

2.3 Steady State Analysis for Infinite MPM

In order to study the long-term behavior of a MPM we need to ensure that the underlying CTMC is *ergodic* [22] which is equivalent to the unique existence of the steady-state distribution π. For a MPM with finite state space, the only requirement for ergodicity is irreducibility, i.e., that each state is reachable by any other state. Unfortunately, many systems like the repressilator model from Example 1 are not finite since there exists no a priory bound for the population counts. Moreover, in Section 3 we will need information about the steady state distribution. We therefore propose to use the results from previous work [8], where we show how to prove ergodicity and compute a finite subset $\mathcal{C}$ of the state space which contains at least $1 - \epsilon$ of the total steady state probability mass for a specified $\epsilon > 0$. In addition, we approximate π inside $\mathcal{C}$ by considering the infinitesimal generator

$$\bar{\mathbf{Q}} = (q_{\mathbf{xy}})_{\mathbf{x},\mathbf{y}\in\mathcal{C}} - diag\left((q_{\mathbf{xy}})_{\mathbf{x},\mathbf{y}\in\mathcal{C}} \cdot \mathbf{e}\right), \tag{4}$$

where transitions of $\mathbf{Q}$ leaving set $\mathcal{C}$ are redirected to the state where the set is left. This is a valid abstraction since using drift arguments as in [8], with high probability the system returns to the set $\mathcal{C}$ quickly. In the following we will assume precise estimates of the steady state probabilities. For details, we refer to [8] in which error bounds are proven.

2.4   Transient Analysis for Infinite MPM

In Section 3 we will have to combine transient and steady-state analysis of MPMs with potentially unbounded state space. If we had to compute the transient distribution of such MPMs exactly, we would have to handle infinitely many states after a non-zero amount of time, since we might have $\pi_{\mathbf{x}}(t) > 0$ for all reachable states $\mathbf{x} \in S$ after $t > 0$ time units. At each time point t, we therefore concentrate on a subset of the state space which only contains states with *significant* probability mass, i.e., states $\mathbf{x} \in S$ with $\pi_{\mathbf{x}}(t) > \delta$ for some $\delta > 0$. The transient analysis algorithm inspired by ideas from [13,9] is shown for completeness in Algorithm 1 and uses a sub-algorithm to advance the probability mass in time. In order to keep the presentation of the paper focused, we only show the Euler method (cf. Algorithm 2) as an example. The actual implementation that was used in this paper is based on the more stable and accurate Runge-Kutta 4th-order method to solve the system of ODEs that governs the behavior of the

---

Algorithm 1. transient$(\mathbf{Q}, \pi(0), t)$	**Algorithm 2.** advance$(\mathbf{Q}, W, \mathbf{p}, h)$
1: $t' \leftarrow 0$; $e \leftarrow 0$	1: $W' \leftarrow \emptyset$
2: $W \leftarrow \{\mathbf{x} \mid \pi_{\mathbf{x}}(0) > 0\}$	2: $\mathbf{p}' \leftarrow$ new HashMap$(S,[0,1])$
3: $\mathbf{p} \leftarrow$ new HashMap$(S,[0,1])$	3: **for** $\mathbf{x} \in W$ **do**
4: $\forall \mathbf{x}$ with $\pi_{\mathbf{x}}(0) > 0$: $\mathbf{p}(\mathbf{x}) \leftarrow \pi_{\mathbf{x}}(0)$	4: **for** $\mathbf{y}$ with $q_{\mathbf{xy}} > 0$ **do**
5: **while** $t' < t$ **do**	5: $\mathbf{p}'(\mathbf{y}) \leftarrow \mathbf{p}'(\mathbf{y}) + q_{\mathbf{xy}} \cdot \mathbf{p}(\mathbf{x}) \cdot h$
6: choose appropriate h	6: $W' \leftarrow W' \cup \{\mathbf{y}\}$
7: $\Delta t \leftarrow \min(t - t', h)$	7: **end for**
8: $[W', \mathbf{p}] \leftarrow$ advance$(\mathbf{Q}, W, \mathbf{p}, \Delta t)$	8: $\mathbf{p}'(\mathbf{x}) \leftarrow \mathbf{p}'(\mathbf{x}) + q_{\mathbf{xx}} \cdot \mathbf{p}(\mathbf{x}) \cdot h$
9: $W \leftarrow \{\mathbf{x} \mid \mathbf{x} \in W' \wedge \mathbf{p}(\mathbf{x}) \geq \delta\}$	9: **end for**
10: $e \leftarrow e + \sum_{\mathbf{x}\in W'\setminus W} \mathbf{p}(\mathbf{x})$	10: **return** $[W \cup W', \mathbf{p}']$
11: remove keys $\mathbf{x} \in W' \setminus W$ in $\mathbf{p}$	
12: $t' \leftarrow t' + \min(t - t', h)$	
13: **end while**	
14: **return** $[\mathbf{p}, e]$	

state probabilities according to the Kolmogorov differential equations (2). The algorithm relies on a fast method to compute the possible successor states of a state, which is the case with transition classes. The time step h should be chosen with care. In our implementation, this value is adapted in each step as $h = 0.5 \cdot u^{-1}$ where u is the current maximal exit rate, i.e., the maximal sum of outgoing rates over all states. Note that concerning the implementation of hash maps from states to probabilities, we rely on the property that whenever a key is not found, the hash map returns 0. Furthermore, our implementation avoids redundant storage of values e.g. by maintaining a single consistent hash map mapping significant states to their current transient probabilities. Also, in addition to storing the probabilities of significant states, it caches the graph-structure of reachable states and their connecting transitions in order to prevent the repeated computation of successor states and transition rates.

3   Defining and Analyzing Oscillatory Behavior

In order to motivate the use of stochastic modeling of certain systems in contrast to traditional techniques based on continuous-deterministic solutions, we will introduce another example, the *3-way oscillator*.

Example 2 (3-way oscillator).
The *3-way oscillator* is another synthetic gene regulatory network which is based

Table 2. The 3-way oscillator model

(a) Chemical reactions.

$$R_1 : A + B \xrightarrow{\tau_A} 2B$$
$$R_2 : B + C \xrightarrow{\tau_B} 2C$$
$$R_3 : C + A \xrightarrow{\tau_C} 2A$$
$$R_4 : A \xrightarrow{\nu_A} B$$
$$R_5 : B \xrightarrow{\nu_B} C$$
$$R_6 : C \xrightarrow{\nu_C} A$$

(b) Transition classes.

$$\alpha_1(\mathbf{x}) = \tau_A \cdot \mathbf{x}_1 \cdot \mathbf{x}_2, \qquad \mathbf{v}_1 = \mathbf{e}_2,$$
$$\alpha_2(\mathbf{x}) = \tau_B \cdot \mathbf{x}_2 \cdot \mathbf{x}_3, \qquad \mathbf{v}_2 = \mathbf{e}_3,$$
$$\alpha_3(\mathbf{x}) = \tau_C \cdot \mathbf{x}_1 \cdot \mathbf{x}_3, \qquad \mathbf{v}_3 = \mathbf{e}_1,$$
$$\alpha_4(\mathbf{x}) = \nu_A \cdot \mathbf{x}_1, \qquad \mathbf{v}_4 = \mathbf{e}_2,$$
$$\alpha_5(\mathbf{x}) = \nu_B \cdot \mathbf{x}_2, \qquad \mathbf{v}_5 = \mathbf{e}_3,$$
$$\alpha_6(\mathbf{x}) = \nu_C \cdot \mathbf{x}_3, \qquad \mathbf{v}_6 = \mathbf{e}_1.$$

on a cyclic Lotka-Volterra model [20] inspired by real world examples of bacterial populations [15,16]. It consists of three chemical species A, B, and C and six reaction types (cf. Table 2a). Reactions R_1, R_2, and R_3 are positive feedback loops, where each species boosts the production of another species in a circular fashion. The remaining three reactions were added to prevent a deadlock situation, where a species becomes totally depleted and the system stops. The corresponding transition classes are stated in Table 2b. For further analysis, we will assume $\tau_A = \tau_B = \tau_C = \nu_A = \nu_B = \nu_C = 1.0$.

3.1 Continuous-Deterministic Solution

At first we would like to use the traditional approach of deriving a continuous-deterministic solution for the system based on the *law of mass action* in chemistry. A phase plot with initial concentration $[A\ B\ C] = [30\ 0\ 0]$ is shown in Figure 1a. As can be seen in the plot, the underlying structure of the deterministic solution is a damped oscillation, i.e., an initial perturbation caused by the asymmetric initial condition is followed by an oscillatory phase with shrinking amplitude until an equilibrium for all species is reached. Here the equilibrium point is $[10\ 10\ 10]$. On the other hand, the state space of the MPM induced by the

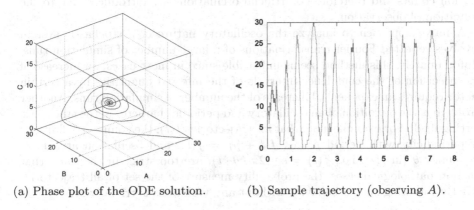

(a) Phase plot of the ODE solution.   (b) Sample trajectory (observing A).

Fig. 1. 3-way oscillator

3-way oscillator's chemical reaction network for initial state $[x_1\ x_2\ x_3] = [30\ 0\ 0]$ is $S = \{[x_1\ x_2\ x_3] \in \mathbb{N}^3 \mid x_1 + x_2 + x_3 = 30\}$ and due to the irreducibility of the underlying CTMC coinciding with ergodicity in the finite case we have for all states $s \in S$ that $\pi_s > 0$. This implies that every state is visited infinitely often which contradicts convergence to any molecule level and further implies everlasting perturbations. Indeed, state $[30\ 0\ 0]$ e.g. has a non-negligible steady state probability of around 0.002 and any simulation run (cf. Figure 1b) of the stochastic system almost surely shows never-ending fluctuations. Actually, the deterministic approach reasoning about expected population counts is only justified in the *thermodynamic limit*, when the number of molecules is very high compared to the considered volume. The treatment of a total number of 30 molecules in our case certainly violates that condition and in the following, we restrict to the case where a deterministic treatment is not justified due to stochastic noise.

3.2 Discrete-Stochastic Approach

Consequently, we propose a method to analyze the stochastic model based on the MPM construction from Section 2.2. Please note that where in the deterministic model we have only to inspect a single solution of the ODE which could be easily treated e.g. by *Fourier analysis*, we have to face an uncountably infinite amount of potential *trajectories* in the stochastic setting.

Definition 3 (Trajectory).
Given a stochastic process $\{X(t)\}_{t \in \mathbb{R}_{\geq 0}}$, *we define a* trajectory *as a function* $f : \mathbb{R}_{\geq 0} \to \mathbb{R}$ *mapping each time point* $t \in \mathbb{R}_{\geq 0}$ *to an observation* $f(t) = X(t) \cdot \mathbf{o}$ *with* $\mathbf{o} \in \mathbb{R}^N$.

The *observation weights* $\mathbf{o}$ allow us to describe several objectives like the observation of a single species i with $\mathbf{o} = \mathbf{e}_i$ or the sum of species i and j via $\mathbf{o} = \mathbf{e}_i + \mathbf{e}_j$. The use of constant weights ensures a linear combination of population counts and therefore no artificial oscillations are introduced due to the definition of observation.

A naive approach to analyze the oscillatory nature of a stochastic system involves repeated Fourier transformations of a large number of simulation runs. Unfortunately, this method seems unfeasible since in most cases, an increase of the precision of the confidence intervals of the inferred quantities by a certain factor requires an exponential increase of the number of simulation runs. Another problem is the definition of an oscillatory and periodic character in the stochastic setting itself. For example, note that for trajectories $f(t)$, the common definitions of periodicity with period p, i.e., $f(t + p) = f(t)$, and oscillation at a fixed frequency θ (in Hz), i.e., $f(t) = \sin(2\pi \cdot \theta \cdot t)$, are too strict in the sense that for non-pathological cases, the probability measure of the set of all trajectories strictly following such a behavior without any violations is trivially zero.

3.3   Oscillatory Behavior and Noisy Periodicity

What we propose instead in order to capture the essence of periodic and oscillatory behavior in the *stochastic setting* is to use a numerical approach. But at first we will formally define the above mentioned concepts.

Definition 4 (Oscillatory behavior).
A MPM is called oscillatory *for observation weights* $\mathbf{o}$ *and amplitude levels* $L, H \in \mathbb{N}$ *with* $H > L$, *if the probability measure of all trajectories visiting intervals* $(-\infty, L)$, $[L, H)$, *and* $[H, \infty)]$ *infinitely often is one*.

Obviously, a MPM with observation weights $\mathbf{o}$ is either *oscillatory* or the probability mass of trajectories with *converging* or *diverging* observations is greater than zero. Assuming a system is oscillatory, we are also interested in the time needed to oscillate once around this interval. We call this duration *noisy period length*. A single period can be split into several events and phases (cf. Figure 2). It starts with crossing the lower bound L (event LE) which is succeeded by a phase where the upper bound H has not been reached yet (phase LC). When this bound is finally reached, the period switches into the HC phase which is ended by another crossing of the lower bound L from below. This is indicated by another LE event and the classification pattern repeats.

Remark 1. In order to simplfiy presentation, here we assume that L and H are chosen such that no transition in the MPM may skip the LC phase. Since we only consider bimolecular reactions which do not alter the observation level by more than one, this is the case if we choose $H - L \geq 2$. Nevertheless, for the case studies in Section 4, we also choose $H = L + 1$, but the resulting special cases were handled in the implementation of the numerical analysis.

3.4 Period Detector Expansion

In the following we will describe how to incorporate the above period classification pattern into a given MPM using a compositional approach.

Definition 5 (Deterministic finite automaton).
A deterministic (non-blocking) finite automaton (DFA) *on set A is a tuple* $(M, m_0, \rightarrow)$, *where M is a finite set of states, $m_0 \in M$ is the initial state, and transition relation* $\rightarrow \subseteq M \times P \times P \times M$ *where P denotes the set of predicates over A. We further demand the transition relation to be deterministic and non-blocking, i.e.,* $\forall m \in M, x \in A . \exists! p, p' \in P, m' \in M . p(x) \wedge p'(x) \wedge (m, p, p', m') \in \rightarrow$.

Definition 6 (Product of MPM and DFA).
The product $\mathcal{M} \otimes \mathcal{D}$ of a MPM $\mathcal{M} = (S, \mathbf{Q}, \pi(0))$ and a DFA $\mathcal{D} = (M, m_0, \rightarrow)$ on S is a MPM $\mathcal{M}' = (S', \mathbf{Q}', \pi'(0))$ where $S' = S \times M$, $\pi'_{[\mathbf{x}\ m]}(0) = \pi_{\mathbf{x}}(0)$ *if* $m = m_0$ *and 0 otherwise, and its infinitesimal generator $\mathbf{Q}'$ defined by*

$$q'_{[\mathbf{x}\ m][\mathbf{y}\ m']} = \begin{cases} q_{\mathbf{xy}} & \text{if } \mathbf{x} \neq \mathbf{y} \wedge eval(\mathbf{x}, \mathbf{y}, m, m'), \\ E([\mathbf{x}\ m]) & \text{if } \mathbf{x} = \mathbf{y} \wedge m = m', \\ 0 & \text{otherwise}, \end{cases}$$

where $E([\mathbf{x}\ m]) = -\sum_{\mathbf{0} \neq \mathbf{x} \vee m'' \neq m} q'_{[\mathbf{x}\ m][\mathbf{0}\ m'']}$ *and predicate* $eval(\mathbf{x}, \mathbf{y}, m, m')$ *is true iff* $\exists p, p' \in P . p(\mathbf{x}) \wedge p'(\mathbf{y}) \wedge (m, p, p', m') \in \rightarrow$. *For a state* $s = [\mathbf{x}\ m]$, *we define* $s_M = m$.

Definition 7 (Period detector expanded MPM). *Given a MPM $\mathcal{M} = (S, \mathbf{Q}, \pi(0))$ with $S \subseteq \mathbb{N}^d$, the period detector expanded MPM (PDMPM) of $\mathcal{M}$ for observation weights $\mathbf{o}$ is the product MPM $\mathcal{M}' = \mathcal{M} \otimes \mathcal{D}^{PD}$ where $\mathcal{D}^{PD}$ denotes the DFA depicted in Figure 3. In order to keep the presentation readable, we have abbreviated some transitions to make the DFA deterministic and non-blocking by the expression* else.

The intuition behind the *period detector expanded* MPM is that it mimics the original behavior but additionally annotates the state space of the MPM by the information of the DFA in which event or phase of an oscillation the system currently is. Note that we do not need an acceptance condition for the DFA.

Theorem 1 (Equivalence preservation).
A MPM $\mathcal{M} = (S, \mathbf{Q}, \pi(0))$ and its product with a DFA $\mathcal{D}$ on S are F-bisimilar [3] written $\mathcal{M} \sim_F \mathcal{M} \otimes \mathcal{D}$ *for* $F = \mathbb{R}$ *with respect to the labeling functions [3]* $l(\mathbf{x}) = \{\mathbf{x} \cdot \mathbf{o}\}$ *on $\mathcal{M}$ and* $l([\mathbf{x}\ m]) = \{\mathbf{x} \cdot \mathbf{o}\}$ *on $\mathcal{M} \otimes \mathcal{D}$.*

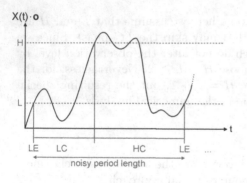

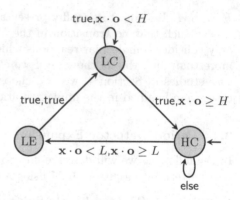

Fig. 2. Events/phases of noisy periods   **Fig. 3.** Period detector DFA

Proof. Let DFA $\mathcal{D} = (M, m_0, \rightarrow)$ and relations $\mathcal{R}_1$ and $\mathcal{R}_2$ be defined as

$$\mathcal{R}_1 = \{(\mathbf{x}, [\mathbf{x}\ m]) \mid \mathbf{x} \in S, m \in M\} \text{ and}$$
$$\mathcal{R}_2 = \{([\mathbf{x}\ m], [\mathbf{x}\ m']) \mid \mathbf{x} \in S, m, m' \in M\}.$$

Then, relation $\mathcal{R} = \mathcal{R}_1 \cup \mathcal{R}_1^{-1} \cup \mathcal{R}_2 \cup id(S)$, where $id(S) = \{(\mathbf{x}, \mathbf{x}) \mid \mathbf{x} \in S\}$ denotes the identity relation on S, is a *F-bisimulation* relation.   □

Theorem 1 ensures, that the MPM and its period detector expanded MPM are equal in the sense that they behave the same with respect to the probability of any observations that can be made starting in any state. In particular, no oscillations are artificially introduced due to the described extension.

3.5   Analysis of the PDMPM

The following theorem shows how the oscillatory character of a MPM can be checked with the help of period detector expansion.

Theorem 2 (Oscillatory character).
Given a MPM with observation weights $\mathbf{o}$ *and amplitude levels* $L, H \in \mathbb{N}$ *with* $H > L$. *If its PDMPM with state space* $S' \subseteq S \times \{LE, LC, HC\}$ *is ergodic and*

$$\forall m \in \{LE, LC, HC\}.\exists \mathbf{x} \in S.\pi([\mathbf{x}\ m]) > 0,$$

where π *denotes the steady state probability distribution of the PDMPM, the MPM is oscillatory.*

Proof. Ergodicity implies positive-recurrence of all states and therefore *divergence* is ruled out. The existence of at least one state $[\mathbf{x}\ m]$ in each phase $m \in \{LE, LC, HC\}$ with positive steady state probability and the construction of the PDMPM imply that each of the observation intervals $(-\infty, L), [L, H), [H, \infty)$ is visited infinitely often contradicting *convergence*.   □

In case of an oscillatory system, we also want to quantify the time needed for oscillations on the long run. We start by defining the time $\mathcal{L}$ needed for the next two LE events in the PDMPM. Since the underlying process is stochastic, $\mathcal{L}$ is a random variable as defined in Equation (5).

$$\mathcal{L}(t) = \min\left[t_2 : \exists t_1 \geq t, t_1 < t_2 . X(t_1)_M = X(t_2)_M = LE \wedge \forall t' \in (t_1, t_2) . X(t')_M \neq LE\right] - t \tag{5}$$

Next, we define the *noisy period length* as the time $\mathcal{L}(t)$ for those states and times where an oscillation just begins, i.e., $X(t)_M = LE$. Our final goal is to approximate the cumulative distribution function (CDF)

$$\lim_{t \to \infty} \mathbf{Pr}[\mathcal{L}(t) \leq T \mid X(t)_M = LE] \tag{6}$$

of the noisy period length on the long run. We will do that in Algorithm 3.

Algorithm 3. nperiod($\mathcal{M} = (S, \mathbf{Q}, \pi(0)), \mathbf{o}, L, H, \Delta, \alpha$)

1: let $\mathcal{M}' = (S' = S \times \{LE, LC, HC\}, \mathbf{Q}', \pi'(0))$ be the PDMPM of $\mathcal{M}$ for $\mathbf{o}$
2: solve $\pi\mathbf{Q}' = 0$ with $\pi\mathbf{e} = 1$
3: $\pi_{[\mathbf{x}\ p]} \leftarrow \pi_{[\mathbf{x}\ LE]}$ if $p = LC$ and 0 otherwise
4: $\pi \leftarrow \pi \cdot (\pi\mathbf{e})^{-1}$
5: $t \leftarrow 0; e_a \leftarrow 0; \mathrm{cdf}(-\Delta) \leftarrow 0$
6: **while** $\sum_{\mathbf{x}} \pi_{[\mathbf{x}\ LE]} < \alpha - e_a$ **do**
7: $\quad [\pi, e] \leftarrow$ transient$(\mathbf{Q}', \pi, \Delta)$ where states $[\mathbf{x}\ LE]$ are made absorbing
8: $\quad \mathrm{cdf}(t) \leftarrow \sum_{\mathbf{x}} \pi_{[\mathbf{x}\ LE]}; e_a \leftarrow e_a + e$
9: $\quad \mathrm{pdf}(t) \leftarrow \mathrm{cdf}(t) - \mathrm{cdf}(t - \Delta); t \leftarrow t + \Delta$
10: **end while**
11: **return** [cdf, pdf]

Theorem 3 (Noisy period length). *Given an oscillatory MPM $X(t)$, Algorithm 3 approximates probability $\lim_{t \to \infty} \mathbf{Pr}[\mathcal{L}(t) \leq T \mid X(t)_M = LE]$ by* cdf(T).

Proof. First, we note that the set of paths satisfying $\mathcal{L}(t) \leq T$ is measurable since the problem can be reduced to bounded reachability where respective proofs as in [3] can be used. Concerning the algorithm, we first compute the steady state distribution π of the PDMPM in line 2 and normalize the sub-distribution $\pi_{[\mathbf{x}\ LE]}$ corresponding to states in the LE event in lines 3 and 4 which resembles the conditioning in Equation (6). The while-loop from lines 6 to 10 performs a transient analysis using this distribution as the initial distribution and states corresponding to a second LE event are made absorbing. Consequently, the total mass in the absorbing states corresponds to the proportion of paths having finished a full oscillatory cycle up to time t. Note that the transient analysis is not exact since we truncate states with probability less than δ (as described in Algorithm 1). Therefore, in addition, we compute the accumulated error in e_a (line 8). We stop iteration as soon as a threshold α of the total initial probability mass minus the accumulated error has been absorbed. In line 8, we keep track of the time and the absorbed mass which gives the CDF quantized by the time step

Δ. Taking finite differences of the CDF finally gives an approximation of the probability density function (PDF) of the noisy period length. The algorithm terminates since each period will finally end with probability one due to the construction of the PDMPM, the ergodicity implied by its oscillatory character and the states corresponding to the end of an oscillation being made absorbing. Note that in line 3 we take states in the LC phase instead of the LE event for the initial distribution. The reason is that this way we do not have to distinguish between the first and second LE event. Consequently, if a state s with $M(s) = LE$ is entered, a full period has been performed. This is justified if $H - L \geq 2$ and only a maximal increase of one in the observation level per transition can be made (bimolecular reactions), since the LE event is left in any case after one step. We restricted to that case in order to simplify the presentation (cf. Remark 1). In our implementation, we have separate annotations for the first and second LE event. □

The total error probability mass $\theta = 1 - \alpha + e_a$ can be used for bounding the period length. If the support of the steady state distribution has to be truncated (cf. Section 2.3), error ϵ has to be added to θ as well. The error θ can be controlled by increasing α. A choice of $\delta = 10^{-20}$ usually results in a negligible error e_a [9]. The time complexity of the steady state computation is $\mathcal{O}(n^3)$, where n is the number of states and the complexity of the CDF/PDF computation using truncation based transient analysis is $\mathcal{O}(u \cdot t \cdot n)$, where t is the maximal period length of interest, u is the maximum exit rate encountered and n is the maximum number of states with significant probability mass until time t.

4   Numerical Results

Finally, we will show the numerical results of applying the presented methods to two case studies from systems biology. All computations were performed on an Intel Core i5 2.66 GHz machine with 8 GB of RAM. In all experiments, we used thresholds $\delta = 10^{-20}$, $\alpha = 0.9999$, and chose $\Delta = 0.5 \cdot u^{-1}$, where u is the maximum exit rate (cf. Section 2.4) over the complete time course.

4.1   Three-Way Oscillator

First, we will analyze the *3-way oscillator* as described in Table 2. We identify the count of molecules of type A (B, C) with variable $\mathbf{x}_1$ $(\mathbf{x}_2, \mathbf{x}_3)$. We choose initial state $(\mathbf{x}_1, \mathbf{x}_2, \mathbf{x}_3) = (30, 0, 0)$ with probability one and rates $\tau_A = \tau_B = \tau_C = \nu_A = \nu_B = \nu_C = 1.0$. As argued in Section 3, the resulting state space is finite. The mean population counts in steady state are $(\overline{\mathbf{x}_1}, \overline{\mathbf{x}_2}, \overline{\mathbf{x}_3}) = (10, 10, 10)$ and due to symmetry we are only interested in species A, i.e., $\mathbf{o} = e_1$. As can be seen in the sample trace in Figure 1b, the oscillations are around this mean value. Consequently, we took $L = \overline{\mathbf{x}_1} - \frac{a}{2}$ and $H = \overline{\mathbf{x}_1} + \frac{a}{2}$ for the interval bounds and varied the amplitude $a \in \{2, 4, 6, 8, 10, 12, 14, 16, 18\}$.

The system is oscillatory for all those amplitude levels and the results of the noisy period length analysis are depicted in Figure 4 with the computation

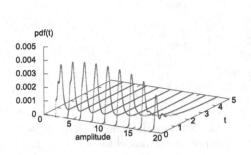

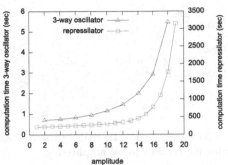

Fig. 4. Noisy period length PDF of the 3-way oscillator model for several amplitudes

Fig. 5. Computation times for computing the noisy period length PDF

times presented in Figure 5. Most likely, the (noisy) period length of the 3-way oscillator is around 0.5 time units and period lengths of 5 or more time units are rare, even in the case of full amplitudes ($a = 18$). This coincides with the observations that can be made from the sample trajectory in Figure 1b.

4.2   Repressilator

The other case study studies the *repressilator* model as described in Table 1 with parameter set $\rho = \rho_A = \rho_B = 10.0$, $\delta = \delta_A = \delta_B = 1.0$, $\beta = \beta_A = \beta_B = 0.05$, and $\nu = \nu_A = \nu_B = 0.2$. We identify molecule counts of species A (B) with variable x_1 (x_2) and represent the active gene G_A (G_B) via variable x_3 (x_4) and inactive gene $\overline{G_A}$ ($\overline{G_B}$) via x_5 (x_6). In contrast to the preceding model, the state space $S = \{x \in \mathbb{N}^2 \times \{0,1\}^4 \mid x_3 = 1 - x_5 \wedge x_4 = 1 - x_6\}$ is not finite for initial state $(0, 0, 1, 1, 0, 0)$. Consequently, geometric bounds for 90% of the steady state probability mass were computed according to [8] which resulted in upper bounds of 32 molecules for both protein species.

Sample traces of the repressilator model (cf. Figure 6) reveal that unlike the oscillation around a mean value, the repressilator with the specified parameters

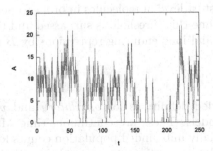

Fig. 6. Sample trajectory of the repressilator model (observing A)

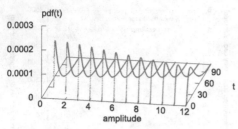
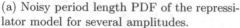

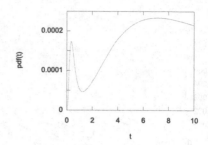

(a) Noisy period length PDF of the repressilator model for several amplitudes.

(b) Noisy period length PDF of the
repressilator model for $a = 1$.

Fig. 7. Results for the repressilator model

has a rather peak-like oscillation pattern, i.e., periods start at the zero level, reach
a maximum peak level and finally return to the zero level again. Therefore, we
choose $L = 1$ and $H = L + a$ for varying a with $1 \leq a \leq 19$ (cf. Remark 1).
The system is oscillatory for all amplitude values and the results of the noisy
period length analysis are depicted in Figure 7a with the respective computation
times presented in Figure 5. The majority of periods have durations of less than
80 time units and the larger the amplitudes, the larger also the period length
becomes, since peaks of higher amplitudes become more rare. An interesting
phenomenon of the repressilator can be witnessed for an amplitude of $a = 1$
(cf. Figure 7b) where we set the smallest constraint on the minimal amplitude.
While more than 98% of the oscillations have a period length of 1.23 time units
or more, a small amount of around 1.87% of the oscillations only lasts for 1.23
time units or less as can be seen by the first peak in probability in Figure 7b.
This bi-modality of the probability distribution can be explained by two effects.
The smaller peak (until $t = 1.23$) occurs since there is little time to build up a
significant amount of A molecules. Consequently, the chance of the A molecules
repressing gene G_B is small and therefore the amount of B molecules grows as
well and finally species B may win the competition to represses its competitor,
gene G_A. Since the degradation rate δ of the molecules is high compared to
the gene unbinding rate ν, it is very likely that all A molecules degrade until
the unbinding event happens and the oscillatory cycle ends. So each oscillation
must first cross a kick-start level of molecules in order to perform a longer cycle.
However, most of the time this threshold is surpassed and the oscillation is only
ended by spontaneous and long enduring repressions by B molecules.

5   Conclusion

In this work we provided definitions for *oscillatory* and *periodic* behavior for
Markov population models. These are continuous-time Markov chains, where
states represent potentially unbounded population counts for several population
types. We further developed an efficient way to check whether a system is oscillatory and also provided a numerical algorithm to approximate the probability

distribution of the period length. We finally applied our techniques to two well-known case studies from systems biology. For future work we plan to extend the approach to capture vanishing oscillations found in damped oscillators as well.

References

1. Andrei, O., Calder, M.: Trend-based analysis of a population model of the akap scaffold protein. TCS Biology 14 (2012)
2. Arkin, A., Ross, J., McAdams, H.: Stochastic kinetic analysis of developmental pathway bifurcation in phage λ-infected escherichia coli cells. Genetics 149, 1633–1648 (1998)
3. Baier, C., Hermanns, H., Haverkort, B., Katoen, J.-P.: Model-checking algorithms for continuous-time Markov chains. IEEE Transactions on Software Engineering 29(6), 524–541 (2003)
4. Ballarini, P., Guerriero, M.L.: Query-based verification of qualitative trends and oscillations in biochemical systems. TCS 411(20), 2019–2036 (2010)
5. Ballarini, P., Mardare, R., Mura, I.: Analysing biochemical oscillation through probabilistic model checking. ENTCS 229(1), 3–19 (2009)
6. Barkai, N., Leibler, S.: Biological rhythms: Circadian clocks limited by noise. Nature 403, 267–268 (2000)
7. Bortolussi, L., Policriti, A.: The importance of being (a little bit) discrete. ENTCS 229(1), 75–92 (2009)
8. Dayar, T., Hermanns, H., Spieler, D., Wolf, V.: Bounding the equilibrium distribution of Markov population models. NLAA (2011)
9. Didier, F., Henzinger, T.A., Mateescu, M., Wolf, V.: Fast adaptive uniformization of the chemical master equation. In: Proc. of HIBI, pp. 118–127. IEEE Computer Society, Washington, DC (2009)
10. Elowitz, M.B., Leibler, S.: A synthetic oscillatory network of transcriptional regulators. Nature 403(6767), 335–338 (2000)
11. Gillespie, D.T.: Exact stochastic simulation of coupled chemical reactions. The Journal of Physical Chemistry 81(25), 2340–2361 (1977)
12. Gillespie, D.T.: A rigorous derivation of the chemical master equation. Physica A 188, 404–425 (1992)
13. Henzinger, T.A., Mateescu, M., Wolf, V.: Sliding window abstraction for infinite Markov chains. In: Bouajjani, A., Maler, O. (eds.) CAV 2009. LNCS, vol. 5643, pp. 337–352. Springer, Heidelberg (2009)
14. Júlvez, J., Kwiatkowska, M., Norman, G., Parker, D.: A systematic approach to evaluate sustained stochastic oscillations. In: Proc. BICoB. ISCA (2011)
15. Kerr, B., Riley, M.A., Feldman, M.W., Bohannan, B.J.M.: Local dispersal promotes biodiversity in a real-life game of rock-paper-scissors. Nature 418 (2002)
16. Kirkup, B.C., Riley, M.A.: Antibiotic-mediated antagonism leads to a bacterial game of rock-paper-scissors in vivo. Nature 428 (2004)
17. Maroto, M., Monk, N.A.M.: Cellular Oscillatory Mechanisms. Advances in Experimental Medicine and Biology, vol. 641. Springer (2009)
18. Meyer, K., Wiegand, K., Ward, D., Moustakas, A.: Satchmo: A spatial simulation model of growth, competition, and mortality in cycling savanna patches. Ecological Modelling 209, 377–391 (2007)

19. Perko, L.: Differential Equations and Dynamical Systems. Texts in Applied Mathematics. Springer (2000)
20. Reichenbach, T., Mobilia, M., Frey, E.: Coexistence versus extinction in the stochastic cyclic lotka-volterra model. Phys. Rev. E 74, 051907 (2006)
21. Spieler, D.: Model checking of oscillatory and noisy periodic behavior in Markovian population models. Technical report, Saarland University (2009), Master thesis available at `http://mosi.cs.uni-saarland.de/?page_id=93`
22. Stewart, W.J.: Introduction to the numerical solution of Markov chains. Princeton University Press (1994)

Model Checking Markov Population Models by Central Limit Approximation

Luca Bortolussi[1] and Roberta Lanciani[2]

[1] Department of Mathematics and Geosciences
University of Trieste, Italy
CNR/ISTI, Pisa, Italy
luca@dmi.units.it
[2] IMT Lucca, Italy
roberta.lanciani@imtlucca.it

Abstract. In this paper we investigate the use of Central Limit Approximation of Continuous Time Markov Chains to verify collective properties of large population models, describing the interaction of many similar individual agents. More precisely, we specify properties in terms of individual agents by means of deterministic timed automata with a single global clock (which cannot be reset), and then use the Central Limit Approximation to estimate the probability that a given fraction of agents satisfies the local specification.

Keywords: Stochastic model checking, fluid approximation, central limit approximation, linear noise approximation, deterministic timed automata, continuous stochastic logic.

1 Introduction

Science and technology face the increasing need of understanding, designing, and controlling large scale complex systems, ranging from biological systems to artificial systems like large computer networks, smart cities, and smart grids. Most of these systems are characterised by a large number of entities that interact in intricate ways to produce the complex gamma of observed behaviours. Mathematical and computational modelling of such *population processes* plays an important role in this challenge. Computational techniques, in particular, are needed to describe, store, validate and analyse such models, which can seldom be treated analytically.

Quantitative Formal Methods (QFM) are very promising in this respect because of their mix of algorithms and formal specification languages for models and properties, resulting in advanced analysis tools like model checking. The class of systems previously mentioned is usually subject to noisy dynamics, so that stochastic processes, like Continuous Time Markov Chains (CTMC [4]), play a predominant role in modelling them. *Stochastic Model Checking* (SMC [4]) can then be used to analyse such CTMCs, building on an established theory and widely used and well-engineered software tools [17]. However, SMC suffers

K. Joshi et al. (Eds.): QEST 2013, LNCS 8054, pp. 123–138, 2013.

severely from the curse of state space explosion, an issue that hampers its applicability when large population models have to be taken into account. The same problem affects other standard analysis techniques for CTMCs, like transient analysis and steady state computation, which are at the heart of SMC algorithms [4]. Indeed, the most successful applications of SMC to population models up to now are either based on statistics [14], or on coarse grained abstractions of the original model [13].

A different class of methods to tackle state space explosion is that of *Fluid Approximation* (FA, [10, 20]), which consists in approximating the collective stochastic dynamics of population processes with a simpler, deterministic one, given by a set of Ordinary Differential Equations (ODE). This operation can be justified invoking the law of large numbers. FA has received attention recently also in the area of QFM, as a tool to approximate the (average) transient evolution of Stochastic Process Algebra (SPA) models, see e.g. [7, 19].

As far as SMC is concerned, fluid approximation has entered the arena only recently [5, 11, 12]. In [5], the authors exploit FA to construct an approximate model of a single individual agent in a (large) population, and check efficiently CSL properties for such an individual. A similar approach is taken in [11], restricting to path properties specified by Deterministic Finite Automata (DFA). In [11, 12], the authors look also at global properties concerned with fraction of agents satisfying local specifications, using moment closure techniques to find approximate bounds on the associated probabilities.

In this paper, we continue along this direction, focussing on the lifting of local specifications to the global level, but using a different FA tool to provide an estimate of the global probabilities involved: the *Central Limit Approximation* (CLA [10]), also known as Linear Noise Approximation [20]. In this respect, our approach complements that of [11, 12]. We also consider a richer class of path properties, expressed by Deterministic Timed Automata (DTA, [8]) with 1 global clock (i.e. a clock referring to the global model time). Hence, this work goes in the direction of merging the approaches of [5] and [11,12] in the light of the logics asCSL [2] or CSL-TA [9], in which until path properties of CSL are replaced by DFA or DTA specifications. The link between local and global properties, with exclusive focus on average collective properites estimated using the fluid limit, has been discussed before in a logical setting in [16].

The paper is organised as follows: in Section 2, we introduce population models by means of a simple automata-based modelling language. In Section 3, we discuss the DTA specification of local properties and their lifting to the global level. In Section 4, we discuss how to combine a population model and a DTA specification into a larger sequence of population models, which is the key step of the algorithm of Section 6, based on the Central Limit Approximation (introduced in Section 5). Finally, in Section 7 we discuss the quality of the approximation using the main example of the paper, a network epidemic model, and, in Section 8, we draw the final conclusions.

2   Population Models

In this section, we introduce an automata-based formalism to specify *Markovian population models* consisting of large collections of interacting components, or *agents*. Each component is a finite state machine, instance of an agent class $\mathscr{A}$ that defines its (finite) state space and its (finite) set of *local* transitions.

Definition 1 (Agent class). *An agent class $\mathscr{A}$ is a pair (S, E) where $S = \{1, \ldots, n\}$ is the state space of the agent and $E = \{\epsilon_1, \ldots, \epsilon_m\}$ is the set of local transitions of the form $\epsilon_i = s_i \xrightarrow{\alpha_i} s_i'$, $i \in \{1, \ldots, m\}$, where α_i is the transition label, taken from the label set $\mathscr{L}$.*

For simplicity, we require that from the same state $s \in S$, there cannot exist two outgoing transitions having the same label. An agent belonging to the class $\mathscr{A} = (S, E)$ is identified by a random variable $Y(t) \in S$, denoting the state of the agent at time t, and the initial state $Y(0) \in S$.

In the following, we consider populations of N agents $Y_k^{(N)}$, $k \in \{1, \ldots, N\}$, all belonging to the same class $\mathscr{A} = (S, E)$ with $S = \{1, \ldots, n\}$ We further make the classical assumption that agents in the same state are indistinguishable, hence the state of the population model can be described by *collective* or *counting* variables $\mathbf{X}^{(N)} = (X_1^{(N)}, \ldots, X_n^{(N)})$, $X_j^{(N)} \in \{0, \ldots, N\}$, defined by $X_j^{(N)} = \sum_{k=1}^{N} \mathbb{1}\{Y_k^{(N)} = j\}$. The initial state $\mathbf{x}_0^{(N)}$ is given by $\mathbf{x}_0^{(N)} = \mathbf{X}^{(N)}(0)$, and the counting variables satisfy the conservation relation $\sum_{j \in S} X_j^{(N)} = N$. To complete the definition of a population model, we need to specify its *global* transitions, describing all possible events that can change the state of the system.

Definition 2 (Population model). *A population model $\mathcal{X}^{(N)}$ of size N is a tuple $\mathcal{X}^{(N)} = (\mathscr{A}, \mathcal{T}^{(N)}, \mathbf{x}_0^{(N)})$, where:*

- *$\mathscr{A}$ is an agent class, as in Definition 1;*
- *$\mathcal{T}^{(N)} = \{\tau_1, \ldots, \tau_\ell\}$ is the set of global transitions of the form $\tau_i = (\mathbb{S}_i, f_i^{(N)})$, where:*
  - *$\mathbb{S}_i = \{s_1 \xrightarrow{\alpha_1} s_1', \ldots, s_p \xrightarrow{\alpha_p} s_p'\}$ is the (finite) set of local transitions synchronized by τ_i;*
  - *$f_i^{(N)} : \mathbb{R}^n \longrightarrow \mathbb{R}_{\geq 0}$ is the (Lipschitz continuous) global rate function.*
- *$\mathbf{x}_0^{(N)}$ is the initial state.*

The rate $f_i^{(N)}$ gives the expected frequency of transition τ_i as a function of the state of the system. We assume $f_i^{(N)}$ equal to zero if there are not enough agents available to perform the transition. The synchronization set $\mathbb{S}_i$, instead, specifies how many agents are involved in the transition τ_i and how they change state: when τ_i occurs, we see the local transitions $s_1 \xrightarrow{\alpha_1} s_1', \ldots, s_p \xrightarrow{\alpha_p} s_p'$ fire at the (local) level of the p agents involved in τ_i. For simplicity, within $\mathbb{S}_i$ we require $s_j \neq s_k$ for $j \neq k$, i.e. agents in the same state cannot be synchronized.

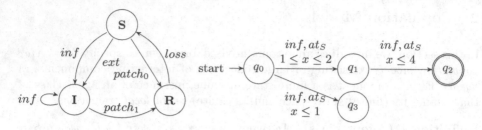

Fig. 1. Left: The automaton representation of a network node. Right: The 1gDTA specifications discussed in Example 1 of Section 3.

Remark 1. The population models we introduced have three main restrictions: (i) there is a single class of agents, (ii) the population is constant, and (iii) synchronising agents must be in different local states. Restrictions (i) and (iii) can be easily dropped, at the price of a heavier notation and of a more involved combinatorics in the definition of the rate functions of the synchronised models, cf. Section 4. Restriction (ii) can be removed as well, as the approximations we will use do not rely on such an assumption. However, extra care has to be put in treating local properties, as discussed in [5].

Given a population model $\mathcal{X}^{(N)} = (\mathscr{A}, \mathcal{T}^{(N)}, \mathbf{x}_0^{(N)})$ and a global transition $\tau = (\mathbb{S}_\tau, f_\tau^{(N)}) \in \mathcal{T}^{(N)}$ with $\mathbb{S}_\tau = \{s_1 \xrightarrow{\alpha_1} s_1', \ldots, s_p \xrightarrow{\alpha_p} s_p'\}$, we encode the net change in $\mathbf{X}^{(N)}$ due to τ in the *update vector* $\mathbf{v}_\tau = \sum_{i=1}^p (\mathbf{e}_{s_i} - \mathbf{e}_{s_i'})$, where $\mathbf{e}_{s_i}$ is the vector that is equal to 1 in position s_i and zero elsewhere.

The CTMC $\mathbf{X}^{(N)}(t)$ associated with $\mathcal{X}^{(N)}$ has state space $\mathcal{S}^{(N)} = \{(z_1, \ldots, z_n) \in \mathbb{N}^n \mid \sum_{i=1}^n z_i = N\}$, initial probability distribution concentrated on $\mathbf{x}_0^{(N)}$, and *infinitesimal generator matrix* $\mathbf{Q}$ defined for $\mathbf{x}, \mathbf{x}' \in \mathcal{S}^{(N)}$, $\mathbf{x} \neq \mathbf{x}'$, by $q_{\mathbf{x}, \mathbf{x}'} = \sum_{\tau \in \mathcal{T} \mid \mathbf{v}_\tau = \mathbf{x}' - \mathbf{x}} f_\tau(\mathbf{x})$.

2.1   Running Example

In order to illustrate the method of the paper, we consider a simple example of a worm epidemic in a peer-to-peer network composed of N nodes (see e.g. [15] for mean field analysis of network epidemics). Each node is modelled by the simple agent shown in Figure 1, which has three states: susceptible to infection (S), infected (I), and patched/immune to infection (R). The contagion of a susceptible node can occur due to an event external to the network (*ext*), like the reception of an infected email, or by file sharing with an infected node within the network (*inf*). Nodes can also be patched, at different rates, depending if they are infected (*patch_1*) or not (*patch_0*). A patched node remains immune from the worm for some time, until immunity is lost (*loss*), modelling for instance the appearance of a new version of the worm.

The agent class $\mathscr{A}_{node} = (S_{node}, E_{node})$ of the network node can be easily reconstructed form the automaton representation in Figure 1. The population model $\mathcal{X}_{net}^{(N)} = (\mathcal{A}_{node}, \mathcal{T}^{(N)}, \mathbf{x}_0^{(N)})$ with population variables $\mathbf{X} = (X_S, X_I, X_R)$ is then obtained by specifying transitions and initial conditions. The latter is simply a network of susceptible nodes, $\mathbf{x}_0^{(N)} = (N, 0, 0)$, while the former is given by five global transitions, $\tau_{ext}, \tau_{loss}, \tau_{patch_0}, \tau_{patch_1}, \tau_{inf} \in \mathcal{T}^{(N)}$. For example, the external infection is defined by $\tau_{ext} = (\{S \xrightarrow{ext} I\}, f_{ext})$, where the synchronisation set specifies that only one susceptible node is involved and changes state from S to I at a rate given by $f_{ext}(\mathbf{X}) = \kappa_{ext} X_S$, corresponding to a rate of infection κ_{ext} per node. The transitions $\tau_{loss}, \tau_{patch_0}, \tau_{patch_1}$ have a similar format, while the internal infection is described by $\tau_{inf} = (\{I \xrightarrow{inf} I, S \xrightarrow{inf} I\}, f_{inf})$ and involves one S-node and one I-node. Furthermore, in this case of τ_{inf}, we assume that an infected node sends infectious messages at rate κ_{inf} to a random node, giving a classical density dependent rate function $f_{inf}(\mathbf{X}) = \frac{1}{N} \kappa_{inf} X_S X_I$ [1].

3 Individual and Collective Properties

We introduce now the class of properties considered in the paper. We distinguish two levels of properties: *local properties*, describing the behaviour of individual agents, and *global properties*, describing the collective behaviour of agents with respect to a local property of interest. In this classification, our approach is similar to [12, 16].

In particular, we are concerned with *time-bounded* local properties specified by Deterministic Timed Automata (DTA). This restriction to finite time horizons is justified because the analysis of steady state properties is always problematic in the context of fluid approximation (see [5, 6, 12] for further discussion on this point).

The global property layer, instead, allows us to specify queries about the fraction of agents that satisfies a given local specification. In particular, given a (local and time-bounded) path property φ, we want to compute the probability that the fraction of agents that satisfies φ at time T is smaller or larger than a threshold α. This will be captured by a proper operator, that can then be combined to specify more complex global queries, as in [16].

Let us fix a population model composed of N agents belonging to a class $\mathcal{A} = (S, E)$. We consider local path properties specified by *1-global-clock Deterministic Timed Automata* (1gDTA), which are DTAs with a single clock variable $x \in \mathbb{R}_{\geq 0}$, called *global clock*, that is never reset. We call $\mathcal{V}$ the set of *valuations of x*, i.e. functions $\eta : \{x\} \longrightarrow \mathbb{R}^{\geq 0}$ that assign a nonnegative real-value to the global clock x, and CC the set of *clock constraints*, which are positive boolean combinations of basic clock constraints of the form $x \leq a$ or $x \geq a$, where $a \subset \mathbb{Q}^{\geq 0}$. We write $\eta(x) \models_{CC} c$ if and only if $c \in CC$ is satisfied when the clock variable takes the value $\eta(x)$. In addition to actions and clock constraints, we also label the edges of 1gDTA by a boolean formula, interpreted on the states $s \in S$ of agent $\mathcal{A}$, similarly to asCSL [2] and CSL-TA [9]. Let Γ_S be the set of these *(atomic) state propositions over S*, and $\mathcal{B}(\Gamma_S)$ the set of boolean combinations over Γ_S. We use the letter φ to range over formulae in $\mathcal{B}(\Gamma_S)$ and we denote by $\models_{\Gamma_S}$ the satisfaction relation over $\mathcal{B}(\Gamma_S)$-formulae. In this way, a local transition $s \xrightarrow{\alpha_\tau} s'$ matches an edge with label α, c, φ in the 1gDTA if and only if the action name is the same, the clock constraint c is satisfied and the $\mathcal{B}(\Gamma_S)$-formulae holds on the initial state s, i.e. $\alpha_\tau = \alpha$, $\eta(x) \models_{CC} c$, and $s \models_{\Gamma_S} \varphi$.

Definition 3 (1-global-clock DTA). *A 1gDTA is specified by the tuple $\mathcal{T} = (\mathcal{L}, \Gamma_S, Q, q_0, F, \rightarrow)$ where $\mathcal{L}$ is the label set of $\mathcal{A}$; Γ_S is the set of atomic state propositions; Q is the (finite) set of states of the DTA, with initial state $q_0 \in Q$; $F \subseteq Q$ is the set of final (or accepting) states, and $\rightarrow \subseteq Q \times \mathcal{L} \times \mathcal{B}(\Gamma_S) \times CC \times Q$ is the edge relation, where $(q, \alpha, \varphi, c, q') \in \rightarrow$ is usually denoted by $q \xrightarrow{\alpha, \varphi, c} q'$. Moreover, $\mathcal{T}$ satisfies:*

- *(determinism) for each $q \in Q$, $\alpha \in \mathcal{L}$, $s \in S$ and clock valuation $\eta(x) \in \mathbb{R}_{\geq 0}$, there is exactly one edge $q \xrightarrow{\alpha, \varphi, c} q'$ such that $s \models_{\Gamma_S} \varphi$ and $\eta(x) \models_{CC} c$;*
- *(absorption) the final states F are all absorbing.*

When we write a 1gDTA, we stick to the convention that all non-specified edges are self-loops on the automata states.

Example 1. As an example, consider the agent class of the network epidemic model of Section 2.1, and the 1gDTA specification of Figure 1 right, where the formula at_S is true in local state S and false in states I and R. The automaton describes the local property stating that an agent is infected by internal contact twice, the first infection happening between time 1 and 2, and the second infection happening before time 4. The sink state q_3 is used to discard agents being infected for the first time before time 1. The use of the state formula at_S allows us to focus only on agents that are infected, ignoring agents that spread the contagion.

An individual agent in a population model satisfies the local property specified by a 1gDTA $\mathcal{T}$ at time T if, feeding to $\mathcal{T}$ the agent trajectory up to time T, we reach a final state. This can be formalised in a standard way, see for instance [8,9]. In order to lift these local specifications to the collective level, we count the number of agents that satisfy the 1gDTA $\mathcal{T}$ at time T. More specifically, we check if the fraction of agents satisfying $\mathcal{T}$ is included in the interval $[a, b]$, which we write as $\mathcal{T}(T) \in [a, b]$, where the bounds a, b are specified in terms of fraction of agents or population density (the number of agents divided by the total population size). To verify the random event $\mathcal{T}(T) \in [a, b]$, we compute its probability, which is then compared with a given threshold. The atomic global properties can be combined together by boolean operators, as in [16], to define more expressive queries.

Definition 4 (Syntax of global properties). *Given a population model $\mathcal{X}^{(N)}$, a collective/global property on $\mathcal{X}^{(N)}$ is given by the following syntax:*

$$\Psi = \texttt{true} \mid \mathbb{P}_{\bowtie p}(\mathcal{T}(T) \in [a, b]) \mid \neg\Psi \mid \Psi_1 \wedge \Psi_2,$$

where $\mathbb{P}_{\bowtie p}(\mathcal{T}(T) \in [a, b])$ is true if and only if $q \bowtie p$, for $\bowtie \in \{<, \leq, \geq, >\}$, with q being the probability that at time T the number of agents that satisfies the local path property $\mathcal{T}$ is contained in the interval $[a, b]$.

As an example, consider again the 1gDTA property $\mathscr{T}$ of Figure 1 right. The atomic global property $\mathbb{P}_{\geq 0.8}(\mathscr{T}(4) \leq \frac{1}{3})$ specifies that, with probability at least 0.8, no more than one third of network nodes will be infected twice in the first 4 time units by an internal contact, with the first infection happening between time 1 and 2.

Remark 2. In addition to path properties specified by 1gDTA, we could have considered state properties in the style of CSL-TA [9]. This can be done at the price of dealing with nesting of path and state properties, which for local specifications rises issues of time-dependency of truth values similar to those discussed in [5]. We leave this for future work.

Remark 3. The fact that final states are absorbing implies that we are looking for properties in which an accepting state of the 1gDTA must be reached at a time instant within $[0, T]$. Punctual properties, looking at satisfaction exactly at time T, can be obtained by dropping the absorbing condition in Definition 3.

4   Synchronisation of Agents and Properties

In this section, we present the model checking procedure for the verification of global atomic properties. We aim at approximating such probabilities by means of central limit results [10, 20]. The first step is to synchronize the agent and the property, constructing an extended Markov population model in which the state space of each agent is combined with the specific path property we are observing. The Central Limit Approximation is then applied to the so obtained model.

The main difficulty in this procedure is the presence of time constraints in the path property specification. However, thanks to the restriction to a single global clock, we can partition the time interval of interest into a finite set of subintervals, within which no clock constraint changes status. Thus, in each subinterval, we can remove the clock constraints, deleting all the edges that cannot fire being their clock constraint false. In this way, we generate a sequence of Deterministic Finite Automata (DFA), that are then combined with the local model $\mathscr{A}$ by a standard product of automata. Then, we construct the population models associated with such a local model (paying attention to the rates) and we obtain a *sequence* of population CTMC models to which we apply the Central Limit Approximation.

Synchronisation of Local Properties

Let $\mathscr{A} = (S, E)$ be an agent class, $\mathscr{T} = (\mathscr{L}, \Gamma_S, Q, q_0, F, \rightarrow)$ be a local path property, and $T > 0$ be the time horizon.

First Step: Uniqueness of Transition Labels. We define a new agent class $\bar{\mathscr{A}} = (S, \bar{E})$ by renaming the local transitions in E to make their label unique. This allows us to remove edge formulae in $\mathscr{T}$, simplifying the product construction. In particular, if there exist $s_1 \xrightarrow{\alpha} s_1', \ldots, s_m \xrightarrow{\alpha} s_m' \in E$ having the same label α, we rename them by $\alpha_{s_1}, \ldots, \alpha_{s_m}$, obtaining $s_1 \xrightarrow{\alpha_{s_1}} s_1', \ldots, s_m \xrightarrow{\alpha_{s_m}} s_m' \in \bar{E}$. The 1gDTA $\mathscr{T}$ is updated accordingly, by substituting each edge $q \xrightarrow{\alpha, \varphi, c} q'$ with the set of edges $q \xrightarrow{\alpha_{s_i}, \varphi, c} q'$, for $i = 1, \ldots, m$. We call $\mathscr{L}$ the label set of $\bar{\mathscr{A}}$.

Second Step: Removal of State Conditions. We remove from the edge relation of $\mathscr{T}$ all the edges $q \xrightarrow{\alpha_{s_i}, \varphi, c} q'$ such that $s_i \not\models_{\Gamma_S} \varphi$, where s_i is the source state of the (now unique) transition of $\bar{\mathscr{A}}$ labeled by α_{s_i}. At this point, the information carried by state propositions becomes redundant, thus we drop them, writing $q \xrightarrow{\alpha_{s_i}, c} q'$ in place of $q \xrightarrow{\alpha_{s_i}, \varphi, c} q'$.

Third Step: Removal of Clock Constraints. Let $t_1, \ldots, t_k$ be the ordered sequence of constants (smaller than T) appearing in the clock constraints of the edges of $\mathscr{T}$. We extend this sequence by letting $t_0 = 0$ and $t_{k+1} = T$. Let $I_j = [t_{j-1}, t_j]$, $j = 1, \ldots, k+1$, be the j-th sub-interval of $[0, T]$ identified by such a sequence. For each I_j, we define a Deterministic Finite Automaton (DFA), $\mathscr{D}_{I_j} = (\mathscr{L}, Q, q_0, F, \rightarrow_j)$, whose edge relation $\rightarrow_j$ is obtained from that of $\mathscr{T}$ by selecting only the edges for which the clock constraints are satisfied in I_j, and dropping the clock constraint. Hence, from $q \xrightarrow{\alpha_{s_i}, c} q'$ such that $\eta(x) \models_{CC} c$ whenever $\eta(x) \in (t_{j-1}, t_j)$, we obtain the DFA edge $(q, \alpha_{s_i}, q') \in \rightarrow_j$, denoted also by $q \xrightarrow{\alpha_{s_i}}_j q'$.

Fourth Step: Synchronization. To keep track of the behaviour of the agents with respect to the property specified by $\mathscr{T}$, we synchronize the agent class $\bar{\mathscr{A}} = (S, \bar{E})$ with each DFA $\mathscr{D}_{I_j}$ through the standard product of automata. The sequence of deterministic automata obtained in this procedure is called the *agent class associated with the local property $\mathscr{T}$*.

Definition 5 (Agent class associated with the local property $\mathscr{T}$). *The agent class $\mathscr{P}$ associated with the local property $\mathscr{T}$ is the sequence $\mathscr{P} = (\mathscr{P}_{I_1}, \ldots, \mathscr{P}_{I_{k+1}})$ of deterministic automata $\mathscr{P}_{I_j} = (\hat{S}, \hat{E}_j)$, $j = 1, \ldots, k+1$, where $\hat{S} = S \times Q$ is the state space and $\hat{E}_j$ is the set of local transitions $\epsilon_i^j = (s, q) \xrightarrow{\alpha_s} (s', q')$, such that $s \xrightarrow{\alpha_s} s'$ is a local transition in $\bar{\mathscr{A}}$ and $q \xrightarrow{\alpha_s} q'$ is an edge in $\mathscr{D}_{I_j}$.*

Synchronisation of Global Properties

The population model $\mathcal{X}^{(N)} = (\mathscr{A}, \mathcal{T}^{(N)}, \mathbf{x}_0^{(N)})$ has to be updated to follow the new specifications at the local level. We do this by defining the population model associated with the local property $\mathscr{T}$ as a sequence $\boldsymbol{\mathcal{X}}^{(N)} = (\mathcal{X}_{I_1}^{(N)}, \ldots, \mathcal{X}_{I_k}^{(N)})$ of population models. Since the agent states are synchronized with the property automaton, each transition in the population model needs to be replicated many times to account for all possible combinations of the extended local state space. Furthermore, we also need to take care of rate functions in order not to change the global rate. Fix the j-th element $\mathscr{P}_{I_j}$ in the agent class $\mathscr{P}$ associated with the property $\mathscr{T}$. The state space of $\mathscr{P}_{I_j}$ is $\hat{S} \times Q$, hence to construct the global model we need nm counting variables ($n = |S|$, $m = |Q|$), where $X_{s,q}$ counts how many agents are in the local state (s, q). Let $\tau = (\mathbb{S}_\tau, f^{(N)}) \in \mathcal{T}^{(N)}$ be a global transition, apply the relabeling of action labels, according to step 1 above, and focus on the synchronisation set $\mathbb{S}_\tau = \{s_1 \xrightarrow{\alpha_{s_1}} s'_1, \ldots, s_k \xrightarrow{\alpha_{s_k}} s'_k\}$. We need to consider all possible ways of associating states of Q with the different states $s_1, \ldots, s_k$ in $\mathbb{S}_\tau$. Indeed, each choice $(q_1, \ldots, q_k) \in Q^k$ generates a different transition in $\mathcal{X}_{I_j}^{(N)}$, with synchronization set $\mathbb{S}_{\tau,r} = \{(s_1, q_1) \xrightarrow{\alpha_{s_1}} (s'_1, q'_1), \ldots, (s_k, q_k) \xrightarrow{\alpha_{s_k}} (s'_k, q'_k)\}$, where q'_i is the unique state of Q such that $q_i \xrightarrow{\alpha_{s_i}} q'_i$. The rate function $f_r^{(N)}$ associated with this instance of τ is a fraction of the total rate function $f^{(N)}$ of τ. Moreover, for all $s_i \xrightarrow{\alpha_{s_i}} s'_i \in \mathbb{S}_\tau$, $f_r^{(N)}$ is proportional to the fraction of agents that before the synchronisation were in s_i and are now in state (s_i, q_i), i.e. X_{s_i,q_i} divided by $X_{s_i} = \sum_{q \in Q} X_{s_i,q}$. Formally,

$$f_r^{(N)}(\mathbf{X}) = \prod_{s_i \xrightarrow{\alpha_{s_i}} s'_i \in \mathbb{S}_\tau} \left(\frac{X_{s_i,q_i}}{\sum_{q \in Q} X_{s_i,q}}\right) f^{(N)}(\widetilde{\mathbf{X}}), \tag{1}$$

where $\widetilde{\mathbf{X}} = (X_1, \ldots, X_n)$ with $X_s = \sum_{r=1}^m X_{s,r}$. Due to the restrictions enforced in Definition 2, summing up the rates $f_r^{(N)}(\mathbf{X})$ for all possible choices of $(q_1, \ldots, q_k) \in Q^k$, we obtain $f^{(N)}(\widetilde{\mathbf{X}})$.[1]

Definition 6 (Population model associated with a local property). *The population model associated with the local property $\mathcal{T}$ is the sequence $\mathcal{X}^{(N)} = (\mathcal{X}_{I_1}^{(N)}, \ldots, \mathcal{X}_{I_k}^{(N)})$. The elements $\mathcal{X}_{I_j}^{(N)} = (\mathscr{P}_{I_j}, \mathcal{T}_j^{(N)})$ are such that $\mathscr{P}_{I_j}$ is the j-th element of the agent class associated with $\mathcal{T}$ and $\mathcal{T}_j^{(N)}$ is the set of global transitions of the form $\tau_i^j = (\mathbb{S}_i^j, f_{j,i}^{(N)})$, as defined above.*[2]

5   Central Limit Approximation

Given a population model $\mathcal{X}^{(N)} = (\mathscr{A}, \mathcal{T}^{(N)}, \mathbf{x}_0^{(N)})$, the Fluid and Central Limit Approximations provide an estimation of the stochastic dynamics of $\mathcal{X}^{(N)}$, exact in the limit of an *infinite* population. In particular, we consider an infinite sequence $(\mathcal{X}^{(N)})_{N \in \mathbb{N}}$ of population models, all sharing the same structure, for increasing population size $N \in \mathbb{N}$ (e.g. the network models $(\mathcal{X}_{net}^{(N)})_{N \in \mathbb{N}}$ with an increasing number of network nodes). To compare the dynamics of the models in the sequence, we consider the *normalised counting variables* $\hat{\mathbf{X}} = \frac{1}{N}\mathbf{X}$ (known also as *population densities* or *occupancy measures*, see [6] for further details)

---

[1] If we drop the restrictions discussed in Remark 1, This will still be true, the only difference being a more complex definition of the coefficient of $f^{(N)}(\widetilde{\mathbf{X}})$ in (1).

[2] Initial conditions of population models in $\mathcal{X}^{(N)}$ are dropped, as they are not required in the following. The initial condition at time zero is obtained from that of $\mathcal{X}^{(N)}$ by letting $(x_0)_{s,q_0} = (x_0)_s$, where q_0 the initial state of $\mathcal{T}$ and $s \in S$.

and we define the *normalized population models* $\hat{\mathcal{X}}^{(N)} = (\mathscr{A}, \hat{\mathcal{T}}^{(N)}, \hat{\mathbf{x}}_0^{(N)})$, obtained from $\mathcal{X}^{(N)}$ by making the rate functions depend on the normalised variables and rescaling the initial conditions. For simplicity, we assume that the rate function of each transition $\tau \in \hat{\mathcal{T}}^{(N)}$ satisfies the *density dependent condition* $\frac{1}{N} f_\tau^{(N)}(\hat{\mathbf{X}}) = f_\tau(\hat{\mathbf{X}})$ for some Lipschitz function $f_\tau : \mathbb{R}^n \longrightarrow \mathbb{R}_{\geq 0}$, i.e. rates on normalised variables are independent of N. Also the *drift* $\mathbf{F}$ of $\mathcal{X}^{(N)}$, that is the mean instantaneous change of the normalised variables, is given by $\mathbf{F}(\hat{\mathbf{X}}) = \sum_{\tau \in \hat{\mathcal{T}}^{(N)}} \mathbf{v}_\tau f_\tau(\hat{\mathbf{X}})$ and, thus, is independent of N. The unique solution[3] $\boldsymbol{\Phi} : \mathbb{R}_{\geq 0} \longrightarrow \mathbb{R}^n$ of the differential equation $\frac{d\boldsymbol{\Phi}(t)}{dt} = \mathbf{F}(\boldsymbol{\Phi}(t))$, given $\boldsymbol{\Phi}(0) = \hat{\mathbf{x}}_0^{(N)}$, is the *Fluid Approximation* of the CTMC $\hat{\mathbf{X}}^{(N)}(t)$ associated with $\hat{\mathcal{X}}^{(N)}$ and has been successfully used to describe the collective behaviour of complex systems with large populations [6]. The correctness of this approximation in the limit of an infinite population is guaranteed by the Kurtz Theorem [6,10], which states that $\sup_{t \in [0,T]} \|\hat{\mathbf{X}}^{(N)}(t) - \boldsymbol{\Phi}(t)\|$ converges to zero (almost surely) as N goes to infinity.

While the Fluid Approximation correctly describes the transient collective behaviour for very large populations, it is less accurate when one has to deal with a *mesoscopic* system, meaning a system with a population in the order of hundreds of individuals and whose dynamics results to be intrinsically probabilistic. Indeed, the (stochastic) behaviour of single agents becomes increasingly relevant as the size of the population decreases. The technique of *Central Limit Approximation* (CLA), also known as *Linear Noise Approximation*, provides an alternative and more accurate estimation of the stochastic dynamics of mesoscopic systems. In particular, in the CLA, the probabilistic fluctuations about the average deterministic behaviour (described by the fluid limit) are approximated by a Gaussian process.

First, we define the process $\mathbf{Z}^{(N)}(t) := N^{\frac{1}{2}} \left(\hat{\mathbf{X}}^{(N)}(t) - \boldsymbol{\Phi}(t) \right)$, capturing the rescaled fluctuations of the Markov chain around the fluid limit. Then, by relying on convergence results for Brownian motion, one shows that $\mathbf{Z}^{(N)}(t)$, for large population *sizes*, can be approximated [10,20] by the Gaussian process[4] $\{\mathbf{Z}(t) \in \mathbb{R}^n \mid t \in \mathbb{R}\}$ (*independent of* N), whose mean $\mathbf{E}[t]$ and covariance $\mathbf{C}[t]$ are given by

$$\begin{cases} \frac{\partial \mathbf{E}[t]}{\partial t} = \mathbf{J}_\mathbf{F}(\boldsymbol{\Phi}(t)) \mathbf{E}[t] \\ \mathbf{E}[0] = 0 \end{cases} \tag{2}$$

and

$$\begin{cases} \frac{\partial \mathbf{C}[t]}{\partial t} = \mathbf{J}_\mathbf{F}(\boldsymbol{\Phi}(t)) \mathbf{C}[t] + \mathbf{C}[t] \mathbf{J}_\mathbf{F}^T(\boldsymbol{\Phi}(t)) + \mathbf{G}(\boldsymbol{\Phi}(t)) \\ \mathbf{C}[0] = 0, \end{cases} \tag{3}$$

where $\mathbf{J}_\mathbf{F}(\boldsymbol{\Phi}(t))$ denotes the Jacobian of the limit drift $\mathbf{F}$ calculated along the deterministic fluid limit $\boldsymbol{\Phi} : \mathbb{R}_{\geq 0} \longrightarrow \mathbb{R}^n$, and $\mathbf{G}(\hat{\mathbf{X}}) = \sum_{\tau \in \hat{\mathcal{T}}^{(N)}} \mathbf{v}_\tau \mathbf{v}_\tau^T f_\tau(\hat{\mathbf{X}})$ is

---

[3] The solution exists and is unique because F is Lipschitz continuous, as each f_τ is.

[4] A Gaussian process $\mathbf{Z}(t)$ is characterised by the fact that the joint distribution of $\mathbf{Z}(t_1), \ldots, \mathbf{Z}(t_k)$ is a multivariate normal distribution for any $t_1, \ldots, t_k$.

called the *diffusion* term. The nature of the approximation of $\mathbf{Z}^{(N)}(t)$ by $\mathbf{Z}(t)$ is captured in the following theorem [10].

Theorem 1. *Let $\mathbf{Z}(t)$ be the Gaussian process with mean (2) and covariance (3) and $\mathbf{Z}^{(N)}(t)$ be the random variable given by $\mathbf{Z}^{(N)}(t) := N^{\frac{1}{2}}\left(\hat{\mathbf{X}}^{(N)}(t) - \boldsymbol{\Phi}(t)\right)$. Assume that $\lim_{N \to \infty} \mathbf{Z}^{(N)}(0) = \mathbf{Z}(0)$. Then, $\mathbf{Z}^{(N)}(t)$ converges in distribution to $\mathbf{Z}(t)$ $(\mathbf{Z}^{(N)}(t) \Rightarrow \mathbf{Z}(t))$.*

The *Central Limit Approximation* then approximates the normalized CTMC $\hat{\mathbf{X}}^{(N)}(t) = \boldsymbol{\Phi}(t) + N^{-\frac{1}{2}}\mathbf{Z}^{(N)}(t)$ associated with $\hat{\mathcal{X}}^{(N)}$ by the stochastic process

$$\boldsymbol{\Phi}(t) + N^{-\frac{1}{2}}\mathbf{Z}(t). \tag{4}$$

Theorem 1 guarantees its asymptotic correctness in the limit of an infinite population.

6 Computing the Probability of Collective Properties

Consider a population model $\mathcal{X}^{(N)}$, for a fixed population size N, and a global property $\mathbb{P}_{\bowtie p}(\mathcal{T}(T) \in [a, b])$. In order to verify the latter, we need to compute the probability $\mathbb{P}(\mathcal{T}(T) \in [a, b])$ that, at time T, the fraction of agents satisfying the local specification $\mathcal{T}$ is contained in $[a, b]$. This probability can be computed exploiting the construction of Section 4, according to which we obtain a sequence of population models $\mathcal{X}^{(N)} = (\mathcal{X}_{I_1}^{(N)}, \ldots, \mathcal{X}_{I_k}^{(N)})$, synchronising local agents with the sequence of deterministic automata associated with $\mathcal{T}$. In such construction we identified a sequence of times $0 = t_0, t_1, \ldots, t_k = T$ and in each interval $I_j = [t_{j-1}, t_j]$ the satisfaction of clock constraints does not change.

Therefore, in order to compute $\mathbb{P}(\mathcal{T}(T) \in [a, b])$, we can rely on *transient analysis* algorithms for CTMCs [3]: first we compute the probability distribution at time t_1 for the first population model $\mathcal{X}_{I_1}^{(N)}$; then we use this result as the initial distribution for the CTMC associated with the population model $\mathcal{X}_{I_2}^{(N)}$ and we compute its probability distribution at time t_2; and so on, until we obtain the probability distribution for $\mathcal{X}_{I_k}^{(N)}$ at time $t_k = T$. Once we have this result, we can find the desired probability by summing the probability of all those states $\mathbf{X} \in \mathcal{S}^{(N)}$ such that $\sum_{s \in S, q \in F} \hat{X}_{s,q} \in [a, b]$.

Unfortunately, this approach suffers from state space explosion, which is severe even for a population size of few hundreds of individuals. Furthermore, for these population levels we cannot either rely on the Fluid Approximation, as it would only give us an estimate of the average of the counting variables, while we need information about their distribution. It is here that the Central Limit Approximation enters the picture.

The idea is simply to compute the average and covariance matrix of the approximating Gaussian Process by solving the ODEs shown at the end of the previous section. In doing this, we have to take proper care of the different population models associated with the time intervals I_j. Then, we integrate the

Gaussian density of the approximating distribution at time T to estimate of the probability $\mathbb{P}(\mathscr{T}(T) \in [a,b])$. The justification of this approach is in Theorem 1, which guarantees that the estimated probability is asymptotically correct, but in practice, we can obtain good approximations also for relatively small populations, in the order of hundreds of individuals.

Verification Algorithm

The *input* of the verification algorithm is:

- an agent class $\mathscr{A} = (S, E)$ and a population model $\mathcal{X}^{(N)} = (\mathscr{A}, \mathcal{T}^{(N)}, \mathbf{x}_0^{(N)})$;
- a local property specified by a 1gDTA $\mathscr{T} = (\mathscr{L}, \varGamma_S, Q, q_0, F, \rightarrow)$;
- a global property $\mathbb{P}_{\bowtie p}(\mathscr{T}(T) \in [a,b])$ with time horizon $T > 0$.

The *steps* of the algorithm are:

1. **Construction of the Population Model Associated with $\mathscr{T}$.** Construct the *normalised* population model $\hat{\mathcal{X}}^{(N)} = (\hat{\mathcal{X}}_{I_1}^{(N)}, \ldots, \hat{\mathcal{X}}_{I_k}^{(N)})$ associated with $\mathscr{T}$ according to the recipe of Section 4. Then modify it by adding to its vector of counting variables $\hat{\mathbf{X}}^{(N)}$ a new variable $\hat{X}_{Final}$ that keeps track of the fraction of agents entering one of the final states (s, q), $q \in F$.

2. **Integration of the Central Limit Equations.** For each $j = 1, \ldots, k$, generate and solve numerically the system of ODEs that describes the fluid limit $\boldsymbol{\varPhi}_j(t)$ and the Gaussian covariance $\mathbf{C}_j[\mathbf{Z}(t)]$ for the population model $\mathcal{X}_{I_j}^{(N)}$ in the interval $I_j = [t_{j-1}, t_j]$, with initial conditions $\boldsymbol{\varPhi}_j(t_{j-1}) = \boldsymbol{\varPhi}_{j-1}(t_{j-1})$ and $\mathbf{C}_j[\mathbf{Z}(t_{j-1})] = \mathbf{C}_{j-1}[\mathbf{Z}(t_{j-1})]$ for $j > 1$, and $\boldsymbol{\varPhi}_1(0) = \mathbf{x}_0$, $\mathbf{C}_1[\mathbf{Z}(0)] = Id$. Define the population mean as $\mathbf{E}^{(N)}[\mathbf{X}(t)] = N\boldsymbol{\varPhi}_j(t)$ and the population covariance as $\mathbf{C}^{(N)}[\mathbf{X}(t)] = N\mathbf{C}_j[\mathbf{Z}(t)]$, for $t \in I_j$. Finally, identify the component $E_{Final}^{(N)}[\mathbf{X}(t)]$ and the diagonal entry $C_{Final}^{(N)}[\mathbf{X}(t)]$ corresponding to X_{Final}.

3. **Computation of the Probability.** Let $g(x \mid \mu, \sigma^2)$ be the probability density of a Gaussian distribution with mean μ and variance σ^2. Then, approximate $\mathbb{P}(\mathscr{T}(T) \in [a,b])$ by

$$\tilde{P}_{\mathscr{T}}^{(N)}(T) = \int_{Na}^{Nb} g(x \mid E_{Final}^{(N)}[\mathbf{X}(t)], C_{Final}^{(N)}[\mathbf{X}(t)]) \mathrm{d}x,$$

and compare the result with the probability bound $\bowtie p$.

The asymptotic correctness of this procedure is captured in the next theorem, whose proof is a straightforward consequence of Theorem 1. We denote by $P_{\mathscr{T}}^{(N)}(T)$ the exact value of $\mathbb{P}(\mathscr{T}(T) \in [a,b])$ and by $\tilde{P}_{\mathscr{T}}^{(N)}(T)$ the approximate value computed by the Central Limit Approximation.

Theorem 2. *Under the hypothesis of Theorem 1, it holds that* $\lim_{N \to \infty} \|P_{\mathscr{T}}^{(N)}(T) - \tilde{P}_{\mathscr{T}}^{(N)}(T)\| = 0.$ ∎

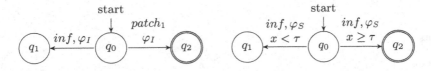

Fig. 2. The 1gDTA specifications experimentally analysed in Section 7

Remark 4. The introduction of the counting variable X_{Final} is needed to correctly capture the variance in entering one of the final states of the property. Indeed, it holds that $X_{Final} = \sum_{s \in S, q \in F} X_{s,q}$, and in principle we could have applied the CLA to the model without X_{Final}, using the fact that the sum of Gaussian variables is Gaussian (with mean and variance given by the sum of means and variances of the addends). In doing this, though, we overestimate the variance of X_{Final}, because we implicitly take into account the dynamics within the final components. The introduction of X_{Final}, instead, avoids this problem, as its variance depends only on the events that allow the agents to enter one of the final states.

7   Experimental Analysis

We discuss now the quality of the Central Limit Approximation for mesoscopic populations from an experimental perspective. We present a detailed investigation of the behaviour of the example describing a network epidemics introduced in Section 2.

We consider two local properties in terms of the 1gDTAs shown in Figure 2. The first property $\mathscr{T}_1$ has no clock constraints on the edges of the automaton, therefore the 1gDTA reduces to a DFA. The property is satisfied if an infected node is patched before being able to infect other nodes in the network, thus checking the effectiveness of the antivirus deploy strategy. The second property $\mathscr{T}_2$, instead, is properly timed. It is satisfied when a susceptible node is infected by an internal infection after the first τ units of time. The corresponding global properties that we consider are $\mathbb{P}(\mathscr{T}_1(T) \geq \alpha_1)$ and $\mathbb{P}(\mathscr{T}_2(T) \geq \alpha_2)$.

In Figure 3, we show the probability of the two global properties as a function of the time horizon T, for different values of N and a specific configuration of parameters. The CLA is compared with a statistical estimate, obtained from 10000 simulation runs. As we can see, the accuracy in the transient phase increases rapidly with N, and the estimate is very good for both properties already for $N = 100$.

Furthermore, in order to check more extensively the quality of the approximation also as a function of the system parameters, we ran the following experiment. We considered five different values of N ($N = 20, 50, 100, 200, 500$). For each of these values, we randomly chose 20 different combinations of parameter values, sampling uniformly from: $\kappa_{inf} \in [0.05, 5]$, $\kappa_{patch_1} \in [0.02, 2]$, $\kappa_{loss} \in [0.01, 1]$, $\kappa_{ext} \in [0.05, 5]$, $\kappa_{patch_0} \in [0.001, 0.1]$, $\alpha_1 \in [0.1, 0.95]$, $\alpha_2 \in [0.1, 0.3]$. For each

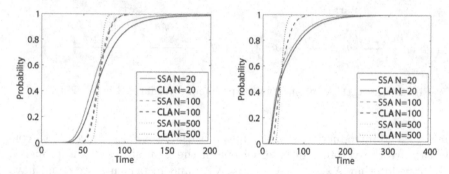

Fig. 3. Comparison of Central Limit Approximation (CLA) and a statistical estimate (using the Gillespie algorithm, SSA) of the path probabilities of the 1gDTA properties of Figure 2 computed on the network epidemic model for different values of the population size N. Parameters of the model are $\kappa_{inf} = 0.05$, $\kappa_{patch_1} = 0.02$, $\kappa_{loss} = 0.01$, $\kappa_{ext} = 0.05$, $\kappa_{patch_0} = 0.001$, $\alpha_1 = 0.5$, $\alpha_2 = 0.2$.

such a parameter set, we compared the CLA of the probability of each global property with a statistical estimate (from 5000 runs), measuring the error in a grid of 1000 equi-spaced time points. We then computed the maximum error and the average error. In Table 1, we report the mean and maximum values of these quantities over the 20 runs, for each considered value of N. We also report the error at the final time of the simulation, when the probability has stabilised to its limit value.[5] It can be seen that both the average and the maximum errors decrease with N, as expected, and are already quite small for $N = 100$ (for the first property, the maximum difference in the path probability for all runs is of the order of 0.06, while the average error is 0.003). For $N = 500$, the CLA is practically indistinguishable from the (estimated) true probability. For the second property, the errors are slightly worse, but still reasonably small.

Finally, we considered the problem of understanding what are the most important aspects that determine the error. To this end, we regressed the observed error against the following features: estimated probability value by CLA, error in the predicted average and variance of X_{Final} (between the CLA and the statistical estimates), and statistical estimates of the mean, variance, skewness and kurtosis of X_{Final}. We used Gaussian Process regression with Adaptive Relevance Detection (GP-ADR, [18]), which performs a regularised regression searching the best fit on an infinite dimensional subspace of continuous functions, and permitted us to identify the most relevant features by learning the hyperparameters of the kernel function. We used both a squared exponential kernel, a quadratic kernel, and a combination of the two, with a training set of 500 points, selected randomly from the experiments performed. The mean prediction error on a test set of other 500 points (independently of N) is around 0.015 for all the considered kernels. Furthermore, GP-ADR selected as most relevant

---

[5] For this model, we can extend the analysis to steady state, as the fluid limit has a unique, globally attracting steady state. This is not possible in general, cf. [6].

Table 1. Maximum and mean of the maximum error (max(err), $\mathbb{E}[\text{max err}]$) for each parameter configuration, maximum and mean of the average error with respect to time (max($\mathbb{E}[\text{err}]$), $\mathbb{E}[\mathbb{E}[\text{err}]]$) for each parameter configuration, maximum and average error at the final time horizon T (max(err(T)), $\mathbb{E}[\text{err}(T)]$), for each parameter configuration. Data is shown as a function of the network size N. Top: First property. Bottom: second property.

N	max(err)	$\mathbb{E}[\text{max err}]$	max($\mathbb{E}[\text{err}]$)	$\mathbb{E}[\mathbb{E}[\text{err}]]$	max(err(T))	$\mathbb{E}[\text{err}(T)]$
20	0.1336	0.0420	0.0491	0.0094	0.0442	0.0037
50	0.0866	0.0366	0.0631	0.0067	0.0128	0.0018
100	0.0611	0.0266	0.0249	0.0030	0.0307	0.0017
200	0.0504	0.0191	0.0055	0.0003	0.0033	0.0002
500	0.0336	0.0120	0.0024	0.0003	0.0002	9.5e-6

N	max(err)	$\mathbb{E}[\text{max err}]$	max($\mathbb{E}[\text{err}]$)	$\mathbb{E}[\mathbb{E}[\text{err}]]$	max(err(T))	$\mathbb{E}[\text{err}(T)]$
20	0.2478	0.1173	0.1552	0.0450	0.1662	0.0448
50	0.2216	0.0767	0.1233	0.0340	0.1337	0.0361
100	0.1380	0.0620	0.0887	0.0216	0.0979	0.0208
200	0.1365	0.0538	0.0716	0.0053	0.0779	0.0162
500	0.1187	0.0398	0.0585	0.0100	0.0725	0.0108

the quadratic kernel, and in particular the following two features: the estimated probability and the error in the mean of X_{Final}. This suggests that moment closure techniques improving the prediction of the average can possibly reduce the error of the method.

8 Conclusions

In this paper we considered population models and properties of individual agents specified by DTAs. We introduced a method based on the central limit theorem for CTMCs to approximate the collective probability with which a given fraction of agents satisfies the local specification. To our knowledge, this has been the first attempt of using central limit results for verification of CTMC properties. The correctness of our method is guaranteed by a convergence result and validated experimentally on a network epidemics model.

For future work, we plan to generalise the class of local specification and the central limit algorithm to more complex DTAs (following [8]), considering also state properties like in CSL-TA [9] (thus extending and connecting with the approach of [5]). We will also investigate the speed of convergence of the approximation in Theorem 2 in order to possibly compare it with fluid limit results (see e.g. [6]). Furthermore, we will also study the connection between the approximation error and the topology of the phase space of the fluid ODEs. We also plan to explore the use of CLA to check purely collective properties, expressed in a temporal logic such as CSL or MiTL, with atoms being inequalities on population variables. Finally, we plan to release a prototype implementation.

Acknowledgements. This research has been partially funded by the EU-FET project QUANTICOL (nr. 600708) and by FRA-UniTS.

References

1. Andersson, H., Britton, T.: Stochastic Epidemic Models and Their Statistical Analysis. Springer (2000)
2. Baier, C., Cloth, L., Haverkort, B.R., Kuntz, M., Siegle, M.: Model checking markov chains with actions and state labels. IEEE Trans. Software Eng. 33(4), 209–224 (2007)
3. Baier, C., Haverkort, B., Hermanns, H., Katoen, J.P.: Model checking continuous-time Markov chains by transient analysis. In: Emerson, E.A., Sistla, A.P. (eds.) CAV 2000. LNCS, vol. 1855, pp. 358–372. Springer, Heidelberg (2000)
4. Baier, C., Katoen, J.P.: Principles of Model Checking. MIT Press (2008)
5. Bortolussi, L., Hillston, J.: Fluid model checking. In: Koutny, M., Ulidowski, I. (eds.) CONCUR 2012. LNCS, vol. 7454, pp. 333–347. Springer, Heidelberg (2012)
6. Bortolussi, L., Hillston, J., Latella, D., Massink, M.: Continuous approximation of collective systems behaviour: A tutorial. Perf. Eval (2013)
7. Bortolussi, L., Policriti, A.: Dynamical systems and stochastic programming: To ordinary differential equations and back. Trans. Comp. Sys. Bio. XI (2009)
8. Chen, T., Han, T., Katoen, J.P., Mereacre, A.: Model checking of continuous-time Markov chains against timed automata specifications. Logical Methods in Computer Science 7(1) (2011)
9. Donatelli, S., Haddad, S., Sproston, J.: Model checking timed and stochastic properties with csl$^{\{TA\}}$. IEEE Trans. Software Eng. 35(2), 224–240 (2009)
10. Ethier, S.N., Kurtz, T.G.: Markov Processes: Characterization and Convergence. Wiley (2005)
11. Hayden, R.A., Bradley, J.T., Clark, A.: Performance specification and evaluation with unified stochastic probes and fluid analysis. IEEE Trans. Software Eng. 39(1), 97–118 (2013)
12. Hayden, R.A., Stefanek, A., Bradley, J.T.: Fluid computation of passage-time distributions in large Markov models. Theor. Comput. Sci. 413(1), 106–141 (2012)
13. Heath, J., Kwiatkowska, M., Norman, G., Parker, D., Tymchyshyn, O.: Probabilistic model checking of complex biological pathways. Theor. Comput. Sci (2007)
14. Jha, S.K., Clarke, E.M., Langmead, C.J., Legay, A., Platzer, A., Zuliani, P.: A bayesian approach to model checking biological systems. In: Degano, P., Gorrieri, R. (eds.) CMSB 2009. LNCS, vol. 5688, pp. 218–234. Springer, Heidelberg (2009)
15. Kolesnichenko, A., Remke, A., de Boer, P.-T., Haverkort, B.R.: Comparison of the mean-field approach and simulation in a peer-to-peer botnet case study. In: Thomas, N. (ed.) EPEW 2011. LNCS, vol. 6977, pp. 133–147. Springer, Heidelberg (2011)
16. Kolesnichenko, A., Remke, A., de Boer, P.T., Haverkort, B.R.: A logic for model-checking of mean-field models. In: Proc. of DSN (2013)
17. Kwiatkowska, M., Norman, G., Parker, D.: PRISM 4.0: Verification of probabilistic real-time systems. In: Gopalakrishnan, G., Qadeer, S. (eds.) CAV 2011. LNCS, vol. 6806, pp. 585–591. Springer, Heidelberg (2011)
18. Rasmussen, C.E., Williams, C.K.I.: Gaussian Processes for Machine Learning. MIT Press (2006)
19. Tribastone, M., Gilmore, S., Hillston, J.: Scalable differential analysis of process algebra models. IEEE Trans. Software Eng. 38(1), 205–219 (2012)
20. Van Kampen, N.G.: Stochastic Processes in Physics and Chemistry. Elsevier (1992)

Fluid Limit for the Machine Repairman Model with Phase-Type Distributions

Laura Aspirot[1], Ernesto Mordecki[1], and Gerardo Rubino[2,*]

[1] Universidad de la República, Montevideo, Uruguay
[2] INRIA, Rennes, France

We consider the Machine Repairman Model with N working units that break randomly and independently according to a phase-type distribution. Broken units go to one repairman where the repair time also follows a phase-type distribution. We are interested in the behavior of the number of working units when N is large. For this purpose, we explore the fluid limit of this stochastic process appropriately scaled by dividing it by N.

This problem presents two main difficulties: two different time scales and discontinuous transition rates. Different time scales appear because, since there is only one repairman, the phase at the repairman changes at a rate of order N, whereas the total scaled number of working units changes at a rate of order 1. Then, the repairman changes N times faster than, for example, the total number of working units in the system, so in the fluid limit the behavior at the repairman is averaged. In addition transition rates are discontinuous because of idle periods at the repairman, and hinders the limit description by an ODE.

We prove that the multidimensional Markovian process describing the system evolution converges to a deterministic process with piecewise smooth trajectories. We analyze the deterministic system by studying its fixed points, and we find three different behaviors depending only on the expected values of the phase-type distributions involved. We also find that in each case the stationary behavior of the scaled system converges to the unique fixed point that is a global attractor. Proofs rely on martingale theorems, properties of phase-type distributions and on characteristics of piecewise smooth dynamical systems. We also illustrate these results with numerical simulations.

1 Introduction

The Machine Repairman Model. The Machine Repairman Model (MRM) is a basic Markovian queue representing a finite number N of machines that can fail independently and, then, be repaired by a repair facility. The latter, in the basic model, is composed of a single repairing server with a waiting room for failed machines managed in FIFO order, in case the repairing server is busy when units fail. In Kendall's notation, this is the $M/M/1//N$ model, specifying that lifetimes and repair times are exponentially distributed. This model is well known

* This paper received partial support from the STIC-AmSud project "AMMA" and project FCE-2-2011-1-6739.

K. Joshi et al. (Eds.): QEST 2013, LNCS 8054, pp. 139–154, 2013.
© Springer-Verlag Berlin Heidelberg 2013

and widely studied in queuing theory and in many applications, as for example in telecommunications or in reliability. Almost all these studies look at the queue in equilibrium. The model is a precursor of the development of queuing network theory, motivated first in computer science. In particular, Scherr from IBM used it in 1972 for analyzing the S360 OS (see [1]). Many extensions of the basic model have been studied, considering more than one repairing server, different queuing disciplines, and other probability distributions for the life-time or for the repair time. We refer to [2] for further reference on the problem.

Fluid Limits. Fluid limits is a widely developed technique that proves very useful for the study of large Markov systems. Many of these systems, under a suitable scaling, have a deterministic limit given by an ordinary differential equation (ODE). As an example, let us consider the fluid limit for a $M/M/1$ queue [3] with arrival rate λ and service rate μ. Let $X(t)$ be the number of units at time t and let $\widehat{X}^N(t) = X(Nt)/N$ be the scaled number of units. Time is accelerated by a factor N, and the initial state is also scaled by the same factor. If the scaled initial condition converges with N, then the process $\widehat{X}^N$ can be approximated, for large N, by the deterministic solution to $\dot{x} = \lambda - \mu$ if $x > 0$, $\dot{x} = 0$ if $x = 0$. For $\lambda < \mu$ the equation defines a piecewise smooth dynamical system, with a solution for the initial condition $x(0)$ that is smooth on $[0, x(0)/(\mu - \lambda))$ and $(x(0)/(\mu - \lambda), \infty)$. If the initial condition is 0, the solution remains at zero.

Other examples from [3] are the $M/M/\infty$ queue and the $M/M/N/N$ queue. Let λN be the arrival rate and μ the service rate in both cases and let $\widetilde{X}^N$ be the number of units in the system. The scaling is different from the $M/M/1$, as time is not scaled, only the arrival rate is accelerated, and the total service rate scales with the number of units in the system. The scaled number of units $X^N = \widetilde{X}^N/N$ converges to the solution to $\dot{x} = \lambda - \mu x$ for the $M/M/\infty$ queue, and to the solution to $\dot{x} = \lambda - \mu x$, if $x < 1$, $\dot{x} = 0$ if $x = 1$, in the $M/M/N/N$ model. In the last case, we find again a piecewise smooth dynamical system, that converges exponentially fast to $\rho = \lambda/\mu$ if $\rho < 1$ and to 1 if $\rho \geq 1$.

Looking for fluid limits is a suitable approach to repairman problems, as shown in [4], where the MRM model with two repair facilities, studied by Iglehart and Lemoine in [5,6], is analyzed using these tools. In [5] there are N operating units subject to exponential failures with parameter λ. Failures are of type 1 (resp. 2) with probability p (resp. $1 - p = q$). If failure is of type i ($i = 1, 2$) the unit goes to repair facility i that has s_N^i exponential servers, each one with exponential service rate μ_i. The goal is to study the stationary distribution when $s_i^N \sim N s_i$ as $N \to \infty$, $i = 1, 2$. The behavior of the system is characterized in terms of the parameter set that defines the model. In addition, the case with spares is presented in [6]. The original approach consists in approximating the number of units in each repair facility by binomial random variables, and then proving for them a law of large numbers and a central limit theorem. Kurtz, in [4] studies the same model with a fluid limit approach, proves convergence to a deterministic system, and goes a step beyond, considering the rate of this approximation through a central limit theorem-type result. The same discussion as in [5] in terms of the different parameter set follows from the study of the ODE's fixed point.

Contributions. In this paper we analyze a repairman problem with N working units that break randomly and independently according to a phase-type distribution. Broken units go to one repairman where the repair time also follows a phase-type distribution (that is, a $PH/PH/1//N$ model). We consider a scaled system, dividing the number of broken units and the number of working units in each phase by the total number of units N. The scaled process has a deterministic limit when N goes to infinity. The first problem that the model presents is that there are two time scales: the repairman changes its phase at a rate of order N, whereas the total scaled number of working units changes at a rate of order 1. Another problem is that transition rates are discontinuous because of idle periods at the repairman (this second issue is also present in the models $M/M/1$ and $M/M/N/N$ described above).

In our main result we prove that the scaled Markovian multi-dimensional process describing the system dynamics converges to the solution of an ODE as $N \to \infty$. The convergence is in probability and takes place uniformly in compact time intervals (usually denoted *u.c.p.* convergence), and the deterministic limit, the solution to the ODE, is only picewise smooth. We analyze the properties of this limit, and we prove the convergence in probability of the system in stationary regime to the ODE's fixed point. We also find that this fixed point only depends on the repair time by its mean. As a matter of fact, recall that when in equilibrium, if the repair times are exponentially distributed, the distribution of the number of broken machines has the insensitivity property with respect to the life-time distribution (only the latter's mean appears in the former). For an example of what happens here, see the end of Section 4.

Related Work. Fluid limits, density dependent population process, approximation by differential equations of Markov chains, are all widely developed objects. As a general reference we refer to the monograph by Ethier and Kurtz [7] and references therein. The main approach there consists in a random change of time that allows to write the original Markov chain as a sum of independent unit Poisson processes evaluated at random times. Darling and Norris [8] present a survey about approximation of Markov chains by differential equations with an approach based on martingales. However, [8] does not deal with discontinuous transition rates. We refer to the books by Shwartz and Weiss [9], and Robert [3] for extensive analysis of the $M/M/1$ and the $M/M/\infty$ queues, including deterministic limits, asymptotic distributions and large deviations results. In particular, in [9] the discontinuous transition rates and different time scales are considered. The latter situation is also considered in [10] and [11]. We mostly follow the approach of [12] to deal with discontinuous transition rates, which considers hybrid limits for continuous time Markov chains with discontinuous transition rates, with examples in queuing and epidemic models. Discontinuous transition rates are also studied in [13,14]. Paper [13] analyzes queuing networks with batch services and batch arrivals, that lead to fluid limits represented by

ODEs with discontinuous right hand sides. Paper [14] models optical packet switches, where the queuing model lead to ODEs with discontinuous right hand sides, and where they consider both exponential and phase-type distributions for packet lengths. Convergence to the fixed points is studied in several works (e.g. [9,13,15]). However there are counterexamples where there is no convergence of invariant distributions to fixed points [15]. There are general results with quite strong hypotheses as in [16], where reversibility is assumed in order to prove convergence to the fixed point.

Organization of the Paper. In Section 2 we present our model. In Section 3 we compute the drift, and we describe the ODE that defines the fluid limit. In Section 4 we state our main results about convergence and description of the fluid limit. In Section 5 we show several numerical examples that illustrates the results and we conclude in Section 6. Proofs are provided in Section 7.

2   Model

We consider N identical units that work independently, as part of some system, that randomly fail and that get repaired. Broken units go to a repairman with one server, where the repair time is a random variable with a phase-type distribution. After being repaired units start working again. A given unit's life-times are independent identically distributed random variables also with phase-type distribution. We want to describe the number of working units in each phase before failure and the number of broken units in the system. We consider the system for large N, with the repair time scaled by N. This means that the repair time per unit decreases as N increases. We describe the limit behavior of the system when N goes to infinity. The assumption of phase-type distributions allows to represent a wide variety of systems, as phase-type distributions well approximate many positive distributions, allowing, at the same time, to exploit properties of exponential distributions and Markov structure. Concerning the repairing facility, we consider a single server with the service time also scaled according with the number of units, and we find a different behavior that for the model scaled both in the number of units and the number of servers.

Phase-Type Distributions. A phase-type distribution with k phases is the distribution of the time to absorption in a finite Markov chain with $k + 1$ states, where one state is absorbing and the remaining k states are transient. With an appropriate numbering of the states, the transient Markov chain has infinitesimal generator $\widehat{M} = \left(\begin{array}{c|c} M & m \\ \hline 0 & 0 \end{array} \right)$, where M is a $k \times k$ matrix, and $m = -M\mathbb{1}$, with $\mathbb{1}$ the column vector of ones in $\mathbb{R}^k$. The initial distribution for the transient Markov chain is a column vector $(r, 0) \in \mathbb{R}^{k+1}$, where r is the initial distribution among the transient states. We represent this phase-type distribution by (k, r, M). We refer to [17] for further background about phase-type distributions.

Variables. We describe the distributions and variables involved in the model. All vectors are column vectors.

Repair Time. The repair time follows a phase-type distribution (m, p, NA), with m phases, matrix NA (where A is a fixed matrix and N is the scaling factor) and initial distribution p. We denote $Na = N(a_1, \ldots, a_m) = -NA\mathbb{1}$.

Life-Time. The life-time for each unit is phase-type (n, q, B), with n phases, matrix B and initial distribution q. We denote $b = (b_1, \ldots, b_n) = -B\mathbb{1}$.

Working Units. $\widetilde{X}_i^N(t)$ is the number of units working in phase i at time t, for $i = 1, \ldots, n$, and $\widetilde{X}^N = (\widetilde{X}_1^N, \ldots, \widetilde{X}_n^N)$.

Repairman State. $\widetilde{Z}_i^N(t)$ is number of units being repaired in phase i for $i = 1, \ldots, m$ ($\widetilde{Z}_i^N(t)$ is zero or one), and $\widetilde{Z}^N = (\widetilde{Z}_1^N, \ldots, \widetilde{Z}_m^N)$.

Waiting Queue. $\widetilde{Y}^N(t)$ is the number of broken units waiting to be repaired.

Scaling. We consider the scaling: $X^N = \dfrac{1}{N}\widetilde{Y}^N$, $Y^N = \dfrac{1}{N}\widetilde{Y}^N$, $Z^N = \dfrac{1}{N}\widetilde{Z}^N$.

Note that $\mathbb{1}^T \widetilde{Z}^N(t) + \chi_{\{\mathbb{1}^T \widetilde{X}^N(t) = N\}} = 1$, where $\chi_\mathcal{P}$ is the indicator function of the predicate $\mathcal{P}$. That means that units are all working, or there is one unit being repaired at the server. In addition, $\mathbb{1}^T \widetilde{X}^N(t) + \widetilde{Y}^N(t) + \sum_{i=1}^m \widetilde{Z}_i^N(t) = N$.

Model Dynamics. $\widetilde{U}^N = \left(\widetilde{X}_1^N, \ldots, \widetilde{X}_n^N, \widetilde{Y}^N, \widetilde{Z}_1^N, \ldots, \widetilde{Z}_m^N\right)$ is a Markov chain.

We denote by $e^i \in \mathbb{R}^{n+m+1}$ the vector $e^i = (e_1^i, \ldots, e_{n+m+1}^i)$ with $e_i^i = 1$ and $e_j^i = 0$ for $i \neq j$, $i, j = 1, \ldots, n+m+1$. We describe the possible transitions for this Markov chain from a vector $\widetilde{u}$ in the state space, with its corresponding transition rates. The vector $\widetilde{u} = (\widetilde{x}, \widetilde{y}, \widetilde{z})$ with $\widetilde{x} = (\widetilde{x}_1, \ldots, \widetilde{x}_n)$, $\widetilde{z} = (\widetilde{z}_1, \ldots, \widetilde{z}_m)$, with $\widetilde{x}_i \in \{0, 1, \ldots, N\}$ for all $i = 1, \ldots n$, $\widetilde{y} \in \{0, 1, \ldots, N\}$ and $\widetilde{z}_i \in \{0, 1\}$ for all $i = 1, \ldots m$.

A working unit in phase i changes to phase j. For $i, j = 1, \ldots, n$, transition $e^j - e^i$ occurs with rate $b_{ij}\widetilde{x}_i$.

A working unit in phase i breaks and goes to the buffer. The unit goes to the buffer because there is one unit in service. For $i = 1, \ldots, n$, transition $e^{n+1} - e^i$ occurs at rate $b_i\widetilde{x}_i\chi_{\{\mathbb{1}^T\widetilde{x}<N\}}$.

A working unit in phase i breaks and starts being repaired. The unit starts being repaired because the repairman is idle, at phase j. For $i = 1, \ldots, n$, $j = 1, \ldots, m$, transition $e^{n+1+j} - e^i$ occurs at rate $b_i p_j \widetilde{x}_i \chi_{\{\mathbb{1}^T\widetilde{x}=N\}}$.

A unit that is being repaired in phase i changes to phase j. For $i, j = 1, \ldots, m$, transition $e^{n+1+j} - e^{n+1+i}$ occurs at rate $Na_{ij}\widetilde{z}_i$.

A unit that is being repaired in phase i ends its service and starts working at phase j with the buffer empty. If the buffer is empty, nobody starts being served and for $j = 1, \ldots, n$, $i = 1, \ldots, m$, the transition $e^j - e^{n+1+i}$ occurs at rate $Na_i q_j \widetilde{z}_i \chi_{\{\widetilde{y}=0\}}$.

A unit that is being repaired in phase i ends its service and starts working at phase j with nonempty buffer. If the buffer is nonempty a unit in the buffer starts being served in phase k at the same time, then for $j = 1, \ldots, n$ and $i, k = 1, \ldots, m$, the transition $e^j + e^{n+1+k} - e^{n+1} - e^{n+1+i}$ occurs at rate $Na_i q_j p_k \widetilde{z}_i \chi_{\{\widetilde{y}>0\}}$.

3   Drift Computation and Description of the Limit

In order to understand an summarize the dynamics of the stochastic process we compute the drift for our model. We compute it and we analyze the behavior of the ODE that will define the limit. For this purpose we also present a brief description of ODEs with discontinuous right hand sides.

Let us recall that for a Markov chain $V \in \mathbb{R}^d$, with transition rates $r_v(x)$ from x to $x + v$, the drift is defined by $\beta(x) = \sum_v v r_v(x)$, where the sum is in all possibles values of v. One possible representation of a Markov chain is in terms of the drift, where in a general way $V(t) = V(0) + \int_0^t \beta(V(s))ds + M(t)$, with $M(t)$ a martingale. One approach to establish a fluid limit is to exploit this decomposition for the scaled process, to prove that there is a deterministic limit for the integral term and to prove that the martingale term converges to 0.

We write down the drift β of the scaled process $U^N = \widetilde{U}^N/N$, evaluated at $u = (x, y, z)$ with $x = (x_1, \ldots, x_n)$, $z = (z_1, \ldots, z_m)$, with $\widetilde{x}_i \in \{0, 1/N, \ldots, 1\}$ for all $i = 1, \ldots n$, $\widetilde{y} \in \{0, 1/N, \ldots, 1\}$ and $\widetilde{z}_i \in \{0, 1/N\}$ for all $i = 1, \ldots m$. Let $\beta = (\beta_1, \ldots, \beta_{n+m+1})$. For $i = 1, \ldots, n$ we have the following equations:

$$\beta_i(u) = \sum_{j=1}^n b_{ji}x_j + q_i \sum_{j=1}^m a_j N z_j.$$

Let us call β_X the first n coordinates of the drift. In matrix notation:

$$\beta_X(u) = B^T x + a^T N z q.$$

For $i = n + 1$ the drift equation (the $(n+1)$th coordinate of the drift) is:

$$\beta_{n+1}(u) = \sum_{j=1}^n b_j x_j \chi_{\{\mathbf{1}^T x < 1\}} - \sum_{i=1}^m a_i N z_i \chi_{\{y > 0\}}$$

and in matrix notation (we also call this coordinate β_Y):

$$\beta_Y(u) = b^T x \chi_{\{\mathbf{1}^T x < 1\}} - a^T N z \chi_{\{y > 0\}}.$$

For $k = n + 1 + i$, with $i = 1, \ldots, m$ we have:

$$\beta_k(u) = p_i \chi_{\{\mathbf{1}^T x = 1\}} \sum_{j=1}^n b_j x_j + \sum_{j=1}^m N z_j \left(a_{ji} + p_i \chi_{\{y > 0\}} a_j \right).$$

In matrix notation (we call these coordinates of the drift β_Z) we have:

$$\beta_Z(u) = b^T x \chi_{\{\mathbf{1}^T x = 1\}} p + A^T N z + a^T N z \chi_{\{y > 0\}} p.$$

We call the drift $\beta(u) = \beta(x, y, z)$.

$$\beta_X(x, y, z) = B^T x + a^T N z q, \tag{1}$$

$$\beta_Y(x, y, z) = b^T x \chi_{\{\mathbf{1}^T x < 1\}} - a^T N z \chi_{\{y > 0\}}, \tag{2}$$

$$\beta_Z(x, y, z) = b^T x \chi_{\{\mathbf{1}^T x = 1\}} p + A^T N z + a^T N z \chi_{\{y > 0\}} p. \tag{3}$$

These equations suggest the ODE that should verify the deterministic limits (x, y) of (X^N, Y^N), if they exist. However, the drift depends on the values of Nz. The process $\widetilde{Z}^N$ varies at a rate of order N whereas the processes X^N and Y^N vary at a rate of order 1. So, we can assume that when N goes to infinity and for a fixed time the process $\widetilde{Z}^N$ has reached its stationary regime and then the limit of the last m coordinates of the drift is negligible. With this argument the candidate to the ODE defining the fluid limit is obtained by replacing in equations (1) and (2) Nz by $\widetilde{z}$, the solution to the n-dimensional equation

$$b^T x \chi_{\{\mathbb{1}^T x = 1\}} p + A^T \widetilde{z} + a^T \widetilde{z} \chi_{\{y > 0\}} p = 0.$$

Solving the last equation (multiplying by $\mathbb{1}^T$, by $\mathbb{1}^T \left(A^T\right)^{-1}$, and using the relationship $\mathbb{1}^T \widetilde{z} = \chi_{\{\mathbb{1}^T x < 1\}}$) we obtain $a^T \widetilde{z} = \mu \chi_{\{\mathbb{1}^T x < 1\}}$, with $1/\mu = -\mathbb{1}^T \left(A^T\right)^{-1} p$, the mean time before absorption for the transient Markov chain defining the phase-type repair time distribution. We refer to [17] for properties of phase-type distributions. As we want to obtain an ODE for x, the candidate to ODE's vector field is $F(x) = B^T x + \mu \chi_{\{\mathbb{1}^T x < 1\}} q$. We observe that the equation $\dot{x} = B^T x + \mu q$ is valid when $\mathbb{1}^T x < 1$, or, in the border $\mathbb{1}^T x = 1$, when the vector field $B^T x + \mu q$ points towards the region $\mathbb{1}^T x < 1$, that is $\mathbb{1}^T (B^T x + \mu q) < 0$. Using that $B\mathbb{1} = -b$ the condition is $b^T x > \mu$. When $\mathbb{1}^T x = 1$ and $b^T x \leq \mu$ the equation presents what is called *sliding motion*. We follow the presentation of this topic in [12]. What happens is that the deterministic system has trajectories in the border surface $\mathbb{1}^T x = 1$. The vector field that drives the equation in the border is $G(x)$, where $\mathbb{1}^T G(x) = 0$ and $G(x)$ is a linear combination of $B^T x + \mu q$ and $B^T x$ (the vectors fields corresponding to the drift in the interior and in the border). Then $G(x) = (1 - \phi(x))(B^T x + \mu q) + \phi(x) B^T x$ with $\mathbb{1}^T G(x) = 0$ that leads to $\phi(x) = 1 - b^T x / \mu$ and then, computing $G(x)$,

$$\dot{x} = \begin{cases} B^T x + \mu q, & \text{if } b^T x > \mu \text{ or } \mathbb{1}^T x < 1, \\ B^T x + b^T x q, & \text{if } b^T x \leq \mu \text{ and } \mathbb{1}^T x = 1, \end{cases} \tag{4}$$

4   Main Results

In this section we state our main results. Proofs are presented in Section 7. First we show that the scaled stochastic process (X^N, Y^N) converges to the deterministic piecewise smooth dynamical system (x, y). Processes (X^N, Y^N) and (x, y) are multidimensional (they live in $\mathbb{R}^{n+1}$), as the number of phases for working units is n. Convergence is in probability, uniformly in compact time intervals (*u.c.p.* convergence). From the calculus of the drift for the stochastic processes in Section 3 we have that the limit processes is driven by the vector field $B^T x + \mu q$ in the interior $\mathbb{1}^T x < 1$ and by the vector field $B^T x$ in the border $\mathbb{1}^T x = 1$. Very close to the border $\mathbb{1}^T x = 1$, when $b^T x \leq \mu$ the vector field $B^T x + \mu q$ points outside the region $\mathbb{1}^T x < 1$, but if we consider the vector field induced by transitions in the border, when $b^T x \leq \mu$ we should have a vector field $B^T x$ that points towards the region $\mathbb{1}^T x < 1$. Because of this, the processes driven by those vector fields present a sliding motion, that means that, when $b^T x \leq \mu$, the trajectory remains in the border $\mathbb{1}^T x = 1$ driven by a linear combination of both fields. So, we must first define the piecewise smooth dynamical system (x, y), where $y = 1 - \mathbb{1}^T x$ and x is the solution (in the sense of Filippov) of the differential equation with discontinuous right hand side (4).

Lemma 1. *The differential equation (4) has a unique solution for each initial condition x_0, with $\mathbb{1}^T x_0 \leq 1$.*

Once our limit candidate is defined, we state the following theorem.

Theorem 1. *Let $\lim_{N \to \infty} X^N(0) = x_0$ in probability, with x_0 deterministic. Then for all $T > 0$*

$$\lim_{N \to \infty} \sup_{[0,T]} \left\| (X^N(t), Y^N(t)) - (x(t), y(t)) \right\| = 0,$$

in probability. The process (x, y) is defined by $y = 1 - \mathbb{1}^T x$ and x the solution to equation (4) with initial condition x_0.

The main contribution here is the study of different time scales, one at the repairman and one for life-times, and the averaging phenomena at the repairman. We treat this problem, together with the problem of discontinuous transition rates. The problem of different time scales has been addressed in other contexts (e.g. [11,10]).

Let us study the behavior of the system defined by equation (4) by studying its fixed points. We observe that $1/\mu$ is the mean of the phase-type distribution (m, p, A). We define $1/\lambda = -\mathbb{1}^T \left(B^T\right)^{-1} q$, the expected value of a phase-type distribution (n, q, B), and

$$\rho = \frac{\mu}{\lambda}. \tag{5}$$

We identify three different behaviors, that we call (using the same definitions that in [3] for the $M/M/N/N$ queue) sub-critical when $\rho < 1$, critical when $\rho = 1$ and super-critical when $\rho > 1$.

Sub-critical case, $\rho < 1$. The mean repair time per unit $1/\mu$ is greater than the mean life-time, so we find an equilibrium with a positive number of broken units in the system. When we compute the fixed points in equation (4) the fixed point is an interior point in $\mathbb{1}^T x \leq 1$, and it is a global attractor.

Super-critical case, $\rho > 1$. When $\rho > 1$, intuitively the repairman is more effective, and we have an equilibrium with all the units (in the deterministic approximation) working. The fixed point is also a global attractor, and it is in the border $\mathbb{1}^T x = 1$.

Critical case, $\rho = 1$. In this case the fixed points for the equation in the interior and in the border coincide, giving a fixed point in the border that is a global attractor.

We state these results in the following lemma.

Lemma 2. *There are three different behaviors for equation (4):*

$\rho < 1$ *(sub-critical). There is a unique fixed point x^* that is a global attractor and verifies $\mathbb{1}^T x^* = \rho < 1$:*

$$x^* = -\mu (B^T)^{-1} q, \tag{6}$$

$\rho > 1$ *(super-critical). There is a unique fixed point x^* that is a global attractor and verifies $\mathbb{1}^T x^* = 1$ and $b^T x^* < \mu$:*

$$x^* = -\lambda (B^T)^{-1} q, \tag{7}$$

$\rho = 1$ *(critical). There is unique fixed point x^* given by equations (6) or (7). It is a global attractor and verifies $\mathbb{1}^T x^* = 1$ and $b^T x^* = \mu = \lambda$.*

Theorem 2. *The system in stationary regime $X^N(\infty)$ converges in probability to x^*, when $N \to \infty$, where x^* is the unique fixed point of equation (4).*

We observe that the fixed point depends only on the mean repair-time but on matrix B. That means that different life-time distribution with the same mean lead to different stationary behaviors.

5   Numerical Examples

We consider the repairman problem with $N = 100$ units, for different phase type time distributions and for different values of ρ. The parameters are defined in Section 2 and ρ is defined in equation (5). As initial condition we fix the total number of working units and sample the number in each phase according to the phase type initial distribution. We show the scaled number of working units in each phase. We illustrate the convergence to the ODE's fixed point x^* and the sliding motion. The parameters for each example (figure) are given in Table 1.

Exponencial life-time and hypoexponential repair time. In Figure 1 we consider exponential life-time and hypoexponential (sum of independent exponentials). We represent for each parameter set the evolution with time of the stochastic process X^N and we show the convergence to the fixed point x^*.

Hypoexponential life and repair time, sliding motion. In Figure 2 we consider two phases both in the life and repair times (both distributions are hypoexponential). At the left we represent, for each parameter set, the evolution with time of the stochastic processes X_1^N and X_2^N and the ODE's solution (x_1, x_2). We also show the convergence to the fixed point $x^* = (x_1^*, x_2^*)$. At the right we represent, for each parameter set, the trajectory of process X^N, the trajectory of the ODE's solution x and the fixed point. Depending on the initial condition, we find sliding motion (as shown in the bottom figures), but, as we have the same parameters, both trajectories converge to the fixed point.

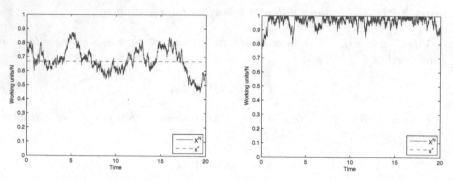

Fig. 1. Exponential life-time and hypoexponential repair time. In the right we show the sliding motion.

Table 1. Parameters for Figures 1, 2 and 3

	Fig.1(left)	Fig.1(right)	Fig.2(top)	Fig.2(bottom)	Fig.3(top)	Fig.3(bottom)
A	$\begin{pmatrix} -1 & 1 \\ 0 & -2 \end{pmatrix}$	$\begin{pmatrix} -2 & 2 \\ 0 & -3 \end{pmatrix}$	$\begin{pmatrix} -1 & 1 \\ 0 & -2 \end{pmatrix}$	$\begin{pmatrix} -1 & 1 \\ 0 & -2 \end{pmatrix}$	-0.5	-1.5
B	-1	-1	$\begin{pmatrix} -2 & 2 \\ 0 & -3 \end{pmatrix}$	$\begin{pmatrix} -2 & 2 \\ 0 & -3 \end{pmatrix}$	$\begin{pmatrix} -3 & 3 & 0 \\ 0 & -2 & 2 \\ 0 & 0 & -5 \end{pmatrix}$	$\begin{pmatrix} -3 & 3 & 0 \\ 0 & -2 & 2 \\ 0 & 0 & -5 \end{pmatrix}$
p	$(1,0)$	$(1,0)$	$(1,0)$	$(1,0)$	1	1
q	1	1	$(1,0)$	$(1,0)$	$(1,0,0)$	$(1,0,0)$
$X^N(0)$	$0.75N$	$0.75N$	$0.75N$	N	$0.75N$	$0.75N$
ρ	0.6667	1.2	0.5556	0.5556	0.5167	1.55

Hypoexponential life-time with three phases and exponential repair time. In Figure 3 we consider three phases (hypoexponential) in the life-time and exponential repair time. At the left we represent the evolution with time of the stochastic processes X_1^N, X_2^N, X_3^N and we show the convergence to the fixed point $x^* = (x_1^*, x_2^*, x_3^*)$. At the right we represent the trajectory of process X^N and the fixed point. The example at the top has $\rho < 1$. The example at the bottom has $\rho > 1$, so we find sliding motion in the plane $\mathbb{1}^T x = 1$ and the fixed point is also in the same plane.

6   Conclusions

In this paper we find a deterministic fluid limit for a Machine Repairman Model with phase type distributions. The fluid limit is a deterministic process with piecewise smooth trajectories, that presents three different behaviors depending only on the expected values of the phase type distributions involved. The stationary behavior of the scaled system converges to a fixed point that only depends on the mean time between failures and on the mean repair time at the

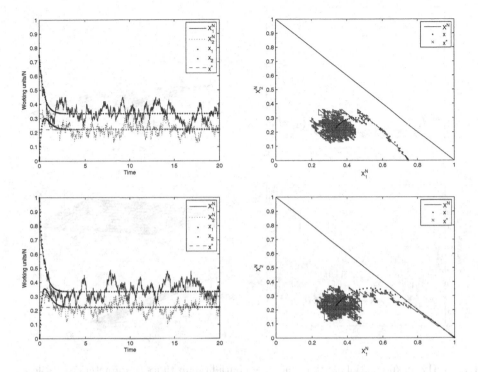

Fig. 2. Hypoexponential life and repair time (parameters in Table 1). The bottom figure shows the sliding motion.

repairman. There are characteristics of the system that hinder us from using classical results from fluid limits or density dependent population processes, as the presence of two different time scales (one for the working units and one for the repairman) and discontinuities in the transition rates due to idle time at the repairman. Concerning the different time scales we find an averaging result, where the phase type distribution at the repairman is represented in the limit only by its mean value. The behavior of the system due to idle times is similar to the behavior of the $M/M/N/N$ queue if we consider exponential distributions instead of phase type ones. The phase type distribution at the failure time adds more dimensions to the problem. This leads to different behaviors for the deterministic system than for the $M/M/N/N$ queue.

Another situation to be addressed in future work is to consider a general distribution at the repairman, as with this scaling it seems to be some insensitivity property where in the limit the behavior depends on the repair time only by its mean. However, the variability at life-time leads to different fluid limits so another possible topic of study is the impact of the life-time distribution on performance measures. Last, we also intend to analyze the asymptotic distribution of the difference between the scaled system and its deterministic limit.

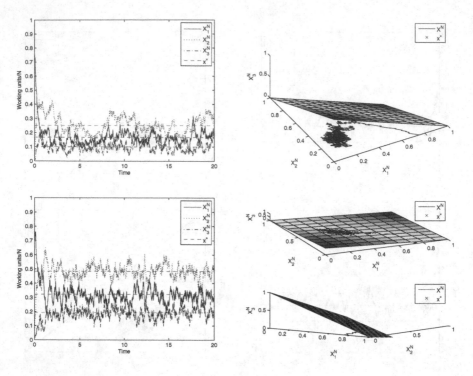

Fig. 3. Hypoexponential life-time and exponential repair times (parameters in Table 1). The three bottom figures shows the sliding motion and the fixed point in the plane $\mathbb{1}^T x = 1$ at the right.

7   Proofs

In what follows we give some proofs of results stated in section 4. First we address the existence of solutions to equation (4) and we prove Lemma 1. For a differential equation $\dot{x} = F(x)$, with F discontinuous, solutions are defined in the set of absolutely continuous functions, instead of differentiable functions as in the classical case. The ODE is defined in the region $\{\mathbb{1}^T x \leq 1\}$ by $B^T x + \mu q$ in the region $\{\mathbb{1}^T x < 1\}$ and $B^T x$ in the region $\{\mathbb{1}^T x = 1\}$. In order to consider the framework of differential equations with discontinuous right hand sides, we extend the definition of the ODE. We define the equation by two continuous fields, $F_1(x) = B^T x + \mu q$ in the region $R_1 = \{\mathbb{1}^T x < 1\}$, $F_2(x) = B^T x$ in the region $R_2 = \{\mathbb{1}^T x > 1\}$ with a region $H = \{x \in \mathbb{R}^n : \mathbb{1}^T x = 1\}$ where the field is discontinuous. The field $B^T x$ always point towards R_1. The field $B^T x + \mu q$ points toward R_1 when $b^T x > \mu$ and point towards R_2 when $b^T x < \mu$. We find that the equation presents *transversal crossing* in H for $b^T x > \mu$ and a *stable sliding motion* in H for $b^T x < \mu$ (defined as in [12]). Transversal crossing occurs when both vector fields point towards R_1 and trajectories from R_2 crosses H. If trajectories start in R_1 or in H, with $b^T x > \mu$ they go into R_1. Stable sliding motion occurs as F_1 points towards R_2 and F_2 points towards F_1 (in H, for $b^T x < \mu$). The ODE has trajectories in the border surface H. The vector field that drives the equation is $G(x)$, with $G = F_1$ in the interior and in the border, when $b^T x < \mu$, $G(x)$ verifies $1\!\!1^T G(x) = 0$ and it is a linear combination of $F_1(x) = B^T x + \mu q$ and $F_2(x) = B^T x$, leading to equation (4). When $b^T x + \mu$, $B^T x + \mu q$ is tangential to H, whereas $B^T x$ points toward R_1, so the trajectories go into R_1. It is called *first order exit condition* of sliding motion.

Proof (Lemma 1). From [12], in order to prove the existence of solutions we need to verify that the field F defined as $F_1(x) = B^T x + \mu q$ in R_1 and $F_2(x) = B^T x$ in R_2 is continuous in each closure $\bar{R}_1$ and $\bar{R}_2$. In addition, if we consider the normal vector to H, $1\!\!1$, we can verify that (except in the region $\{b^T x = \mu\}$) we have $1\!\!1^T F_1(x) > 0$ and $1\!\!1^T F_2(x) < 0$. These conditions mean that there is a stable sliding motion, where the solution belongs to H, driven by G, the linear combination of F_1 and F_2. In addition, unique solutions are also defined for initial conditions in H.   □

Now we address the proof of Theorem 1. We recall the definitions of the drift and the vector fields. Let $U^N = (X^N, Y^N, Z^N)$ and $u = (x, y, 0)$. We have that $(x(t), y(t)) = (x_0, y_0) + \int_0^t G(x(s), y(s)) ds$.

Proof (Theorem 1). First, we observe that $1\!\!1^T X^N(t) + Y^N(t) + 1\!\!1^T Z^N(t) = 1$, and that $\lim_{N \to +\infty} \sup_{[0,T]} Z^N(t) = 0$. Then, in order to prove that

$$\lim_{N \to +\infty} \sup_{[0,T]} \left\| (X^N(t), Y^N(t)) - (x(t), y(t)) \right\| = 0$$

in probability, we only need to prove that $\lim_{N \to +\infty} \sup_{[0,T]} \left\| X^N(t) - x(t) \right\| = 0$ in probability. In the proof of this theorem we follow the approach of [18].

$$\sup_{[0,T]} \left\| X^N(t) - x(t) \right\| \leq \left\| X^N(0) - x(0) \right\| \tag{8}$$

$$+ \sup_{[0,T]} \left\| X^N(t) - X^N(0) - \int_0^t \beta_X \left(U^N(s) \right) ds \right\| \tag{9}$$

$$+ \sup_{[0,T]} \left\| \int_0^t \beta_X \left(U^N(s) \right) ds - \int_0^t F \left(X^N(s) \right) ds \right\| \tag{10}$$

$$+ \sup_{[0,T]} \left\| \int_0^t F \left(X^N(s) \right) ds - \int_0^t G \left(X^N(s) \right) ds \right\| \tag{11}$$

$$+ \sup_{[0,T]} \left\| \int_0^t G \left(X^N(s) \right) ds - \int_0^t G \left(x(s) \right) ds \right\| \tag{12}$$

We want to prove that (9), (10) and (11) converge to 0 in probability. So, provided that the initial condition $X^N(0)$ converges to $x(0)$, we have that, with probability that tends to 1 with N,

$$\sup_{[0,T]} \left\| X^N(t) - x(t) \right\| \leq \varepsilon + \sup_{[0,T]} \left\| \int_0^t G \left(X^N(s) \right) ds - \int_0^t G \left(x(s) \right) ds \right\| .$$

Using that G is piecewise linear and Gronwall inequality, we obtain the bound $\sup_{[0,T]} \left\| X^N(t) - x(t) \right\| \leq \varepsilon e^{KT}$, which leads to $\sup_{[0,T]} \left\| X^N(t) - x(t) \right\| \to 0$ in probability. We study the convergence of (9), (10) and (11). To show the convergence of (9), we first notice that

$$(9) \leq \left\| U^N(t) - U^N(0) - \int_0^t \beta \left(U^N(s) \right) ds \right\|.$$

Convergence follows from the representation of the process as the initial condition plus the integral of the drift plus a martingale term. The martingale term goes to 0 with N because of the scaling. Let us define for a Markov chain $V \in \mathbb{R}^d$, with transition rates $r_v(x)$ from x to $x + v$, $\alpha(x) = \sum_v \|v\|^2 r_v(x)$. Let us also call α the corresponding object for U^N. Convergence of (9) can be then proved using Proposition 8.7 in [8], that states that

$$E \left(\sup_{[0,T]} \left\| U^N(t) - U^N(0) - \int_0^t \beta \left(U^N(s) \right) ds \right\|^2 \right) \leq 4 \int_0^T \alpha \left(U^N(s) \right) ds.$$

As for our scaling $\sup_{[0,T]} \alpha \left(U^N(t) \right) \sim \mathcal{O}(1/N)$, convergence holds.

To prove that (10) converges to 0 in probability we consider the last m coordinates of (9), corresponding to the phases at the repairman. As we have proved that (9) converges to 0 in probability, we conclude that

$$\int_0^t \beta_Z \left(U^N(s) \right) ds = \int_0^t \left(b^T X^N(s) \chi_{\{\mathbf{1}^T X^N(s)=1\}} p + A^T \widetilde{Z}^N(s) + a^T \widetilde{Z}^N(s) \chi_{\{Y^N(s)>0\}} p \right) ds$$

converges to 0 in probability. Multiplying by $\mu \mathbf{1}^T (A^T)^{-1}$,

$$\int_0^t \left(-b^T X^N(s) \chi_{\{\mathbf{1}^T X^N(s)=1\}} + \mu \chi_{\{\mathbf{1}^T X^N(s)<1\}} - a^T \widetilde{Z}^N(s) \chi_{\{Y^N(s)>0\}} \right) ds \quad (13)$$

goes to 0 in probability. In addition, $\mathbf{1}^T \beta_X + \beta_Y + \mathbf{1}^T \beta_Z = 0$. Then, as $\int_0^t \beta_Z \left(U^N(s) \right) ds$ converges to 0 in probability, $\int_0^t \left(\mathbf{1}^T \beta_X \left(U^N(s) \right) + \beta_Y \left(U^N(s) \right) \right) ds$ also converges to 0 in probability and it is equal to

$$\int_0^t \left(-b^t X^N(s) \chi_{\{\mathbf{1}^T X^N(s)=1\}} + a^T \widetilde{Z}^N(s) \chi_{\{Y^N(s)=0\}} \right) ds. \quad (14)$$

Then, considering the sum of equations (13) and (14), we obtain

$$\lim_{N \to +\infty} (10) = \lim_{N \to +\infty} \sup_{[0,T]} \int_0^t \left(a^T \widetilde{Z}^N(s) q - \mu \chi_{\{\mathbf{1}^T X^N(s)<1\}} q \right) ds = 0.$$

The convergence of (11) can be proved by approximating the continuous process in the border by a discrete process with the same jumps. As our model verifies the hypotheses of [12], the same proof that in Lemma 3 of [18] holds.   □

Proof (Lemma 2). First, we compute the fixed points of both fields: $B^T x + \mu q$ in $\mathbb{R}^n$, and $(B^T + qb^T)x$ in $\{\mathbb{1}^T x = 1\}$. We will exploit the linearity of the field in each region where it is continuous. Then we discuss in terms of ρ.

The fixed point for $B^T x + \mu q$ is $x_1^* = -\mu(B^T)^{-1}q$. We recall that B^T is regular due to the properties of phase type distributions. In addition, the eigenvalues of B^T are all negative (since B^T has the same eigenvalues than B). Matrix B has a negative diagonal, and the sum of all non diagonal entries per row (that are all positive) is less than or equal to the absolute value of the diagonal element. This is because the sum of each row of $\widehat{B}$ is 0 and the last column (that does not belong to B) has non-negative entries. Then, considering the field in $\mathbb{R}^n$, we have that $x_1^* = -\mu(B^T)^{-1}q$ is a global attractor. In addition, as $\mathbb{1}^T x_1^* = \rho$, we have that x_1^* is an interior point of R_1 iff $\rho < 1$, x_1^* is an interior point of R_2 iff $\rho > 1$, and $x_1^* \in H \cap \{b^T x = \mu\}$ iff $\rho = 1$. We observe that when $\rho \geq 1$, as x_1^* is the unique fixed point of $B^T x + \mu q$, and the vector field points outside R_1, the solution of $\dot{x} = B^T x + \mu q$ is pushed towards H.

Now we consider the fixed point of $(B^T + qb^T)x$ in $\{\mathbb{1}^T x = 1\}$. As $b = -B\mathbb{1}$, we have that $(B^T + qb^T)x = 0$ iff $(I - q\mathbb{1}^T)B^T x = 0$, where I is the identity matrix. As B^T is invertible, if v is an eigenvector of $q\mathbb{1}^T$ with eigenvalue 1, $x_1^* = (B^T)^{-1}v$ is a fixed point. Matrix $q\mathbb{1}^T$ has eigenvalues 0 and 1 and, as $q\mathbb{1}^T$ has range 1, the dimensions of the corresponding eigenspaces are respectively $n - 1$ and 1. Then there is a one-dimensional space with eigenvalue 1, that intersects $\{\mathbb{1}^T x = 1\}$, giving the fixed point $x_2^* = -\lambda(B^T)^{-1}q$. Since $\mathbb{1}^T x_2^* = 1$, we have that $x_2^* \in H$. In addition $x_2^* \in H \cap \{b^T x < \mu\}$ iff $\rho > 1$, $x_2^* \in H \cap \{b^T x > \mu\}$ iff $\rho < 1$ and $x_2^* \in H \cap \{b^T x = \mu\}$ iff $\rho = 1$. We also have that, restricted to H, x_2^* is a global attractor, so in the case of $\rho \leq 1$, the solution in H is pushed to the region $\{b^t x > \mu\}$, where the solution is again driven by the field $B^T x + \mu q$, so the solution that starts in H does not remain in H for $\rho < 1$. $\square$

Proof (Theorem 2). To prove the convergence in stationary regime, we use the results in [14]. In Theorem 5 in [14] it is proved convergence in stationary regime to the fixed point for piecewise smooth dynamical systems. This is an extension of a general result in [15]. The hypotheses needed are that the fixed point is a unique global attractor and regularity assumptions for the trajectories. We use Lemma 2 to characterize the fixed point. are verified in our case, where solutions are piecewise linear and the discontinuity surface H is a hyperplane. Then applying Theorem 5 in [14] we obtain Theorem 2. $\square$

References

1. Lavenberg, S.: Computer performance modeling handbook. Notes and reports in computer science and applied mathematics. Academic Press (1983)
2. Haque, L., Armstrong, M.J.: A survey of the machine interference problem. European Journal of Operational Research 179(2), 469–482 (2007)
3. Robert, P.: Stochastic Networks and Queues. Stochastic Modelling and Applied Probability Series. Springer, New York (2003)

4. Kurtz, T.G.: Representation and approximation of counting processes. In: Fleming, W.H., Gorostiza, L.G. (eds.) Advances in Filtering and Optimal Stochastic Control. LNCIS, vol. 42, pp. 177–191. Springer, Berlin (1982)

5. Iglehart, D.L., Lemoine, A.J.: Approximations for the repairman problem with two repair facilities, I: No spares. Advances in Applied Probability 5(3), 595–613 (1973)

6. Iglehart, D.L., Lemoine, A.J.: Approximations for the repairman problem with two repair facilities, II: Spares. Advances in Appl. Probability 6, 147–158 (1974)

7. Ethier, S.N., Kurtz, T.G.: Markov processes: Characterization and convergence. Wiley Series in Probability and Statistics. John Wiley & Sons Inc., New York (1986)

8. Darling, R.W.R., Norris, J.R.: Differential equation approximations for Markov chains. Probab. Surv. 5, 37–79 (2008)

9. Shwartz, A., Weiss, A.: Large deviations for performance analysis. Stochastic Modeling Series. Chapman & Hall, London (1995)

10. Ayesta, U., Erausquin, M., Jonckheere, M., Verloop, I.M.: Scheduling in a random environment: Stability and asymptotic optimality. IEEE/ACM Trans. Netw. 21(1), 258–271 (2013)

11. Ball, K., Kurtz, T.G., Popovic, L., Rempala, G.: Asymptotic analysis of multiscale approximations to reaction networks. The Annals of Applied Probability 16(4), 1925–1961 (2005)

12. Bortolussi, L.: Hybrid limits of continuous time markov chains. In: Proceedings of Eighth International Conference on Quantitative Evaluation of Systems (QEST), pp. 3–12 (September 2011)

13. Bortolussi, L., Tribastone, M.: Fluid limits of queueing networks with batches. In: Proceedings of the Third Joint WOSP/SIPEW International Conference on Performance Engineering, ICPE 2012, pp. 45–56. ACM, New York (2012)

14. Houdt, B.V., Bortolussi, L.: Fluid limit of an asynchronous optical packet switch with shared per link full range wavelength conversion. In: SIGMETRICS 2012, pp. 113–124 (2012)

15. Benaïm, M., Le Boudec, J.Y.: A class of mean field interaction models for computer and communication systems. Perform. Eval. 65(11-12), 823–838 (2008)

16. Le Boudec, J.Y.: The Stationary Behaviour of Fluid Limits of Reversible Processes is Concentrated on Stationary Points. Technical report (2010)

17. Asmussen, S., Albrecher, H.: Ruin probabilities, 2nd edn. Advanced Series on Statistical Science & Applied Probability, vol. 14. World Scientific Publishing Co. Pte. Ltd., Hackensack (2010)

18. Bortolussi, L.: Supplementary material of Hybrid Limits of Continuous Time Markov Chains, http://www.dmi.units.it/~bortolu/files/qest2011supp.pdf (2011) (accesed March 15, 2013)

Tulip: Model Checking Probabilistic Systems Using Expectation Maximisation Algorithm

Rastislav Lenhardt

Department of Computer Science, University of Oxford, United Kingdom

Abstract. We describe a novel tool for model checking ω-regular specifications on interval Markov chains, recursive interval Markov chains and interval stochastic context-free grammars. The core of the tool is an iterative expectation maximisation procedure to compute values for the unknown probabilities in a parametrised system, which maximises the probability of satisfying the specification. The tool supports specifications given as LTL formulas or unambiguous Büchi automata.

1 Introduction

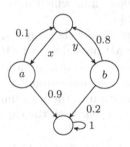

Interval Markov chain $\mathcal{M}$

Interval Markov chains (IMCs) generalise ordinary Markov chains by allowing undetermined transition probabilities that are constrained to intervals [7]. IMCs arise naturally in the modelling and verification of probabilistic systems. They are useful for modelling systems in which some transition probabilities depend on an unknown environment, are only approximately known, or are parameters that can be controlled. Consider the interval Markov chain $\mathcal{M}$ with undefined transition probabilities, represented by variables $x, y \in [0, 1]$, shown on the left. For example, we can optimise $\mathcal{M}$ with respect to the LTL formula $\varphi \stackrel{\text{def}}{=} \Diamond a \wedge \Diamond b$.

IMCs can be also seen as a type of Markov decision process. Valuations of their undetermined transition probabilities can correspondingly be seen as history-free stochastic schedulers. This enforced history-independence makes them different. Here we consider the problem of computing the maximum probability that an IMC can satisfy specification, which is given as an automaton or as a Linear Temporal Logic (LTL) formula. We consider also recursive IMCs. They are extension of ordinary recursive Markov chains [4], where we allow transition probabilities to be intervals. Since there is a straightforward translation from stochastic context-free grammar (SCFG) to recursive IMCs, we can handle also model checking SCFG with rules having interval probabilities. Popular applications of SCFG include e.g. describing the secondary structure of tRNA (see Figure 1) and natural language processing. Interval probabilities provide a more realistic model than fixed values, because the probabilities in a model are often only approximately known, being themselves obtained as a result of learning. We refer the reader for the full complexity analysis of the problem to [2].

K. Joshi et al. (Eds.): QEST 2013, LNCS 8054, pp. 155–159, 2013.

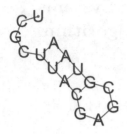

Secondary structure of RNA on the left can be represented by a string, where matching brackets indicate connections: UCGC(U(U(A(CGAGCG)U)A)A)A). We can then use LTL to query this string generated by the rules of SCFG with interval ranges. Some of them can be e.g.: $S \xrightarrow{[0.19, 0.2]} HT$, $H \xrightarrow{[0.38, 0.4]} (cHg)$, $H \xrightarrow{[0.5, 0.51]} (uHt)$, $H \xrightarrow{[0.1, 0.11]} g$, ...

Fig. 1. Secondary structure of RNA

2   Tulip

Tulip is a web application available at http://tulip.lenhardt.co.uk, along with several built-in examples and documentation. There is no installation needed. The tool accepts as input either a labelled interval Markov chain, SCFG or a labelled recursive IMC, along with properties specified either by LTL formulas or by unambiguous Büchi automata. Tulip can translate LTL formulas directly to unambiguous automata. We take advantage of this translation, which outperforms the traditional approach via deterministic automata [2].

Tulip brings a novel algorithmic approach to solving the model checking problem in practice by using the *expectation-maximisation* (EM) procedure, which is ubiquitous in machine learning. Indeed, our algorithm can be seen as a variant of the classical Baum-Welch procedure [1]. It performs a specified number of iterations of the EM algorithm, and outputs an approximation to the maximum probability with which the model satisfies the property, together with the values within the intervals for which the maximum is achieved.

EM Algorithm. We start with initial feasible values within intervals, which we plug into a cross product of chain and automaton. In each iteration, we improve these values by computing (i) the expected number of times to visit a given state before reaching an accepting strongly connected component (SCC) and (ii) the probability of reaching an accepting SCC given a starting state. This allows us to compute for each interval, the expected number of times a run takes this edge before it is clear it will be accepted. For all intervals leaving the same vertex of an IMC we set their new values to be proportional to these expectations. In case of IMCs, see [2] for more details about the update procedure, convergence, the choice of initial refinement and handling the cases, where new improved values violates the given ranges for intervals.

Optimisations. Firstly, we identify reachable states of the cross-product. Then, in the non-recursive case, we use probabilistic bisimulation and a heuristic to collapse together vertices v and v' whenever we can determine that starting in v any path will pass through v' with probability one. Finally, we use the Sparse Newton method implementation from [8] to solve systems of linear equations (or non-linear equations in the recursive case). We have chosen this method because for our type of equations it (i) converges, as shown in [4] (ii) has the best performance among many compared in [8] (iii) allows us to run the whole algorithm in linear time.

Problems Tulip Can Solve. (i) Model checking IMCs and recursive IMCs against LTL or automata (ii) Model checking ordinary Markov chains and recursive Markov chains (RMCs). Note that PReMo [8], the only tool for model-checking RMCs and SCFG, does not support intervals, nor LTL specification (iii) Finding optimal positional schedulers for MDPs and recursive MDPs given an LTL specification. Note that the latter problem is undecidable for schedulers with memory [5] (iv) Perform *model repair*, by adjusting transition probabilities to meet the specification (v) Model check stochastic context free grammars also with interval ranges for probabilities.

3   Experiments

We evaluate the performance of Tulip using a single core of 1.7 GHz Intel Core i5 CPU and show the impact of our optimisations on several examples which are described on our website. They cover a range of scenarios, including finding mixed strategies in some economic games and evaluating properties specifying competing goals. One of the larger examples we consider is a *Bounded Retransmission Protocol* [3], where we model-check the property that *the sender does not report a successful transmission*. We include also examples covering interval stochastic context-free grammars and recursive IMCs.

Examples: *Rendezvous in the Park* (R), *Predicting Football* (F), *Bounded Retransmission Protocol* (B), *Natural Language Processing* (N), *Secondary Structure of RNA* (S).

	R	F	B	N	S
Example is recursive	no	no	no	yes	yes
Size of interval Markov chain	5	22	1767	78	57
Initial automaton size	6	82	5	19	37
Automaton after bisimulation	6	29	3	8	9
Naive cross product	30	638	5301	624	513
Product with reachable states only	20	171	1767	330	459
Product after collapse	7	41	610	-	-
Product after bisimulation	6	15	544	-	-
Iterations for 5-decimal digit precision	14	12	1	1	1
Overall running time (in seconds)	0.013	0.024	1.241	0.495	0.729

For all our examples fewer than 15 iterations are sufficient to attain a precision up to five decimal places. Each iteration of our EM algorithm runs in practice in linear time in the number of vertices and edges of the cross product. It is because all parts of the algorithm run in linear time: finding reachable vertices, strongly connected components, building systems of equations and also solving iteratively these equations. It is confirmed by the example we consider in Figure 2, where we model a *Probabilistic Broadcast Protocol* [6], a synchronous variant with message collision.

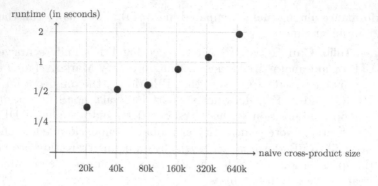

Fig. 2. Runtime of one iteration for different sizes of Probabilistic Broadcast Protocol. Note that both axes grow exponentially and that we use k to denote thousands.

4 Conclusions and Future Work

We have shown that Tulip can be used to model check temporal logic properties of many different systems. In many cases, there were no tools available before. The linear running time in the size of the cross product means that Tulip already scales well to large systems. To make it even more useful in practice and overcome memory limitations, we plan to improve the running time of the optimisations and include symbolic algorithms as in the symbolic engine of PRISM. They usually make algorithm a little slower, but much less memory is sufficient. Another extension we are working on is to support a richer set of properties. We would like to support finding the values within intervals for which the expected number of steps to reach the accepting set is minimised. That would allow us, for example, to find the optimal parameters for the runs of Las Vegas algorithms.

References

1. Baum, L.E., Petrie, T., Soules, G., Weiss, N.: A maximization technique occurring in the statistical analysis of probabilistic functions of markov chains. The Annals of Mathematical Statistics 41(1), 164–171 (1970)
2. Benedikt, M., Lenhardt, R., Worrell, J.: LTL model checking of interval markov chains. In: Piterman, N., Smolka, S.A. (eds.) TACAS 2013 (ETAPS 2013). LNCS, vol. 7795, pp. 32–46. Springer, Heidelberg (2013)
3. D'Argenio, P.R., Jeannet, B., Jensen, H.E., Larsen, K.G.: Reachability analysis of probabilistic systems by successive refinements. In: de Luca, L., Gilmore, S. (eds.) PAPM-PROBMIV 2001. LNCS, vol. 2165, pp. 39–56. Springer, Heidelberg (2001)
4. Etessami, K., Yannakakis, M.: Recursive markov chains, stochastic grammars, and monotone systems of nonlinear equations. In: Diekert, V., Durand, B. (eds.) STACS 2005. LNCS, vol. 3404, pp. 340–352. Springer, Heidelberg (2005)

5. Etessami, K., Yannakakis, M.: Recursive markov decision processes and recursive stochastic games. In: Caires, L., Italiano, G.F., Monteiro, L., Palamidessi, C., Yung, M. (eds.) ICALP 2005. LNCS, vol. 3580, pp. 891–903. Springer, Heidelberg (2005)
6. Fehnker, A., Gao, P.: Formal verification and simulation for performance analysis for probabilistic broadcast protocols. In: Kunz, T., Ravi, S.S. (eds.) ADHOC-NOW 2006. LNCS, vol. 4104, pp. 128–141. Springer, Heidelberg (2006)
7. Jonsson, B., Larsen, K.G.: Specification and refinement of probabilistic processes. In: LICS (1991)
8. Wojtczak, D., Etessami, K.: PReMo: An analyzer for probabilistic recursive models. In: Grumberg, O., Huth, M. (eds.) TACAS 2007. LNCS, vol. 4424, pp. 66–71. Springer, Heidelberg (2007)

PLASMA-lab: A Flexible, Distributable Statistical Model Checking Library

Benoît Boyer, Kevin Corre, Axel Legay, and Sean Sedwards

INRIA Rennes – Bretagne Atlantique

Abstract. We present PLASMA-lab, a statistical model checking (SMC) library that provides the functionality to create custom statistical model checkers based on arbitrary discrete event modelling languages. PLASMA-lab is written in Java for maximum cross-platform compatibility and has already been incorporated in various performance-critical software and embedded hardware platforms. Users need only implement a few simple methods in a simulator class to take advantage of our efficient SMC algorithms.

PLASMA-lab may be instantiated from the command line or from within other software. We have constructed a graphical user interface (GUI) that exposes the functionality of PLASMA-lab and facilitates its use as a standalone application with multiple 'drop-in' modelling languages. The GUI adds the notion of projects and experiments, and implements a simple, practical means of distributing simulations using remote clients.

Background and Motivation

Statistical model checking (SMC) is a form of probabilistic model checking that employs Monte Carlo methods to avoid the state explosion problem. SMC uses a number of independent simulation traces of a discrete event model to estimate the probability of a property. The traces may be generated on different machines, so SMC can efficiently exploit parallel computation (see Fig. 2). Reachable states are generated on-the-fly and the length of simulations is only weakly related to the size of the state space. Hence SMC tends to scale polynomially with respect to system description (see Fig. 1). Properties may be specified in bounded versions of the same temporal logics used in probabilistic model checking. Since SMC is thus applied to finite traces, it is also possible to use logics and functions that would otherwise be intractable or undecidable.

SMC abstracts the probabilistic model checking problem to one of estimating the parameter of a Bernoulli random variable with well defined confidence (e.g., using a Chernoff bound). The complexity of the estimation problem with respect to confidence is largely independent of the total number of possible traces. Hence SMC may also be applied to stochastic models with continuous states.

Dedicated SMC tools, such as YMER[1], VESPA, APMC[2] and COSMOS[3], have been joined by statistical extensions of established tools such as PRISM[4]

---

[1] www.tempastic.org/ymer   [2] sylvain.berbiqui.org/apmc
[3] www.lsv.ens-cachan.fr/~barbot/cosmos/   [4] www.prismmodelchecker.org

K. Joshi et al. (Eds.): QEST 2013, LNCS 8054, pp. 160–164, 2013.

and UPPAAL[5]. In the case of UPPAAL-SMC, this has required the definition of stochastic timed semantics. The tool MRMC[6] has both numerical and statistical functionality, but takes as input a low level textual description of a Markov chain. Many other tools are available or under development, with most using a single high level modelling language related to a specific semantics. Our previous tool [2] suffered the same limitation, prompting us to develop a radically new tool with modular architecture.

PLASMA-lab

PLASMA-lab [3] is an efficient SMC library written in Java, featuring a customisable simulator class. This allows SMC functionality to be added to existing domain-specific modelling platforms, such as DESYRE[7], and allows rapid prototyping of formal verification solutions using, e.g., Scilab[8] and MATLAB[9]. High performance standalone model checkers can also be constructed with PLASMA-lab by including a suitable language parser in the simulator class. PLASMA-lab's integrated development environment facilitates distributed simulation and can work with multiple user-defined language plug-ins.

Properties. PLASMA-lab accepts properties described in a form of bounded linear temporal logic (BLTL) extended with custom temporal operators based on concepts such as *minimum*, *maximum* and *mean* of a variable over time.

Model Checking Modes. PLASMA-lab offers three basic modes of model checking: simple Monte Carlo, Monte Carlo using a Chernoff confidence bound and sequential hypothesis testing. There is also a simulation mode for debugging. Rare event model checking modes, such as importance sampling and importance splitting, can be implemented as part of the simulator class when the modelling semantics support them.

- Monte Carlo: the user explicitly specifies the number of simulations that PLASMA-lab must use to estimate the probability of a property.
- Chernoff: the user specifies an absolute error ε and a probability δ. PLASMA-lab calculates the number of simulations required to ensure that the resulting estimate is within $\pm\varepsilon$ of the correct value with minimum probability δ.
- Sequential: PLASMA-lab adopts the sequential hypothesis ratio test of [4] to verify that the probability of a property is above a user-specified threshold. The user also specifies a level of indifference and parameters to control errors of Types I and II. The number of simulations is not specified a priori: simulations are performed as necessary. See [4] for details.

---

[5] www.uppaal.org   [6] www.mrmc-tool.org   [7] www.ales.eu.com   [8] www.scilab.org
[9] www.mathworks.com

Usage. PLASMA-lab may be invoked from the command line or embedded in other software as a library. PLASMA-lab is provided as a pre-compiled jar file (plasmalab.jar) and a source template (Simulator.java) to create the simulator class. The minimum requirement is to implement the methods newTrace() and nextState(), that initiate a new simulation and advance the simulation by one step, respectively. Language parsers are typically invoked in the constructor.

Graphical User Interface. The GUI provides an integrated development environment (IDE) to facilitate the use of PLASMA-lab as a standalone statistical model checker with multiple 'drop-in' modelling languages. To demonstrate this, we have included a biochemical language and a language based on reactive modules. The website [3] includes other examples. The GUI implements the notion of a project file, that links the description of a model to a specific modelling language simulator and a set of associated properties and experiments. The GUI also provides 2D and 3D graphical output of results and implements a distributed algorithm that will work with any of its associated modelling languages.

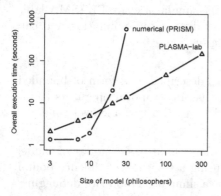

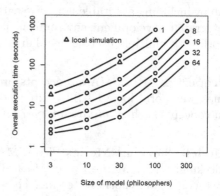

Fig. 1. Exponential scaling of numerical model checking vs. linear scaling of PLASMA-lab SMC, considering a fairness property of the probabilistic dining philosophers protocol.

Fig. 2. Scaling of PLASMA-lab distributed algorithm applied to dining philosophers. Numbers are quantity of simulation nodes. Local simulation scaling is shown for reference.

Distributed Algorithm. The administrative time needed to distribute SMC on parallel computing architectures is often a deterrent. To overcome this, the PLASMA-lab GUI implements a simple and robust client-server architecture, based on Java Remote Method Invocation (RMI) using IPv4/6 protocols. The algorithm will work on dedicated clusters and grids, but can also take advantage of ad hoc networks of heterogeneous computers. The minimum requirement is that the IP address of the GUI is available to the clients. PLASMA-lab implements the SMC distribution algorithm of [4], which avoids the statistical bias that might otherwise occur from load balancing. Distributed performance in illustrated in Fig. 2. The user selects the distributed mode via the GUI and publishes the IP address of the instance of PLASMA-lab GUI that is acting as server. Clients (instances of the PLASMA-lab service application) willing to participate respond by sending a message to the published IP address. The server sends an encapsulated version of the model and property to each of the participating clients, which then wait to be told how many simulations to perform. When sufficient clients are available, the user initiates the analysis by causing the server to broadcast the simulation requirements to each client.

Applications

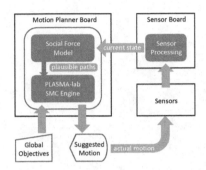

PLASMA-lab has been applied to problems from, e.g., systems biology, rare events, performance, reliability, motion planning and systems of systems [3]. PLASMA-lab is the focus of ongoing collaborations with companies Dassault, Thales, IBM, and EADS. PLASMA-lab is also used by several European projects. The following examples relate to the DALi[10] and DANSE[11] projects.

Fig. 3. Control loop of DALi motion planner

Motion Planning. PLASMA-lab is used by the DALi project in a novel motion planning application of SMC. DALi aims to develop an autonomous device to help those with impaired ability to negotiate complex crowded environments (e.g. shopping malls). High level constraints and the objectives of the user are expressed in temporal logic, while low level behaviour is predicted by the 'social force model' [1].

PLASMA-lab was integrated with MATLAB to develop the prototype algorithm. The final version is implemented directly in C on embedded hardware and finds the optimum trajectory in a fraction of a second. PLASMA-lab improves the social force model's ability to avoid collisions by a factor of five.

Systems of Systems. The DANSE project is concerned with the design and analysis of 'systems of systems' (SoS). SoS feature a dynamicity of configurations that introduces significant additional complexity (the state and state space of the model are not necessarily known a priori). PLASMA-lab is now an integral part of the DANSE software platform, using a Simulator class that wraps the DESYRE[12] hybrid simulation engine to make dynamicity transparent to SMC.

References

1. Helbing, D., Molnár, P.: Social force model for pedestrian dynamics. Phys. Rev. E 51, 4282–4286 (1995)

10 www.ict-dali.eu   11 www.danse-ip.eu   12 www.ales.eu.com

2. Jegourel, C., Legay, A., Sedwards, S.: A Platform for High Performance Statistical Model Checking – PLASMA. In: Flanagan, C., König, B. (eds.) TACAS 2012. LNCS, vol. 7214, pp. 498–503. Springer, Heidelberg (2012)
3. PLASMA-lab project page, https://project.inria.fr/plasma-lab/
4. Younes, H.L.S.: Verification and planning for stochastic processes with asynchronous events. PhD thesis, Carnegie Mellon University (2005)

STRONG: A Trajectory-Based Verification Toolbox for Hybrid Systems

Yi Deng[1], Akshay Rajhans[2], and A. Agung Julius[1]

[1] ECSE Department, Rensselaer Polytechnic Institute
[2] ECE Department, Carnegie Mellon University

Abstract. We present STRONG, a MATLAB toolbox for hybrid system verification. The toolbox addresses the problem of reachability/safety verification for bounded time. It simulates a finite number of trajectories and computes robust neighborhoods around their initial states such that any trajectory starting from these robust neighborhoods follows the same sequence of locations as the simulated trajectory does and avoids the unsafe set if the simulated trajectory does. Numerical simulation and computation of robust neighborhoods for linear dynamics scale well with the size of the problem. Moreover, the computation can be readily parallelized because the nominal trajectories can be simulated independently of each other. This paper showcases key features and functionalities of the toolbox using some examples.

1 Introduction

The problem of safety verification using reachability analysis, i.e., finding out whether the trajectories of a system reach a goal set and/or avoid an unsafe set, has received a lot of attention particularly in the hybrid systems community. The different approaches from the literature can be roughly classified into two types: state-space exploration techniques and construction of certificate-based guarantees. Despite recent progress, the applicability of these formal techniques still remains limited due the state-explosion problem and due to challenges in coming up with the right certificates necessary. On the other hand, in practice, simulation remains a widely-used approach for analyzing systems despite it being incomplete and informal. To bridge the divide between simulation and verification, tools that combine simulation with some formal analysis are recently being developed [3,2]. In the similar spirit, we have been developing STRONG (System Testing using RObust Neighborhood Generation)[1], a Matlab toolbox for trajectory-based reachability/safety verification of hybrid systems.

Our approach combines simulation and formal verification. By simulating trajectories from a finite number of initial points within a compact set of initial conditions, we can obtain reachability and safety properties for the entire set of initial conditions [5].

---

[1] The toolbox and the supporting examples can be downloaded at http://dengy3.myrpi.org/strong.html . Preliminary work on the tool was done at University of Pennsylvania as a part of the masters thesis [6].

K. Joshi et al. (Eds.): QEST 2013, LNCS 8054, pp. 165–168, 2013.
© Springer-Verlag Berlin Heidelberg 2013

2   Features and Functionalities

Model Consistency Checking. The toolbox has the ability to detect and correct certain kinds of ill-posedness in the model. It tracks the validity of state flow to detect common mistakes such as the reset state after a discrete transition falling out of the invariant of the new location. It can also detect a particular case of Zeno behaviors (infinite jumps in finite time between two neighboring locations), and correct the model by replacing such transitions with sliding modes.

Trajectory Simulation. Simulating a trajectory for a given initial condition is one of the main functionalities of the toolbox and forms a basis for the verification. The toolbox uses MATLAB's ode45 solver as a default for numerical integration. For every trajectory, the tool gathers all the information including the continuous evolution, transition events (e.g. unsafe), and event times. Trajectories of linear as well as nonlinear dynamics can be simulated.

Robustness Computation. The trajectory robustness and upper/lower time difference bounds between a simulated trajectory and its neighbors can be computed automatically for each continuous segment within every discrete location visited by the simulated trajectory. For linear dynamics, the computation involves solving a Lyapunov equation, for which we use standard convex optimization tools. Automatic verification of nonlinear systems is still under development. A trajectory and its robust ball of initial states can be visualized for any two specified dimensions.

Initial Set Coverage. Using the robustness analysis of a single trajectory, we can ascertain that the portion of the initial set covered by a robust ball around the chosen initial state leads to trajectories with the same safety and reachability properties as the simulated one. The final goal is to cover a given compact initial set as much as possible using simulated trajectories and their robust neighborhoods. Doing this in an effective way involves smartly choosing initial states for the simulated trajectories and assessing current coverage. Currently, the coverage strategy implemented is to generate random points as initial states, and an unbiased estimator [1] is used to evaluate the percentage of the covered initial set, which has a precision independent of the dimension of the state space.

The procedure of property verification for an initial set is parallelized in the tool. As shown in the adjacent figure, trajectory simulations and robustness computation, which form the majority of the computation, can be performed independently and the initial set can be covered in a highly parallel manner.

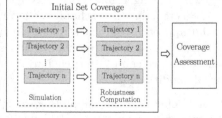

3 Examples

Demos of all examples can be found in the toolbox. Readers are recommended to view them for specific usage of commands. Here we present the examples briefly, and summarize the results as well as the performance of the tool in a table. For details, we refer the reader to the tool user guide available on the tool webpage.

Navigation Benchmark. Consider the navigation benchmark problem from [4,5]. The system state vector x is comprised of position variables (x_1, x_2) and velocity variables (v_1, v_2). As shown in the plot, a simulated trajectory reaches four locations $(\ell_2, \ell_5, \ell_2, \ell_3)$ and no unsafe state ever reached. Any initial state within the robust ball leads to a safe trajectory that will reach $(\ell_2, \ell_5, \ell_2, \ell_3)$ with upper and lower bounds on each transition time.

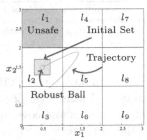

Automotive Cruise Control. In the automotive cruise control example from [7], v is the vehicle's velocity and r is the distance to another vehicle. The original dynamics has chattering, so we invoke our model consistency checking feature to automatically incorporate sliding modes. After that, normal verification can be performed. As shown in the plot, part of the simulated trajectory is along a guard, where we have inserted a location with sliding dynamics.

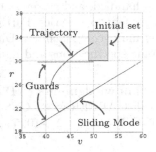

A High-dimensional Example. To demonstrate the scalability of our tool with the system dimension, we use a finite-element model to describe the heat-flow phenomenon along a rod as a 60th-order differential equation. A critical element is to be protected from being under- or over-heated ($\neg(10 \leq T(C) \leq 30)$) by injecting a hot/cold flux into the ends of the rod.

We simulate and verify the 60-dimensional system under two different initial conditions. For the first case shown in the adjacent figure, although several elements start from a relatively high temperature, safety can been maintained as the maximum $T(C)$, which occurs at time $t = 32$ is less than the unsafe threshold. On the other hand, in the second case depicted, the higher (worse) initial temperature of some elements results in unsafe temperature $T(C)$ at time $t = 33$. Properties of two different initial sets are verified as in the summary table.

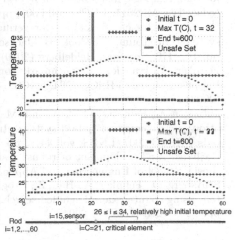

problem	Navigation Benchmark		Cruise Control	Heat Flow	
end time	3		20	600	
dimension	4		2	60	
initial set	$0.3 \leq x_1 \leq 0.7$ $1.3 \leq x_2 \leq 1.7$ $-2.1 \leq v_1 \leq -1.9$ $-2.1 \leq v_2 \leq -1.9$	$0.3 \leq x_1 \leq 0.7$ $1.3 \leq x_2 \leq 1.7$ $1.9 \leq v_1 \leq 2.1$ $0.9 \leq v_2 \leq 1.1$	$49 \leq x_1 \leq 53$ $30 \leq x_2 \leq 35$	For $26 \leq i \leq 34$, $35.8 \leq x_i \leq 36.2$; otherwise, $x_i = 27$.	For $26 \leq i \leq 34$, $39.8 \leq x_i \leq 40.2$; otherwise, $x_i = 27$.
results[1]	t 0.7s / # traj. 1 / cvg 100%	unsafe[2]	t 75.4s / # traj. 1500 / cvg 100%	t 60.1s / # traj. 50 / cvg 100%	unsafe[2]

[1] t = computation time on a 3.40 GHz Inter Xeon CPU, 16GB RAM, 4 cores; # traj. = number of trajectories tested; cvg = coverage assessment ($\pm 2\%$, $Pr > 99\%$).

[2] An unsafe trajectory is detected. The initial set cannot have uniform reachability and safety property.

To summarize, the STRONG toolbox is developed for bounded time reachability and safety verification of hybrid systems. Based on the idea of robust test generation and coverage, the tool computes a mathematically proven bound on the trajectory divergence and provides formal verification for the covered initial states. The tool does not use gridding; high-dimensional problems can be handled, and systems that are robustly safe can be verified with potentially very few trajectories. Further speed up can be achieved by using parallelization on multi-core machines. Directions for future work include supporting temporal logic specifications and handling stochastic system models.

Acknowledgments. YD and AAJ would like to acknowledge the support of NSF CAREER grant CNS-0953976. AR would like to acknowledge the support of NSF grants CNS-1035800 and CCF-0926181.

References

1. Afshari, S.: Coverage assessment criteria for approximate bisimulation theory and introduction of computer games in hybrid systems safety/reachability design. Master's thesis, Rensselaer Polytechnic Institute, NY (2010)
2. Annpureddy, Y., Liu, C., Fainekos, G., Sankaranarayanan, S.: S-TaLiRo: A tool for temporal logic falsification for hybrid systems. In: Abdulla, P.A., Leino, K.R.M. (eds.) TACAS 2011. LNCS, vol. 6605, pp. 254–257. Springer, Heidelberg (2011)
3. Donzé, A.: Breach, A toolbox for verification and parameter synthesis of hybrid systems. In: Touili, T., Cook, B., Jackson, P. (eds.) CAV 2010. LNCS, vol. 6174, pp. 167–170. Springer, Heidelberg (2010)
4. Fehnker, A., Ivančić, F.: Benchmarks for hybrid systems verification. In: Alur, R., Pappas, G.J. (eds.) HSCC 2004. LNCS, vol. 2993, pp. 326–341. Springer, Heidelberg (2004)
5. Julius, A.A., Fainekos, G.E., Anand, M., Lee, I., Pappas, G.J.: Robust test generation and coverage for hybrid systems. In: Bemporad, A., Bicchi, A., Buttazzo, G. (eds.) HSCC 2007. LNCS, vol. 4416, pp. 329–342. Springer, Heidelberg (2007)
6. Rajhans, A.: Development of robust testing toolbox for hybrid systems. Master's thesis, School of Engineering and Applied Science, Univ. of Pennsylvania (2007)
7. Stursberg, O., Fehnker, A., Han, Z., Krogh, B.H.: Verification of a cruise control system using counterexample-guided search. In: Control Engineering Practice, vol. 12, pp. 1269–1278 (October 2004)

PEPERCORN: Inferring Performance Models from Location Tracking Data

Nikolas Anastasiou and William Knottenbelt

Department of Computing
Imperial College London
South Kensington Campus
London SW7 2AZ
{na405,wjk}@doc.ic.ac.uk

Abstract. Stochastic performance models are widely used to analyse the performance of systems that process customers and resources. However, the construction of such models is traditionally manual and therefore expensive, intrusive and prone to human error. In this paper we introduce PEPERCORN, a Petri Net Performance Model (PNPM) construction tool, which, given a dataset of raw location tracking traces obtained from a customer-processing system, automatically formulates and parameterises a corresponding Coloured Generalised Stochastic Petri Net (CGSPN) performance model.

Keywords: Performance Modelling, Location Tracking, Data Mining, Coloured Generalised Stochastic Petri Nets.

1 Introduction

Performance modelling and analysis facilitates the understanding of customer and resource flow in complex physical customer-processing systems, such as hospitals, airports and car assembly lines. The accurate formulation and parameterisation of a performance model is critical to the validity of subsequent analysis. Yet the construction of such a model usually requires the availability of large amounts of data. Current data-gathering techniques, such as time and motion studies, involve tedious manual tasks that are not only time consuming, but may also be inaccurate and disrupt the system's natural flow.

The increasing adoption of real time location systems (RTLSs) has led to an abundance of low-level data describing the fine grained flow of customers and resources in customer-processing systems. In this paper, we introduce PEPERCORN, a tool which exploits the availability of such data in order to automatically infer and construct the PNPM of the underlying system and thus, provide insights regarding the system's high-level operations and performance. Constructed models can be visualised and/or analysed in PIPE2, the platform-independent Petri Net editor [4].

K. Joshi et al. (Eds.): QEST 2013, LNCS 8054, pp. 169–172, 2013.

2 PEPERCORN

PEPERCORN is a Java-based implementation of our earlier work [1–3] which presented a methodology, based on a four-stage data processing pipeline (cf. Figure 1), that allows the automated construction of CGSPN performance models from high-precision location tracking data. Key assumptions include static service areas with single-server service semantics and random service discipline.

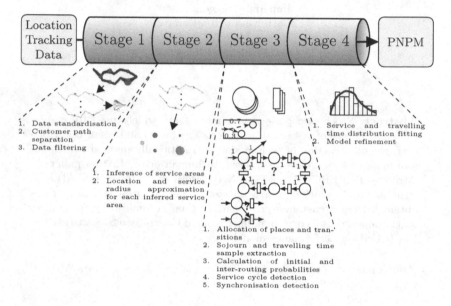

Fig. 1. The four-stage data processing pipeline that is implemented by PEPERCORN

The basic actions provided by PEPERCORN's main menu bar are shown in Figure 2. In order to construct a PNPM using PEPERCORN, the user first opens a file containing raw location tracking updates retrieved from a particular customer-processing system. PEPERCORN currently supports location tracking data obtained from a Ubisense UWB-based RTLS and synthetic data generated by the location-aware queueing network simulator LocTrackJINQS [6]. This data is a stream of tuples of the form (tag id, type, x, y, time, stderr). The tag id field denotes the monitoring tag's unique identifier and type contains the tag's category, e.g. doctor, patient, etc. In the case of multiple customer classes, type can also be used to specify the customer class that the tag belongs to. time denotes the timestamp of the location update, i.e. the time when the location update was recorded by the RTLS, and x, y specify the location of the tag at that particular instance. stderr is the expected deviation between the tag's recorded location and actual location.

Once a file has been imported, the user can initiate the data processing pipeline through the 'Process File' toolbar button. The user may wish to adjust some of the pipeline's default parameters or disable the synchronisation

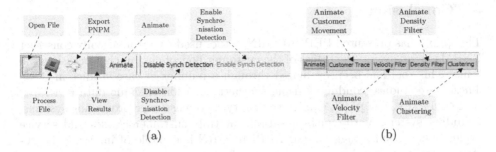

(a)   (b)

Fig. 2. Figure 2(a) shows the basic actions provided by PEPERCORN's toolbar. Figure 2(b) shows additional animation actions, once the 'Animate' button is pressed.

detection mechanism before processing commences. A user can also animate the processing phases of the second pipeline stage (cf. Figure 2(b)).

During data processing only one mandatory input is required: the *Eps* value used by the DBSCAN [5] clustering algorithm (cf. Figure 3(a)). This value defines the area of the neighbourhood around each point in the dataset for which the density is measured[1]. When processing is completed, the user can export the constructed PNPM as an XML file (a custom variation of PNML) so it can be visualised and/or analysed in PIPE2 (cf. Figure 3(b)). PEPERCORN also allows users to examine key quantitative results, such as the service and travelling time distribution fits (obtained by interfacing with the G-FIT tool [7]), the compatibility of the extracted time samples with the fitted distribution, and the inferred service area locations.

(a)   (b)

Fig. 3. Figure 3(a) shows the *Eps* selection dialog. Figure 3(b) shows the constructed PNPM as visualised in PIPE2 (in compact transition form).

---

[1] DBSCAN requires a second parameter in order to determine density-connected points: the value of the minimum number of points (*MinPts*) that must lie within the neighbourhood (defined by *Eps*) of each point in the dataset. However, we set the value of *MinPts* to be equal to four in all cases after the suggestion of [5].

3   Conclusion

This paper has presented PEPERCORN, a tool used to automatically construct PNPMs by analysing the traces of the customer flow of customer-processing systems. This tool has been evaluated through a number of case studies in [1–3]. These case studies, conducted using synthetic location tracking data generated by LOCTRACKJINQS [6], employ several types of customer-processing systems, including systems with synchronisation, multiple customer classes and service cycles. Their results suggest that PEPERCORN is capable of inferring the abstract structure, stochastic features and high-level customer flow of complex systems, at least when synthetic location tracking data is used.

The constructed models can be used to provide insights into the system's performance through the computation of end-to-end response time distributions and to identify bottlenecks not likely to be discovered by a manual process. At this stage PEPERCORN is particularly suitable for small-scale indoor customer-processing systems characterised by complex processes which are difficult to capture via manually collected data.

References

1. Anastasiou, N., Horng, T.-C., Knottenbelt, W.: Deriving Generalised Stochastic Petri Net performance models from High-Precision Location Tracking Data. In: Proc. 5th Intl. Conference on Performance Evaluation Methodologies and Tools, VALUETOOLS 2011 (2011)
2. Anastasiou, N., Knottenbelt, W.: Deriving Coloured Generalised Stochastic Petri Net Performance Models from High-Precision Location Tracking Data. In: Proc. 4th ACM/SPEC International Conference on Performance Engineering (2013)
3. Anastasiou, N., Knottenbelt, W., Marin, A.: Automatic Synchronisation Detection in Petri Net Performance Models Derived from Location Tracking Data. In: Thomas, N. (ed.) EPEW 2011. LNCS, vol. 6977, pp. 29–41. Springer, Heidelberg (2011)
4. Dingle, N.J., Knottenbelt, W.J., Suto, T.: PIPE2: A tool for the Performance Evaluation of Generalised Stochastic Petri Nets. ACM SIGMETRICS Performance Evaluation Review 36(4), 34–39 (2009)
5. Ester, M., Kriegel, H.-P., Sander, J., Xu, X.: A density-based algorithm for discovering clusters in large spatial databases with noise. In: Proc. 2nd Intl. Conf. on Knowledge Discovery and Data Mining, KDD 1996 (1996)
6. Horng, T.-C., Anastasiou, N., Knottenbelt, W.: LocTrackJINQS: An Extensible Location-aware Simulation Tool for Multiclass Queueing Networks. In: Proc. 5th Intl. Workshop on Practical Applications of Stochastic Modelling (2011)
7. Thümmler, A., Buchholz, P., Telek, M.: A Novel Approach for Phase-Type Fitting with the EM Algorithm. IEEE Transactions on Dependable and Secure Computing 3, 245–258 (2005)

ADTool: Security Analysis
with Attack–Defense Trees*

Barbara Kordy, Piotr Kordy, Sjouke Mauw, and Patrick Schweitzer

University of Luxembourg, SnT
{barbara.kordy,piotr.kordy,sjouke.mauw,patrick.schweitzer}@uni.lu

Abstract. ADTool is free, open source software assisting graphical modeling and quantitative analysis of security, using attack–defense trees. The main features of ADTool are easy creation, efficient editing, and automated bottom-up evaluation of security-relevant measures. The tool also supports the usage of attack trees, protection trees and defense trees, which are all particular instances of attack–defense trees.

1 Background and Motivation

Attack–defense trees (ADTrees) extend and improve the well-known formalism of attack trees, by including not only the actions of an attacker, but also possible counteractions of a defender. Since interactions between an attacker and a defender are modeled explicitly in ADTrees, the extended formalism allows for a more thorough and accurate security analysis compared to regular attack trees. This paper presents ADTool software [7] which supports quantitative and qualitative security assessment using attack–defense trees.

Theoretical foundations of the ADTree methodology, including a graphical and a term-based syntax as well as numerous formal semantics, have been introduced in [6]. A mathematical framework for quantitative evaluation of ADTrees is based on the notion of attributes, which allow us to formalize and specify relevant security metrics. Standard quantitative analysis of ADTrees relies on a step-wise computation procedure. Numerical values are assigned to all atomic actions, represented by the non-refined nodes. The values for the remaining nodes, including the root of the tree, are deduced automatically in a bottom-up way. This bottom-up algorithm makes use of attribute domains which specify operators to be used while calculating values for different node configurations.

The practical use of the ADTree methodology requires dedicated tool support. Lack of such support may result in numerous modeling difficulties and computational errors. On the one hand, there exist a number of commercial software applications for attack tree-like modeling, including SecurITree[1] and

---

* The research leading to the results presented in this work received funding from the Fonds National de la Recherche Luxembourg under the grants C08/IS/26 and PHD-09-167 and the European Commission's Seventh Framework Programme (FP7/2007-2013) under grant agreement number 318003 (TREsPASS).

[1] http://www.amenaza.com/

K. Joshi et al. (Eds.): QEST 2013, LNCS 8054, pp. 173–176, 2013.

AttackTree+[2]. However, these are closed source tools and their use is not free of charge. On the other hand, existing academic software, such as SeaMonster[3], does not support quantitative analysis and uniformly integrated defenses.

The above observations motivated the development of ADTool, which

- is a free and open source application supporting qualitative and quantitative analysis of tree-based models integrating attack and defense components;
- is based on well-founded formal framework;
- guides the user in constructing well-formed and visually appealing models;
- facilitates sharing, management and updating of the models;
- automates computation of security related parameters.

This paper provides a brief overview of the main features and practical capabilities of ADTool. For a more detailed description we refer the reader to an extended and illustrated version of this article [5] and to the ADTool manual available at http://satoss.uni.lu/software/adtool/manual.pdf.

2 Main Features of ADTool

ADTool is guiding the user in constructing models that comply with the graphical ADTree language. All options that allow to modify or refine the models can be accessed via a user-friendly GUI of the application.

ADTool uses an improved version of Walker's algorithm [2] to produce trees having an appealing layout. Furthermore, when an ADTree is built, the corresponding attack–defense term (ADTerm), is immediately displayed. ADTerms form a compact, algebraic representation of ADTrees. The shortest tree edit distance algorithm [3] implemented in ADTool ensures that when an ADTerm is modified, the corresponding ADTree is adapted accordingly.

ADTool provides advanced features for model manipulation and management. Folding, expanding and zooming options make the analysis of large models possible. Temporarily hiding parts of a tree permits users to focus on the displayed components. This is highly appreciated during industrial meetings and presentations. ADTrees created with ADTool can be saved as special .adt files, which enables their reuse and modification. Models can also be exported to vector graphics files (pdf), raster graphics files (png, jpeg) and LaTeX files (tex). Resulting figures can be used as illustrations in presentations, research papers and posters. A dedicated option makes it possible to print trees on a specified number of pages, which enhances readability of large-scale models.

The bottom-up algorithm for evaluation of attributes on ADTrees has been implemented in ADTool. Supported measures include: attributes based on real values (e.g., time, cost, probability), attributes based on levels (e.g., required skill level, reachability of the goal in less than k units of time), and Boolean properties (e.g., satisfiability of a scenario). The implemented measures can be

---

[2] http://www.isograph-software.com/2011/software/attacktree/
[3] http://sourceforge.net/projects/seamonster/

computed from the point of view of an attacker (e.g., the cost of an attack), of a defender (e.g., the cost of defending a system), or relate to both of them (e.g., overall maximum power consumption). Using different attribute domains allows us to distinguish between actions executed sequentially or in parallel.

After a user selects an attribute, the tool decorates the ADTree with default values representing the worst case scenario, e.g., infinite cost or maximal required skill level. The user then customizes the inputs for the relevant non-refined nodes and the linear bottom-up algorithm computes the values of the remaining nodes. Input values can be modified directly on the tree or using an overview table which is particularly helpful in case of large models. The tool ensures that the provided values are consistent and belong to a specified value domain. This is especially important when several specialists supply values for different parts of the tree.

The tool has been extensively tested and has proven to be able to easily handle realistic models containing a few thousand nodes. The computations using ADTool are performed instantaneously. The limiting factor is the graphical display of ADTrees. For trees of more than ten thousand nodes, a delay of about five seconds occurs when a new node is added. This is due to the recalculation of the positions of some nodes.

3   Implementation Characteristics

The application has been written in a modular way with a clear distinction between the GUI and the Implementation Model. An overview of the ADTool architecture is depicted in Figure 1. The Implementation Model consists of the

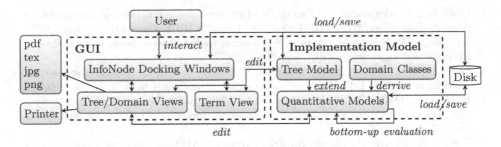

Fig. 1. An overview of the ADTool architecture

Tree Model (which stores the basic tree structure), Domain Classes (defining the implemented attribute domains), and Quantitative Models (which are derived from Domain Classes and contain inserted and computed values). The functionality of the tool can easily be extended by defining new attributes. For this purpose, a new Domain Class needs to be created and compiled. Domain Classes have been designed to be simple, in order to make it possible for a user with minimal knowledge of Java to add a new domain. Due to the use of Java reflection, no recompilation or other modifications of the program are required after adding a new Domain Class.

ADTool runs on all common operating systems (Windows, Linux, Mac OS). The program is written in Java and it depends on the following free libraries: abego TreeLayout[4], implementing an efficient and customizable tree layout algorithm in Java, and InfoNode Docking Windows[5], a pure Java Swing based docking windows framework, allowing to set up windows in a flexible way and to save and restore their layout. ADTool is available for download and as an online application at http://satoss.uni.lu/software/adtool/.

4   Conclusion and Future Work

ADTool provides security consultants as well as academic researchers with a rigorous but user-friendly application that supports security analysis using ADTrees. It integrates two crucial modeling aspects: the creation of security models and their quantitative analysis. From a formal perspective, attack trees [8], protection trees [4], and defense trees [1] are instances of ADTrees. Thus, ADTool can also be employed to automate and facilitate the usage of all these formalisms.

We are currently working on combining the ADTree methodology with Bayesian Networks, to make probabilistic reasoning about scenarios involving dependent actions possible. Related theoretical findings and newly identified features will be implemented in the next versions of ADTool.

References

1. Bistarelli, S., Fioravanti, F., Peretti, P.: Defense Trees for Economic Evaluation of Security Investments. In: ARES 2006, pp. 416–423. IEEE Computer Society (2006)
2. Buchheim, C., Jünger, M., Leipert, S.: Drawing rooted trees in linear time. Software: Practice and Experience 36(6), 651–665 (2006)
3. Demaine, E.D., Mozes, S., Rossman, B., Weimann, O.: An Optimal Decomposition Algorithm for Tree Edit Distance. ACM Trans. Algorithms 6(1), 2:1–2:19 (2009)
4. Edge, K.S., Dalton II, G.C., Raines, R.A., Mills, R.F.: Using Attack and Protection Trees to Analyze Threats and Defenses to Homeland Security. In: MILCOM, pp. 1–7. IEEE (2006)
5. Kordy, B., Kordy, P., Mauw, S., Schweitzer, P.: ADTool: Security Analysis with Attack–Defense Trees (Extended Version). CoRR abs/1305.6829 (2013)
6. Kordy, B., Mauw, S., Radomirović, S., Schweitzer, P.: Attack–Defense Trees. Journal of Logic and Computation pp. 1–33 (2012), http://logcom.oxfordjournals.org/content/early/2012/06/21/logcom.exs029.short?rss=1
7. Kordy, P., Schweitzer, P.: ADTool (2012), http://satoss.uni.lu/software/adtool
8. Mauw, S., Oostdijk, M.: Foundations of Attack Trees. In: Won, D., Kim, S. (eds.) ICISC 2005. LNCS, vol. 3935, pp. 186–198. Springer, Heidelberg (2006)

---

[4] http://code.google.com/p/treelayout/
[5] http://www.infonode.net/index.html?idw

SAT-Based Analysis and Quantification of Information Flow in Programs

Vladimir Klebanov[1], Norbert Manthey[2], and Christian Muise[3]

[1] Karlsruhe Institute of Technology (KIT)
Am Fasanengarten 5, 76131 Karlsruhe, Germany
klebanov@kit.edu
[2] Knowledge Representation and Reasoning Group
Technische Universität Dresden, 01062 Dresden, Germany
norbert@iccl.tu-dresden.de
[3] Department of Computer Science
University of Toronto, Toronto, Canada
cjmuise@cs.toronto.edu

Abstract. Quantitative information flow analysis (QIF) is a portfolio of security techniques quantifying the flow of confidential information to public ports. In this paper, we advance the state of the art in QIF for imperative programs. We present both an abstract formulation of the analysis in terms of verification condition generation, logical projection and model counting, and an efficient concrete implementation targeting ANSI C programs. The implementation combines various novel and existing SAT-based tools for bounded model checking, #SAT solving in presence of projection, and SAT preprocessing. We evaluate the technique on synthetic and semi-realistic benchmarks.

1 Introduction

Quantitative information flow analysis (QIF) is a collection of techniques for security assessment of software. The research in QIF is motivated by the observation that it is not feasible to completely prevent information leaks (i.e., the flow of confidential information to public ports) in realistic systems. Instead, practical security analysis demands a measure of leaked information in order to decide if a leak is tolerable.

QIF techniques have been applied to a variety of problems. Deciding whether a PIN generation algorithm produces PINs that are hard to guess [1], or whether a particular image transformation is a secure anonymization mechanism (cf. Figure 1) [17] are examples of QIF applications. While the information-theoretical foundations of QIF in deterministic programs are relatively well-understood, practical analysis techniques and tools are still under development.

So far, QIF analyses have been typically described *operationally*, i.e., with focus on algorithm development. One contribution of this paper is an *abstract* formulation, describing a whole class of QIF analyses in terms of verification

K. Joshi et al. (Eds.): QEST 2013, LNCS 8054, pp. 177–192, 2013.

condition generation, logical projection (most notably), and model enumeration/counting. This view facilitates understanding and comparison of existing approaches and better connects QIF to the existing body of work in these areas.

Inspired by this connection is another contribution of this paper: a novel combination of SAT preprocessing, projection, and counting, resulting in a QIF analysis that is more efficient than its predecessors. Our toolchain for analysis of C programs consists of an off-the-shelf bounded model checker CBMC [6], a propositional formula preprocessor that we developed previously, and three tools for propositional model enumeration/counting under projection that we developed (resp. extended) for this paper.

The discerning properties of our analysis are: (1) The implemented analysis is general-purpose, i.e., it is not taylored to a particular software application domain. No restrictions on the shape of the indistinguishability relation are posed (in contrast to [2,14,13]). The implementation supports almost all of ANSI C by virtue of using CBMC. (2) The analysis is not compositional, but it is fully automated—the only required user input is the program under analysis. Loops are handled by bounded unwinding, which is computationally expensive, but fully automatic and complete (with unwinding assertion checking). (3) The analysis supports measuring both the conditional min-entropy (counting the number of program outputs) and conditional Shannon entropy (counting output preimage sizes). (4) The analysis is, conceptually, sound and precise (in contrast to [17,18,21]). Of course, QIF is a hard problem, so computational constraints may force the user to settle for merely deriving (more or less tight) leak bounds as program complexity increases. (5) The analysis outpferforms comparable previous approaches both for theoretical reasons (e.g., it avoids computationally expensive program self-composition used in [11,12,2,13]) and practically due to the use of a number of well-connected novel and existing state-of-the-art techniques and tools for propositional reasoning.

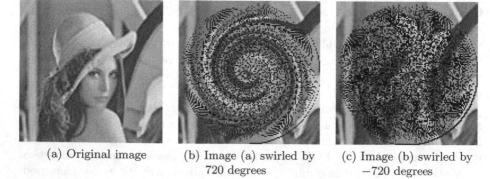

(a) Original image   (b) Image (a) swirled by   (c) Image (b) swirled by
                              720 degrees   −720 degrees

Fig. 1. Anonymization by image swirling. Details in Section 6.3.

2   QIF Basics and Technical Preliminaries

Programs, States, and Transition Relation. A *program state* is a semantical structure assigning values to mutable program vocabulary of a program p. Let S be the set of all (program) states for p. A program p induces a *transition relation* $\rho_p \subseteq S \times S$ on states as follows: $(s, s') \in \rho_p$ iff p started in state s terminates in a state s'. A security analysis may sometimes wish to focus on a particular set of initial states $S_I \subseteq S$. In this case $\rho_p \subseteq S_I \times S$.

We only consider programs that are written in a deterministic (read: sequential) programming language and are terminating, i.e., we require that ρ_p is a total function. The termination requirement is enforced by model checking (see Section 4). We call a pair of an initial and a final state $(s, s') \in \rho_p$ a *run* of p.

Unless stated otherwise, we establish the convention that the program takes its input in the variable I and produces its output in the variable O. The shorthand phrase *value of I* resp. O is to be understood as referring to the value in the initial resp. final state of a given run of p. Whenever necessary, I and O are silently lifted to be vectors (with $I \cap O = \emptyset$). A treatment of C structs as program output is shown in Section 6.2.

More amenable to reasoning is a description of ρ_p by a logical formula with two free variables I and O. We denote such formula as $\langle p \rangle(I, O)$ or, later, simply $\langle p \rangle$. The formula $\langle p \rangle(I_0, O_0)$ evaluates to TRUE iff p started with the input denoted by I_0 terminates with output denoted by O_0.

Attacker Model and Indistinguishability Relation. We assume that the attacker knows the program p, and that the input I is secret and the output O is public. The attacker has observed the value of the output O in a final state of some run of p and wants to learn something about the value of I in the initial state. It is the goal of QIF analysis to measure p's vulnerability to such an attack.

In the above attacker model, each program induces a partition on secret inputs $\approx_p$ called the *indistinguishability relation*. Each block in this partition is a set corresponding to some output value of the program and containing exactly the input values leading to this output. Formally, $\approx_p = \{\rho_p^{-1}(s') \mid s' \in \rho_p \circ S_I\}$. One also speaks of blocks in $\approx_p$ as *preimages* of program outputs. For example, if I is an unsigned 32-bit integer, then the program `if(I==42) O=1 else O=0;` induces $\approx_p = \{\{0, \ldots, 41, 43, \ldots, 2^{32} - 1\}, \{42\}\}$.

Intuitively, an attacker can discern secret inputs from different blocks but not within one block. Secure programs have a coarse $\approx_p$, while insecure a fine one. If $\approx_p$ is identity (very fine), then all blocks are singleton sets, and each output corresponds uniquely to a secret input: the attacker has perfect knowledge. Conversely, the coarsest indistinguishability relation $\approx_p = S_I \times S_I$ with only one block means that the attacker learns nothing about the secret inputs by observing program outputs (a scenario known as "non-interference").

Sometimes, a more powerful attacker is considered who can observe multiple runs while partially choosing the program inputs (so called *low inputs*). In this case, the indistinguishability relation becomes parametrized by a set of actualized low inputs L. If the set L is small, the QIF problem can be reduced to the no-low-input case by calculating the cartesian product of outputs for each low input $l \in L$. If the set L is large, other approaches (typically based on computationally more expensive self-composition) must be used.

Quantitative Security Measures. Given the number and sizes of blocks in $\approx_p$, it is possible to compute a range of security measures summarizing information flow (leakage) in a program. The leaked information is the difference between the attacker's initial uncertainty about the secret inputs and the remaining (a.k.a. residual) uncertainty after observing the output of the program [23].

It should be noted that different security measures have different properties and are appropriate for different scenarios. It may also be necessary to consider several measures in order to give dependable operational guarantees. We focus on two popular measures and refer to [23] for an in-depth discussion.

For quantification purposes, we interpret I and O as random variables ranging over S_I and S respectively. The program p restricts the values of I and O that can occur simultaneously. We assume that I follows a uniform distribution, i.e., that all secret inputs are equally likely. If this is not the case, techniques exist for reducing the analysis to a uniform case [1].

Under these assumptions, and given $\approx_p = \{C_1, \dots, C_n\}$ (n is, thus, the total number of possible distinct outputs of p), the following measures can be computed [23,2]:

$$H_\infty(I|O) = \log_2 \frac{|S_I|}{n} \qquad \text{and} \qquad H(I|O) = \frac{1}{|S_I|} \sum_{i=1}^{n} |C_i| \log_2 |C_i|$$

where the *conditional min-entropy* $H_\infty(I|O)$ is a measure in bit reflecting the probability of correctly determining I in a single guess after observing O, and the *conditional Shannon entropy* $H(I|O)$ is a lower bound in bit on the expected message length needed to communicate the remaining secret about I after observing O.

3   Analysis, Abstractly

In this section, we formulate our QIF framework in abstract logical terms. Our implementation, described later, is based on propositional logic, but other logics supporting model generation (e.g., QF_ABV) could be used just as well. We assume that logical formulas are built from usual logical connectives ($\wedge$, $\vee$, $\neg$, etc.) and user-defined vocabulary Σ. In propositional logic, Σ is a set of propositional variables. A *model* is a logic-specific semantical structure used to give meaning to user-defined vocabulary of a formula. In propositional logic, a model $m \colon \Sigma \to \{\text{TRUE}, \text{FALSE}\}$ is a map assigning every variable in Σ a truth value. In general, a given model m can be homomorphically extended to give a truth value to a formula Φ according to standard rules for logical connectives. We call a model m a *model of* Φ, if m assigns Φ the value TRUE. A formula Φ is *satisfiable* if it has at least one model, and *unsatisfiable* otherwise.

Definition 1. *We build our analysis from a number of abstract operators, which we define below, using the following designations. Σ and Δ are vocabularies with $\Delta \subseteq \Sigma$. A Σ-entity (i.e., formula or model) is an entity defined (only) over vocabulary from Σ. In the following, Φ is a Σ-formula, Ψ is a Δ-formula, i is an integer, m is a Σ-model, m_1 is a Δ-model, M is a set of models, p is a program, I and O are program variables.*

Expression	Meaning
$\Phi := \langle p \rangle$	formula encoding the behaviors of p (i.e., its transition relation or the set of traces)
$\Delta := \langle I \rangle, \Delta := \langle O \rangle$	vocabulary denoting in $\langle p \rangle$ the input and output variables of p (while p is implied)
$m := model(\Phi)$	some model satisfying Φ. If Φ is unsatisfiable, the result is a special value $\bot$.
$M := models(\Phi)$	the set of all models satisfying Φ. If Φ is unsatisfiable, the result is the empty set $\emptyset$.
$i := count(\Phi)$	$i := |models(\Phi)|$ (number of models satisfying Φ)
$m_1 := m\big|_\Delta$	the Δ-model that coincides with the Σ-model m on the vocabulary Δ
$\Psi := \Phi\big|_\Delta$	the strongest Δ-formula that, when interpreted as a Σ-formula, is entailed by Φ (projection of Φ on Δ). $models(\Phi\big|_\Delta) = \{ m\big|_\Delta \mid m \in models(\Phi) \}$
$\Psi := \Delta \simeq m_1$	a Δ-formula that is true in m_1 and false in all other Δ-models.
$\Psi := \Delta \not\simeq m_1$	$\neg(\Delta \simeq m_1)$, a Δ-formula that is false in m_1 and true in all other Δ-models.

The most interesting operator in the list above is projection. It makes the formula $\Phi\big|_\Delta$ say the same things about Δ as Φ does—but nothing else. Projection allows isolating aspects of program behavior along syntactical boundaries. For instance, the formula $\langle p \rangle\big|_{\langle O \rangle}$ describes (just) the set of outputs that are compatible with the behavior of program p. Orthogonally, the formulas $\Delta \simeq m_1$ and $\Delta \not\simeq m_1$ allow—when conjoined with $\langle p \rangle$—selecting or rejecting particular runs of the program. These formulas are easier to illustrate if the underlying logic is first-order; an implementation in propositional logic is given later. For instance, the first-order formula $\langle p \rangle \wedge \langle O \rangle = 5$ would be a particular instance of $\langle p \rangle \wedge \langle O \rangle \simeq m_1$ (for m_1 where $\langle O \rangle$ has the value 5) and describe all those runs of p that terminate with $O = 5$. Employing projection, we can describe the set of inputs that produce the output $O = 5$ by $(\langle p \rangle \wedge \langle O \rangle = 5)\big|_I$. In this light, we now formulate a general result:

Proposition 1.

$$H_\infty(I|O) = log_2 \frac{|S_I|}{count(\langle p \rangle\big|_{\langle O \rangle})} \quad and \quad H(I|O) = \frac{1}{|S_I|} \sum_{o \in M} |C(o)| \, log_2 |C(o)|$$

where $M = models(\langle p \rangle\big|_{\langle O \rangle})$ and $|C(o)| = count((\langle p \rangle \wedge \langle O \rangle \simeq o)\big|_{\langle I \rangle})$.

We note that computing $H(I|O)$ requires model enumeration *and* counting, while $H_\infty(I|O)$ only requires counting. We also note that since searching for models is computationally expensive, determining the residual min-entropy is easier the more secure the program of a given complexity is (fewer blocks in $\approx_p$). This does not hold for the Shannon entropy, as there is a tension between the number and size of blocks in $\approx_p$ (fewer blocks entail larger block sizes and vice versa).

4   From Program to Transition Relation with Bounded Model Checking

SAT-Based Bounded Model Checking. To implement the $\langle \cdot \rangle$ operator for translating programs into (propositional) logic, we use the SAT-based model checker CBMC [6] for C programs. CBMC is a very mature and popular verification tool supporting almost all ANSI C language features, including pointer constructs, dynamic memory allocation, recursion, and the float and double data types [6]. A similar, if less mature, tool for Java is JForge [8].

Given a C program p and a specification *spec* (given by `assert` statements in the code), CBMC generates a formula $\langle p \rangle \wedge \neg \langle spec \rangle$ in propositional logic, where $\langle p \rangle$ encodes the behaviors of the program p, and $\neg \langle spec \rangle$ encodes the behaviors that a specification-compliant program should not exhibit. This *verification condition* $\langle p \rangle \wedge \neg \langle spec \rangle$ is passed to a SAT solver. If it is unsatisfiable, then the program is correct w.r.t. the specification; otherwise, any model of $\langle p \rangle \wedge \neg \langle spec \rangle$ describes a violation of the specification.

During CBMC operation, functions are inlined and loops are unwound to the user-specified depth. CBMC warns the user if the unwinding depth is insufficient to cover all of the program behaviors (this is known as *unwinding assertion* checking). The unwound program is transformed into the static-single-assignment (SSA) form. In this form, statements can be interpreted as equations over bit vectors. The equations are combined and reduced to a formula of propositional logic in a process resembling synthesis of arithmetic circuits. The formula is flattened into conjunctive normal form (CNF) $\bigwedge_i \bigvee_j L_{i,j}$, where each literal $L_{i,j}$ is either a propositional variable or its negation. The formula can be exchanged with other tools by means of a standard DIMACS format.

Translating Programs into Logic. First, we carry out a preliminary verification pass, during which we incrementally increase the unwinding depth until CBMC reports no more unwinding assertion violations. This ensures that all program behaviors are covered and also that the program terminates for all inputs. In the main CBMC pass, we augment the program with the specification `assert(0);` (i.e., an assertion that is never fulfilled) before each return statement and make CBMC export the verification condition formula $\langle p \rangle \wedge \neg \langle spec \rangle$. The specification reduces the $\neg \langle spec \rangle$ conjunct to *true*, leaving the desired $\langle p \rangle$. The process may consume large amounts of memory but it is not a computational bottleneck as long as the unwinding depth is reasonable. The runtimes in our examples ranged from instantaneous to under a minute.

Identifying Program Variables in the Transition Relation Formula.
Internally, CBMC represents each program variable bit-wise according to its
type (and machine architecture). For example, the initial value of a **char**-typed
program variable is represented by 8 propositional variables. More precisely,
CBMC tracks the evolution of each program variable over a series of *time frames*
(relative to each variable). Typically, we are only interested in time frame one
for I (i.e., initial state) and the highest time frame for O (final state). The
mapping from program variables and time frames to sets of propositional variables is embedded as comments in the CBMC-generated formula (lines starting
with c, at the bottom of the DIMACS file). These comments have the following
structure: c *function_id* :: *prg_var_id* ! *thr_nr* @ *rec_depth* # *time prop_var_list*.
Thus, c c::main::1::I!0@1#1 1 2 3 4 5 6 7 8 means that the variable I in
function **main** in thread 0 at recursion depth 1 during time frame 1 is represented
by propositional variables $v_1, \ldots, v_8$. We extract this information with a simple
parser.

5   Model Enumeration and Counting

In this section, we present two conceptually different approaches and three tools that
we developed to implement $models(\Phi|_\Delta)$ resp.
$count(\Phi|_\Delta)$.

input : Σ-Formula Φ,
        projection
          scope $\Delta \subseteq \Sigma$
output: $models(\Phi|_\Delta)$

$M \leftarrow \emptyset$
$m \leftarrow model(\Phi)$
while $m \neq \bot$ **do**
   | $M \leftarrow M \cup m|_\Delta$
   | $\Phi \leftarrow \Phi \wedge (\Delta \not\approx m|_\Delta)$
   | $m \leftarrow model(\Phi)$
end
return M

5.1   Iterative Model Enumeration/Counting

Proposition 2. *The algorithm shown in Figure 2 implements model enumeration of a formula under projection.*

We have implemented this projection-capable
version of a well-known model enumeration algorithm in a tool named SHARPCDCL[1]. The
basic $model(\Phi)$-finding functionality is offered
by the SAT solver MINISAT [9]. Implementing
model projection $m|_\Delta$ is trivial, as one simply

Fig. 2. An algorithm for
enumerating $models(\Phi|_\Delta)$

restricts the domain of the mapping m to the scope Δ. The formula $\Delta \not\approx m|_\Delta$
can be constructed as $\bigvee_{v \in \Delta} flip(v, m)$, where $flip(v, m) \equiv v$, if $m(v) = \text{FALSE}$,
and $flip(v, m) = \neg v$, if $m(v) = \text{TRUE}$. This way, the truth value of at least one
variable in $m|_\Delta$ must flip in order to satisfy $\Delta \not\approx m|_\Delta$. Conjoining this formula
(which is already in CNF) with Φ ensures that the current model m will not be
found again. After the loop terminates (or SHARPCDCL is interrupted), the set
resp. number of found models is returned.

---

[1] Available at http://tools.computational-logic.org/

5.2   Model Counting via Compilation to d-DNNF

State-of-the-art deterministic #SAT solvers implement $count(\cdot)$ via compilation of the formula to Deterministic Decomposable Negation-Normal Form (d-DNNF). We have extended two such tools[2], SHARPSAT [25] and DSHARP [20], with projection capabilities—something that has not been available in #SAT solvers so far. While iterative model enumeration/counting works better on large formulas with few models, d-DNNF-based #SAT solvers are useful to analyze smaller formulas with a large number of models. Empirical evidence is presented in Section 6. Below, we briefly sketch the necessary theoretical results for integrating model counting and projection.

Definition 2. *A formula in d-DNNF is a rooted tree such that:*

- *The label of each leaf node is either true, false, or a literal (i.e., negation can only appear attached to variables), while the label of each internal node is either a conjunction ($\wedge$) or a disjunction ($\vee$).*
- *Decomposability holds: any two children c_i and c_j of a conjunctive node share no vocabulary: $\Sigma_{c_i} \cap \Sigma_{c_j} = \emptyset$.*
- *Determinism holds: let $\Phi(n)$ be the formula represented by the subtrees rooted at node n. For any two children d_i and d_j of a disjunctive node, $\Phi(d_i)$ and $\Phi(d_j)$ must be contradictory, i.e., $\Phi(d_i) \wedge \Phi(d_j)$ is unsatisfiable.*

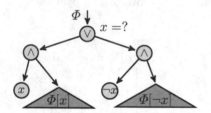

Fig. 3. A typical d-DNNF fragment

A propositional formula is typically compiled to d-DNNF by an exhaustive DPLL-style algorithm alternating systematic case distinctions (decisions) and unit propagation. Each decision gives rise to an $\vee$-node as per equality $\Phi = \Phi[v] \vee \Phi[\neg v]$. Figure 3 illustrates such a decision on variable x. The #SAT solvers also employ a number of optimizations (e.g., subtree caching, clause learning, etc.), but these are of no interest here. After a (computationally hard) compilation to d-DNNF, both projection computation and model counting—though not in combination—can be carried out in linear time [7].

Proposition 3. *If Φ is a Σ-formula in d-DNNF, then $\Phi|_\Delta$ can be computed in polynomial time by replacing every satisfiable $(\Sigma \setminus \Delta)$-subtree in Φ by true, and every unsatisfiable $(\Sigma \setminus \Delta)$-subtree in Φ by false.*

This is a direct consequence of [7, Theorems 3 and 9]. Unfortunately, projection can destroy determinism (e.g., if the nodes x and $\neg x$ in Figure 3 are removed), making later model counting impossible. Yet, it is easy to see that:

---

[2] Available at http://formal.iti.kit.edu/~klebanov/software/

Proposition 4. *Determinism is retained during projection of a d-DNNF formula Φ on scope Δ, if every subtree rooted at an $\vee$-node associated with a decision on variable $v \in \Sigma \setminus \Delta$ only contains variables from $\Sigma \setminus \Delta$.*

In other words, losing determinism in a subtree is not harmful, if the whole subtree is bound to be removed.

We enforce projection determinism by modifying the variable selection heuristic of the #SAT solvers' d-DNNF compilers to always perform decisions on variables from Δ first. Further, we implemented the satisfiability check of Proposition 3 by integrating MINISAT into DSHARP. We omitted a similar check in projecting SHARPSAT; the latter can thus report a result that is higher than the actual model count (though this never happened in our benchmarks). When computing min-entropy, such overapproximation entails an error on the conservative side.

5.3   Boosting Counting Performance with Formula Preprocessing

In [10], model counting has been improved by a few preprocessing techniques, namely unit propagation, equivalence reduction and hyper binary resolution. In general, any equivalence-preserving preprocessing technique can be applied before model counting, because the set of models does not change. However, there are also many powerful preprocessing techniques that are merely satisfiability-preserving but not equivalence-preserving, such as variable elimination or blocked clause elimination. For general model counting, these techniques cannot be applied. The situation changes when projection is involved.

Proposition 5. *Let Φ be a propositional Σ-formula and $\Delta \subseteq \Sigma$ a projection scope. Applying satisfiability-preserving preprocessing on $\Sigma \setminus \Delta$ in Φ does not change the set of models of the projection $\Phi|_\Delta$.*

We use the propositional preprocessor COPROCESSOR 2^3 [15,16], which we developed earlier, for equivalence-preserving simplification and scope-restricted satisfiability-preserving simplification. While the more advanced simplification techniques are not always beneficial, preprocessing boosts model counting performance in most cases, as shown in the next section. The benchmark results are given for default settings. The exact set of applied techniques can be configured by the user.

6   Benchmarks and Evaluation

6.1   Synthetic Benchmarks

A number of microbenchmarks for general-purpose QIF have appeared in [2] and [21]. The collection has later been consolidated and extended in [18]. It is valuable as it is quite varied and targets different bottlenecks in QIF analyses.

---

[3] Available at `http://tools.computational-logic.org/`

```
O = ((I >> 16) ^ I);      if (I == R1) O = R1;       O = 0;
O = O & 0xffff;           else if (I==R2) O = R2;     for (i = 0; i < N; i++) {
O = O | (O << 16);        ...                            m = 1 << (31-i);
                          else if (I==R9) O = R9;        if (O + m <= I) O += m;
                          else O = R10;               }
```

(a) Mix and duplicate   (b) Ten random outputs   (c) Binary search

Fig. 4. Benchmarks from [21] and [18]

The only drawback is that the majority of the benchmarks no longer pose a challenge. Below we report results on five benchmarks (out of eleven total) that are still difficult or interesting in some sense.

Table 1 summarizes the performance results. The experiments were performed on a machine with an Intel Core i7 860 2.80GHz CPU. We have included the timings published in [18] for comparison, though no hardware description is available in that paper. The code is presented in Figure 4. Unless noted otherwise, the variable I is the secret input, O is the observable output, and the type of variables is uint32_t (32-bit unsigned integer).

The mix and duplicate benchmark (Figure 4a)—we cite [21]—"combines the two halves of its input word with XOR, and then duplicates these 16 bits in both the upper and lower halves of its output", leaking 16 bit of the secret. "[This leak] is too large to be effectively measured exhaustively, too small for effective sampling, and too uniformly distributed for range queries, so only our probabilistic #SAT strategy gives an accurate estimate" [21]. We can see that neither iterative model enumeration nor modern precise #SAT solvers have difficulties with this benchmark.

The ten-random benchmark (Figure 4b) is essentially a program with ten outputs that do not follow a particular pattern. On this benchmark, the two-bit abstraction from [18] overapproximates the leak.

In the sum benchmark O = I1 + I2 + I3;, we increase the difficulty compared to [18], dropping the restriction of the summands to the range 0–10. Instead, we

Table 1. Benchmark runtimes (seconds)

w/PP=with preprocessing, t/o=timeout at 1h, *=result overapproximates number of models. Preprocessing time was negligible in all cases.

| | | Iterative enum. | | Precise #SAT | | Overapprox. #SAT | | |
| | | sharp-CDCL | w/PP | Dsharp | w/PP | sharp-SAT | w/PP | |
Benchmark	models							[18]
mix-n-dup	2^{16}	16.2	< 0.1	< 0.1	< 0.1	< 0.1	< 0.1	1.3
ten-random	10	< 0.1	< 0.1	< 0.1	< 0.1	< 0.1	< 0.1	4.6*
sum-three-32	2^{32}	t/o	t/o	t/o	< 0.1	t/o	< 0.1	n/a
bin-search-16	2^{16}	6.9	9.6	166.6	39.3	12.9	5.7	6.4
bin-search-32	2^{32}	t/o	t/o	t/o	48.2	t/o	9.8	55.5

```
1  int atalk_getname(struct socket *sock, struct sockaddr *uaddr,
2                    int *uaddr_len, int peer)
3  {
4      struct sockaddr_at sat;
5      struct sock *sk = sock->sk;
6      struct atalk_sock *at = at_sk(sk);
7      int err;
8
9      // lock_sock(sk);
10     err = -ENOBUFS;
11     if (sock_flag(sk, SOCK_ZAPPED)) if (atalk_autobind(sk) < 0) goto out;
12
13     *uaddr_len = sizeof(struct sockaddr_at);
14     // memset(&sat.sat_zero, 0, sizeof(sat.sat_zero)); // leak patch
15
16     if (peer) {      err = -ENOTCONN;
17                      if (sk->sk_state != TCP_ESTABLISHED) goto out;
18                      sat.sat_addr.s_net  = at->dest_net;
19                      sat.sat_addr.s_node = at->dest_node;
20                      sat.sat_port        = at->dest_port;
21
22     } else {         sat.sat_addr.s_net  = at->src_net;
23                      sat.sat_addr.s_node = at->src_node;
24                      sat.sat_port        = at->src_port;
25     }
26
27     err = 0;
28     sat.sat_family = AF_APPLETALK;
29     memcpy(uaddr, &sat, sizeof(sat));
30
31 out:
32     // release_sock(sk);
33     unsigned char 0; int i;
34     for (i=0; i<sizeof(struct sockaddr_at); i++) 0=((char *)uaddr)[i];
35     assert(0);
36     return err;
37 }
```

Fig. 5. AppleTalk driver function leaking kernel memory (CVE 2009-3002)

consider the sum of three arbitrary 32-bit secret values (variable type int32_t). Unsurprisingly, iterative enumeration of models is ineffective, while preprocessing quickly simplifies $\langle p \rangle|_O$ to *true*, corresponding to 2^{32} outputs.

The binary search benchmark (Figure 4c) is valuable, because it is parametric and can help assess analysis scalability. The program leaks the most significant N bit of the secret by repeated dichotomy. We note the improved rates of slowdown between N = 16 and N = 32 with our tools.

6.2   Linux Kernel

This benchmark has originally appeared in [12], where the authors analyze a number of vulnerabilities previously found in the Linux kernel. The goal is to measure the amount of unsanitized kernel memory leaking to applications in the userland. The value of the benchmark stems less from the operational significance of the leak size, but rather from its origin in actual systems software. We revisit the most complex example presented in the above paper, a leak in the atalk_getname routine in the AppleTalk driver (Figure 5).

The leak is as follows. The kernel allocates a 16-byte structure sat (the secret input) and initializes it. Later, the content of the structure is copied to userland. Due to a programming error, parts of the structure are not initialized properly. The official patch fixing the bug is shown in line 14.

In order to deal with an output that is a C struct, we introduce an auxiliary variable O and a loop reading the structure before the return statement (lines 33–34). The observable output is then the last sizeof(struct sockaddr_at) values of O. This is possible since CBMC actually encodes the full trace of variable values rather than merely the initial and the final states.

We have used the code from net/appletalk/ddp.c of the Linux kernel 3.4.28 (minus the bugfix). The only simplification that we performed was to remove the locking calls in line 9 and 32, caused by technical difficulties with the code organization of the kernel. It took the analysis in [12] one hour and 39 minutes to find at least 64 blocks in $\approx_p$ of the function—a time explained by an extreme form of self-composition exploring the function behavior 64 times. With SHARP-CDCL, finding 64 blocks was instantaneous, while finding 65536 blocks (i.e., a 16-bit leak) took 20 seconds resp. 14 seconds with preprocessing. The full size of the leak is too large to be established precisely.

6.3   Image Anonymization

This benchmark has originally appeared in [17], where the authors assess the effect of several image anonymization techniques on a 125×125 pixel test image. While effective leakage bounds could be established for blurring or pixelation, no useful bounds (in either direction) could be established for image swirling demonstrated in Figure 1. While, the analysis in [17] sacrifices soundness and precision for scalability (cf. next section), we test our tools by establishing precise leakage in a variant of this application.

As in [17], the source code is derived from the SwirlImage() function of the popular ImageMagick[4] image manipulation suite. We deviate from [17] by discounting target image interpolation. This simplification eliminates the influence of image data, and leaves us with a function that merely transforms a pair of integer coordinates into another pair (Figure 6). The coordinates that are not reachable by the swirling transformation appear as black pixels in the illustration in Figure 1b.

Two things should be noted about analyzing code with non-integral data types. First, CBMC approximates data types such as double with a 16+16 bit fixed-point representation. We think, it is reasonable to assume that this precision is sufficient in this example. Second, modern general-purpose processors typically implement mathematical functions such as sine, cosine, and square root in hardware. For the analysis, we have used their software counterparts from the popular Freely Distributable C Math Library FDLIBM[5]. Thus, the main function shown in Figure 6 accounts for only 23 out of 379 total lines of

---

[4] http://imagemagick.org/. The SwirlImage() function is in fx.c.
[5] http://www.netlib.org/fdlibm/

```
1  int main(int argc, char **argv) {
2
3      unsigned char x,y; // secret inputs
4      __CPROVER_assume(x>=0 && x<125); // range limit
5      __CPROVER_assume(y>=0 && y<125);
6      unsigned char newx, newy; // observable outputs
7
8      double center = (double) 125/2.0;
9      double radius = center;
10     double degrees = (double) (3.141593*720.0/180.0);
11     double deltay = (double) (y-center);
12     double deltax = (double) (x-center);
13     double distance = deltax*deltax + deltay*deltay;
14
15     if (distance < radius*radius) {
16         double factor=1.0-sqrt((double) distance)/radius;
17         double d = (double) (degrees*factor*factor);
18         double sine=sin(d);
19         double cosine=cos(d);
20
21         newx = ((cosine*deltax-sine*deltay)+center);
22         newy = ((sine*deltax+cosine*deltay)+center);
23     } else { newx = x; newy = y; }
24     assert(0);
25     return 0;
26  }
```

Fig. 6. Image anonymization main function

analyzed code (about 6%). The code contains four loops (maximal unwinding depth 32). A total of 2079 bitvector equations encode the transition relation $\langle p \rangle$.

It took SHARPCDCL 4h58m to find all 12228 blocks in $\approx_p$, which corresponds to a leak of 13.58 out of 13.93 bit, if one measures min-entropy. In other words, swirling is not a good anonymization technique. The deanonymization in Figure 1c actually underestimates the leakage, as the unswirling transformation used is also lossy.

Measuring the residual Shannon entropy of the secret was quite more costly. It took SHARPCDCL 5h23m to find all six inputs in the preimage of a single output (newx=87,newy=62), and this was only possible as we used—in this case only—MINISAT's randomized variable selection heuristic, which can sometimes produce much faster SAT solver runs at the price of generally unpredictable performance.

Altogether, while we do obtain proof of the secret leaking almost completely, we clearly cannot claim a practical benefit of our analysis in this case. The reasons for this are twofold. First, the analyzed code is too large, and we see this benchmark as marking the frontier of what is barely possible with current technology. Second, the size of the secret is too small, making simple exhaustive simulation an attractive alternative

A new perspective opens if our toolchain were used as part of probabilistic QIF for large secrets (and on programs of more appropriate size). Köpf and Rybalchenko show in [14] that it is sufficient to randomly choose $\frac{(\log_2 |S_I|)^2}{(1-P)\delta^2}$ input samples and measure the respective size of the enclosing block in $\approx_p$, in order to probabilistically estimate the residual Shannon entropy to a degree of precision δ and a confidence level $P \in [0, 1)$. As the size of the secret increases, the polylogarithmic probabilistic approach remains feasible in contrast to exhaustive simulation.

7   Related Work

We survey most recent and relevant works in the field; a further survey of QIF models and techniques is available in [19].

Backes et al. [2] describe a precise QIF analysis for programs with affine indistinguishability relations based on self-composition and Barvinok's counting algorithm. This was later extended in [14] to improve scalability. In order to maintain automation, the latter approach gives up precise computation of the leak and opts for an approximative characterization, deriving lower and upper bounds on residual min-entropy, as well as probabilistic bounds on residual Shannon entropy. A different extension based on symbolic Barvinok counting was proposed by Klebanov in [13].

Heusser and Malacaria have developed two relevant QIF approaches: [12] and [11]. The former encodes detection of (small) leaks as a pure model checking problem via self-composition in CBMC. The latter one and our SAT-based analysis are quite similar in spirit, though [11] builds on expensive self-composition, supplementing it with a model enumerator and a #SAT solver.

The following three approaches compute leakage bounds by approximating the projection $\langle p \rangle|_{\langle O \rangle}$ with a series of entailment queries on $\langle p \rangle$, followed by precise or approximative model counting.

Newsome et al. [21] use a series of SMT entailment queries to identify and narrow down in-/feasible output ranges. Such approximations of $\langle p \rangle|_{\langle O \rangle}$ are amenable to simple model counting. The approach is complemented by sampling and probabilistic counting. Models generated by the SMT solver are used to identify the presence of a feasible output within a range, but this procedure is not leveraged fully as an output-finding technique.

Phan et al. [22] encode a full binary search for feasible outputs (models of $\langle p \rangle|_{\langle O \rangle}$) in a bounded model checker. This approach is precise, but requires in practice more than one call to the underlying solver to find a single feasible output. It is useful when the program verification system does not expose the underlying logical representation or when the used solver cannot generate models.

Meng and Smith [18] use "two-bit-pattern" SMT entailment queries to calculate a propositional overapproximation (w.r.t. the number of models) of $\langle p \rangle|_{\langle O \rangle}$ and count its instances with a #SAT solver of the computer algebra system Mathematica.

McCamant and Ernst combine in [17] a dynamic bitwise taint analysis with static analysis to derive bounds on information leakage in C programs. The technique has been applied to large programs used in practice. On the other hand, it only measures leakage along one or a few selected program paths, leaving it to the user to supply "representative" inputs.

Another tool for dynamic analysis is reported by Chatzikokolakis in [5]. The tool automatically derives bounds of information leakage in terms of mutual information and capacity from trial runs of the system, which is treated as a black box.

The theoretical hardness of QIF has been shown by Terauchi et al. in [27,24]. As with other hard problems (e.g., SAT), these results do not preclude the existence of efficient analyses for individual instances or subclasses of the problem.

8 Conclusion

We presented a unifying abstract formulation of a class of QIF analyses for imperative programs and an instance of this class outperforming previous comparable approaches. We demonstrated that logical projection is a useful framework for understanding and implementing QIF. In the future, we are interested in exploring more advanced projection computation techniques [3,4,26].

Though our implementation is not a single tool, all its components are available publicly. A part of the performance improvement is due to advances in the underlying reasoning technology, which have been fueled by regular SAT competitions and associated benchmark collections. Maintaining and extending a set of canonical benchmarks would benefit the QIF field as well.

Acknowledgments. This work was in part supported by the German National Science Foundation (DFG) under the priority programme 1496 "Reliably Secure Software Systems – RS3." The authors would like to thank Christoph Wernhard for his comments on projection computation.

References

1. Backes, M., Berg, M., Köpf, B.: Non-uniform distributions in quantitative information-flow. In: ASIACCS 2011, pp. 367–375. ACM (2011)
2. Backes, M., Köpf, B., Rybalchenko, A.: Automatic discovery and quantification of information leaks. In: S&P 2009, pp. 141–153. IEEE Computer Society (2009)
3. Brauer, J., King, A.: Approximate quantifier elimination for propositional boolean formulae. In: Bobaru, M., Havelund, K., Holzmann, G.J., Joshi, R. (eds.) NFM 2011. LNCS, vol. 6617, pp. 73–88. Springer, Heidelberg (2011)
4. Brauer, J., King, A., Kriener, J.: Existential quantification as incremental SAT. In: Gopalakrishnan, G., Qadeer, S. (eds.) CAV 2011. LNCS, vol. 6806, pp. 191–207. Springer, Heidelberg (2011)
5. Chatzikokolakis, K., Chothia, T., Guha, A.: Statistical measurement of information leakage. In: Esparza, J., Majumdar, R. (eds.) TACAS 2010. LNCS, vol. 6015, pp. 300 404. Springer, Heidelberg (2010)
6. Clarke, E., Kroning, D., Lerda, F.: A tool for checking ANSI-C programs. In: Jensen, K., Podelski, A. (eds.) TACAS 2004. LNCS, vol. 2988, pp. 168–176. Springer, Heidelberg (2004)
7. Darwiche, A.: Decomposable negation normal form. J. ACM 48(4), 608–647 (2001)
8. Dennis, G., Chang, F.S.-H., Jackson, D.: Modular verification of code with SAT. In: ISSTA 2006, pp. 109–120. ACM (2006)

9. Eén, N., Sörensson, N.: An extensible SAT-solver. In: Giunchiglia, E., Tacchella, A. (eds.) SAT 2003. LNCS, vol. 2919, pp. 502–518. Springer, Heidelberg (2004)

10. Guo, Q., Sang, J., He, Y.-M.: Effective preprocessing in #SAT. In: ICMV 2011. SPIE (2011)

11. Heusser, J., Malacaria, P.: Applied quantitative information flow and statistical databases. In: Degano, P., Guttman, J.D. (eds.) FAST 2009. LNCS, vol. 5983, pp. 96–110. Springer, Heidelberg (2010)

12. Heusser, J., Malacaria, P.: Quantifying information leaks in software. In: ACSAC 2010, pp. 261–269. ACM (2010)

13. Klebanov, V.: Precise quantitative information flow analysis using symbolic model counting. In: Martinelli, F., Nielson, F. (eds.) Proceedings of the International Workshop on Quantitative Aspects in Security Assurance, QASA (2012)

14. Köpf, B., Rybalchenko, A.: Approximation and randomization for quantitative information-flow analysis. In: CSF 2010, pp. 3–14. IEEE Computer Society, Washington, DC (2010)

15. Manthey, N.: Coprocessor 2.0 – A flexible CNF simplifier. In: Cimatti, A., Sebastiani, R. (eds.) SAT 2012. LNCS, vol. 7317, pp. 436–441. Springer, Heidelberg (2012)

16. Manthey, N., Heule, M.J.H., Biere, A.: Automated reencoding of boolean formulas. In: Proceedings of Haifa Verification Conference 2012 (2012)

17. McCamant, S., Ernst, M.D.: Quantitative information flow as network flow capacity. In: PLDI 2008, pp. 193–205. ACM (2008)

18. Meng, Z., Smith, G.: Calculating bounds on information leakage using two-bit patterns. In: PLAS 2011, pp. 1–12. ACM (2011)

19. Mu, C.: Quantitative information flow for security: a survey. Technical Report TR-08-06, Department of Computer Science, King's College London (2008), http://www.dcs.kcl.ac.uk/technical-reports/papers/TR-08-06.pdf (updated 2010)

20. Muise, C., McIlraith, S.A., Beck, J.C., Hsu, E.I.: DSHARP: Fast d-DNNF compilation with sharpSAT. In: Kosseim, L., Inkpen, D. (eds.) Canadian AI 2012. LNCS, vol. 7310, pp. 356–361. Springer, Heidelberg (2012)

21. Newsome, J., McCamant, S., Song, D.: Measuring channel capacity to distinguish undue influence. In: PLAS 2009, pp. 73–85. ACM, New York (2009)

22. Phan, Q.-S., Malacaria, P., Tkachuk, O., Păsăreanu, C.S.: Symbolic quantitative information flow. In: Mehlitz, P., Rungta, N., Visser, W. (eds.) Proceedings, Java Pathfinder Workshop, pp. 1–5 (2012)

23. Smith, G.: On the foundations of quantitative information flow. In: de Alfaro, L. (ed.) FOSSACS 2009. LNCS, vol. 5504, pp. 288–302. Springer, Heidelberg (2009)

24. Terauchi, T., Aiken, A.: Secure information flow as a safety problem. In: Hankin, C., Siveroni, I. (eds.) SAS 2005. LNCS, vol. 3672, pp. 352–367. Springer, Heidelberg (2005)

25. Thurley, M.: sharpSAT – counting models with advanced component caching and implicit BCP. In: Biere, A., Gomes, C.P. (eds.) SAT 2006. LNCS, vol. 4121, pp. 424–429. Springer, Heidelberg (2006)

26. Wernhard, C.: Tableaux for projection computation and knowledge compilation. In: Giese, M., Waaler, A. (eds.) TABLEAUX 2009. LNCS (LNAI), vol. 5607, pp. 325–340. Springer, Heidelberg (2009)

27. Yasuoka, H., Terauchi, T.: Quantitative information flow – verification hardness and possibilities. In: CSF 2010, pp. 15–27. IEEE Computer Society (2010)

PRINSYS—On a Quest for Probabilistic Loop Invariants*

Friedrich Gretz[1,2], Joost-Pieter Katoen[1], and Annabelle McIver[2]

[1] RWTH Aachen University, Germany
lastname@cs.rwth-aachen.de
[2] Macquarie University, Australia
firstname.lastname@mq.edu.au

Abstract. PRINSYS (pronounced "princess") is a new software-tool for probabilistic invariant synthesis. In this paper we discuss its implementation and improvements of the methodology which was set out in previous work. In particular we have substantially simplified the method and generalised it to non-linear programs and invariants. PRINSYS follows a constraint-based approach. A given parameterised loop annotation is speculatively placed in the program. The tool returns a formula that captures precisely the invariant instances of the given candidate. Our approach is sound and complete. PRINSYS's applicability is evaluated on several examples. We believe the tool contributes to the successful analysis of sequential probabilistic programs with infinite-domain variables and parameters.

Keywords: invariant generation, probabilistic programs, non-linear constraint solving.

1 Introduction

Motivation. Probabilistic programs are pivotal in different application fields like security, privacy [2]—several probabilistic protocols (e.g. onion-routing) aim to ensure privacy, and there is an increasing interest in the topic, partly driven by the social-media world—and cryptography [1] as well as quantum computing [13]. Such programs are single threaded and typically consist of a small number of code lines, but are hard to understand and analyse. The two major reasons for their complexity are the occurrence of program variables with unbounded domains, and parameters. Such parameters can be either loop bounds, number of participants (in a protocol), or probabilistic choices where the parameters range over concrete probabilities. For example, the following simple program generates a sample x according to a geometric distribution with parameter p. In every loop iteration, the variable x is increased by one with probability $1-p$ and *flip* is set to one with probability p, where p is an unknown real value from

---

* This work is partially funded by the DFG Research Training Group Algosyn, the EU FP7 Project CARP (Correct and Efficient Accelerator Programming), and the EU MEALS exchange project with Latin America.

K. Joshi et al. (Eds.): QEST 2013, LNCS 8054, pp. 193–208, 2013.

Listing 1. $x \sim \text{geom}(p)$

```
x := 0;
flip := 0;
while (flip = 0) {
    ( flip := 1 [p] x := x+1 );
}
```

the range $(0, 1)$. The occurrence of unbounded variables and parameters comes at a price, namely that probabilistic programs in general cannot be analysed automatically by model-checking tools such as PRISM [10], PARAM [6], PASS [5] or APEX [9].

Approach. Instead we resort to deductive techniques. Recall that one of the main approaches to the verification of sequential programs rests on the pioneering work of Floyd, Hoare, and Dijkstra in which annotations are associated with control points in the program. Whereas the annotations for sequential programs are qualitative and can be expressed in predicate logic, quantitative annotations are needed to reason about probabilistic program correctness. McIver and Morgan [11] have extended the method of Floyd, Hoare, and Dijkstra to probabilistic programs by making the annotations real- rather than Boolean-valued expressions in the program variables. Using these methods we can prove that in the above program the average value of x is $\frac{1-p}{p}$. Annotating a probabilistic program with such expressions is non-trivial and undecidable in general. The main reason is the occurrence of loops. This all boils down to the question on how to establish a loop invariant. It is known that this is a notorious hard problem for traditional programs. For probabilistic programs it is even more difficult as loop invariants are quantitative—so-called probabilistic loop invariants. Variables do no longer have a value, but have a certain value with a given likelihood. Finding an invariant is hard and requires both ingenuity as well as involved computations to check that a given expression is indeed invariant. Recently, Katoen *et al.* [7] have proposed a technique for finding linear invariants for linear probabilistic programs. Linearity refers to the fact that right-hand sides of assignments and guards are linear expressions in the program variables (and parameters). This technique is based on speculatively annotating a loop with a template (in fact a linear inequality) and using constraint solving techniques to distill all parameters for which the template is indeed a loop invariant.

Contributions of this paper. The contributions of this paper are manifold. First and foremost, this paper presents PRINSYS (pronounce "princess"), a novel tool for supporting the semi-automated generation of probabilistic invariants of pGCL[1] programs. This publicly available tool implements the technique

---
[1] pGCL extends Dijkstra's guarded command language with a probabilistic choice operator.

advocated in [7], i.e., automatically computes the constraints under which a user-provided template is invariant, saving the user from tedious and error prone calculations. To the best of our knowledge, it is the first tool for synthesizing probabilistic invariants. Secondly, we show that the theory in [7] can be considerably simplified. In particular, we show that the usage of Motzkin's transposition theorem (a generalisation of Farkas' lemma) to turn an existentially quantified formula into a universally quantified one, is not needed. As a result, PRINSYS allows arbitrary formulas in templates and program guards. This allows for polynomial invariant templates and non-linear program expressions. So, an immediate consequence of this simplification is that the restriction to linear programs and linear invariants can be dropped. This is more of theoretical interest than of practical interest, as polynomial invariants—as for the traditional, non-probabilistic setting—are hard to synthesize in practice. Finally, we present some applications of the tool such as proving the equivalence of two programs computing a sample from $X-Y$ where X and Y are both geometrically distributed, and the generation of a fair coin from a biased one. We evaluate the experiments and give directions for future research.

Organization of the paper. Section 2 provides the preliminaries such as pGCL, probabilistic invariants, and expectations. Section 3 presents the steps of our approach and the simplification of [7]. Section 4 provides three examples to give insight about what PRINSYS can establish. Section 5 evaluates the tool and approach, whereas Sect. 6 concludes the paper and provides pointers to future work.

2   Background

When probabilistic programs are executed they determine a probability distribution over final values of program variables. For instance, on termination of

$$(x := 1 \; [0.75] \; x := 2);$$

the final value of x is 1 with probability $\frac{3}{4}$ or 2 with probability $1 - \frac{3}{4} = \frac{1}{4}$. An alternative way to characterise that probabilistic behaviour is to consider the expected values over random variables with respect to that distribution. For example, to determine the probability that x is set to 1, we can compute the expected value of the random variable "x is 1" which is $\frac{3}{4} \cdot 1 + \frac{1}{4} \cdot 0 = \frac{3}{4}$. Similarly, to determine the average value of x, we compute the expected value of the random variable "x" which is $\frac{3}{4} \cdot 1 + \frac{1}{4} \cdot 2 = \frac{5}{4}$. More generally, rather than a distribution centred approach, we take an "expectation transformer" [11] approach. We annotate probabilistic programs with *expectations*.

Expectations. Expectations map program states to non-negative real values. They generalise Hoare's predicates for non-probabilistic programs towards real-valued functions. Intuitively, implication between predicates is generalised to pointwise inequality between expectations. For convenience we use square brackets to link Boolean truth values to numbers and by convention [true] $= 1$ and [false] $= 0$. In the example above, we call "x" the *post-expectation* and $\frac{5}{4}$ its *pre-expectation*. Thus the annotated program is $\langle \frac{5}{4} \rangle$ $(x := 1 \; [0.75] \; x := 2);$ $\langle x \rangle$.

The formal mechanism for computing pre-expectations for a given program and post-expectation is the expectation transformer semantics [11]. Expectation transformers are the quantitative pendant to Dijkstra's predicate transformers. McIver and Morgan extend Dijkstra's concept and introduce a function wp(*prog,post*) which based on the program *prog* determines the *greatest pre-expectation* for any given post-expectation *post*. A summary of pGCL's expectation transformer semantics is given in Table 1 where f is a given post-expectation. From an operational perspective, pGCL programs can be viewed as (infinite state) MDPs with a reward structure induced by the given post-expectation f. Then the greatest pre-expectation can be computed as the expected cummulative reward on that model [4].

Table 1. Syntax and expectation transformer semantics of pGCL

syntax *prog*	semantics wp(*prog,f*)
skip	f
abort	0
x := E	$f[x/E]$
P ; Q	$\mathrm{wp}(P, \mathrm{wp}(Q, f))$
if (G) { P } **else** { Q }	$[G] \cdot \mathrm{wp}(P, f) + [\neg G] \cdot \mathrm{wp}(Q, f)$
P [] Q	$\min\{\mathrm{wp}(P, f), \mathrm{wp}(Q, f)\}$
P [p] Q	$p \cdot \mathrm{wp}(P, f) + (1 - p) \cdot \mathrm{wp}(Q, f)$
while (G) { P }	$\mu X.([G] \cdot \mathrm{wp}(P, X) + [\neg G] \cdot f)$

For loop-free programs, the pre-expectation is simply given by syntactic rules. However, loops pose a problem because their expectation over final values is given in terms of a least fixed point (over the domain of expectations with the ordering $\leq$, a pointwise ordering on expectations).

Invariants. Using special expectations which we call *invariants* we can avoid the calculation of a loop's fixed point. Assume we are given two expectations *pre* and *post* and we want to show that *pre* is a lower bound on the loop's actual pre-expectation, i.e.

$$pre \leq \mathrm{wp}(\mathbf{while}(G)\{body\}, post) \ .$$

Instead of computing the greatest pre-expectation wp(**while**(G){*body*}, *post*) directly, it is more practical to divide this problem into simpler subtasks:

1. find an expectation $\mathcal{I}$ such that

$$pre \leq \mathcal{I} \quad \text{and} \quad \mathcal{I} \cdot [\neg G] \leq post \, ,$$

2. show $\mathcal{I}$ is *invariant*[2], that is $\mathcal{I} \cdot [G] \leq \text{wlp}(body, \mathcal{I})$
3. show $\mathcal{I}$ is sound, that is $\mathcal{I} \leq \text{wp}(\textbf{while}(G)\{body\}, \mathcal{I} \cdot [\neg G])$

Points 2. and 3. may seem odd as they resemble the original problem of proving an inequality between an expectation and the greatest pre-expectation of a loop. However they are easier than the original problem, because in 2. the greatest pre-expectation can be explicitly computed because *body* is a loop-free program. In order to guarantee soundness (point 3.) the loop must terminate with probability one and the invariant $\mathcal{I}$ has to additionally meet one of the following sufficient conditions [11]:

– from every initial state of the loop only a finite state space is reachable
– or $\mathcal{I}$ is bounded above by some fixed constant
– or $\text{wp}(body, \mathcal{I} \cdot [G])$ tends to zero as the number of iterations tends to infinity.

Remark 1. It is an open problem to give the *necessary and sufficient* conditions for soundness.

Put all together this proves the inequality above as

$$pre \leq \mathcal{I} \leq \text{wp}(\textbf{while}(G)\{body\}, \mathcal{I} \cdot [\neg G]) \leq^3 \text{wp}(\textbf{while}(G)\{body\}, post) \, .$$

Example 1 (Application of invariants.). Consider the program *prog* in Lst. 2. On each iteration of the loop it sets x to -1 with probability 0.15, to 0 with probability 0.5 and to 1 with probability 0.35. We would like to prove that the probability to terminate in a state where $x = 1$ is 0.7 or equivalently

$$\text{wp}(prog, [x = 1]) = 0.7 \, .$$

Instead of computing the least fixed point of the loop wrt. post-expectation $[x = 1]$, we can show that $\mathcal{I} = [x = 0] \cdot 0.7 + [x = 1]$ is invariant. If the loop terminates, we can establish:

$$[\neg G] \cdot \mathcal{I} = [x \neq 0] \cdot [x = 0] \cdot 0.7 + [x = 1]$$
$$= [x = 1] \, .$$

At the beginning of the program the initialisation of x transforms the invariant to:

$$\text{wp}(x := 0, \mathcal{I}) = [0 = 0] \cdot 0.7 + [0 = 1]$$
$$= 0.7 \, .$$

---

[2] wlp is the "liberal" version of wp. Both expectation transformers coincide for almost surely terminating programs. Since in this paper we do not consider nested loops, i.e. *body* is loop-free (and hence surely terminates), we do not discuss the theoretical differences between wp and wlp here.

[3] wp is monotonic in its second argument [11].

Listing 2. A simple loop

```
x := 0;
while (x=0) {
    (x := 0;) [0.5] { (x := -1 [0.3] x := 1); }
}
```

In this way we obtain the annotation

$$\langle 0.7 \rangle \; x := 0; \; \langle \mathcal{I} \rangle \; \mathbf{while}(x = 0)\{\ldots\} \; \langle [x \neq 0] \cdot [\mathcal{I}] = [x = 1] \rangle$$

as desired. It is sound because the program obviously terminates with probability one and $\mathcal{I}$ is bounded.

The crucial point in determining a pre-expectation of a program is to discover the necessary loop invariants for each loop. Checking soundness and carrying out subsequent calculations for the other program constructs turns out be easy in practice. In the following section we explain our approach to finding invariants step by step.

3   Our Approach

To explain the steps carried out by PRINSYS we revisit the geometric distribution program from Lst. 1. In the next section, we will view it in a broader context.

Template. Consider the loop:

$$\mathbf{while} \; (\mathit{flip} = 0)\{ \; (\mathit{flip} := 1 \; [p] \; x := x+1); \; \}$$

and an expectation

$$\mathcal{T}_\alpha = [x \geq 0] \cdot x + [x \geq 0 \wedge \mathit{flip} = 0] \cdot \alpha$$

where α is an unknown (real) parameter. We call $\mathcal{T}_\alpha$ a *template*. Replacing α by a real value yields an *instance* of the template. Depending on this value, some instances may satisfy the invariance condition $\mathcal{T}_\alpha \cdot [G] \leq \mathrm{wlp}(\mathit{body}, \mathcal{T}_\alpha)$.

Goal. PRINSYS gives a characterisation of all invariant instances of a given template. This characterisation is a formula which is true for all admissible values of the template parameters, α in our example. It is important to stress that this method is *complete* in the sense that for any given template the resulting formula captures *precisely* the invariant instances.

Workflow. Stage 1: After parsing the program text and template, PRINSYS traverses the generated control flow graph of the program and computes:

$$\text{wp}(\textit{flip} := 1 \ [p] \ x := x{+}1, \ \mathcal{T}_\alpha)$$
$$= [x \geq 0] \cdot px + (1 - p) \cdot ([x + 1 \geq 0] \cdot (x + 1) + [x + 1 \geq 0 \wedge \textit{flip} = 0] \cdot \alpha) \ .$$

For details, cf. Table 1. After expanding this expression, the invariance condition amounts to:

$$\overbrace{[x \geq 0 \wedge \textit{flip} = 0] \cdot (x + \alpha)}^{\mathcal{T}_\alpha \cdot [G]} \leq \quad [x \geq 0] \cdot px$$
$$+ [x + 1 \geq 0] \cdot ((1 - p)x - p + 1)$$
$$\underbrace{+ [x + 1 \geq 0 \wedge \textit{flip} = 0] \cdot (1 - p)\alpha}_{\text{wlp}(\textit{body}, \mathcal{T}_\alpha)} \ .$$

Our goal is to find all α such that the point-wise inequality is satisfied, i.e. it holds for every x and every *flip*. This can be done by pairwise comparison of the summands on the left-hand side and the right-hand side. But summands may overlap. This makes it necessary to rewrite the expectations in disjoint normal form (DNF).

Theorem 1 (Transformation to DNF [7]). *Given an expectation of the form*

$$f = [P_1] \cdot w_1 + \ldots + [P_n] \cdot w_n.$$

Then an equivalent expectation in DNF can be written as:

$$\sum_{I \in \mathcal{P}(\overline{n}) \backslash \emptyset} \left(\left[\bigwedge_{i \in I} P_i \wedge \neg \left(\bigwedge_{j \in \mathcal{P}(\overline{n}) \backslash I} P_j \right) \right] \cdot \left(\sum_{i \in I} w_i \right) \right)$$

where $\overline{n}$ is the index set $\{1, \ldots, n\}$ and $\mathcal{P}(\cdot)$ denotes the power set.

The left-hand side of the inequality for the example program above is already in DNF as there is only one summand. We apply the transformation to the right-hand side expression. The result is an expectation with 15 summands. For better readability we only show the summands that are not trivially zero:

$$[x + 1 \geq 0 \wedge x < 0 \wedge \textit{flip} = 0)] \cdot ((1 - p)x + (1 - p)\alpha - p + 1)$$
$$+ [x \geq 0 \wedge \textit{flip} = 0)] \cdot (x + (1 - p)\alpha - p + 1)$$
$$+ [x + 1 \geq 0 \wedge x < 0 \wedge \textit{flip} \neq 0] \cdot ((1 - p)x - p + 1)$$
$$+ [x \geq 0 \wedge \textit{flip} \neq 0] \cdot (x - p + 1) \ .$$

The following theorem provides a straightforward encoding of the inequality as a first-order formula.

Theorem 2. *Given two expectations over variables* $x_1, \ldots, x_n$ *in disjoint-normal form*

$$f = [P_1] \cdot u_1 + \ldots + [P_M] \cdot u_M , \qquad g = [Q_1] \cdot w_1 + \ldots + [Q_K] \cdot w_K .$$

The inequality $f \leq g$ *holds if and only if*

$$\forall x_1, \ldots, x_n \in \mathbb{R} : \bigwedge_{m \in \overline{M}} \bigwedge_{k \in \overline{K}} (P_m \wedge Q_k \Rightarrow (u_m - w_k \leq 0))$$

$$\wedge \bigwedge_{m \in \overline{M}} \left(P_m \wedge \left(\bigwedge_{k \in \overline{K}} \neg Q_k \right) \Rightarrow u_m \leq 0 \right)$$

$$\wedge \bigwedge_{k \in \overline{K}} \left(Q_k \wedge \left(\bigwedge_{m \in \overline{M}} \neg P_m \right) \Rightarrow 0 \leq w_k \right)$$

holds, where $\overline{X}$ *is the set of indices* $\{1, 2, \ldots, X\}$.

The idea is that we consider individual summands on the left-hand and right-hand side of the inequality and compare their values. It may also be the case that for some evaluations, all predicates on the right-hand side are false and hence the expectation is zero (i.e., the zero function). Then it must be ensured that no summand is greater than zero on the left-hand side. Conversely, if none of the predicates on the left-hand side are satisfied, the summands on the right-hand side may be no less than zero.

Theorem 2 originally appears in [7] where the last case is omitted because expectations are assumed to be non-negative by definition. However it is crucial to encode such informal assumptions in the formula as the tools are not aware of such expectation properties and instead treat them as usual functions over real values. This issue remained undiscovered until its implementation in PRINSYS caused incorrect results. The lesson learned is that bridging the gap between an idea and a working implementation requires more than "just" coding.

Continuing our example, the (simplified) first-order formula obtained is:

$$\forall x, \textit{flip} : (\alpha p + p - 1 \leq 0 \vee \textit{flip} \neq 0 \vee x < 0)$$
$$\wedge (\alpha p - \alpha + px + p - x - 1 \leq 0 \vee \textit{flip} \neq 0 \vee x + 1 < 0 \vee x \geq 0)$$
$$\wedge (\textit{flip} = 0 \vee px + p - x - 1 \leq 0 \vee x + 1 < 0 \vee x \geq 0)$$
$$\wedge (\textit{flip} = 0 \vee p - x - 1 \leq 0 \vee x < 0) .$$

The calculation of this formula by PRINSYS concludes the first stage.

Stage 2: The formula is passed to REDLOG which simplifies the formula by quantifier elimination. Sometimes the result returned by REDLOG still contains redundant information and can be further reduced by its built-in simplifiers or by the SLFQ tool. In the end the user is presented a formula that characterises all αs that make $\mathcal{T}_\alpha$ invariant:

$$\alpha p + p - 1 \geq 0 .$$

Listing 3. Annotated program from Lst. 1

```
⟨ 1−p / p ⟩
x  :=  0;
flip  :=  0;
⟨[x ≥ 0] · x + [x ≥ 0 ∧ flip = 0] · 1−p/p ⟩
while (flip = 0) {
    ( flip := 1 [p] x := x+1 );
}
⟨x⟩
```

We pick the greatest admissible α and obtain an invariant:

$$\mathcal{T}_{\frac{1-p}{p}} = [x \geq 0] \cdot x + [x \geq 0 \wedge \mathit{flip} = 0] \cdot \frac{1-p}{p} \ .$$

This can be used to prove that the program in Lst. 1 has an average outcome of $\frac{1-p}{p}$ which indeed is the mean of a geometric distribution with parameter p. The annotated program now looks as follows: The soundness of our invariant is given because there is always a non-zero probability to exit the loop, cf. definition of invariants above.

Figure 1 pictures the described workflow of PRINSYS.

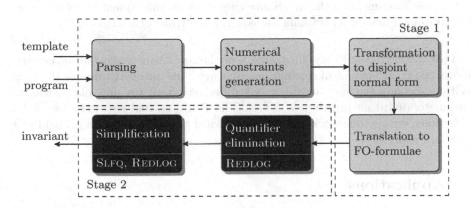

Fig. 1. Tool chain workflow

New Insights. There are major differences with the approach sketched in [7]. In PRINSYS we skip the additional step of translating the universally quantified formula into an existential one using the Motzkin's transposition theorem. This step turns out to be not necessary. In fact it complicates matters as the existential formula will have more quantified variables which is bad for quantifier

Listing 4.

```
c := IC; // capital c (is set to some InitialCapital)
b := 1; // initially bet one unit
rounds := 0; //number of rounds played (survived)
while (b > 0){
    {// win with probability p
    c := c+b;
    b := 0;}
    [p]
    {// lose with probability 1-p
    c := c-b;
    b := 2*b;}
    rounds := rounds+1;
}
```

elimination. Furthermore, Motzkin's transposition theorem requires the universally quantified formula to be in a particular shape. Our implementation however does not have these restrictions and allows arbitrary predicates in the program's guards and in templates. Also the template and program do not have to be linear (theoretically at least) because REDLOG and SLFQ can work with polynomials. Moreover the invariant generation method remains complete in this case. This is because starting with the invariance condition all subsequent steps to obtain the simplified first-order formula are equivalence transformations.

This section has not only illustrated how the tool-chain works but also clearly shows the great amount of calculations that are done automatically for the user. Within seconds the user may try out different templates and play with the parameters until an invariant is found. The PRINSYS tool saves the user a lot of tedious, error-prone work and pushes forward the automation of probabilistic program analysis.

4 Applications

This section presents three examples, for simplicity all based on our running example of the geometric distribution, that illustrate the possibilities of the PRINSYS approach. Let us start with a relatively simple example.

Martingale Betting Strategy. Another variant of the geometric distribution appears in the following program, which models a gambler with infinite resources who is playing according to the martingale strategy. Note that this program has two unbounded variables. Using the same template as before, we discover that $\frac{1}{p}$

| Listing 5. | Listing 6. |

```
x := 0;                              x := 0;
flip := 0;                           (flip := 0 [0.5] flip := 1);
while (flip = 0) {                    if (flip = 0) {
    (x := x+1 [p] flip := 1);            while (flip = 0) {
}                                            (x := x+1 [p] flip := 1);
flip := 0;                               }
while (flip = 0) {                   } else {
    (x := x-1 [q] flip := 1);            flip := 0;
}                                        while (flip = 0) {
                                             x := x-1;
                                             (skip [q] flip := 1);
                                         }
                                     }
```

is the expected number of rounds played before the gambler stops. The expectation differs from what we have computed for the program in Lst. 1 because here the counter is increased also on the last iteration before the loop terminates.

Geometric Distribution. This example is taken from [8] where amongst others it has been shown that the two programs in Lst. 5 and Lst. 6 are equivalent for $p = \frac{1}{2}$ and $q = \frac{2}{3}$. The proof in [8] relies on language equivalence checking of probabilistic automata. Here, we show how the techniques supported by PRINSYS can be used to show that both programs are equivalent for any p and q satisfying $q = \frac{1}{2-p}$. Let us explain the example in more detail. The aim is to generate a sample x according to the distribution $X-Y$ where X is geometrically distributed with parameter $1-p$ and Y is geometrically distributed with $1-q$.

Although it is not common to say that a distribution has a parameter $1-p$, it is natural in the context of these programs where x is manipulated with probability p and the loop is terminated with the remaining probability. The difference between the programs in Lst. 5 and Lst. 6 is that the first uses two loops in sequence whereas the latter needs only one out of two loops. Our goal is to determine when the two programs are equivalent, in the sense that they compute the same value for x on average.

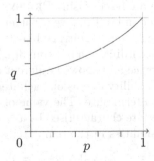

Fig. 2. Pairs (p, q) for which the programs in List. 5 and List. 6 produce the same x on average.

Listing 7. x is set to zero or one, each with probability 0.5

```
x := 0; // stores outcome of first biased coin flip
y := 0; // stores outcome of second biased coin flip

while (x−y = 0) {
    (x := 0 [p] x := 1);
    (y := 0 [p] y := 1);
}
```

The PRINSYS tool generates invariants for single loops, so we consider each loop separately. Using the template $\mathcal{T}_\alpha = [x \geq 0] \cdot x + [x \geq 0 \wedge \mathit{flip} = 0] \cdot \alpha$ from our running example, PRINSYS yields the following invariants:

- $\mathcal{I}_{11} = x + [\mathit{flip} = 0] \cdot \frac{p}{1-p}$,
- $\mathcal{I}_{12} = x + [\mathit{flip} = 0] \cdot \left(-\frac{q}{1-q}\right)$,
- $\mathcal{I}_{21} = \mathcal{I}_{11}$ and
- $\mathcal{I}_{22} = x + [\mathit{flip} = 0] \cdot \left(-\frac{1}{1-q}\right)$,

where $\mathcal{I}_{ij}$ is the invariant of the j-th loop in program i, $i, j \in \{1, 2\}$. With these invariants we can easily derive the expected value of x, which is $\frac{p}{1-p} - \frac{q}{1-q}$ and $\frac{p}{2(1-p)} - \frac{1}{2(1-q)}$ for the program in List. 5 and List. 6, respectively. The two programs thus are equivalent whenever these two expectations coincide; e. g. this is the case for $p = \frac{1}{2}$ and $q = \frac{2}{3}$ as discussed in [8]. Figure 2 visualises our result: for every point (p, q) on the graph the two programs are equivalent. This result cannot be obtained using the techniques in [8]; to the best of our knowledge there are no other automated techniques that can establish this.

Generating a Fair Coin from a Biased Coin. In [7], Hurd's algorithm to generate a sample according to a biased coin flip using only fair coin flips has been analysed. Using PRINSYS the calculations can be automated. This was elaborated in [3]. Here we consider an algorithm for the opposite problem. Using a coin with some arbitrary bias $0 < p < 1$, the algorithm in Lst. 7 generates a sample according to a fair coin flip. The loop terminates when the biased coin was flipped twice and showed different outcomes. Obviously the program terminates with probability one as on each iteration of the loop there is a constant positive chance to terminate. The value of x is taken as the outcome. The two possible outcomes are characterised by $x = 0 \wedge y = 1$ and $x = 1 \wedge y = 0$. We encode these two possibilities in the template:

$$[x = 0 \wedge y - 1 = 0] \cdot (\alpha) + [x - 1 = 0 \wedge y = 0] \cdot (\beta)$$

PRINSYS returns one constraint:

$$\alpha p^2 - \alpha p + \beta p^2 - \beta p \leq 0$$

As before we look for the maximum value, hence we consider equality with zero. The equation simplifies to $\alpha = -\beta$ because we know that $0 < p < 1$. Hence $[x = 0 \wedge y - 1 = 0] - [x - 1 = 0 \wedge y = 0]$ is invariant[4] which, together with almost sure termination, gives us

$$\text{wp}(prog, [x = 0 \wedge y - 1 = 0] - [x - 1 = 0 \wedge y = 0])$$
$$= \text{wp}(prog, [x = 0 \wedge y - 1 = 0]) - \text{wp}(prog, [x - 1 = 0 \wedge y = 0])$$
$$= 0 . \tag{1}$$

where $prog$ is the entire program from Lst 7. The previous argument about almost sure termination and possible outcomes shows that

$$\text{wp}(prog, [x = 0 \wedge y - 1 = 0] + [x - 1 = 0 \wedge y = 0])$$
$$= \text{wp}(prog, [x = 0 \wedge y - 1 = 0]) + \text{wp}(prog, [x - 1 = 0 \wedge y = 0])$$
$$= 1 . \tag{2}$$

The unique solution to (1) and (2) is

$$\text{wp}(prog, [x = 0 \wedge y - 1 = 0])$$
$$= \text{wp}(prog, [x - 1 = 0 \wedge y = 0])$$
$$= 0.5 .$$

This concludes the proof that x is distributed evenly for any p satisfying $0 < p < 1$.

5   Evaluation

We have seen three pGCL programs that were variants of the geometric distribution. Our approach allows us to exploit their common structure and enables us to calculate the expectation of these programs using the same template although they compute different (mean) values. Since our method does not rely on numerical calculation we are able to parameterise the programs and provide very general results. In particular we could decide when two programs have the same expectation depending on their parametric distributions. Another handy feature of reasoning with expectation-transformer wp is that we can exploit its properties as well. For example, the reasoning is modular with respect to sequential composition. That means we can compute the pre-expectation for individual loops and then add the results when we put the loops in sequence. The last example demonstrates yet another use of invariants. Instead of deriving a bound on the pre-expectation we have shown how an invariant may give constraints on the pre-expectation. Together with termination these constraints produced the sought pre-expectation. This exemplifies that invariants are not just a particular

---

[4] We pick $\alpha = 1$ and $\beta = -1$ but in fact any non-zero pair of values $\alpha = -\beta$ would result in the same argument.

way to compute an expectation but rather they describe the behaviour of the program and can be used in different ways.

Together with the three (other) examples discussed in [4,7] we have a set of interesting programs which we can analyse with the help of PRINSYS. Note, that our examples do not make use of the non-deterministic choice statement. This is because the algorithms we focused on do not need it, however PRINSYS also supports non-deterministic pGCL programs. There is no commonly accepted benchmark suite that we can compare against as this area of research has not spawned many tools yet. We refrain from giving a table that shows for each program the state space size, the number of discovered invariants or running times. This is because the beauty of this approach is exactly that the number of states does not matter. In fact all programs that generate (a variant) of the geometric distribution have an infinite set of reachable states! The number of discovered invariants cannot be really be given as, first of all the result depends on the template provided and second we get a characterisation of *all* invariant instances of a template. Since we reason over the reals there are uncountably many.

The runtime of PRINSYS depends on the size of the expressions that we have to handle. This means that if we have many choices in the loop (i.e. there are many paths in the control flow graph) this will blow up the size of wp($body, \mathcal{T}$). The same is true for templates that have many summands. Finally, the external tools used by PRINSYS affect the overall running time. Their execution time cannot be predicted exactly but experience shows that the final simplification step takes considerably longer the more parameters we allow in the template. The overall runtime for the presented examples lies within a second on a laptop computer.

Since there is no software that could be easily adapted to support our methods, PRINSYS was developed from scratch. It was recently redesigned to be more extensible and easier to maintain as we hope that future developments in the area of constraint-based methods will use our work as a basis. From the user's point of view, the usability was substantially increased with the introduction of a graphical user interface that allows an intuitive interaction.

Programs and templates considered in our examples are linear. This means all guards, assignments or terms are linear in the program variables. As pointed out earlier, our approach per se allows polynomial expressions as well. To see to what extent this applies in practice we have tried to generate polynomial invariants for variants of a bounded random walk, cf. Lst. 8. The goal is here to estimate the number of steps taken before x hits its lower bound zero or upper bound M where M is a fixed parameter. Surprisingly quantifier elimination works reasonably fast for formulas with polynomials but the returned quantifier-free formula is very big. The lack of powerful simplification methods makes it difficult to find a concise representation of the formula that describes all invariant instances of the template. REDLOG's simplifier might increase the formula size or not terminate at all, whereas SLFQ hits the memory bound quickly and crashes, even if the allocated memory is increased maximally.

Listing 8. Bounded random walk

```
counter := 0;
while (x > 0 and x–M < 0){
    (x := x+1 [p] x := x−1);
    counter := counter+1;
}
```

6   Conclusion

We have presented a new software tool called PRINSYS for probabilistic invariant generation. Its functionality was explained and its merits were assessed in the discussion. Also implementation details that deviate from the theoretic description of the method in [7] were pointed out. During our evaluation we have reached the next challenge, that is to extend invariant generation to polynomial templates. Related work, e.g. [12] suggests a workaround to find polynomial invariants for non-probabilistic programs. This comes at the price that they sacrifice completeness and limit the class of systems permitted. In the future we would like to work out a similar approximate invariant generation method for probabilistic systems and evaluate it within PRINSYS.

References

1. Barthe, G., Grégoire, B., Béguelin, S.Z.: Probabilistic relational Hoare logics for computer-aided security proofs. In: Gibbons, J., Nogueira, P. (eds.) MPC 2012. LNCS, vol. 7342, pp. 1–6. Springer, Heidelberg (2012)
2. Barthe, G., Köpf, B., Olmedo, F., Béguelin, S.Z.: Probabilistic relational reasoning for differential privacy. In: Symp. on Principles of Programming Languages (POPL), pp. 97–110. ACM (2012)
3. Gretz, F.: Invariant Generation for Linear Probabilistic Programs. Master's thesis, RWTH Aachen (2010), http://www-i2.informatik.rwth-aachen.de/i2/gretz/
4. Gretz, F., Katoen, J.P., McIver, A.: Operational versus Weakest Precondition Semantics for the Probabilistic Guarded Command Language. In: QEST, pp. 168–177 (2012)
5. Hahn, E.M., Hermanns, H., Wachter, B., Zhang, L.: PASS: Abstraction refinement for infinite probabilistic models. In: Esparza, J., Majumdar, R. (eds.) TACAS 2010. LNCS, vol. 6015, pp. 353–357. Springer, Heidelberg (2010)
6. Hahn, E.M., Hermanns, H., Zhang, L.: Probabilistic Reachability for Parametric Markov Models. STTT 13(1), 3–19 (2011)
7. Katoen, J.-P., McIver, A.K., Meinicke, L.A., Morgan, C.C.: Linear-Invariant Generation for Probabilistic Programs: In: Cousot, R., Martel, M. (eds.) SAS 2010. LNCS, vol. 6337, pp. 390–406. Springer, Heidelberg (2010)
8. Kiefer, S., Murawski, A.S., Ouaknine, J., Wachter, B., Worrell, J.: On the Complexity of the Equivalence Problem for Probabilistic Automata. In: Birkedal, L. (ed.) FOSSACS 2012. LNCS, vol. 7213, pp. 467–481. Springer, Heidelberg (2012)

9. Kiefer, S., Murawski, A.S., Ouaknine, J., Wachter, B., Worrell, J.: APEX: An Analyzer for Open Probabilistic Programs. In: Madhusudan, P., Seshia, S.A. (eds.) CAV 2012. LNCS, vol. 7358, pp. 693–698. Springer, Heidelberg (2012)
10. Kwiatkowska, M., Norman, G., Parker, D.: PRISM 4.0: Verification of Probabilistic Real-time Systems. In: Gopalakrishnan, G., Qadeer, S. (eds.) CAV 2011. LNCS, vol. 6806, pp. 585–591. Springer, Heidelberg (2011)
11. McIver, A., Morgan, C.: Abstraction, Refinement and Proof For Probabilistic Systems. Monographs in Computer Science. Springer (2004)
12. Sankaranarayanan, S., Sipma, H., Manna, Z.: Non-linear Loop Invariant Generation Using Gröbner Bases. In: POPL, pp. 318–329 (2004)
13. Ying, M.: Floyd-Hoare logic for quantum programs. ACM Trans. Program. Lang. Syst. 33(6), 19 (2011)

Revisiting Weak Simulation
for Substochastic Markov Chains

David N. Jansen[1], Lei Song[2,5], and Lijun Zhang[3,4,5]

[1] Radboud Universiteit, Model-Based System Development,
Nijmegen, The Netherlands
dnjansen@cs.ru.nl
[2] Max-Planck-Institut für Informatik, Saarbrücken, Germany
song@cs.uni-saarland.de
[3] State Key Laboratory of Computer Science, Institute of Software,
Chinese Academy of Sciences, Beijing, China
zhanglj@ios.ac.cn
[4] Technical University of Denmark, DTU Compute, Denmark
[5] Universität des Saarlandes, Saarbrücken, Germany

Abstract. The spectrum of branching-time relations for probabilistic systems has been investigated thoroughly by Baier, Hermanns, Katoen and Wolf (2003, 2005), including weak simulation for systems involving substochastic distributions. Weak simulation was proven to be sound w. r. t. the liveness fragment of the logic $PCTL_{\setminus \mathcal{X}}$, and its completeness was conjectured. We revisit this result and show that soundness does not hold in general, but only for Markov chains without divergence. It is refuted for some systems with substochastic distributions. Moreover, we provide a counterexample to completeness. In this paper, we present a novel definition that is sound for live $PCTL_{\setminus \mathcal{X}}$, and a variant that is both sound and complete.

A long version of this article containing full proofs is available from [11].

1 Introduction

Simulation relations are often used to verify that one system correctly implements another, more abstract system [1]. Simulation relations are therefore used as a basis for abstraction techniques, where the rough idea is to replace the model to be verified by a smaller model and to verify the latter instead of the original one. Dually, simulation relations are also used to refine a high-level specification into a low-level implementation. To be useful for abstraction and refinement, a simulation relation has to show a form of *weak preservation*, i. e., all properties expressible as positive formulas are preserved.

We choose a *liveness* view on simulation, for reasons that will be explained shortly. In this view, an abstract model underapproximates a concrete one, so the latter simulates the former. Every behaviour possible in the abstract model is also possible in the concrete one; i. e., every liveness property ensured by the former also holds in the latter. In a probabilistic context, a liveness property is

K. Joshi et al. (Eds.): QEST 2013, LNCS 8054, pp. 209–224, 2013.

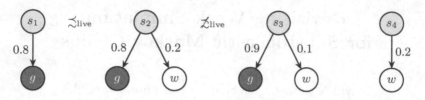

Fig. 1. Without substochastic distributions, simulation degenerates to bisimulation.

a lower bound on the probability of some (good) behaviour. For example, for strong simulation $\precsim$ in labelled Markov processes, $s \precsim t$ iff for all formulas Φ in $\mathcal{L}_\vee$ (a logic for liveness properties), $s \models \Phi$ implies $t \models \Phi$ [8]. The concrete state t satisfies all liveness properties that hold in the abstract state s.

Simulation for fully probabilistic models (without nondeterminism) faces a difficulty: many modelling formalisms require that all probability distributions are stochastic, i.e. the probabilities sum to exactly one. Consider s_2 in Fig. 1. (We use colours to indicate the state labelling: a state can only simulate states with the same colour.) If it is required to reach the goal state ● with probability at least 0.8, such a model cannot leave unspecified what happens with the remaining probability. For example, the wrong state ○ is reached with probability 0.2. As a consequence, s_3 in the same figure, while satisfying the requirement, does not simulate s_2 because the probability to reach ○ from s_3 is not large enough. Simulation degenerates to bisimulation. A solution to this problem is to allow *substochastic* distributions: it is enough if the probabilities sum to *at most* one, so that we can model the requirement like s_1 in Fig. 1. It is not specified what s_1 will do with the remaining probability 0.2. Another interpretation is that with probability 0.2, s_1 will do nothing at all, i.e. it deadlocks. In both interpretations, any model will simulate an unspecified or deadlocking model.

Alternatively, one could have chosen a safety view on simulation, i.e. the abstract model overapproximates the concrete one and every behaviour forbidden by the abstract model is also forbidden in the concrete one. But if we try to model forbidden behaviours by substochastic distributions, we get models like s_4 in Fig. 1, which should express that with probability (at most) 0.2, ○ is reached and with probability (at least) 0.8, any behaviour except entering ○ is acceptable – a much more complex semantics.

In a *weak* simulation relation, only visible steps are compared, while internal computations (called silent steps) are neglected. Weak simulation for Markov chains (including substochastic ones) was introduced in [2,4] and denoted $\precsim_d$. The authors claim that weak simulation is sound w.r.t. the liveness fragment of the logic $\mathrm{PCTL}_{\backslash \mathcal{X}}$. Completeness is conjectured to hold as well. Unfortunately, neither of the properties holds on substochastic DTMCs.

The main problem with soundness is that $\precsim_d$ only compares probabilities under the condition that some visible step is taken. However, if the concrete model deadlocks, nothing visible will happen, nor is there a successor state that could take the required visible step. Completeness is broken in a similar way: A single PCTL path property is not able to express multiple requirements on behaviours, but $\precsim_d$ still requires that the concrete state reached after a silent step can execute all behaviours of the abstract state.

To combat these problems, we base our definition of weak simulation on a notion of weak transition called *derivative*. In a derivative, one does not look too closely at intermediary states reached by silent steps, but concentrates on the visibly reached states. Overall, we get a relation that is sound w.r.t. the liveness fragment of $\mathrm{PCTL}_{\backslash \mathcal{X}}$, and we conjecture its completeness. A variant of the definition is provably sound and complete.

2 Preliminaries

A distribution μ over the set Σ is a function $\mu : \Sigma \to [0, 1]$ satisfying the condition $\mu(\Sigma) \le 1$, where $\mu(T) := \sum_{s \in T} \mu(s)$. We let $Dist(\Sigma)$ denote the set of distributions over Σ. The *support* of μ is the set of states on which μ is non-zero, i.e., $Supp(\mu) = \{s \in \Sigma \mid \mu(s) > 0\}$. We assume that all distributions considered have countable supports; most distributions will even have finite supports.

The distribution μ is called *stochastic* if $\mu(\Sigma) = 1$ and *absorbing* if $\mu(\Sigma) = 0$. Otherwise, i.e. if $0 < \mu(\Sigma) < 1$, we say μ is *substochastic*. Some authors call a substochastic or absorbing distribution a *subdistribution*. We sometimes use an auxiliary outcome $\bot \notin \Sigma$ and set $\mu(\bot) := 1 - \mu(\Sigma)$. Let $\Sigma_\bot$ denote the set $\Sigma \cup \{\bot\}$. $\mathcal{D}_s$ denotes the *Dirac* distribution such that $\mathcal{D}_s(s) = 1$.

For a relation $R \subseteq \Sigma \times \Pi$ (for sets Σ and Π) and some $s \in \Sigma$, we let $R[s]$ denote the set $\{p \in \Pi \mid s\,R\,p\}$. Similarly, $R[S] = \{p \in \Pi \mid \exists s \in S : s\,R\,p\}$.

2.1 Substochastic Discrete-Time Markov Chains

Let AP denote a fixed, finite, nonempty set of atomic propositions.

Definition 1. *A* substochastic discrete-time Markov chain *(sDTMC) is a tuple* $\mathcal{M} = (S, \mathbf{P}, L)$ *where:*

- *S is a finite or countable set of states,*
- *$\mathbf{P} : S \times S \to [0, 1]$ is a subprobability matrix such that for all $s \in S$, $\mathbf{P}(s, \cdot)$ is a distribution over S with finite support,*
- *$L : S \to 2^{AP}$ is a labelling function.*

A state $s \in S$ is called stochastic, absorbing, or substochastic if the distribution $\mathbf{P}(s, \cdot)$ is stochastic, absorbing, or substochastic, respectively. (A state s with $\mathbf{P}(s, s) = 1$ is stochastic.) Intuitively, $\mathbf{P}(s, t)$ denotes the probability of moving from s to t in a single step. For $s \in S$, let $post_\bot(s) := \{t \subset S_\bot \mid \mathbf{P}(s, t) > 0\}$, i.e., the set of successor states of s (including $\bot$ if s is not stochastic). A sDTMC without substochastic states is a *discrete-time Markov chain*. A path π is either an infinite sequence $s_0, s_1 \ldots$ such that $\mathbf{P}(s_i, s_{i+1}) > 0$ for $i = 0, 1, \ldots$, or a finite sequence $s_0, s_1 \ldots s_n$ satisfying $s_n = \bot$ and $\mathbf{P}(s_i, s_{i+1}) > 0$ for $i = 0, 1, \ldots, n - 1$. We use $\pi_i = s_i$ to denote the $(i + 1)$th state, if it exists. A path fragment is a strict prefix of a path. Each state s induces a probability space, whose σ-algebra is generated by *cylinder sets* like $C(s, s_1, \ldots, s_n)$, the set that contains all paths beginning with the path fragment $s, s_1, \ldots, s_n$. The probability measure $Prob_s$ is uniquely determined by: $Prob_s(C(s, s_1, \ldots, s_n)) = \mathbf{P}(s, s_1)\mathbf{P}(s_1, s_2) \cdots \mathbf{P}(s_{n-1}, s_n)$.

For $k \in \mathbb{N}$, $s \in S$, and sets $Tau, G \subseteq S$, let $Prob_s(Tau \, \mathcal{U}^{=k} \, G)$ denote the probability to be in a G-state after exactly k steps and to pass through Tau-states before, if starting in s. Similarly, $Prob_s(Tau \, \mathcal{U}^{\leq k} \, G)$ denotes the probability to reach G after passing through Tau for at most k steps, and $Prob_s(Tau \, \mathcal{U} \, G)$ is an abbreviation for $\lim_{k \to \infty} Prob_s(Tau \, \mathcal{U}^{\leq k} \, G)$. Finally, $Prob_s(\Diamond^{\leq k} G)$ is an abbreviation for $Prob_s(S \, \mathcal{U}^{\leq k} \, G)$.

In the following, we assume given a fixed sDTMC $\mathcal{M} = (S, \mathbf{P}, L)$.

2.2 Probabilistic CTL

We recall briefly the PCTL$_{\backslash \mathcal{X}}$ liveness formulas and their semantics. Details can be found in [4]. The syntax of the PCTL$_{\backslash \mathcal{X}}$ liveness formulas is defined by:

$$\Phi = true \mid false \mid a \mid \neg a \mid \Phi \wedge \Phi \mid \Phi \vee \Phi \mid \mathcal{P}_{>p}(\Phi \, \mathcal{U} \, \Phi) \mid \mathcal{P}_{\geq p}(\Phi \, \mathcal{U} \, \Phi),$$

where $a \in AP$ runs over the atomic propositions and $p \in [0, 1]$. The semantics for *true*, *false*, atomic propositions, negation, conjunction and disjunction are defined as usual. We denote the set of states that satisfy Φ by $Sat(\Phi)$.

A path π satisfies the until formula $\Phi_1 \, \mathcal{U} \, \Phi_2$ if there exists an index i such that π_i exists with $\pi_i \models \Phi_2$, and $\pi_j \models \Phi_1$ for all $j < i$. A state s satisfies the probabilistic formula $\mathcal{P}_{\trianglerighteq p}(\Phi_1 \, \mathcal{U} \, \Phi_2)$ if the probability that a path from s satisfies $\Phi_1 \, \mathcal{U} \, \Phi_2$ meets the bound, i. e. $Prob_s(Sat(\Phi_1) \, \mathcal{U} \, Sat(\Phi_2)) \trianglerighteq p$. We write $\mathcal{P}_{\trianglerighteq p}(\Diamond \Phi_2)$ as an abbreviation for $\mathcal{P}_{\trianglerighteq p}(true \, \mathcal{U} \, \Phi_2)$.

We define a relation $\precsim_{\text{live}}$ by: $s \precsim_{\text{live}} t$ if for all PCTL$_{\backslash \mathcal{X}}$ liveness formulas Φ it holds that $s \models \Phi$ implies $t \models \Phi$. The equivalence relation $\approx_{\text{live}}$ can be defined as the intersection $\precsim_{\text{live}} \cap \succsim_{\text{live}}$. So, $s \approx_{\text{live}} t$ if for all PCTL$_{\backslash \mathcal{X}}$ liveness formulas Φ it holds that $s \models \Phi$ if and only if $t \models \Phi$.[1]

3 Weak Bisimulation and Divergence

Weak bisimulation $\approx_d$ (as defined in [4]) is sound and complete, i. e., it coincides with $\approx_{\text{live}}$, for most sDTMCs; only for infinite sDTMCs with nonzero probability to take infinitely many silent transitions (to diverge), there is a problem:

Example 2. Consider the infinite DTMC in Fig. 2, constructed by Chenyi Zhang and Carroll Morgan [6, Example 3.16]. The probability to diverge, i. e. to take infinitely many transitions within the $\bigcirc$-states, when starting from s'_k, is

$$\prod_{i=k}^{\infty} \frac{i^2 - 1}{i^2} = \lim_{m \to \infty} \prod_{i=k}^{m} \frac{(i-1)(i+1)}{i^2} = \lim_{m \to \infty} \frac{(k-1)(m+1)}{km} = \frac{k-1}{k}.$$

---

[1] Others define $s \approx_{\text{PCTL} \backslash \mathcal{X}} t$ to hold if for *all* PCTL$_{\backslash \mathcal{X}}$ formulas Φ, even those that are not liveness formulas, $s \models \Phi$ iff $t \models \Phi$. However, this relation coincides with $\approx_{\text{live}}$. See Thm. 10.67 in [3, page 813sq.], (c) $\Longleftrightarrow$ (d), for an analogous statement, whose proof can easily be adapted.

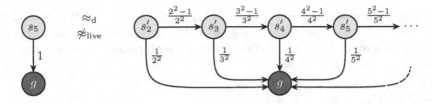

Fig. 2. $\approx_d$ may be unsound for DTMCs that diverge with nonzero probability.

As a result, $Prob_{s'_k}(\diamond\, Sat(\bullet)) = 1 - (k-1)/k = 1/k$ and $s_5 \not\approx_{\text{live}} s'_k$. However, $s_5 \approx_d s'_k$ for all $k \geq 2$: All transitions between $\bigcirc$-states can be considered silent, and then the probability to reach $\bullet$ under the condition to take a visible step agrees between s_5 and s'_k.

Reachability probabilities are often calculated with a linear equation system (Eqn. (6) in [4]). The proof that $\approx_d$ is sound relies on the assumption that it has a unique solution, which holds if the probability of divergence is zero. Generally, the reachability probabilities are the smallest solution, which is always unique because of the Knaster–Tarski fixpoint theorem [15]. So it is enough to restrict the probability of divergence. We propose to change the third condition of the definition:

Definition 3. *The equivalence relation $R \subseteq S \times S$ is a divergence-sensitive weak bisimulation[2] iff for all s, t with $s\,R\,t$:*

1. *$L(s) = L(t)$,*
2. *Let $B := R[s] = R[t]$ be the equivalence class of s and t. If $\mathbf{P}(s, B) < 1$ and $\mathbf{P}(t, B) < 1$, then for all $C \in S/R$ with $C \neq B$:*

$$\frac{\mathbf{P}(s, C)}{1 - \mathbf{P}(s, B)} = \frac{\mathbf{P}(t, C)}{1 - \mathbf{P}(t, B)},$$

3. *$Prob_s(\diamond\, S \setminus B) = Prob_t(\diamond\, S \setminus B)$.*

States s and t are ds-weakly bisimilar, denoted $s \approx t$, if there exists a divergence-sensitive weak bisimulation R with $s\,R\,t$.

Proposition 4. *Divergence-sensitive weak bisimulation $\approx$ is sound and complete for sDTMCs, both countable and finite. On sDTMCs that diverge with probability 0, it coincides with $\approx_d$.*

4   Defects of Original Weak Simulation

We recall the definition of weak simulation [4]. It is based on the notion of weight functions, used to lift a relation $R \subseteq S \times M$ to a relation $\sqsubseteq_R \subseteq Dist(S) \times Dist(M)$. We will first use the definition only for relations $R \subseteq S \times S$. Weight functions were introduced in [12] and adapted in [4] to incorporate substochastic states.

---

[2] The name reminds of divergence-sensitive stutter equivalence [5].

Definition 5 (Weight function). *Let S and M be sets and $R \subseteq S \times M$ be a relation. Let $\sigma \in Dist(S)$ and $\mu \in Dist(M)$ be distributions with at most countable supports. A weight function for (σ, μ) with respect to R is a function $\Delta : S_\perp \times M_\perp \to [0, 1]$ such that*

1. *$\Delta(s, m) > 0$ implies $s \, R \, m$ or $s = \perp$,*
2. *$\sigma(s) = \Delta(s, M_\perp)$ for $s \in S_\perp$, and*
3. *$\mu(m) = \Delta(S_\perp, m)$ for $m \in M_\perp$.*

We write $\sigma \sqsubseteq_R \mu$ if there exists a weight function for (σ, μ) with respect to R.

Note that the support of Δ is a subset of $Supp(\sigma) \times Supp(\mu)$, so it is at most countable. Therefore, the sums in Conds. 2 and 3 have at most a countable number of nonzero summands.

The following equivalent characterisation of the lifting will be useful later. See [16, 9], and a detailed proof can be found in [13, Lemma 1].

Lemma 6. *With the notations of Def. 5, $\sigma \sqsubseteq_R \mu$ iff $\sigma(G) \leq \mu(R[G])$ for all $G \subseteq Supp(\sigma)$.*

To check whether some relation R is a weak simulation, [4] defines, for every pair $s_1 \, R \, s_2$, which successors of s_i are visible and which ones are silent. The functions $\delta_i : S_\perp \to [0, 1]$ below have this task: $\delta_i(s') = 0$ means that the transition $s_i \to s'$ is silent. Then, $K_i := \sum_{u \in S_\perp} \mathbf{P}(s_i, u)\delta_i(u)$ is the probability to take a visible transition from s_i at all. If R is a weak simulation, there should exist a mapping from the visible transitions of s_1 to (a subset of) the visible transitions of s_2. To this end, [4] compares (through the lifting of R) the probabilities to move from s_i to u, under the condition that the transition is visible: $\mathbf{P}(s_i, u \,|\, \text{visible}) := \mathbf{P}(s_i, u)\delta_i(u)/K_i$.

Definition 7 (Weak simulation $\precsim_d$ in [4]). *The relation $R \subseteq S \times S$ is a weak simulation if $s_1 \, R \, s_2$ implies that $L(s_1) = L(s_2)$ and there exist functions $\delta_i : S_\perp \to [0, 1]$ such that, using the sets*

$$U_i = \{u_i \in post_\perp(s_i) \mid \delta_i(u_i) > 0\} \qquad \text{(visible successors)}$$
$$V_i = \{v_i \in post_\perp(s_i) \mid \delta_i(v_i) < 1\} \qquad \text{(silent successors)},$$

the following conditions hold:

1. *$v_1 \, R \, s_2$ for all $v_1 \in V_1 \setminus \{\perp\}$ and $s_1 \, R \, v_2$ for all $v_2 \in V_2 \setminus \{\perp\}$.*
2. *If both $K_1 > 0$ and $K_2 > 0$, then $\mathbf{P}(s_1, \cdot \,|\, \text{visible}) \sqsubseteq_R \mathbf{P}(s_2, \cdot \,|\, \text{visible})$.*
3. *For every $u_1 \in U_1 \setminus \{\perp\}$, $Prob_{s_2}(R[s_1] \, \mathcal{U} \, R[u_1]) > 0$.*

We say that s_2 weakly simulates s_1, denoted $s_1 \precsim_d s_2$, iff there exists a weak simulation R such that $s_1 \, R \, s_2$.

Weak simulation on DTMCs arises as a special case of the above definition, as every DTMC is an sDTMC (where each state is absorbing or stochastic).

Theorem 63 of [4] now states the soundness of $\precsim_d$ w.r.t. live $PCTL_{\setminus \mathcal{X}}$. Namely, that for $s, t \in S$, we have: If $s \precsim_d t$, then for all $PCTL_{\setminus \mathcal{X}}$ liveness formulas Φ, $s \models \Phi$ implies $t \models \Phi$. In the conclusion of [4] it is conjectured that also the converse – completeness of $\precsim_d$ – holds. Unfortunately this is false:

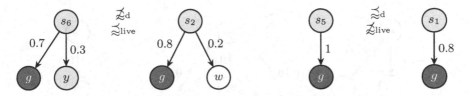

Fig. 3. $\precsim_d$ is not complete.   **Fig. 4.** $\precsim_d$ is not sound.

Example 8. The DTMC depicted in Fig. 3 illustrates that weak simulation is not complete w. r. t. live PCTL$_{\setminus \mathcal{X}}$. Let us prove that for all formulas Φ, $s_6 \models \Phi$ implies $s_2 \models \Phi$. The only formulas for which the proof is not trivial are those that measure the paths in $C(s_6, y)$, say $s_6 \models \mathcal{P}_{\geq 0.3}(\Phi_1 \, \mathcal{U} \, \Phi_2)$ with $y \models \Phi_2$. As s_2 has the same colour as y, also $s_2 \models \Phi_2$, and thus $s_2 \models \mathcal{P}_{\geq 0.3}(\Phi_1 \, \mathcal{U} \, \Phi_2)$.

If it would hold that $s_6 \precsim_d s_2$, then $\delta_1(g) = \delta_2(g) = \delta_2(w) = 1$. So, $U_1 = \{g\}$ or $\{g, y\}$, $K_1 \geq 0.7$ and $K_2 = 1$, therefore a weight function Δ would exist. However, as $g \not\precsim_d w$ and $y \not\precsim_d w$, it satisfies $0 = \Delta(U_1, w) = \mathbf{P}(s_2, w \,|\, \text{visible}) = \mathbf{P}(s_2, w)\delta_2(w)/K_2 = 0.2$. Contradiction!

Even worse: the relation $\precsim_d$ is not sound on sDTMCs.

Example 9. The sDTMC in Fig. 4 illustrates that weak simulation is not sound w. r. t. live PCTL$_{\setminus \mathcal{X}}$. Namely, $s_5 \precsim_d s_1$, because we can choose $\delta_1(g) = \delta_2(g) = 1$ and $\delta_2(\bot) = 0$. Then, the sets U_i and V_i are: $U_1 = U_2 = \{g\}$, $V_1 = \emptyset$, $V_2 = \{\bot\}$, and $K_1 = 1$, $K_2 = 0.8$. The conditions hold trivially.

Now consider the formula $\Phi := \mathcal{P}_{>0.9}(\bigcirc \, \mathcal{U} \, \bullet)$, which states that the probability to reach $\bullet$-states is greater than 0.9. Obviously, the probability to reach $\bullet$-states from s_5 is 1, and from s_1 is 0.8, thus $s_5 \models \Phi$ but $s_1 \not\models \Phi$.

The problem went undetected because the proof of Thm. 63 in [4] allows a nice intuition with just one wrong detail: one constructs an intermediary sDTMC that contains states $\langle s, t, 1 \rangle$ and $\langle s, t, 2 \rangle$ for every state pair $s \precsim_d t$, defined in a way that it is easy to see $s \approx_d \langle s, t, 1 \rangle \precsim_{\text{live}} \langle s, t, 2 \rangle \approx_d t$. If $K_1 > 0$ and $K_2 > 0$, the new state $\langle s, t, 1 \rangle$ has $1/(1 + M)$ times the original transitions of s (for some carefully selected constant $M \in \mathbb{R}_{\geq 0}$) and moves to states bisimilar to s with probability $M/(1 + M)$, so that $s \approx_d \langle s, t, 1 \rangle$ follows immediately. The bisimilar states have the form $\langle s, v_2, 1 \rangle$ for $v_2 \in V_2$ – except that there is no state $\langle s, \bot, 1 \rangle$. This is problematic if $M > 0$ (which is equivalent to $K_2 < 1$) and $\bot \in V_2$. Note that $K_2 < 1$ follows from $\bot \in V_2$.

In terms of the example, for any silent step $s_1 \to v_2$, the reached state satisfies $s_5 \precsim_d v_2$ and therefore $\mathbf{P}(s_5, \bullet) \leq Prob_{v_2}(\Diamond \, \bullet)$ – except for $v_2 = \bot$.

As the proof also relies on the soundness of $\approx_d$, it does not work for sDTMCs that may diverge. In particular, also $s_5 \precsim_d s'_k$, similar to Example 2.

Lemma 10. $\precsim_d$ *is sound on sDTMCs that diverge with probability 0, if no state pair* $s \precsim_d t$ *requires a choice of* δ_1 *and* δ_2 *such that* $K_1 > 0$, $K_2 > 0$ *and* $\bot \in V_2$.

For DTMCs without substochastic states, always $\bot \notin V_2$. So, $\precsim_d$ is sound if the simulating sDTMC is not substochastic and almost surely does not diverge.

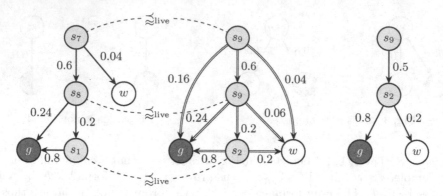

Fig. 5. Some sDTMCs illustrating the weak simulation relation.

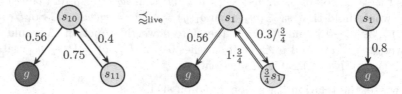

Fig. 6. Rescaling states: $\frac{3}{4}$ of s_1 is used to simulate s_{11}.

5   A New Notion of Weak Simulation

Before we come to an improved definition, let us give three motivating examples.

Example 11. This example illustrates which kinds of delaying or stuttering are needed for weak simulation.

Consider the sDTMCs on the left and right of Fig. 5. To simulate the transitions of s_7, state s_9 has to *delay* or to stutter with probability 0.6, and with the remaining probability, it moves on, so that it reaches a $\bigcirc$-state with probability $(1 - 0.6) \cdot 0.5 \cdot 0.2 = 0.04$. Note that we cannot simulate the transition $s_7 \to s_8$ by $s_9 \to s_2$ because the probability of the latter is lower than of the former.

Now consider $s_8 \precsim_{\text{live}} s_9$. Here, the transition to s_1 cannot be simulated by delaying in s_9 because the probability to reach a $\bullet$-state from the latter is too small. We therefore choose to delay in state s_2 instead with probability 0.2, so we reach a $\bullet$-state with probability $(0.5 - 0.2) \cdot 0.8 = 0.24$.

In our definition, we use *derivatives*, a kind of weak transition, to describe these delays systematically. In the center of Fig. 5, we show the weak transitions with double lines; see Example 22 below for the exact definitions of the derivative. State s_9 is drawn twice because we use two different derivatives to simulate s_7 and s_8, respectively.

Example 12. Sometimes, we have to rescale a part of the derivative.

Now consider state s_{10} in Fig. 6. The probability to reach g from s_{10} satisfies $Prob_{s_{10}}(\diamond \bullet) = 0.56 + 0.4 \cdot 0.75 \cdot Prob_{s_{10}}(\diamond \bullet)$, so it is 0.8. We conclude $s_{10} \precsim_{\text{live}} s_1$.

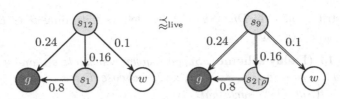

Fig. 7. Splitting states: $s_{12} \gtrsim_{\text{live}} s_9$. A part of s_2 is used to simulate s_1.

How can we find a derivative of s_1 to simulate $\mathbf{P}(s_{10}, \cdot)$? The naïve choice would be to delay in s_1 with probability 0.4, corresponding to $s_{10} \xrightarrow{0.4} s_{11} \gtrsim_{\text{live}} s_1$. But then, the probability to go to g can be at most $\mathbf{P}(s_1, g) \cdot (1 - 0.4) = 0.8 \cdot 0.6 = 0.48$, which is too small for $s_{10} \xrightarrow{0.56} g$. The point here is that s_1 oversimulates s_{11}; it would be enough to use $\frac{3}{4}$ of s_1. Our definitions allow to *rescale* this part of the derivative, so that enough probability mass is left to simulate the transition $s_{10} \to g$. The correct derivative therefore only delays in s_1 with probability 0.3; this corresponds to moving to "$\frac{3}{4}$ of s_1" with probability 0.4. The derivative then moves on to g with probability $\mathbf{P}(s_1, g) \cdot (1 - 0.3) = 0.56$, the required value. We draw the incomplete state as a partial eclipse.

Now, one might think that these two ideas – delaying and rescaling – provide enough liberty to define a new notion of weak simulation. However, we have to generalise rescaling slightly:.

Example 13. Consider state s_{12} in Fig. 7. One can show that $s_{12} \gtrsim_{\text{live}} s_9$. However, if we try to find a (rescaled) derivative, we get that the derivative is not allowed to delay in s_9 nor in s_2, because otherwise, the probability to get to w in one step by the simulating derivative would become too small.

The solution is to move to s_2 (a state that can simulate s_{12}) and rescale that state selectively: $0.5 \cdot \mathbf{P}(s_2, \cdot)$ is split into two substates with transition distributions $\sigma := \{(g, 0.24), (w, 0.1)\}$ and $\rho' := \{(g, 0.16)\}$, respectively. The first is used to simulate the transitions from s_{12} to g and w. In order to simulate the transition from s_{12} to s_1, we delay in the part of s_2 that has been split off. We denote this substate (rescaled appropriately) as $s_{2\restriction\rho}$.

We now introduce the concept of substates formally as follows:

Definition 14 (Substate). *A substate of $s \in S$ is a pair $(s, \sigma) \in S \times Dist(S)$ such that $\sigma \le \mathbf{P}(s, \cdot)$ (pointwise). We write this pair as $s_{\restriction\sigma}$. We extend $\mathbf{P}$ (in the first argument) to substates by setting $\mathbf{P}(s_{\restriction\sigma}, \cdot) := \sigma$. Let $Sub(T)$, for any $T \subseteq S$, denote the set of all substates $s_{\restriction\sigma}$ with $s \in T$.*

We will often write the improper substate $s_{\restriction\mathbf{P}(s, \cdot)}$ as $s_\restriction$.

We adapt the notion of *derivatives*, weak transitions, introduced in [7] to our state-based setting. We will assume given a set $Tau \subseteq S$; transitions between states in Tau are regarded as silent steps or τ steps. It typically contains the start state together with states that should not be distinguished from it.

For a distribution $\nu \in Dist(Sub(S))$, we let its *flattening* $\overline{\nu} \in Dist(S)$ be
$\overline{\nu} := \sum_{u_{\upharpoonright\upsilon} \in Sub(S)} \nu(u_{\upharpoonright\upsilon})\upsilon$.

Definition 15 (Delay scheme). *Suppose given a substate $s_{\upharpoonright\sigma}$ and a set $Tau \subseteq S$. For every $t \in Tau$ and $i \in \mathbb{N}_1 = \{1, 2, \ldots\}$, choose distributions $\mu_{t,i}^{\rightarrow} \le \mathbf{P}(t, \cdot)$ and $\mu_{t,i}^{\times} \in Dist(Sub(\{t\}))$ such that*

$$\mu_{t,i}^{\rightarrow}(S) + \mu_{t,i}^{\times}(Sub(S)) \le 1 \qquad and \qquad (1)$$

$$\mu_{t,i}^{\rightarrow} + \overline{\mu_{t,i}^{\times}} \le \mathbf{P}(t, \cdot). \qquad (2)$$

Similarly, choose $\mu_{s_{\upharpoonright\sigma},0}^{\rightarrow} \le \sigma$ and $\mu_{s_{\upharpoonright\sigma},0}^{\times} \in Dist(Sub(\{s\}))$, such that

$$\mu_{s_{\upharpoonright\sigma},0}^{\rightarrow}(S) + \mu_{s_{\upharpoonright\sigma},0}^{\times}(Sub(S)) \le 1,$$

$$\mu_{s_{\upharpoonright\sigma},0}^{\rightarrow} + \overline{\mu_{s_{\upharpoonright\sigma},0}^{\times}} \le \sigma, \quad and$$

$$if\ s \notin Tau,\ then\ \mu_{s_{\upharpoonright\sigma},0}^{\times}(s_{\upharpoonright\sigma}) = 1. \qquad (3)$$

This choice $(\mu_{t,i}^{\rightarrow}, \mu_{t,i}^{\times})_{t \in Tau, i \in \mathbb{N}_1}, \mu_{s_{\upharpoonright\sigma},0}^{\rightarrow}, \mu_{s_{\upharpoonright\sigma},0}^{\times}$ is a delay scheme.

The idea behind the scheme is: Whenever $t \in Tau$ is visited (after i transitions), we will either choose to continue with the probabilities indicated by $\mu_{t,i}^{\rightarrow}$ or to stop in (a substate of) t with the probabilities indicated by $\mu_{t,i}^{\times}$. The conditions ensure that the total probability is at most 1 and the probability to reach any successor of t, either directly or via a delay state, does not increase over $Prob_t(Tau\ \mathcal{U}\ \cdot)$. For technical reasons, the counter i is added; however, one can often choose $\mu_{t,i}^{\rightarrow}$ and $\mu_{t,i}^{\times}$ independent from i.

Definition 16 (Derivative). *Suppose given a substate $s_{\upharpoonright\sigma}$, a set $Tau \subseteq S$, and a delay scheme $(\mu_{t,i}^{\rightarrow}, \mu_{t,i}^{\times})_{t \in Tau, i \in \mathbb{N}_1}, \mu_{s_{\upharpoonright\sigma},0}^{\rightarrow}, \mu_{s_{\upharpoonright\sigma},0}^{\times}$. We extend $\mu_{t,i}^{\times}$ to S by setting $\mu_{t,i}^{\times}(t_{\upharpoonright}) := 1$ for $t \notin Tau$. Let $\nu_i^{\rightarrow} \in Dist(S)$ and $\nu_i^{\times} \in Dist(Sub(S))$, for every $i \ge 0$, be as follows:*

$$\nu_0^{\rightarrow} := \mu_{s_{\upharpoonright\sigma},0}^{\rightarrow} \qquad\qquad \nu_{i+1}^{\rightarrow} := \sum_{t \in Tau} \nu_i^{\rightarrow}(t)\mu_{t,i+1}^{\rightarrow}$$

$$\nu_0^{\times} := \mu_{s_{\upharpoonright\sigma},0}^{\times} \qquad\qquad \nu_{i+1}^{\times} := \sum_{t \in S} \nu_i^{\rightarrow}(t)\mu_{t,i+1}^{\times}.$$

The distribution

$$\nu := \sum_{i=0}^{\infty} \nu_i^{\times} \ \in Dist(Sub(S)) \qquad (4)$$

is a derivative *of $s_{\upharpoonright\sigma}$. (The support of ν may be countable.) We write $s_{\upharpoonright\sigma} \xoverset{Tau} \Longrightarrow \nu$ if ν is a derivative of $s_{\upharpoonright\sigma}$.*

The following lemma shows that our definition of derivative exactly models the reachability probabilities: A derivative cannot exceed the probability to reach a set of states; concretely, given a set of states $G \subseteq S$, $\nu(Sub(G))$ is at most the probability to reach G.

Lemma 17. *Suppose given a substate $s_{\upharpoonright\sigma}$, sets $Tau \subseteq S$ and $G \subseteq S$, and a derivative $s_{\upharpoonright\sigma} \overset{Tau}{\Longrightarrow} \nu$. Then, $Prob_{s_{\upharpoonright\sigma}}(Tau \, \mathcal{U} \, G) \geq \nu(Sub(G))$.*
Equality holds if the delay scheme satisfies, for all $i \in \mathbb{N}_1$,

$$\mu^{\rightarrow}_{s_{\upharpoonright\sigma},0} = \sigma \text{ if } s \in Tau \setminus G \qquad\qquad \mu^{\rightarrow}_{t,i} = \mathbf{P}(t, \cdot) \text{ if } t \in Tau \setminus G$$

$$\mu^{\times}_{s_{\upharpoonright\sigma},0}(Sub(S)) = 1 \text{ if } s \in G \qquad\qquad \mu^{\times}_{t,i}(Sub(S)) = 1 \qquad \text{if } t \in G.$$

Proof sketch. We first prove equality under the mentioned conditions. Later, we will show that a condition violation does not increase $\nu(Sub(G))$.

One can prove by induction over $k \geq 0$ the following stronger statements:

1. $Prob_{s_{\upharpoonright\sigma}}(Tau \, \mathcal{U}^{\leq k} \, G) = \sum_{i=0}^{k} \nu^{\times}_i(Sub(G))$.
2. $Prob_{s_{\upharpoonright\sigma}}([Tau \setminus G] \, \mathcal{U}^{=k+1} \, \cdot) = \nu^{\rightarrow}_k$.

The lemma then follows from Statement 1 by taking the limit $k \to \infty$.

To prove that violating the equality conditions does not increase $\nu(Sub(G))$, assume that we reduce $\mu^{\rightarrow}_{t,i+1}$ below $\mathbf{P}(t, \cdot)$ for some $t \in Tau \setminus G$. Then we can see immediately from Def. 15 that $\nu^{\rightarrow}_{i+1}$ and $\nu^{\times}_{i+1}$ will not increase. Even if we now have room to set $\mu^{\times}_{t,i+1}$ to some nonzero value, it still holds that $\mu^{\times}_{t,i+1}(Sub(G)) = 0$, so $\nu(Sub(G))$ is not affected.

If some other equality condition is violated, one can argue similarly that $\nu(Sub(G))$ will not increase. $\qquad\square$

From the above lemma, we derive a corollary that provides the heart of the soundness proof:

Corollary 18. $s \models \mathcal{P}_{\unrhd p}(\Phi \, \mathcal{U} \, \Psi)$ *iff there exists a derivative* $s_{\upharpoonright} \overset{Sat(\Phi)}{\Longrightarrow} \nu$ *such that $\nu(Sub(Sat(\Psi))) \unrhd p$.*

Remark 19. Note that in Defs. 15 and 16, we allowed as an atypical case that $s \notin Tau$. The reason for this now becomes clear: we can apply Corollary 18 even if $s \not\models \Phi$. To make sure that Lemma 17 holds even then, Cond. (3) in Def. 15 was added. – Additionally, we do not require that Tau be an R-upset (an R-upward closed set), i. e. it may happen that $R[Tau] \not\subseteq Sub(Tau)$.

Definition 20 (Weak simulation). *Suppose given a relation $R \subseteq S \times Sub(S)$. We let $R[s]^{\text{St}} := \{s' \mid s \, R \, s'_{\upharpoonright}\}$. The relation R is a weak simulation if $s \, R \, t_{\upharpoonright\tau}$ implies that $L(s) = L(t)$ and there exists $t_{\upharpoonright\tau} \overset{R[s]^{\text{St}}}{\Longrightarrow} \nu$ such that $\mathbf{P}(s, \cdot) \sqsubseteq_R \nu$. We say that $t_{\upharpoonright\tau}$ weakly simulates s, denoted as $s \precsim t_{\upharpoonright\tau}$, iff there exists a weak simulation R such that $s \, R \, t_{\upharpoonright\tau}$.*

Let us first apply our definition of weak simulation to the examples above.

Example 21. The pathological examples in Figs. 2–4 are handled correctly:

$s_5 \precsim s'_{k\upharpoonright}$: From Lemma 17, we conclude that any derivative $s'_{k\upharpoonright} \overset{R[s_5]^{\text{St}}}{\Longrightarrow} \nu$ satisfies $\nu(Sub(\{g\})) \leq 1/k$. But $\mathbf{P}(s_5, \cdot) \sqsubseteq_R \nu$ (for any sensible R) would imply, according to Lemma 6, $1 = \mathbf{P}(s_5, \{g\}) \leq \nu(Sub(\{g\})) \leq 1/k$. Contradiction!

$s_6 \gtrsim s_{2\uparrow}$: Let $R := \{(s_6, s_{2\uparrow}), (g, g_\uparrow), (y, s_{2\uparrow 0}), (y, s_{2\uparrow})\}$. We simulate y by s_2, rescaled to no transitions at all. We have to prove that R is a weak simulation. Obviously, the labellings are compatible ($L(s_6) = L(s_2)$ etc.), and the proof for $g \, R \, g_\uparrow$ is trivial.

Let us have a look at $s_6 \, R \, s_{2\uparrow}$. Here, $Tau = R[s_6]^{St} = \{s_2\}$, so our choice of delay scheme only consists of $\mu^{\rightarrow}_{s_{2\uparrow},0} := 0.7\mathcal{D}_g$ and $\mu^{\times}_{s_{2\uparrow},0} := 0.3\mathcal{D}_{s_{2\uparrow 0}}$. (Choices $\mu^{\rightarrow}_{s_2,i}$ for $i > 0$ are irrelevant.) This delay scheme satisfies the conditions; note, in particular, that we have dropped the probability to reach w, so that the total probability to go anywhere is ≤ 1. The derivative is constructed by:

$$\vec{\nu_0} = 0.7\mathcal{D}_g \qquad\qquad \vec{\nu_1} = \mathbf{0}$$
$$\nu_0^{\times} = 0.3\mathcal{D}_{s_{2\uparrow 0}} \qquad\qquad \nu_1^{\times} = 0.7\mathcal{D}_{g_\uparrow}$$

So, $s_{2\uparrow} \xrightarrow{R[s_6]^{St}} 0.3\mathcal{D}_{s_{2\uparrow 0}} + 0.7\mathcal{D}_{g_\uparrow} =: \nu$, and to show $\mathbf{P}(s_6, \cdot) \sqsubseteq_R \nu$, we can use the weight function $\Delta : S_\perp \times Sub(S)_\perp \rightarrow [0, 1]$ with:

$$\Delta(g, g_\uparrow) = 0.7 \qquad\qquad \Delta(y, s_{2\uparrow 0}) = 0.3$$

and $\Delta(s, t_{\uparrow\tau}) = 0$ otherwise.

For the other pairs in R, the proof that they satisfy the conditions of weak simulation is easy.

$s_5 \not\gtrsim s_{1\uparrow}$: Similar to $s_5 \not\gtrsim s'_{k\uparrow}$, all derivatives $s_{1\uparrow} \xrightarrow{R[s_5]^{St}} \nu$ satisfy $\nu(Sub(\{g\})) \leq 0.8$. Again, $1 \leq 0.8$ would follow. Contradiction!

Example 22. Reconsider s_7 and s_9 in Fig. 5. We are going to prove that $s_7 \gtrsim s_{9\uparrow}$. Let $R = \{(s_7, s_{9\uparrow}), (s_7, s_{2\uparrow}), (s_8, s_{9\uparrow}), (s_8, s_{2\uparrow}), (s_1, s_{2\uparrow}), (g, g_\uparrow), (w, w_\uparrow)\}$. Let us look at $s_8 \, R \, s_{9\uparrow}$ first. $Tau = R[s_8]^{St} = \{s_9, s_2\}$. We choose the delay scheme

$$\mu^{\rightarrow}_{s_{9\uparrow},0} := \mathbf{P}(s_9, \cdot) \qquad\qquad \mu^{\rightarrow}_{s_2,1} := 0.6\mathbf{P}(s_2, \cdot)$$
$$\mu^{\times}_{s_{9\uparrow},0} := \mathbf{0} \qquad\qquad \mu^{\times}_{s_2,1} := 0.4\mathcal{D}_{s_{2\uparrow}},$$

as suggested by Fig. 5. As the derivative of $s_{9\uparrow}$, we get $s_{9\uparrow} \xrightarrow{R[s_8]^{St}} 0.2\mathcal{D}_{s_{2\uparrow}} + 0.24\mathcal{D}_{g_\uparrow} + 0.06\mathcal{D}_{w_\uparrow} =: \nu$. Then, we have to prove $\mathbf{P}(s_8, \cdot) \sqsubseteq_R \nu$. The weight function $\Delta : S_\perp \times Sub(S)_\perp \rightarrow [0, 1]$ that witnesses this relation is

$$\Delta(g, g_\uparrow) = 0.24 \qquad\qquad \Delta(\perp, w_\uparrow) = 0.06$$
$$\Delta(s_1, s_{2\uparrow}) = 0.2 \qquad\qquad \Delta(\perp, \perp) = 0.5$$

and $\Delta(s, t_{\uparrow\tau}) = 0$ otherwise.

For the proof of $s_7 \, R \, s_{9\uparrow}$, one has to define a derivative according to the same principles; this is left to the reader.

Now let us find a derivative for $s_8 \, R \, s_{2\uparrow}$. Here, $Tau = \{s_9, s_2\}$ again, but $\mu^{\rightarrow}_{s_9,i}$ and $\mu^{\times}_{s_9,i}$ are irrelevant, as s_9 is not reachable from s_2. For $\mu^{\rightarrow}_{s_{2\uparrow},0}$ and $\mu^{\times}_{s_{2\uparrow},0}$, we can choose between several values, as $s_{2\uparrow}$ oversimulates s_8. For example, let $\mu_{s_2\uparrow,0}^{\rightarrow} := 0.55\mathbf{P}(s_2, \cdot)$ and $\mu_{s_2\uparrow,0}^{\times} := 0.4\mathcal{D}_{s_2\uparrow}$. This will lead to $s_{2\uparrow} \xRightarrow{R[ss]^{St}} 0.4\mathcal{D}_{s_2\uparrow} + 0.44\mathcal{D}_{g\uparrow} + 0.11\mathcal{D}_{w\uparrow}$.

The proof for $s_1 R s_{2\uparrow}$ is even easier, as $\mathbf{P}(s_1, \cdot) \sqsubseteq_R \mathbf{P}(s_2, \cdot^{St})$.

So, every pair in R satisfies the requirements, and R is a weak simulation.

Example 23. Now let us prove that $s_{12} \gtrsim s_{9\uparrow}$. Let $R = \{(s_{12}, s_{9\uparrow}), (s_{12}, s_{2\uparrow}), (s_1, s_{2\uparrow 0.8\mathcal{D}_g}), (s_1, s_{2\uparrow}), (g, g_\uparrow), (w, w_\uparrow)\}$.

First, look at $s_{12} R s_{9\uparrow}$. Here, $Tau = R[s_{12}]^{St} = \{s_9, s_2\}$. We choose the delay scheme

$$\mu_{s_9\uparrow,0}^{\rightarrow} := \mathbf{P}(s_9, \cdot) \qquad\qquad \mu_{s_2,1}^{\rightarrow} := 0.48\mathcal{D}_g + 0.2\mathcal{D}_w$$

$$\mu_{s_9\uparrow,0}^{\times} := \mathbf{0} \qquad\qquad \mu_{s_2,1}^{\times} := 0.32\mathcal{D}_{s_2\uparrow 0.8\mathcal{D}_g}.$$

The conditions for delay schemes are satisfied; in particular, we have $\mu_{s_2,1}^{\rightarrow}(S) + \mu_{s_2,1}^{\times}(Sub(S)) = 0.68 + 0.32 \le 1$ and $\mu_{s_2,1}^{\rightarrow} + \overline{\mu_{s_2,1}^{\times}} = \mu_{s_2,1}^{\rightarrow} + \mu_{s_2,1}^{\times}(s_{2\uparrow 0.8\mathcal{D}_g})0.8\mathcal{D}_g = (0.48 + 0.32 \cdot 0.8)\mathcal{D}_g + 0.2\mathcal{D}_w \le \mathbf{P}(s_2, \cdot)$. For the derivative of $s_{9\uparrow}$, we get

$$\nu_0^{\rightarrow} = \mathbf{P}(s_9, \cdot) \quad \nu_1^{\rightarrow} = 0.5 \cdot [0.48\mathcal{D}_g + 0.2\mathcal{D}_w] \quad \nu_2^{\rightarrow} = \mathbf{0}$$

$$\nu_0^{\times} = \mathbf{0} \quad \nu_1^{\times} = 0.5 \cdot 0.32\mathcal{D}_{s_2\uparrow 0.8\mathcal{D}_g} \quad \nu_2^{\times} = 0.24\mathcal{D}_{g\uparrow} + 0.1\mathcal{D}_{w\uparrow}$$

and therefore, we have $s_{9\uparrow} \xRightarrow{R[s_{12}]^{St}} 0.16\mathcal{D}_{s_2\uparrow 0.8\mathcal{D}_g} + 0.24\mathcal{D}_{g\uparrow} + 0.1\mathcal{D}_{w\uparrow} =: \nu$. The weight function that witnesses $\mathbf{P}(s_{12}, \cdot) \sqsubseteq_R \nu$ is straightforward.

The other pairs in R are easy to handle. Therefore, R is a weak simulation.

6   Soundness and Completeness

In this section we prove the soundness of weak simulation with respect to PCTL$_{\setminus \mathcal{X}}$ and give a fragmentary proof of its completeness.

Lemma 24. *The relation $R \subseteq S \times Sub(S)$ is a weak simulation iff $s R t_{\upharpoonright T}$ implies that $L(s) = L(t)$ and for any set $Tau \subseteq S$, whenever $s_\uparrow \xRightarrow{Tau} \mu$ (with a delay scheme that never delays, i. e. $\mu(Sub(Tau)) = 0$), there exists $t_{\upharpoonright T} \xRightarrow{R[Tau]^{St}} \nu$ such that $\mu^{St} \sqsubseteq_R \nu$.*

Proof sketch. The "if" direction is almost trivial; let us concentrate on the "only if" direction. Let $s_\uparrow \xRightarrow{Tau}_n \mu_n$ denote the partial derivative: instead of summing $\sum_{i=0}^{\infty} \nu_i^{\times}$ in (4) of Def. 16, we let $\mu_n := \sum_{i=0}^{n} \nu_i^{\times}$. Then $\mu = \lim_{n\to\infty} \mu_n$.

One first proves by induction on n that for any $s R t_{\upharpoonright T}$, $Tau \subseteq S$, and $s_\uparrow \xRightarrow{Tau}_n \mu_n^{(s,t_{\upharpoonright T})}$ with $\mu_n^{(s,t_{\upharpoonright T})}(Sub(Tau)) = 0$, there exists $t_{\upharpoonright T} \xRightarrow{R[Tau]^{St}} \nu_n^{(s,t_{\upharpoonright T})}$ such that $(\mu_n^{(s,t_{\upharpoonright T})})^{St} \sqsubseteq_R \nu_n^{(s,t_{\upharpoonright T})}$, and additionally that $\nu_{n-1}^{(s,t_{\upharpoonright\downarrow})} \le \nu_n^{(s,t_{\upharpoonright\downarrow})}$.

Now assume given a derivative $s_\uparrow \xRightarrow{Tau} \mu$ that never delays, and let μ_n be the corresponding partial derivatives. The above induction gives us, for every n, a derivative $t_{\upharpoonright T} \xRightarrow{R[Tau]^{St}} \nu_n$ such that $\mu_n^{St} \sqsubseteq_R \nu_n$. Taking the limit on both sides, we get $\mu^{St} = \lim_{n\to\infty} \mu_n^{St} \sqsubseteq_R \lim_{n\to\infty} \nu_n =: \nu$. This ν is the derivative that we were required to construct. $\qquad\square$

Theorem 25 ($\precsim$ is sound). $s \precsim t_\uparrow$ *implies for all* $PCTL_{\backslash \mathcal{X}}$ *liveness formulas* Φ, $s \models \Phi$ *implies* $t \models \Phi$.

Proof. We need to prove that $s \precsim t_\uparrow$ implies $s \precsim_{\text{live}} t$. Suppose that $s \precsim t_\uparrow$ and $s \models \Phi$, where Φ is a $PCTL_{\backslash \mathcal{X}}$ liveness formula. Our goal is to prove that $t \models \Phi$. This can be done by induction on the structure of Φ. The cases $true, a, \neg a, \Phi_1 \wedge \Phi_2$ and $\Phi_1 \vee \Phi_2$ are standard, so we omit them here.

The remaining case is the probabilistic operator, namely $\Phi = \mathcal{P}_{\unrhd p}(\Phi_1 \, \mathcal{U} \, \Phi_2)$. Let $Tau := Sat(\Phi_1)$ and $G := Sat(\Phi_2)$. According to Corollary 18, there exists $s_\uparrow \xrightarrow{Tau} \mu$ such that $\mu(Sub(G)) \unrhd p$. We use w. l. o. g. a delay scheme for μ that satisfies the equality conditions in Lemma 17 and $\mu_t^\times(s'_\uparrow) = 1$ for all $s' \in G$. Note that this implies that $s_\uparrow \xrightarrow{Tau \backslash G} \mu$ is also a derivative, and it never delays. By Lemma 24 there exists $t_\uparrow \xrightarrow{\precsim[Tau \backslash G]^{\text{St}}} \nu$ such that $\mu^{\text{St}} \sqsubseteq_{\precsim} \nu$, which indicates that $\nu(Sub(G)) \geq \mu(Sub(G)) \unrhd p$. As $\precsim[Tau \backslash G]^{\text{St}} \subseteq \precsim[Tau]^{\text{St}} \subseteq Tau = Sat(\Phi_1)$ by induction hypothesis, $t \models \Phi$ by Corollary 18. $\qquad\square$

We also explain why we think that $\precsim$ is complete with respect to $PCTL_{\backslash \mathcal{X}}$.

Conjecture 26 ($\precsim$ is complete). *For* $s, t \in S$, *we have: if* $s \models \Phi$ *implies* $t \models \Phi$ *for all* $PCTL_{\backslash \mathcal{X}}$ *liveness formulas* Φ, *then* $s \precsim t_\uparrow$.

Proof fragment. Let $R \subseteq S \times Sub(S)$ be the following relation: $s \, R \, t_{\upharpoonright \tau}$ if $L(s) = L(t)$ and for all $\precsim_{\text{live}}$-upsets $U_1, U_2 \subseteq S$, we have $Prob_s(U_1 \, \mathcal{U} \, U_2) \leq Prob_{t_{\upharpoonright \tau}}(U_1 \, \mathcal{U} \, U_2)$. We will have to prove two things: First, $\precsim_{\text{live}}$ is a subrelation of R, i.e., $\{(s', t'_\uparrow) \mid s' \precsim_{\text{live}} t'\} \subseteq R$; and second, R is a weak simulation relation.

For the first part, assume to the contrary that there existed a pair of states s', t' such that $s' \precsim_{\text{live}} t'$ but not $s' \, R \, t'_\uparrow$. So there would exist $\precsim_{\text{live}}$-upsets $U_1, U_2 \subseteq S$ with $p := Prob_{s'}(U_1 \, \mathcal{U} \, U_2) > Prob_{t'_\uparrow}(U_1 \, \mathcal{U} \, U_2)$. Both U_1 and U_2 can be described by some live $PCTL_{\backslash \mathcal{X}}$-formula, say Ψ_1 and Ψ_2 with $Sat(\Psi_1) = U_1$ and $Sat(\Psi_2) = U_2$. Obviously, $s' \models \mathcal{P}_{\geq p}(\Psi_1 \, \mathcal{U} \, \Psi_2)$, therefore $t' \models \mathcal{P}_{\geq p}(\Psi_1 \, \mathcal{U} \, \Psi_2)$. So it would follow from the semantics of $\mathcal{P}$ that $p \leq Prob_{t'}(Sat(\Psi_1) \, \mathcal{U} \, Sat(\Psi_2)) = Prob_{t'_\uparrow}(U_1 \, \mathcal{U} \, U_2) < p$. Contradiction!

It is easy to see that

$$\forall G \subseteq S : \exists t_{\upharpoonright \tau} \xrightarrow{R[s]^{\text{St}}} \nu_G : \mathbf{P}(s, G) \leq \nu_G(R[G]), \tag{5}$$

and we would have to prove

$$\exists t_{\upharpoonright \tau} \xrightarrow{R[s]^{\text{St}}} \nu : \forall G \subseteq S : \mathbf{P}(s, G) \leq \nu(R[G]). \tag{6}$$

From Lemma 6, we know that (6) implies $\mathbf{P}(s, \cdot) \sqsubseteq_R \nu$, so R would be a weak simulation, and $\precsim$ would be complete as well.

While swapping two quantifiers like in (5) $\Longrightarrow$ (6) is not allowed in general, we believe that this implication holds because the ν_G are all derivatives. $\qquad\square$

6.1 A Sound and Complete Variant

We now proceed to a slightly modified definition of $\precsim$, which is provably sound and complete. We call this relation Π-weak simulation because it is similar to (5), a Π_2^1-formula in the analytical hierarchy.

Definition 27 (Π-weak simulation). *Suppose given relation* $R \subseteq S \times Sub(S)$. *The relation* R *is a* Π*-weak simulation if* $s \, R \, t_{\upharpoonright\tau}$ *implies that* $L(s) = L(t)$ *and* $\forall G, Tau \subseteq S$, *whenever* $s_{\upharpoonright} \overset{Tau}{\Longrightarrow} \mu$ *(with a delay scheme that never delays, i. e.* $\mu(Sub(Tau)) = 0$*), there exists* $t_{\upharpoonright\tau} \overset{R[Tau]^{St}}{\Longrightarrow} \nu$ *such that* $\mu(Sub(G)) \leq \nu(R[G])$. *We say that* $t_{\upharpoonright\tau}$ Π*-weakly simulates* s, *denoted as* $s \precsim^{\Pi} t_{\upharpoonright\tau}$, *iff there exists a* Π*-weak simulation* R *such that* $s \, R \, t_{\upharpoonright\tau}$.

Theorem 28. $\precsim^{\Pi}$ *is sound w. r. t. PCTL*$_{\setminus \mathcal{X}}$.

Proof. The proof is completely analogous to the proof of Thm. 25.   □

Theorem 29. $\precsim^{\Pi}$ *is complete w. r. t. PCTL*$_{\setminus \mathcal{X}}$.

Proof. Let R be the same relation as in Conjecture 26. We already have shown that $\precsim_{\text{live}}$ is a subrelation of R; it remains to be proven that R is a Π-weak simulation.

Assume given a pair $s \, R \, t_{\upharpoonright\tau}$. Let $G, Tau \subseteq S$, and $s_{\upharpoonright} \overset{Tau}{\Longrightarrow} \mu$ be arbitrary. By Lemma 17, $\mu(Sub(G)) \leq Prob_{s_{\upharpoonright}}(Tau \, \mathcal{U} \, G) \leq Prob_{s_{\upharpoonright}}(\precsim_{\text{live}}[Tau] \, \mathcal{U} \, \precsim_{\text{live}}[G])$. The definition of R, together with Lemma 17, ensures that there exists $t_{\upharpoonright\tau} \overset{\precsim_{\text{live}}[s]}{\Longrightarrow} \nu$ such that $\mu(Sub(G)) \leq Prob_{t_{\upharpoonright\tau}}(\precsim_{\text{live}}[Tau] \, \mathcal{U} \, \precsim_{\text{live}}[G]) = \nu(Sub(\precsim_{\text{live}}[G]))$. One can define ν in such a way that its support only contains improper substates of $\precsim_{\text{live}}[G]$. All these substates are contained in $R[G]$. Obviously $\mu(Sub(G)) \leq \nu(R[G])$.   □

7 Conclusion

In this paper we have redefined the notion of weak simulation for Markov chains such that it is sound with respect to the logical preorder induced by the PCTL$_{\setminus \mathcal{X}}$ liveness properties. Unfortunately, we were unable to prove its completeness; but at least there exists a variant that is provably sound and complete.

Our definition of weak simulation relies on the concept of substates, which are closely related to (bi)simulation defined on distributions instead of states. In [14], probabilistic forward simulation is defined as the coarsest congruence relation preserving probabilistic trace distribution on probabilistic automata; while in [10], weak bisimulation – a symmetric version of probabilistic forward simulation – is introduced for Markov automata (subsuming probabilistic automata). Both relations are defined over distributions. An important difference is that our substates are labelled, i. e. they have a "colour".

We hope that the scientific community can fill in the gap in the proof left by us. Of course one also has to prove that $\precsim$ is a congruence, to find an axiomatisation and an efficient algorithm to abstract a sDTMC – however, we think that the definitions and the completeness proof should be finalised first.

Acknowledgements. The authors are partially supported by DFG/NWO Bilateral Research Programme ROCKS, MT-LAB (a VKR Centre of Excellence) and IDEA4CPS. We thank Holger Hermanns, Verena Wolf and Rob van Glabbeek for their extensive remarks and helpful discussions.

References

1. Abadi, M., Lamport, L.: The existence of refinement mappings. Theoretical Computer Science 82(2), 253–284 (1991)
2. Baier, C., Hermanns, H., Katoen, J.-P., Wolf, V.: Comparative branching-time semantics for Markov chains (extended abstract). In: Amadio, R., Lugiez, D. (eds.) CONCUR 2003. LNCS, vol. 2761, pp. 492–507. Springer, Heidelberg (2003)
3. Baier, C., Katoen, J.-P.: Principles of model checking. MIT Press, Cambridge (2008)
4. Baier, C., Katoen, J.-P., Hermanns, H., Wolf, V.: Comparative branching-time semantics for Markov chains. Information and Computation 200(2), 149–214 (2005)
5. De Nicola, R., Vaandrager, F.: Three logics for branching bisimulation. Journal of the ACM 42(2), 458–487 (1995)
6. Deng, Y., van Glabbeek, R., Hennessy, M., Morgan, C.: Testing finitary probabilistic processes, http://www.cse.unsw.edu.au/~rvg/pub/finitary.pdf, an extended abstract has been published as [7]
7. Deng, Y., van Glabbeek, R., Hennessy, M., Morgan, C.: Testing finitary probabilistic processes (extended abstract). In: Bravetti, M., Zavattaro, G. (eds.) CONCUR 2009. LNCS, vol. 5710, pp. 274–288. Springer, Heidelberg (2009)
8. Desharnais, J., Gupta, V., Jagadeesan, R., Panangaden, P.: Approximating labelled Markov processes. Information and Computation 184(1), 160–200 (2003)
9. Desharnais, J., Laviolette, F., Tracol, M.: Approximate analysis of probabilistic processes: Logic, simulation and games. In: QEST 2008, pp. 264–273. IEEE Computer Society, Los Alamitos (2008)
10. Eisentraut, C., Hermanns, H., Zhang, L.: On probabilistic automata in continuous time. In: 25th Annual IEEE Symposium on Logic in Computer Science: LICS, pp. 342–351. IEEE Computer Society, Los Alamitos (2010)
11. Jansen, D.N., Song, L., Zhang, L.: Revisiting weak simulation for substochastic Markov chains. Tech. Rep. ICIS–R13005, Radboud Universiteit, Nijmegen (2013), http://www.cs.ru.nl/research/reports
12. Jonsson, B., Larsen, K.G.: Specification and refinement of probabilistic processes. In: Sixth Annual IEEE Symposium on Logic in Computer Science (LICS), pp. 266–277. IEEE Computer Society, Los Alamitos (1991)
13. Sack, J., Zhang, L.: A general framework for probabilistic characterizing formulae. In: Kuncak, V., Rybalchenko, A. (eds.) VMCAI 2012. LNCS, vol. 7148, pp. 396–411. Springer, Heidelberg (2012)
14. Segala, R.: Modeling and verification of randomized distributed real-time systems. Ph.D. thesis, Massachusetts Institute of Technology, Cambridge (1996)
15. Tarski, A.: A lattice-theoretical fixpoint theorem and its applications. Pacific Journal of Mathematics 5(2), 285–309 (1955)
16. Zhang, L.: Decision Algorithms for Probabilistic Simulations. Ph.D. thesis, Universität des Saarlandes, Saarbrücken (2008)

A Performance Analysis of System S, S4, and Esper via Two Level Benchmarking

Miyuru Dayarathna[1] and Toyotaro Suzumura[1,2]

[1] Department of Computer Science, Tokyo Institute of Technology, 2-12-1 Ookayama,
Meguro-ku, Tokyo 152-8552, Japan
`dayarathna.m.aa@m.titech.ac.jp`, `suzumura@cs.titech.ac.jp`
[2] IBM Research - Tokyo

Abstract. Data stream processing systems have become popular due to their effectiveness in applications in large scale data stream processing scenarios. This paper compares and contrasts performance characteristics of three stream processing softwares System S, S4, and Esper. We study about which software aspects shape the characteristics of the workloads handled by these software. We use a micro benchmark and different real world stream applications on System S, S4, and Esper to construct 70 different application scenarios. We use job throughput, CPU, Memory consumption, and network utilization of each application scenario as performance metrics. We observed that S4's architectural aspect which instantiates a Processing Element (PE) for each keyed attribute is less efficient compared to the fixed number of PEs used by System S and Esper. Furthermore, all the Esper benchmarks produced more than 150% increased performance in single node compared to S4 benchmarks. S4 and Esper are more portable compared to System S and could be fine tuned for different application scenarios easily. In future we hope to widen our understanding of performance characteristics of these systems by investigating in to the code level profiling.

Keywords: stream processing, data-intensive computing, workload characterization, performance analysis, benchmarking, systems scalability.

1 Introduction

Stream processing [16] (which is also called Complex Event Processing [5]) has emerged as an exciting new filed to support online information processing activities. These software process data on-the-fly, in memory without requiring to store data in secondary storage. There have been extensive studies for characterizing the workload and performance implications of computing systems. However, there has not been sufficient amount of such studies carried out in the area of stream computing. In this paper we work on stream processing software performance characterization using System S, S4, and Esper; which are currently three prominent stream processing software in the field. Decision to

K. Joshi et al. (Eds.): QEST 2013, LNCS 8054, pp. 225–240, 2013.

choose these three software was stimulated due to their unique architectural designs. While System S is developed following a manager and worker model; S4 has a decentralized and symmetric architecture and follows Actors model [9]. Esper is completely different from System S and S4 because it is just a component for stream processing [4]. However, Esper provides a complete software suite for stream processing which has been used by popular software vendors for their event processing back-ends. Furthermore, current implementation of S4's operators are purely based on Java and also Esper is a pure Java library, whereas System S allows for both C/C++ and Java versions. We checked the software's licenses and got confirmed that they allow for publishing performance comparisons.

In achieving the aforementioned objectives we created 70 different experiment scenarios using three real-world application benchmarks and a micro benchmark. Performance characteristics such as Job throughput, CPU usage, Memory Usage, Network I/O were observed in arriving at conclusions about the design of stream processing system architectures.

2   Related Work

There have been several previous studies on characterizing performance of stream processing systems. Mendes *et al.* have conducted a performance evaluation of three event processing systems by running several micro-benchmarks [8]. Their intention was to provide a first insight in to the performance of event processing systems. We try to delve more deep in to the performance characteristics of such systems. They have used Esper similar to us. However, their study has been conducted in single node settings. Different to them, we implemented all the benchmarks using Esper in distributed settings using Java Messaging Service (JMS) [12].

Suzumura *et al.* made a performance study considering ETL (Extract-Transform-Load) scenario of System S [14]. However, our intention is completely different from their work. While they evaluate performance of System S in the context of ETL applications; we aim for identifying the characteristics of stream processing systems in a more general context. The works done by Parekh *et al.* and Zhang *et al.* study methods for characterizing resource usage profiles and characterizing the resource usage of Processing Elements (PEs) respectively [19][11]. Both these works are based on System S and their aim is to provide solid foundation for modeling and predicting resource usage of stream programs which is different from our motivation of performance characterization.

Arasu *et al.* described a stream data management benchmark [3] which has been originally used by members of Aurora [1] and STREAM [15] projects to compare performance characteristics of Data Stream Management Systems (DSMS) (i.e., Stream Processing Systems). Recently Zeitler *et al.* implemented the Linear Road benchmark on SCSQ DSMS [18]. However, in this study our approach does not concentrate on single concrete benchmark. Rather we use a collection of applications which is a distinguishing point of our work from their's. Yet we see Linear Road as a possible avenue for extending our work.

On their paper introducing S4 [9], Neumeyer *et al.* have conducted two experiments (online and offline) on S4 with a Streaming Click-Through Rate computation application. While they conducted their experiments maximum at 20000 input events per second rate; we conduct the experiments with several magnitudes higher input data rates to compare performance of S4 with respect to System S.

3   An Overview of Stream Processing Software

System S, S4, and Esper are three popular stream processing software in use. We provide brief introduction to each of them below.

System S is an operator-based, large-scale distributed data stream processing middleware [7]. The project was initiated in 2003 and is currently under development at IBM Research [17]. System S uses an operator-based programming language called SPADE [6] for defining data flow graphs. SPADE has a set of built-in operators (BIOP) and also supports for creating customized operators (i.e., User Defined Operators (UDOP)) which allows for extending the SPADE language. Communication between operators is specified as streams. SPADE compiler fuses operators into one or more Processing Elements (PEs) during the compilation process. System S Scheduler (see Figure 1 (c)) makes the connection between PEs, when the application is run in a stream processing cluster. Out of the BIOPs used for implementing the sample programs Source creates a stream from data coming from an external source. Sink converts a stream into a flow of tuples that can be used by external components. Functor performs tuple-level manipulations (e.g., filtering, mapping, projection, attribute creation, transformation, etc.). Aggregate groups and summarizes incoming tuples. Split splits a stream into multiple output streams.

S4 is an open source operator-based stream processing system released by Yahoo Inc. in October 2010 [9]. S4 follows Actors model which makes it considerably different from System S. S4 has a decentralized and symmetric architecture where all nodes share the same responsibilities. Furthermore, S4 uses a pluggable architecture (see Figure 1 (a)) that keeps the design generic and customizable to a greater extent.

In S4 the computations are performed by PEs. Communication between the PEs is done in the form of data events. A sequence of events in S4 is defined as a stream. Event emission and consumption is the only mode of communication between PEs. Current version of S4 provides several PEs for standard tasks such as count, aggregate, join, etc. However, custom PEs can be easily created by extending the classes provided by S4 API. All PEs consume exactly the events which correspond to the values on which they are keyed.

Programming model of S4 has been created in such a way that developers write PEs in Java programming language, and the PEs are assembled into applications using Spring framework. Developers need to essentially implement input event handler `processEvent()`, and `output()` which implements the output mechanism [9].

Esper is a software component for stream processing developed by EsperTech Inc. [4]. The software is available in Java with the name Esper, and in.NET as NEsper. However, this paper uses only Esper since we wanted to compare performance of S4 with a Java based stream processing software. One of the important differences of Esper from System S and S4 is that it is just a software library. Therefore, important features of stream processing systems such as distributed processing, fault tolerance, etc. has to be coded manually. However, since Esper is a software library it can be easily integrated with variety of applications such as J2EE web servers, distributed caches, web browsers, etc.

Since Esper is just a software library we implemented the distributed event processing functionality by using Esper with ActiveMQ [12] which is an open source message broker implementation of the Java Messaging Service (JMS) specification.

Software architecture of the three stream processing software is shown in Figure 1.

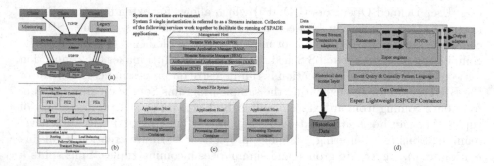

Fig. 1. Software architecture of the three stream processing software. (a) S4 Framework overview (b) A processing node of S4 (c) System S runtime environment (d) Esper's architecture.

4   Methodology and Performance Metrics

Our methodology is based on two level benchmarking. We use a micro-benchmark to get basic characterization of the three software's performance. Next, we use three different stream programs (Application-Specific Benchmarks) which are used for different purposes. These programs, the reasons for choosing them, and the features of their associated data sets are described below.

4.1   Sample Programs and Data Sets

Micro-benchmark. We use a three operator micro-benchmark program to get a basic understanding of the behavior of the stream processing software. This program's structure is shown in Figure 2 (d). It has only one functor (F1) that increments an integer value it receives from the Source operator (S). The result from F1 gets stored in `results.dat` via Sink operator (SI). We chose this program for our study because of its simple nature which allows us to reveal the behavior of the stream processing systems during heavy work loads. We used a synthetic data set of two digit integer values (in CSV format) for our experiment. The data set contained 10 million records.

VWAP. Calculation of Volume-Weighetd Average Price (VWAP) is a real world application scenario of stream processing in the financial services domain. VWAP is calculated as the ratio of the value traded and the volume traded within a specified time horizon. This can be depicted as,

$$VWAP = \frac{\sum_{i=1}^{n}(P_i.V_i)}{\sum_{i=1}^{n}(V_i)}$$

where V_i represents each traded volume and P_i represents the corresponding traded price in a series of n transactions. Figure 2 (a) shows the data flow graph of the VWAP application used for performance evaluation. This application is part of a larger financial trading application described at [2]. The first functor (F1) filters the tuples for valid records and the aggregate operator (AG) keeps a tuple window of 4 based on the ticker id it receives from F1. Furthermore, it calculates the $P_i.V_i$ portion of the VWAP formula for each tuple. AG outputs a new tuple each time it receives a tuple from F1. Finally, F2 calculates the VWAP value and transfers to Sink operator (SI) which stores the result in `result.dat` file. We used a dataset that is available with System S's sample VWAP application; but magnified it to a larger data set of 1 million tuples.

CDR. Call Detail Record (CDR) is a piece of information produced by a telephone exchange containing details of a phone call passed through it. Telecommunication networks process massive amounts of CDR events, another application area for stream processing software systems. Furthermore, we wanted to test how well different stream processing software scale under massive data rates if programmers do not worry about producing optimized version of their codes. Considering these factor, we decided to use a four operator CDR processing application (shown in Figure 2 (b)) in our study to evaluate the three stream processing software. First functor (F1) splits the input tuples read by Source operator to different routes based on a hash value of call station ID (which identifies each user). The Aggregate operator (AG) calculates the total packet count by adding input and output packet values mentioned in the tuple. Finally, the result is stored in the disk via Sink (SI) operators. We used a synthesized data set with 2 million data tuples with each tuple having 22 fields for this experiment. The data set had information of 0.1 million users.

Twitter Topic Counter. As the third application specific benchmark we selected a Twitter hash tag count application. Twitter users can label the key words of their tweets using # (e.g., #car) which indicate to which conversations the messages relate to. As shown in Figure 2 (c) we developed a six operator SPADE program with similar functionality to the Twitter Topic Counter application (here onwards referred to as Twitter application) available with S4 and used both these applications during our performance evaluations. We used a real data set with 90507 tweets gathered from twitter on 4th June 2011 in the period 00:00-01:00 JST for this experiment.

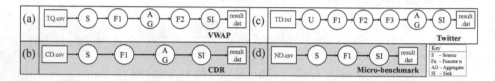

Fig. 2. Data flow graphs of sample applications

4.2 Experimental Setup

The experiments were conducted on 12 compute nodes each with AMD Opteron^{TM} Processor 242, 1600MHz 1MB L2 cache per core, 250GB hard drive. Seven out of twelve nodes had 8GB RAM (Nodes labeled as sa0<n>, n is from 1 to 7) while the remaining five (Nodes labeled as sb0<m>, m is from 1 to 5) had 4GB RAM. From the profiling results we obtained using Oprofile [13], and Nmon [10] we observed that the difference of main memory in the nodes did not affect our experiments. All the nodes were installed with Linux Cent OS release 5.4, IBM Infosphere Streams Version 1.2, and JRE 1.6.0; and were connected via 1Gigabit Ethernet. We used S4 version 0.3.0.0 and Esper version 4.5.0. We set both initial Java heap sizes and Maximum Java heap sizes to 3GB to avoid CPU time being spent on increasing Java heap memory during the experiments.

4.3 Performance Metrics

We used job throughput, CPU usage, network I/O, and memory usage of each sample application (i.e., job) as the metrics for the evaluations.

- Job Throughput : Measures number of input tuples processed by each stream processing system. On System S we measured this in Tuples per second while on S4 and Esper it was measured in Events per second.
- CPU usage (%): Overall CPU utilization during the experiments. This metric provides information on which nodes are busy processing data hence allows for identifying stream processing system's node level bottlenecks.

- CPU usage by process (%): Measures what processes contribute to overall CPU utilization and the amount of their contribution. This allows us to identify where the performance bottlenecks exist (i.e., in stream processing system level or OS level) and which processes are responsible for such behavior.
- Memory Usage (MB) : The amount of main memory (RAM) use during application execution. It is essential to maintain high percentage of free main memory for proper functioning of stream applications. Memory usage is one of the important metrics for this study because unlike most batch processing tasks stream processing keeps all the required data in main memory during the program execution.
- Network I/O (KB/s) : We measured network I/O of each node of stream processing system cluster during experiments to quantitatively understand what kind of communication overhead exists between the nodes.

4.4   Objectives and Methodology

Measurement of throughput was done in two different ways for S4, Esper, and System S. In S4, and Esper we measured time required to process a specific amount of events while in System S we measured time between processing first tuple and a last tuple in order to increase the accuracy of the results.

By measuring throughput our intention was to understand how well the applications scale in both stream processing systems. Scalability of current S4 applications are characterized by S4 runtime. The runtime determines what number of PEs to be created to suit for a particular node allocation under different workloads. We cannot explicitly allocate PEs for different nodes. Hence we utilized the automatic node allocation for S4. However, SPADE allows for allocating a node pool (a collection of hosts in a stream processing cluster) and attaching different operators to specific nodes programmatically. We used this feature to distribute workload of System S jobs among nodes. The Esper applications we developed had to be manually allocated for different nodes giving us more finer grain control over their distributed execution. Furthermore, we augmented the sample programs shown in Figure 2 in such a way that the modified versions allow for attaching more nodes with the job. The objective was to use more nodes than the operators (PEs) available in the data flow graph and allocate multiple operators to each node. E.g., If we used 12 nodes we need a data flow graph with at least 12 operators which is not possible with any sample program layout shown in Figure 2. How the augmentation of data flow graphs of sample applications developed for System S has been carried out is shown in Figure 3.

Apart from the four augmented sample applications we used an optimized version of the CDR application (shown in Figure 3 (c)) in order to observe how introduction of multiple sources affects the throughput of a stream program. However, we used this program only for this purpose but used normal augmented version of CDR application (in Figure 3 (b)) in order to make the comparison a fair one. In order to supply four source files to CDR Optimized data flow graph the data file used for CDR was splitted equally in to four files using Linux's `split` command and resulting files were checked to make sure that the first and last packets were in the correct format. Note that we do not do any optimization of the workload across different nodes in our experiments in order to make a fair comparison between the three systems.

As shown in Figure 3 we introduced a split operator (SP) for each sample program that splitted the data stream from source operator to several sub graphs. The splitting was based on a hash value produced for each incoming packet. We made sure the packets are evenly splitted among sub-graphs by choosing an appropriate filed to hash and by generating the input data file with proper distribution of tuples. Furthermore, we introduced two Functor operators (FS (at the beginning) and FE (at the end)) which recorded the time of receiving start packet and a desired last packet.

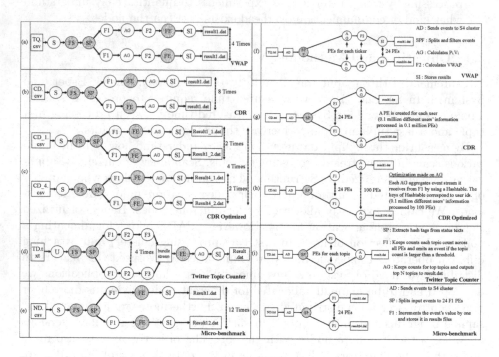

Fig. 3. Augmentation of sample application data flow graphs on System S and S4. Sub-figures (a) to (e) correspond to System S application while (f) to (j) correspond to S4.

In the case of VWAP we used a data set of 1 million data tuples and we attached last packet with trade ticker "GMD". The time it took to receive this packet at one of the four PEs was taken as the total runtime and throughput was calculated. For CDR the input data file had 2 million data tuples. We measured the time it takes to receive the last packet (i.e., 2 millionth packet) by one of the FEs. The same method was employed to measure the runtime of CDR optimized version.

The data flow graph of Twitter application is significantly different from other augmented versions because we wanted to keep the similarity of program design with S4 version of Twitter application. It should be noted that in CDR, CDR Optimized, and Twitter applications measurement of the last packet was done before the Aggregate operators. It is because Aggregate operators output only a subset of tuples they receive. The measurements of the throughput was conducted on the corresponding Aggregate operators on S4 versions; hence this ensured that the measurements were taken on the same operator on both System S and S4.

How the sample applications have been implemented on S4 is shown in Figure 3 (f) to (j). For data flow graphs shown in Figure 3 (g), (h), (i), and (j) the storing of results was done without use of a Sink (SI) PE because we can store the results by using the code of last operator. However, we defined a separate SI PE for VWAP since data output involved an additional step of conversion to CSV format. Furthermore, as shown in Figure 3 (g) and (h) we developed two versions of CDR for S4 since the original version shown in Figure 3 (g) resulted in inefficient use of system resources by spawning 0.1 million PEs. The solution was to reduce the number of PEs to 100 by using a hash table on each PE with user id as the key. All the PEs were distributed equally to all nodes during the experiments that ran on S4.

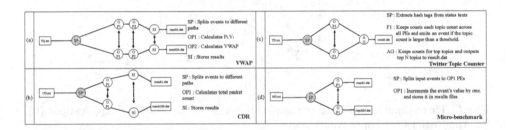

Fig. 4. Augmentation of sample application data flow graphs for Esper

Augmented versions of the Esper applications are shown in Figure 4. Each operator denoted by OP (e.g., OP1) is attached to a JMS server running on the same node.

For each augmented sample program we allocated different number of operators per node as the number of worker nodes involved in the processing vary. Each version of the sample program was run three times for both System S, Esper, and S4. Average value of the running times was taken for calculating the throughput to improve the accuracy of end results. Furthermore, we ran both Nmon and Oprofile on each node associated with the experiment well before one of the experiments during three experiment runs begin. The two daemons were shutdown after the experiment completed keeping enough delay from the experiment end time.

5   Performance Evaluation

5.1   Job Throughput

The throughput results obtained from running the four sample applications and the micro-benchmark based on System S, S4, and Esper are shown in Figure 5 (a), (b), and (c) respectively. Note that a tuple in System S corresponds to an event in S4 and Esper. Since most research literature on System S uses the term tuple to represent a data event we use the same terminology in our work. The performance curves on Figure 5 (b) shows that S4 achieves sub-linear scalability for CDR, VWAP, and micro-benchmark because doubling the number of nodes did not result in a corresponding doubling of the system's performance. We have drawn two different curves for CDR on S4. The curve marked with CDR corresponds to the data flow graph shown in Figure 3 (g) which is a naive unoptimized version while the curve CDR Optimized is obtained by running the optimized program shown in Figure 3 (h). The latter was created to avoid inefficient resource usage as described in Section 4.4. Twitter application indicated super-linear scalability because in each case shown in Figure 5 (b) increasing the number of nodes increased the system performance more than twice.

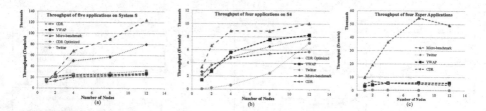

Fig. 5. Throughput comparison of sample applications. (a) System S Applications (b) S4 Applications (c) Esper Applications.

Out of the System S versions of these applications; micro-benchmark and CDR Optimized version produced almost linear throughput curves. CDR, VWAP, and Twitter applications indicated sub-linear scalability (saturated). However, considering the throughput characteristics obtained from optimized version of CDR, it is apparent that having a single source operator in the data flow graph results in such sub-linear scalability.

Esper reported more than 1.5 times higher performance for single node applications compared to S4. This indicates that Esper is much suited for single node event processing scenarios. However, the distributed versions of the Esper applications did not scale well compared to S4 or System S. We believe the reason for such less scalability was because of the overhead associated with object serialization.

In order to get more details of System S, S4, and Esper (on single node) we used Nmon and Oprofiler tools. We wanted to get profile information of Esper on single node setting because it reported best performance on single node. The results are discussed below.

5.2   CPU Usage

CPU utilization of manager nodes and worker nodes of both S4 and System S are shown in Figure 6. We refer the node on which the S4 adapter component run during the experiment as Manager node and the remaining nodes as Worker nodes in the case of an S4 cluster. The node on which Streams Application Manager (SAM) [7] resides is referred as Manager node and the other nodes are taken as Worker nodes.

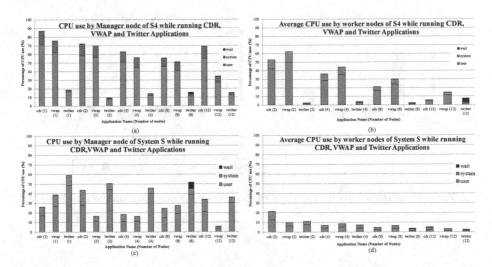

Fig. 6. CPU usage on different nodes of System S and S4 while running CDR, VWAP, and Twitter applications

To increase the legibility of results and to distinguish between application-level and micro-level benchmarks we separately list the results of micro-benchmark experiments' CPU utilization in Figure 7.

It was clear that System S's workers used (in Figure 6 (c), (d)) less CPU compared to S4 with VWAP and CDR applications. However, in the case of Twitter which operated relatively low input data rate S4 workers and manager node reported relatively less CPU usage compared to their System S counter parts. Furthermore, in System S a considerable amount of processing has been performed by the system processes rather than by user processes as can be observed from S4.

Esper on single node reported average CPU usage of 83.4%, 72.1%, 66.2%, 50.8% for VWAP, Twitter, CDR, and micro-benchmark respectively.

We used Oprofile to identify what happens in system and user processes in the case of the CDR and micro-benchmark experiments with 12 nodes for both

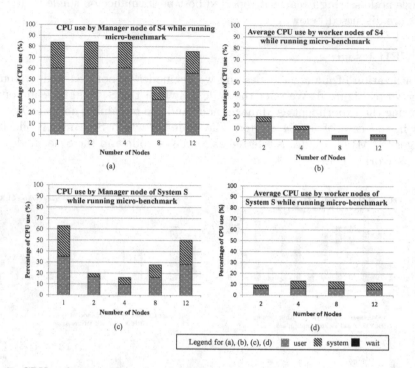

Fig. 7. CPU utilization by nodes when running micro-benchmark on System S and S4

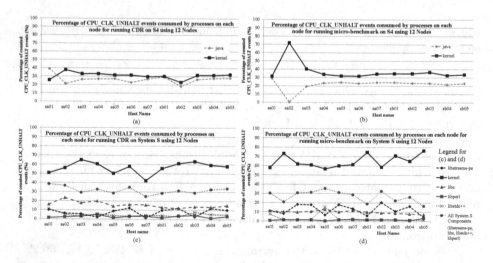

Fig. 8. Breakdown of CPU usage by processes for CDR and Micro-benchmark

System S and S4. The results are shown in Figure 8. Each graph of Figure 8 shows the share of CPU_CLK_UNHALT events used by different processes. It can be observed that in both experiments involved with S4 (Figure 8 (a), (b)), in most of the nodes the Java virtual machine (java) had used slightly less than 30% of CPU_CLK_UNHALT events which is little less than the amount consumed by the operating system's kernel. However, in the case of System S when running both CDR and micro-benchmark the System S components had consumed roughly 20% to 30% CPU_CLK_UNHALT events. However, unlike the case with S4 nodes' kernel consumed 50% to 60% CPU_CLK_UNHALT events. This gives an explanation for why we saw considerable amount of processing conducted by system processes in Figure 6.

5.3 Memory Usage

Memory consumption of different nodes while running the four augmented applications on S4 and System S are shown in Figure 9. It can be observed that worker nodes of System S consumed less memory compared to S4 workers. Twitter application introduced relatively less data transfer rate among the nodes since it only injected 90507 events. This is a reason for why S4 nodes indicate less memory use compared to other applications. Huge amount of memory consumption can be observed on manager node (sa01). It should be noted that the node named sa01 was used as the master node in all the experiments carried out in this paper. Memory usage during Esper run on single node was 7.7GB, 5.5GB, 6.1GB, and 7.3GB micro-benchmark, Twitter, VWAP, and CDR respectively.

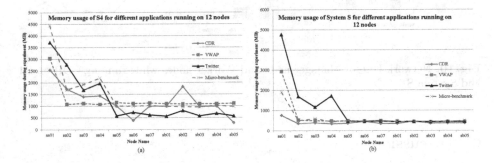

Fig. 9. Memory consumption of sample applications on System S and S4

5.4 Network I/O

The network I/O happening between nodes of a stream processing system is a key criterion that determines its performance. Figure 10 shows the network I/O of different nodes when running CDR application on both S4 and System S. Note that we compare only the external communications happening through Ethernet port (eth0) rather than through the loop back interface (lo). When considering

the `eth0-write` at master nodes and `eth0-read` at worker nodes it is apparent
that the amount of data transferred between the master and the worker nodes is
considerably larger in S4 compared to System S. For the data shown in Figure
10 the average `eth0-write` rate of Manager node of S4 is 60% larger than the
average `eth0-write` rate of System S's Manager node. Moreover, average rate
of `eth0-read` at worker nodes of S4 is 120% larger than average `eth0-read` at
worker nodes of System S. However, both System S and S4 applications used
the same 2 million data set. This indicates that S4's data transfer protocol
consumes more memory bandwidth compared to System S. The network I/O for
eight nodes while running CDR for both S4 and System S (shown in Figure 10
(c) and (f)) confirms the aforementioned fact.

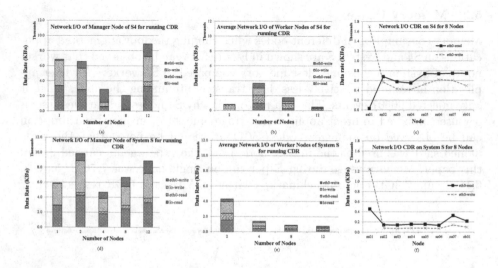

Fig. 10. Network I/O for Manager node and Worker nodes while running CDR on S4
and System S

6   Discussion

The results we obtained by running ten different applications (totaling 70 applications considering 1,2,4,8,12 node scenarios including System S and S4 source
code optimization) on System S, S4, and Esper gave us sufficient insight to their
internals. It became clear from the throughput comparison made on Figure 5
that the stream programming models that allow/require programmers to write
optimized codes should be used carefully to maximize the throughput. E.g.,
A SPADE program written with single source operator might not scale well
in different hardware configurations. Also a S4 application that generates huge
numbers of PEs for incoming events cannot scale well with limited `PEContainer`
queue size. Yet introduction of multiple source operators resulted in a 3.2 times

speedup for CDR application on System S and a 1.34 times speed up for reducing 0.1 million Aggregator PEs of S4 to 100 PEs (See Figure 5) which indicated possible avenues for performance improvements in different stream programming models.

While Java based stream processing system architectures are gaining considerable attention due to their portability, system designers and programmers have to think carefully before choosing the desired solution. E.g., A light weight input event rate job could be easily processed using S4 with few amount of PEs (E.g., Twitter application run on S4). However, a large scale application with high commercial importance such as the VWAP and CDR might produce millions of PEs since S4 dynamically generates PEs for each new data events it receives. As with any other JVM related optimizations, setting of maximum and minimum heap size values plays a key role in determining the performance of Java based stream processing systems. However, in the case of S4 and Esper based stream processing system administrators need to be vigilant about the characteristics of the data handled by their S4/Esper applications properly (See Figure 9 (a)).

Esper applications' throughput results indicate that scalability is one of the key challenges faced by JMS based distributed Esper applications. While there was a slight scalability advantage scaling from one node to two nodes for all the benchmarks (See Figure 5), the performance tend to degrade when the application is scaled to more nodes. We believe the reason for such behavior is slowness in network communication and serialization. However, the Esper microbenchmark application had considerably higher performance compared to its S4 counterpart.

While we observed heavy use of network bandwidth by S4, by using optimized protocols and techniques such as Java New I/O, InfiniBand Remote Direct Memory Access the conditions could be improved.

7   Conclusion

In this paper we presented a performance study on three stream processing software. We used three popular stream processing software: IBM System S, Yahoo S4, and EsperTech's Esper. We ran three application benchmarks covering different domains of stream processing. We used a micro-benchmark to further clarify performance of the software systems. The study used job throughput, CPU usage, memory usage, and network usage of each node as the performance metrics. By analyzing the throughput and profiling results we observed that carefully designed stream applications result in high throughput. Another conclusion we arrived at is that choice of a stream processing system need to be made considering factors such as performance, platform independence, and size of the jobs. Furthermore, we understood the importance of key role played by operating system kernel in stream processing system's performance.

A stream processing system architecture that scales in terms of number of PEs is a further work that is inspired by this work. In future we hope to extend this work to a code level performance study on S4, specially to identify which components, code segments are most resource intensive.

Acknowledgments. This research was supported by the Japan Science and Technology Agency's CREST project titled "Development of System Software Technologies for post-Peta Scale High Performance Computing".

References

1. Abadi, D.J., et al.: Aurora: a new model and architecture for data stream management. The VLDB Journal 12, 120–139 (2003)
2. Andrade, H., et al.: Scale-up strategies for processing high-rate data streams in systems. In: ICDE 2009 (2009)
3. Arasu, A., et al.: Linear road: a stream data management benchmark. In: VLDB 2004, pp. 480–491 (2004)
4. EsperTech. Esper - Complex Event Processing (February 2012), http://esper.codehaus.org/
5. Etzion, O., Niblett, P.: Event Processing in Action (2011)
6. IBM. Ibm infosphere streams version 1.2.0.1: Programming model and language reference (February 2010)
7. IBM. Ibm infosphere streams version 1.2.1: Installation and administration guide (October 2010)
8. Mendes, M.R.N., Bizarro, P., Marques, P.: A performance study of event processing systems. In: Nambiar, R., Poess, M. (eds.) TPCTC 2009. LNCS, vol. 5895, pp. 221–236. Springer, Heidelberg (2009)
9. Neumeyer, L., et al.: S4: Distributed stream computing platform. In: KDCloud 2010 (December 2010)
10. Nmon. nmon for Linux (June 2011), http://nmon.sourceforge.net
11. Parekh, S., et al.: Characterizing, constructing and managing resource usage profiles of systems applications: challenges and experience. In: CIKM 2009, pp. 1177–1186 (2009)
12. Snyder, B., Bosanac, D., Davies, R.: ActiveMQ in Action (2011)
13. SourceForge. OProfile - A System Profiler for Linux (June 2011), http://oprofile.sourceforge.net
14. Suzumura, T., Yasue, T., Onodera, T.: Scalable performance of systems for extract-transform-load processing. In: SYSTOR 2010 (2010)
15. The_STREAM_Group. Stream: The stanford stream data manager. Technical Report 2003-21 (2003)
16. Turaga, D., et al.: Design principles for developing stream processing applications. In: Software: Practice and Experience (August 2010)
17. Wolf, J., Bansal, N., Hildrum, K., Parekh, S., Rajan, D., Wagle, R., Wu, K.-L., Fleischer, L.K.: SODA: An optimizing scheduler for large-scale stream-based distributed computer systems. In: Issarny, V., Schantz, R. (eds.) Middleware 2008. LNCS, vol. 5346, pp. 306–325. Springer, Heidelberg (2008)
18. Zeitler, E., Risch, T.: Scalable splitting of massive data streams. In: Kitagawa, H., Ishikawa, Y., Li, Q., Watanabe, C. (eds.) DASFAA 2010. LNCS, vol. 5982, pp. 184–198. Springer, Heidelberg (2010)
19. Zhang, X.J., et al.: Workload characterization for operator-based distributed stream processing applications. In: DEBS 2010, pp. 235–247 (2010)

Effect of Codeword Placement on the Reliability of Erasure Coded Data Storage Systems

Vinodh Venkatesan and Ilias Iliadis

IBM Research – Zurich, 8803 Rüschlikon, Switzerland
{ven,ili}@zurich.ibm.com

Abstract. Modern data storage systems employ advanced erasure codes to protect data from storage node failures because of their ability to provide high data reliability at high storage efficiency. In contrast to previous studies, we consider the practical case where the length of codewords in an erasure coded system is much smaller than the number of storage nodes in the system. In this case, there exists a large number of possible ways in which different codewords can be stored across the nodes of the system. In this paper, it is shown that a declustered placement of codewords can significantly improve system reliability compared to other placement schemes. A detailed reliability analysis is presented that accounts for the rebuild times involved, the amounts of partially rebuilt data when additional nodes fail during rebuild, and an intelligent rebuild process that attempts to rebuild the most critical codewords first.

1 Introduction

Modern data storage systems are complex in nature consisting of several components of hardware and software. To perform a reliability analysis, we require a model that abstracts the reliability behavior of this complex system and lends itself to theoretical analysis, but at the same time, preserves the core features that affect the system failures and rebuilds. In this article, we develop and describe a relatively simple yet powerful model that captures the essential reliability behavior of an erasure coded data storage system. Using this model, we show the effect of codeword placement on the system reliability.

As an alternative to replication, storage systems employ advanced erasure codes to protect data from storage node failures because of their ability to provide high data reliability as well as high storage efficiency. The use of such erasure codes can be dated back to as early as the 1980s when they were applied in systems with redundant arrays of inexpensive disks (RAID) [1, 2]. When nodes fail, storage systems try to maintain the redundancy through node rebuild processes that use the data from the surviving nodes to reconstruct the lost data in new replacement nodes. There exists a non-zero probability of further node failures during rebuild that can cause the system to lose enough redundant data to make some of the originally stored data irrecoverable. The average amount of time taken by the system to end up in irrecoverable data loss, also known as the mean time to data loss, or MTTDL, is a measure of reliability commonly

K. Joshi et al. (Eds.): QEST 2013, LNCS 8054, pp. 241–257, 2013.
© Springer-Verlag Berlin Heidelberg 2013

used for comparing different coding schemes and studying the effect of various design parameters [3]. The length of codewords in an erasure coded system is typically much smaller than the number of storage nodes in the system (e.g. RAID-6 typically uses a codeword length of 16). This implies that there exist a large number of possible ways in which codewords can be stored across the nodes of the system. However, many reliability analyses in the literature are performed under the assumption that the number of storage nodes is equal to the codeword length [1, 2, 4]. In addition, some of the reliability analyses do not account for the time taken to rebuild [4–7]. For replication-based systems, it is well-known that the MTTDL is significantly affected by the choice of placement of replicas [5, 6, 8–10]. In particular, it is known that a certain replica placement scheme, known as declustered placement, can provide significantly higher reliability than other placement schemes, especially for large storage systems [9]. The intuition behind this is that the declustered placement scheme spreads replicas of data across more number of devices thereby reducing the amount of critical data (i.e., the amount of data with the least number of surviving replicas) when multiple node failures occur. An intelligent rebuild scheme that always rebuilds the critical data first reduces the risk of data losing all its replicas, and hence the risk of data loss in the system.

This paper addresses the following practical questions regarding erasure coded systems. How does the MTTDL of a system depend on the codeword length and the number of parities in the erasure code? For a given codeword length and a given number of parities, how does the codeword placement affect the MTTDL of a system? Do the results on the effect of replica placement on MTTDL in replication-based systems extend to the effect of codeword placement on the MTTDL in erasure coded systems? How does the trade-off between storage efficiency and MTTDL depend on the codeword placement scheme?

The key contributions of this article are the following. We extend previous work in the literature by considering the general case of erasure coded systems, which includes replication-based systems. A new model enhancing previous ones is developed here to evaluate the MTTDL of erasure coded systems. The model developed captures the effect of the various system parameters as well as the effect of various codeword placement schemes. The reliability analysis is detailed, in the sense that it accounts for the rebuild times involved, the amounts of partially rebuilt data when additional nodes fail during rebuild, and and an intelligent rebuild process that attempts to rebuild the most critical codewords first. The validity of the model is confirmed by simulation.

The remainder of this article is organized as follows: Section 3 describes the system model considered. Section 4 describes the methodology of reliability analysis used. Using the methodology described in the previous section, Section 5 evaluates the reliability of clustered and declustered placement schemes. Section 6 provides numerical results and discusses the effect of codeword placement on reliability. Section 7 compares simulation-based MTTDL values with the theoretical predictions. Finally, the paper is concluded in Section 8.

Table 1. Parameters of a storage system

c	amount of data stored on each storage node (bytes)
n	number of storage nodes
$c\mu$	average read-write rebuild bandwidth of a storage node (bytes/s)
$1/\lambda$	mean time to failure of a storage node (s)
$1/\mu$	mean time to read/write c amount of data from/to a node (s)

2 Related Work

A comparison between erasure codes and replication in terms of availability in peer-to-peer systems has been presented in [11]. It has been well-established that erasure codes can provide much higher reliability than replication for the same level of storage efficiency. The trade-off, however, is in the performance as erasure codes may require Galois field arithmetic for encoding and decoding. Therefore, many recent works have laid emphasis on the development of new codes as well as new encoding and decoding techniques to improve the performance of erasure coded systems (see [12] and references therein). Some works have also addressed the reliability assessment of erasure codes through simulation [13]. One thing that is common in all these works is that they essentially consider the case where the codeword length is equal to the number of nodes. In contrast, our work provides a unified framework for assessing the reliability of erasure coded systems where the codeword length may be larger than the number of nodes, in which case, there exist many possible ways of storing each codeword across the nodes in the system. This is a practically relevant case as, for performance reasons, the lengths of the erasure codes used in real storage systems are kept constant and small, whereas the number of nodes in the system grows with the system capacity. For replication-based systems, it was shown that the reliability is significantly affected by the choice of placement of replicas [9, 10, 14]. In this article, we extend these results to a more general case of maximum distance separable (MDS) erasure codes. To the best of our knowledge, this is the first work exploring the space of codeword placement for erasure codes in a homogeneous environment through both theory and simulation, which shows that codeword placement can have a significant impact on reliability.

3 System Model

The storage system is modeled as a collection of n *storage nodes* each of which stores c amount of data. In addition to the space required for the c amount of data that is stored, each node is assumed to have sufficient spare space that may be used for a distributed rebuild process (see Section 3.5) when other nodes fail. The main parameters used in the storage system model are listed in Table 1.

3.1   Storage Node

Each storage node is a fairly complex entity that comprises of disks, memory, processor, network interface, and power supply. Any of these components can fail and lead to the node either becoming temporarily unavailable, or failed. It is assumed that there is some mechanism, such as regular pinging of each node, in place to detect node failures as they occur.

Node Unavailability vs. Node Failure: As noted in [15], more than 90% of the node unavailabilities are transient and do not last for more than 15 minutes. As most of the unavailabilities are transient, a node rebuild process is initiated only if a node stays unavailable for more than 15 minutes [15]. In other words, node unavailabilities lasting longer than a certain amount of time are treated as node failures. The primary focus of this paper is on node failures. Node *unavailabilities* have been observed to exhibit strong correlation that may be due to short power outages in the datacenter, or part of a rolling reboot or upgrade activity at the datacenter management layer [15]. However, there is no indication that correlations exist among node *failures*. Therefore, for ease of analysis, we assume that node failures are independent.

3.2   Redundancy

In erasure coded systems, the user data is divided into blocks (or symbols) of a fixed size (e.g. sector size of 512 bytes) and each set of l blocks is encoded into a set of m $(> l)$ blocks, called a codeword, before storing them on m distinct nodes. In this paper, we consider (l, m)-MDS codes, in which the encoding is done, such that, any subset of l symbols of a codeword can be used to decode the l symbols of user data corresponding to that codeword. Replication-based systems, with a given replication factor r, are a subset of erasure coded systems where the parameters l and m are equal to 1 and r, respectively.

3.3   Codeword Placement

In a large storage system, the number of nodes, n, is typically much larger than the codeword length, m. Therefore, there exist many ways in which a codeword of m blocks can be stored across n nodes.

Clustered Placement: If n is divisible by m, one simple way to place codewords would be to divide the n nodes into disjoint sets, of m nodes each, and store each codeword across the nodes of a particular set. We refer to this type of data placement as *clustered* placement, and each of these disjoint sets of nodes as *clusters*. Note that we disregard the various permutations of the m symbols in each codeword, e.g., we consider both RAID-4 and RAID-5 to have clustered placement, although they have different placement of parities. In such a placement scheme, it can be seen that no cluster stores the redundancies corresponding to the data on another cluster. The entire storage system can essentially be modeled as consisting of n/m independent clusters. Reliability behavior of a cluster under exponential failure and rebuild time distributions is well-known [1, 2, 16]. To the best of our knowledge, all prior work in the reliability analysis of erasure coded systems have solely been for clustered placement (see [12] and references therein).

Declustered Placement: A placement scheme that can potentially offer far higher reliability than the clustered placement scheme, especially as the number of nodes in the system grows, is the *declustered* placement scheme. There exist $\binom{n}{m}$ different ways of placing m symbols of each codeword across n nodes. In this scheme, all these $\binom{n}{m}$ possible ways are equally used to store all the codewords in the system. It can be seen that, in such a placement scheme, when a node fails, the redundancy corresponding to the data on the failed node is equally spread across all the surviving nodes (as opposed to clustered placement in which it is spread only across the surviving nodes of the corresponding cluster). This allows one to use the rebuild read-write bandwidth available at all surviving nodes to do a *distributed* rebuild in parallel, which can be extremely fast when the number of nodes is large. As it turns out, this is one of the main reasons why declustered placement can offer significantly higher reliability than clustered placement for large systems.

Spread Factor: A broader set of placement schemes can be defined using the concept of *spread factor*. For each node in the system, its *redundancy spread factor* is defined as the number of nodes over which the data on that node and its corresponding redundant data are spread. In an erasure coded system, when a node fails, its spread factor determines the number of nodes which have the redundancy corresponding to the lost data, and this in turn determines the degree of parallelism that can be used in rebuilding the data lost by that node. In this paper, we will consider symmetric placement schemes in which the spread factor of each node is the same, denoted by k. Two examples of such symmetric placement schemes are the clustered and declustered placement schemes for which the spread factor, k, is equal to m and n, respectively. A number of different placement schemes can be generated by varying the spread factor, k, between m and n.

3.4   Node Failure

Based on the discussion in Section 3.1, the times to node failures are modeled as independent and identically distributed random variables. Denote the cumulative distribution function of the times to node failure by F_λ, with mean, $1/\lambda$. An interesting result of this paper is that the mean time to data loss of an erasure coded storage system tends to be invariant within a large class of failure time distributions, that includes the exponential distribution and, most importantly, real-world distributions like Weibull and gamma. A similar result has been established earlier for replication-based systems [14].

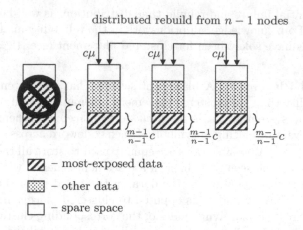

Fig. 1. Distributed rebuild in declustered placement

3.5   Node Rebuild

When storage nodes fail, codewords lose some of their symbols and this leads to a reduction in data redundancy. The system attempts to maintain the redundancy of the system by reconstructing the lost codeword symbols using the surviving symbols of the affected codewords.

Codeword Reconstruction: For a system using an (l, m)-MDS code for redundancy, a simple way to reconstruct a codeword that has lost up to $m - l$ symbols is to read any of its l symbols, decode the original l user data blocks, re-encode these l user data blocks using the (l, m)-MDS code, and recover the lost codeword symbols. The reconstruction process takes an amount of time that depends on the amount of data to read, the time taken for decoding and re-encoding this data, and the amount of data to write. Typically the amount of time taken for decoding and re-encoding this data is much smaller than the time taken to read the required data and write the re-encoded data. It is assumed that the decoding and re-encoding of data is done in a streaming fashion, that is, as the data is being read, the decoding and re-encoding is assumed to be done on-the-fly which converts a stream of input data to a stream of output data. This implies that the time taken for the reconstruction is equal to the time taken to stream the input and output data. Alternative methods of reconstruction based on regenerating codes have been proposed as a solution to reduce the amount of data *transferred* over the storage network during reconstruction (see [17] and references therein). The effect of these methods on the system reliability is outside the scope of this paper and is a subject of further investigation.

Intelligent Rebuild: In an intelligent rebuild process, the system attempts to first recover the codewords of the user data that have the least number of codeword symbols left. These codewords are also referred to as the *most-exposed* codewords. In contrast to intelligent rebuild, one may consider a *blind* rebuild, where lost codeword symbols are being recovered in an order that is not specifically aimed at recovering the codewords with the least number of surviving symbols first. Clearly, such a blind rebuild is more vulnerable to data loss. So, in the remainder of the paper, we consider only intelligent rebuild.

Distributed Rebuild: When a storage node fails, all the codewords that had one of their symbols stored on this node are affected. For a symmetric placement scheme with spread factor k, $m \leq k \leq n$, the surviving symbols of the affected codewords are equally distributed across $k - 1$ other surviving nodes of the system.

For clustered placement, i.e., $k = m$, the surviving symbols are present in the $m - 1$ surviving nodes of the affected cluster. The lost symbols are recovered by reading the required codeword symbols from a set of l nodes of the corresponding surviving cluster. The lost symbols are reconstructed on the fly and directly written to a new replacement node.

For other placement schemes, i.e., $m + 1 \leq k \leq n$, the surviving symbols are present in $k - 1$ ($\geq m$) surviving nodes. Performing a rebuild similar to clustered placement would, in general, degrade reliability for these placement schemes. This is because, although the rebuild time would be the same (as the same amount of data is written to the new replacement node), there are more nodes ($k - 1 > m - 1$) that contain the surviving symbols of the affected codewords The failure of any of these nodes can result in additional symbols of the affected codewords being lost. We therefore consider instead distributed rebuild for these placement schemes as illustrated in Fig. 1. Distributed rebuild involves reading the required codeword symbols from all the $k - 1$ nodes, computing the lost codeword symbols, and writing them to the spare space of these $k - 1$ nodes in such a way that no symbol is written to a node in which another symbol corresponding to the same codeword is already present. Once all lost codeword symbols are recovered, they are transferred to a new replacement node. Due to the parallel nature of distributed rebuild, the rebuild times can be extremely short for large storage systems. Such a distributed rebuild process is in fact used in practical systems [18].

Node Rebuild Bandwidth: During the rebuild process, an average read-write bandwidth of $c\mu$ bytes/s is assumed to be reserved at each node for the rebuild. This implies that the average time required to read (or write) c amount of data from (or to) a node is equal to $1/\mu$. The average rebuild bandwidth is usually only a fraction of the total bandwidth available at each node; the remainder is being used to serve user requests. Denote the cumulative distribution function of the time required to read (or write) c amount of data from (or to) a node by G_μ, and its corresponding probability density function by g_μ.

3.6   Failure and Rebuild Time Distributions

It is known that real-world storage nodes are *generally reliable*, that is, the mean time to read all contents of a node (which is typically of the order of tens of hours) is much smaller than the mean time to failure of a node (which is typically at

least of the order of thousands of hours). So, it follows that generally reliable
nodes satisfy the following condition:

$$1/\mu \ll 1/\lambda, \quad \text{or} \quad \lambda/\mu \ll 1. \tag{1}$$

This assumption is holds for each node. The effect of this assumption on a collection of nodes is captured in the analysis. In the subsequent analysis, this
condition implies that terms involving powers of λ/μ greater than one are negligible compared to λ/μ and can be ignored. Let the cumulative distribution
functions F_λ and G_μ satisfy the following condition:

$$\mu \int_0^\infty F_\lambda(t)(1 - G_\mu(t))dt \ll 1, \quad \text{with} \quad \frac{\lambda}{\mu} \ll 1. \tag{2}$$

The results of this paper are derived for the class of failure and rebuild distributions that satisfy the above condition. In particular, the mean time to data
loss of a system is shown to be insensitive to the failure distributions within
this class. This result is of great importance because it turns out that this condition holds for a wide variety of failure and rebuild distributions, including,
most importantly, distributions that are seen in real-world storage systems [14].
Condition (2) can also be stated in the following alternate way [14]:

$$F_\lambda(t) \ll 1 \text{ when } G_\mu(t) < 1 \text{ and } \lambda \ll \mu, \tag{3}$$
$$\mu(1 - G_\mu(t)) \ll 1 \text{ when } F_\lambda(t) > 0 \text{ and } \mu \gg \lambda. \tag{4}$$

4   Reliability Analysis

The reliability analysis in this article uses a methodology similar to [9, 10, 14]. It
involves a series of approximations, each of which is justified for generally reliable
nodes with failure and rebuild time distributions satisfying (2). The theoretical
estimates of mean times to data loss predicted using this methodology have also
been shown to match with simulations, which avoid all the approximations made
in the methodology, over a wide range of system parameters [9, 10, 14].

4.1   Mean Time to Data Loss (MTTDL)

In an erasure coded system, a data loss is said to have occurred when sufficient
number of blocks of at least one codeword have been lost, rendering the code-
word(s) undecodeable. The average time taken for the system to end up in data
loss, also referred to as the mean time to data loss, or MTTDL, is a commonly
used measure that is useful for assessing trade-offs, for comparing schemes, and
for estimating the effect of the various parameters on the system reliability [3].

At any point of time, the system can be thought to be in one of two modes:
fully-operational mode or *rebuild* mode. During the fully-operational mode, all
data in the system has the original amount of redundancy and there is no active
rebuild process. During the rebuild mode, some data in the system has less than

the original amount of redundancy and there is an active rebuild process that is trying to restore the lost redundancy. A transition from fully-operational mode to rebuild mode occurs when a node fails; we refer to this node failure that causes a transition from the fully-operational mode to the rebuild mode as a *first-node* failure. Following a first-node failure, a complex sequence of rebuilds and subsequent node failures may occur, which eventually lead the system either to irrecoverable data loss, with probability P_{DL}, or back to the original fully-operational mode by restoring all codeword symbols, with probability $1 - P_{DL}$. In other words, the probability of data loss in the rebuild mode, P_{DL}, is defined as follows:

$$P_{DL} := \Pr \left\{ \begin{matrix} \text{data loss occurs before returning} \\ \text{to the fully-operational mode} \end{matrix} \,\middle|\, \text{system enters rebuild mode} \right\} \tag{5}$$

Since the rebuild times are much shorter than the times to failure, when computing the time to data loss, the time spent by the system in rebuild mode can be ignored. If we ignore the rebuild times, the system timeline consists of one first-node failure after another, each of which can end up in data loss with a probability P_{DL}. It can be shown that the mean time between two successive first-node failures, converges to $1/(n\lambda)$ [14] and that the MTTDL is given by the following proposition. In the remainder of this article, by $A \approx B$, we mean $\lim_{\lambda/\mu \to 0} A/B = 1$.

Proposition 1. *Consider a system with generally reliable nodes whose failure and rebuild distributions, F_λ and G_μ, satisfy (2). Its MTTDL is given by*

$$\text{MTTDL} \approx 1/(n\lambda P_{DL}), \tag{6}$$

where P_{DL} is defined in (5). The relative error in the approximation tends to zero as λ/μ tends to zero.

Proof. See [14]. □

4.2   Probability of Data Loss in Rebuild Mode (P_{DL})

This section shows how P_{DL} is estimated so that MTTDL can be obtained using Proposition 1.

Exposure Levels: Consider an erasure coded storage system with an (l, m)-MDS code. Let

$$\tilde{r} := m - l + 1. \tag{7}$$

We model the system as evolving from one exposure level to another as nodes fail and rebuilds complete. At time $t \geq 0$, let $D_j(t)$ be the amount of user data that have lost j symbols of their corresponding codewords, for $0 \leq j \leq \tilde{r}$. At time t, the system is said to be in exposure level e, $0 \leq e \leq \tilde{r}$, if $e = \max_{D_j(t) > 0} j$.

Direct Path Approximation: A path to data loss following a first-node-failure event is a sequence of exposure level transitions that begins in exposure level 1 and ends in exposure level $\tilde{r}$ (data loss) without going back to exposure level 0, that is, for some $j \geq r$, a sequence of $j - 1$ exposure level transitions $e_1 \rightarrow e_2 \rightarrow \cdots \rightarrow e_j$ such that $e_1 = 1$, $e_j = \tilde{r}$, $e_2, \cdots, e_{j-1} \in \{1, \cdots, \tilde{r}-1\}$, and $|e_i - e_{i-1}| = 1$, $\forall i = 2, \cdots, j$. To estimate P_{DL}, we need to estimate the probability of the union of *all* such paths to data loss following a first-node failure. As the set of events that can occur between exposure level 1 and exposure level $\tilde{r}$ is complex, estimating P_{DL} is a non-trivial problem. Therefore, we proceed by considering the direct path of successive transitions from exposure levels 1 to $\tilde{r}$. Denote the probability of the direct path to data loss by $P_{DL,\text{direct}}$, that is,

$$P_{DL,\text{direct}} := \Pr\{\text{exposure level path } 1 \rightarrow 2 \rightarrow \cdots \rightarrow \tilde{r}\}, \tag{8}$$

and approximate P_{DL} by $P_{DL,\text{direct}}$ using the following proposition.

Proposition 2. *Consider a system with generally reliable nodes whose failure and rebuild distributions, F_λ and G_μ, satisfy (2). Its P_{DL} is given by*

$$P_{DL} \approx P_{DL,\text{direct}}, \tag{9}$$

The relative error in the approximation tends to zero as λ/μ tends to zero.

Proof. See [9]. □

4.3   Probability of the Direct Path to Data Loss ($P_{DL,\text{direct}}$)

Consider the direct path to data loss, that is, the path $1 \rightarrow 2 \rightarrow \cdots \rightarrow \tilde{r}$ through the exposure levels. At each exposure level, the *intelligent* rebuild process attempts to rebuild the most-exposed data, that is, the data with the least number of codeword symbols left (see Section 3.5). Let the rebuild times of the most-exposed data at each exposure level in this path be denoted by R_e, $e = 1, \cdots, \tilde{r} - 1$. Let t_e, $e = 2, \cdots, \tilde{r}$, be the times of transitions from exposure level $e - 1$ to e following a first-node failure. Let $\tilde{n}_e$ be the number of nodes in exposure level e whose failure before the rebuild of most-exposed data causes an exposure level transition to level $e + 1$. Denote the time period from t_e until the next failure of node i by $E_{t_e}^{(i)}$. The time, F_e, until the first failure among the $\tilde{n}_{e-1}$ nodes that causes the system to enter exposure level e from $e - 1$, is

$$F_e := \min_{i \in \{1, \cdots, \tilde{n}_{e-1}\}} E_{t_{e-1}}^{(i)}, \quad e = 2, \cdots, \tilde{r}. \tag{10}$$

At exposure level e, let α_e be the fraction of the rebuild time R_e still left when a node failure occurs causing an exposure level transition, that is, let

$$\alpha_e := (R_e - F_{e+1})/R_e, \quad e = 1, \cdots, \tilde{r} - 2. \tag{11}$$

It can be shown that α_e is uniformly distributed in $(0, 1)$ (see [19, Lemma 2]). Now, denote by $1/\mu_e$ the following conditional means of R_e:

$$1/\mu_e := E[R_e | R_{e-1}, \alpha_{e-1}], \quad e = 2, \cdots, r - 1. \tag{12}$$

The actual values of $1/\mu_e$ depend on the codeword placement and this will be further discussed in later sections of this paper. Now, the distribution of R_e given R_{e-1} and α_{e-1} could be modeled in several ways. We consider the model B presented in [14], namely,

$$R_e | R_{e-1}, \alpha_{e-1} = 1/\mu_e \qquad w.p.\, 1 \text{ for } e = 2, \cdots, \tilde{r} - 1. \tag{13}$$

This model assumes that the rebuild time R_e is determined completely by R_{e-1} and α_{e-1} and no new randomness is introduced in the rebuild time of exposure level e. For further discussion on this model see [14]. Under this model, the probability of the direct path to data loss is given by the following proposition.

Proposition 3. *Consider a system with generally reliable nodes whose failure and rebuild distributions, F_λ and G_μ, satisfy (2). Consider the direct path $1 \rightarrow 2 \rightarrow \cdots \rightarrow \tilde{r}$ through the exposure levels in which the rebuild times R_e satisfy (13). The probability of this direct path is given by*

$$P_{DL,direct} \approx \lambda^{\tilde{r}-1} \times \tilde{n}_1 \cdots \tilde{n}_{\tilde{r}-1} \int_{\tau_1=0}^{\infty} \cdots \int_{\tau_{\tilde{r}-1}=0}^{\infty} \int_{a_1=0}^{1} \cdots \int_{a_{\tilde{r}-2}=0}^{1} \left(\tau_1 \cdots \tau_{\tilde{r}-1} g_{\mu_1}(\tau_1) \right.$$

$$\left. \times \delta \left(\tau_2 - \frac{1}{\mu_2} \right) \cdots \delta \left(\tau_{\tilde{r}-1} - \frac{1}{\mu_{\tilde{r}-1}} \right) da_{\tilde{r}-2} \cdots da_1 d\tau_{\tilde{r}-1} \cdots d\tau_1 \right). \tag{14}$$

The relative error in the approximation tends to zero as λ/μ tends to zero.

Proof. See [19, Appendix A]. □

5   Effect of Codeword Placement on Reliability

In this section, we consider different codeword placement schemes as discussed in Section 3.3. We wish to estimate their reliability in terms of their MTTDL and understand how codeword placement affects data reliability. To use the expression (14) for $P_{DL,\text{direct}}$, we need to compute the conditional means of rebuild times in each exposure level, $1/\mu_e$, $e = 1, \cdots, \tilde{r} - 1$, and the number of nodes whose failure can cause a transition to the next exposure level, $\tilde{n}_e$, $e = 1, \cdots, \tilde{r} - 1$. The values of these quantities depend on the underlying codeword placement and the nature of the rebuild process used. Now, denote the kth raw moment of the rebuild distribution G_μ by $M_k(G_\mu)$. The MTTDL of clustered and declustered codeword placement schemes are given by the following propositions.

Proposition 4. *Consider a storage system using clustered codeword placement with generally reliable nodes whose failure and rebuild distributions satisfy (2). Its mean time to data loss is given by*

$$\text{MTTDL}^{clus.} \approx \frac{\mu^{m-l}}{n\lambda^{m-l+1}} \frac{1}{\binom{m-1}{l-1}} \frac{M_1^{m-l}(G_\mu)}{M_{m-l}(G_\mu)}. \tag{15}$$

The relative error in the above approximation tends to zero as λ/μ tends to zero.

Proof. See [19, Appendix B].   □

Proposition 5. *Consider a storage system using declustered codeword place-ment with generally reliable nodes whose failure and rebuild distributions satisfy (2). Its mean time to data loss is given by*

$$\mathrm{MTTDL}^{declus.} \approx \frac{\mu^{m-l}}{n\lambda^{m-l+1}} \frac{(m-l)!}{(l+1)^{m-l}} \frac{M_1^{m-l}\left(G_{\frac{n-1}{l+1}\mu}\right)}{M_{m-l}\left(G_{\frac{n-1}{l+1}\mu}\right)} \prod_{e=1}^{m-l-1}\left(\frac{n-e}{m-e}\right)^{m-l-e}.(16)$$

The relative error in the above approximation tends to zero as λ/μ tends to zero.

Proof. See [19, Appendix C].   □

Remark 1. The expressions for MTTDL obtained in this paper are better ap-proximations for smaller values of λ/μ. This implies that, if simulation-based MTTDL values match the theoretically predicted MTTDL values for a certain value of λ/μ, it will also match for all smaller values of λ/μ. This fact is used in Section 7, where simulations are shown to match theory for values of λ/μ that are much larger than those observed in real-world storage systems, thereby establishing the applicability of the theoretical results to real-world storage systems.

6   Numerical Results

In this section, we compare the MTTDLs of (l, m)-MDS code based systems for clustered and declustered placement schemes for various choice of parameters l and m with the help of figures.

Single Parity Codes: Single parity (l, m)-MDS codes correspond to the case where $m - l = 1$. When $l = 1$, this corresponds to two-way replication. For higher values of l, this corresponds to RAID-5 [1]. It is observed that the MTTDL of single parity codes under both placement schemes are directly proportional to the square of the mean time to node failure, $1/\lambda$, and inversely proportional to the mean time to read all contents of a node during rebuild, $1/\mu$. In addi-tion, the MTTDL values are seen to be independent of the underlying rebuild distribution. The result for clustered placement is well known since the 1980s when the reliability of RAID-5 systems were studied [1]. Fig. 2(a) illustrates the MTTDL behavior of single parity codes with respect to the number of nodes in the system. It is seen that the MTTDL is inversely proportional to the number of nodes for both clustered and declustered placement schemes.

Fig. 2(b) shows how the MTTDL varies as a function of both the codeword length m and the spread factor k for single parity codes, for a given number of nodes n. In Fig. 2(b), clustered placement corresponds to the cases where the spread factor is equal to the codeword length, and declustered placement corresponds to the case where the spread factor is equal to the number of nodes. It is observed that the clustered placement scheme has slightly higher MTTDL

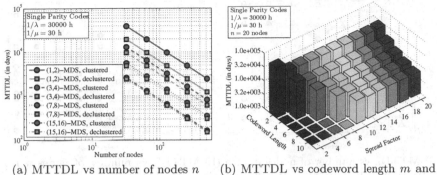

(a) MTTDL vs number of nodes n (b) MTTDL vs codeword length m and
spread factor k for $n = 20$

Fig. 2. MTTDL for single parity codes with $1/\lambda = 30000$ h and $1/\mu = 30$ h

values than other placement schemes, and that increasing the codeword length
decreases the MTTDL.

Double Parity Codes: It is observed that the MTTDL of double parity codes
under both placement schemes are directly proportional to the cube of the mean
time to node failure, $1/\lambda$, and inversely proportional to the square of the mean
time to read all contents of a node during rebuild, $1/\mu$. The result for clustered
placement is well known in the context of RAID-6 systems [2].

In contrast to single parity codes, it is seen that the MTTDL of double parity
codes depends on rebuild distribution. For deterministic rebuild times, the ra-
tios $M_1^2(G_\mu)/M_2(G_\mu)$ and $M_1^2\left(G_{\frac{n-1}{m-1}\mu}\right)/M_2\left(G_{\frac{n-1}{m-1}\mu}\right)$ become one. However,
for random rebuild times, these ratios are upper-bounded by one by Jensen's
inequality. The MTTDL of a system using a $(2,4)$-MDS code is plotted against
the number of nodes in the system for clustered and declustered placements, as
well as for deterministic and exponential rebuild times, in Fig. 3(a). It is observed that the rebuild time distribution scales down the MTTDL, but leaves
the behavior with respect to the number of nodes, n, unaffected.

In contrast to single parity codes, the difference in MTTDL between the two
schemes can be significant, depending on the number of nodes, n, in the system.
This is because, the MTTDL of clustered placement is inversely proportional
to n, whereas the MTTDL of declustered placement is roughly invariant with
respect to n. This is illustrated in Fig. 3(b) in which MTTDL of double parity
codes is plotted against the number of nodes, n, in a log-log scale. The lines
corresponding to clustered placement have a slope of -1, whereas the lines corresponding to declustered placement have a slope of roughly 0. It is also observed
from Fig. 3(b) that longer codes, which are more desirable as they have higher
storage efficiency, can have better MTTDL with declustered placement than
shorter codes with clustered placement for large systems. This is seen, for example, by observing the lines corresponding to $(4,6)$-MDS code with declustered
placement and $(1,3)$-MDS code with clustered placement, for $n > 100$. Just like

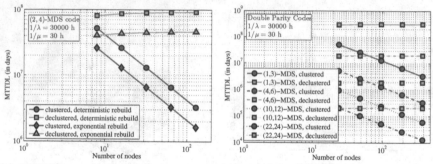

(a) MTTDL vs number of nodes n for a (2,4)-MDS code illustrating the effect of various rebuild distribution

(b) MTTDL vs number of nodes n for double parity codes

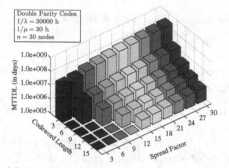

(c) MTTDL vs codeword length m and spread factor k for $n = 30$

Fig. 3. MTTDL for double parity codes with $1/\lambda = 30000$ h and $1/\mu = 30$ h

in the case of single parity codes, the difference in MTTDL between clustered and declustered is observed to be smaller for larger values of the codeword length, m. Fig. 3(c) shows how the MTTDL varies as a function of both the codeword length m and the spread factor k for double parity codes, for a given number of nodes, n.

Codes with Higher Number of Parities: Comparing the MTTDL values of clustered placement in (15) with those of declustered placement in (16), we observe that they are both directly proportional to the $(m-l+1)$th power of the mean time to node failure $1/\lambda$, and inversely proportional to the $(m-l)$ power of the mean time to node rebuild $1/\mu$. This is a general trend in the MTTDL behavior of data storage systems. However, in contrast to clustered placement, which always scales inversely proportional to the number of nodes, the MTTDL of declustered placement is observed to scale differently with the number of nodes for different values of $\tilde{r} = m - l + 1$. In particular, for codes with more than two parities, the MTTDL of declustered placement increases with n. This shows that, by changing the codeword placement scheme, one can influence the scaling of MTTDL with respect to the number of nodes n, resulting in a tremendous improvement in reliability for large storage systems.

Table 2. Range of values of different simulation parameters

Parameter	Meaning	Range
c	amount of data stored on each node	12 TB
n	number of storage nodes	4 to 200
$m - l$	number of parities	1, 2
b	average rebuild bandwidth at each storage node	96 MB/s
$1/\lambda$	mean time to failure of a node	1000 h to 10000 h
$1/\mu$	average time to read/write c amount of data from/to a node during rebuild ($1/\mu = c/b$)	35 h

7   Simulations

Event-driven simulations are used to verify the theoretical estimates of MTTDL of erasure coded systems for two placement schemes, namely, clustered and declustered. The simulations are more involved than the theoretical analysis as they do not make any of the approximations made in theory. Despite this fact, it is found that the theoretical estimates match the simulation results for a wide range of parameters, including the parameters generally observed in practice, thereby validating the applicability of the reliability analysis to real-world storage systems. A detailed description of the simulation method and why the simulations are more realistic than theory can be found in [19].

Simulation Results: Table 2 shows the range of parameters used for the simulations. Typical values for practical systems are used for all parameters, except for the mean times to failure of a node, which have been chosen artificially low (1000 h to 10000 h) to run the simulations fast. The running times of simulations with practical values of the mean times to node failure, which are of the order of 10000 h or higher, are prohibitively high; this is due to the fact that P_{DL} becomes extremely low thereby making the number of first-node-failure events that

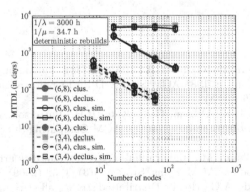

Fig. 4. MTTDL of two different erasure codes with the same storage efficiency for a system with mean time to node failure $1/\lambda = 3000$ h and mean time to read all contents of a node during rebuild $1/\mu = 34.7$ h.

need to be simulated (along with the other complex set of events that restore all lost codeword symbols following each first-node-failure event) extremely high for each run of the simulation. It is seen that, despite the unrealistically low values of mean times to node failure, the simulation-based values are a good match to the theoretical estimates. This observation in conjunction with Remark 1 implies that the theoretical estimates will also be accurate for realistic values of mean times to node failure, $1/\lambda$, which are generally much higher.

Figure 4 shows the comparison between the theoretically-predicted MTTDL values and the simulation-based MTTDL values for systems using $(3, 4)$ and $(6, 8)$ MDS codes. The simulation-based MTTDLs are observed to be in agreement with the theoretical predictions.

8   Conclusion

The reliability of erasure coded systems was studied with a detailed analytical model that accounts for the rebuild times involved, the amounts of partially rebuilt data when additional nodes fail during rebuild, and the fact that modern systems utilize an intelligent rebuild process that rebuilds the most critical codewords first. It was shown that the mean time to data loss of erasure coded systems are practically insensitive to distribution of times to node failure but sensitive to the distribution of node rebuild times. In particular, it was shown that random rebuild times result in lower MTTDL values compared to deterministic rebuild times. The codeword placement scheme, and the rebuild process used, are major factors that influence the scaling of MTTDL with the number of nodes in the system. Declustered codeword placement with distributed rebuild was shown to potentially have significantly larger values of MTTDL compared to clustered codeword placement as the number of nodes in the system increases. Simulations were used to confirm the validity of the theoretical model. Extensions of this work to non-MDS codes and correlated failures are subjects of further investigation.

References

1. Patterson, D.A., Gibson, G., Katz, R.H.: A case for redundant arrays of inexpensive disks (RAID). In: Proc. 1988 ACM SIGMOD Int'l Conference on Management of Data, pp. 109–116 (1988)
2. Chen, P.M., Lee, E.K., Gibson, G.A., Katz, R.H., Patterson, D.A.: RAID: high-performance, reliable secondary storage. ACM Computing Surveys 26(2), 145–185 (1994)
3. Thomasian, A., Blaum, M.: Higher reliability redundant disk arrays: Organization, operation, and coding. ACM Trans. Storage 5(3), 1–59 (2009)
4. Leong, D., Dimakis, A.G., Ho, T.: Distributed storage allocation for high reliability. In: Proc. IEEE Int'l Conference on Communications, pp. 1–6 (2010)
5. Leslie, M., Davies, J., Huffman, T.: A comparison of replication strategies for reliable decentralised storage. Journal of Networks 1(6), 36–44 (2006)

6. Thomasian, A., Blaum, M.: Mirrored disk organization reliability analysis. IEEE Transactions on Computers 55, 1640–1644 (2006)
7. Li, X., Lillibridge, M., Uysal, M.: Reliability analysis of deduplicated and erasure-coded storage. ACM SIGMETRICS Performance Evaluation Review 38(3), 4–9 (2011)
8. Xin, Q., Miller, E.L., Schwarz, T.J.E.: Evaluation of distributed recovery in large-scale storage systems. In: Proc. 13th IEEE Int'l Symposium on High Performance Distributed Computing (HPDC 2004), pp. 172–181 (2004)
9. Venkatesan, V., Iliadis, I., Fragouli, C., Urbanke, R.: Reliability of clustered vs. declustered replica placement in data storage systems. In: Proc. 19th Annual IEEE/ACM Int'l Symposium on Modeling, Analysis, and Simulation of Computer and Telecommunication Systems (MASCOTS 2011), pp. 307–317 (2011)
10. Venkatesan, V., Iliadis, I., Haas, R.: Reliability of data storage systems under network rebuild bandwidth constraints. In: Proc. 20th Annual IEEE Int'l Symposium on Modelling, Analysis, and Simulation of Computer and Telecommunication Systems (MASCOTS 2012), pp. 189–197 (2012)
11. Weatherspoon, H., Kubiatowicz, J.D.: Erasure coding vs. replication: A quantitative comparison. In: Druschel, P., Kaashoek, M.F., Rowstron, A. (eds.) IPTPS 2002. LNCS, vol. 2429, pp. 328–338. Springer, Heidelberg (2002)
12. Plank, J.S., Huang, C.: Tutorial: Erasure coding for storage applications. Slides presented at 11th Usenix Conference on File and Storage Technologies (FAST 2013) (February 2013)
13. Greenan, K.M., Miller, E.L., Wylie, J.: Reliability of flat XOR-based erasure codes on heterogeneous devices. In: Proc. 38th Annual IEEE/IFIP Int'l Conference on Dependable Systems and Networks (DSN 2008), pp. 147–156 (June 2008)
14. Venkatesan, V., Iliadis, I.: A general reliability model for data storage systems. In: Proc. 9th Int'l Conference on Quantitative Evaluation of Systems (QEST 2012), pp. 209–219 (2012)
15. Ford, D., Labelle, F., Popovici, F.I., Stokely, M., Truong, V.A., Barroso, L., Grimes, C., Quinlan, S.: Availability in globally distributed storage systems. In: Proc. 9th USENIX Symposium on Operating Systems Design and Implementation (OSDI 2010), pp. 61–74 (2010)
16. Ramabhadran, S., Pasquale, J.: Analysis of long-running replicated systems. In: Proc. 25th IEEE Int'l Conference on Computer Communications (INFOCOM 2006), pp. 1–9 (2006)
17. Dimakis, A.G., Ramchandran, K., Wu, Y., Suh, C.: A survey on network coding for distributed storage. Proceedings of the IEEE 99(3) (2011)
18. IBM: XiV Storage System Specifications, http://www.xivstorage.com
19. Venkatesan, V., Iliadis, I.: Effect of codeword placement on the reliability of erasure coded data storage systems. Technical Report RZ 3827, IBM Research - Zurich (2012)

Fault-Impact Models Based on Delay and Packet Loss for IEEE 802.11g

Daniel Happ, Philipp Reinecke, and Katinka Wolter

Freie Universität Berlin
Institut für Informatik
Takustraße 9
14195 Berlin, Germany
{daniel.happ,philipp.reinecke,katinka.wolter}@fu-berlin.de

Abstract. In this paper we derive fault-impact models for wireless network traffic as it could be used in the control traffic for smart grid nodes.
We set up experiments using a testbed with 116 nodes which uses the
protocol IEEE 802.11g. We develop models for packet loss, the length
of consecutive packet loss or non-loss as well as for packet transmission time. The latter is a known challenge and we propose a sampling
technique that benefits from the wireless as well as wired connections
between the nodes in the testbed. The data obtained shows similarity
with previous measurements. However, we progress the state of the art
in two ways: we show measurements of packet transmission times and
fit models to those and we provide some more detailed insight in the
data. We find that with increasing link quality, the distributions of lossy
and loss-free periods show major fluctuation. It is shown that in those
cases, phase-type distributions can approximate the data better than
traditional Gilbert models. In addition, the medium access time is also
found to be approximated well with a PH distribution.

1 Introduction

Precise stochastic description of network behaviour is important for various types
of studies. Several experimentation studies on wireless network behaviour have
been conducted in the past [1, 4, 11, 17, 27] and evaluated in terms of standard
analysis techniques such as moments and histograms. These studies provide insight into network behaviour and the traces can be used in detailed simulation
models. However, all experimental results are specific to the environment where
they are sampled. In this paper we aim at providing more general results by
extensive experiments in a testbed with 116 wireless nodes. The network is deployed across three buildings and includes nodes on the outside of buildings as
well as inside. We try to achieve generality by randomly selecting communicating
pairs of nodes. We obtain a wide variety of connections some of which have very
good and some very poor transmission quality.

Existing studies of wireless networks mostly determine the packet loss rate.
Time-synchronisation of wireless networks is difficult and therefore transmission times can normally not be determined. In the testbed we use nodes that

K. Joshi et al. (Eds.): QEST 2013, LNCS 8054, pp. 258–273, 2013.

are connected through several interfaces, one of them being a standard ethernet connection. We use the different interfaces to measure transmission times over the wireless link. This gives us very valuable data which we have not seen published elsewhere before.

As illustrated in e.g. [21, 22], packet loss may affect dependability of higher networking and system layers even with the reliable TCP protocol, which guarantees reliable data transmission. Consequently, methods for reproducing disturbances are required for evaluating the reliability of those systems. Using traces has several disadvantages; in particular, traces are often large, and abstract models can rarely benefit from them. Therefore, a solid model-based description of experimental data is very valuable.

While in the simplest case packet loss may be described by a Bernoulli model, packet loss is often comprised of bursts of elevated loss probability, which can be modelled more closely with Gilbert-Elliot (GE) models [5–7,11,12,17,27–29]. A Gilbert-Elliot model describes the loss process as a Markov Chain with different loss probabilities for each state. We extend the continuous-time Gilbert-Elliot model [23] by using a phase-type distribution [19] for the state transitions. We show that this more complex model fits the data much better, especially when the link quality is high.

The contributions of this paper can be summarised as follows:

- We provide a large data set obtained through extensive randomised experiments in a testbed with different connection characteristics
- We propose a method for sampling transmission times in such a testbed
- We fit extended Gilbert-Elliot models using phase-type fitting to the data
- We show that PH distributions should be considered for modeling packet delay

The paper is organised as follows. In Section 2 we describe the examples motivating our study of packet loss and packet delay characteristics and the metrics we consider. We then discuss related work in Section 3. In Section 4 we introduce our methodology for obtaining accurate measurements and describe the experiments and the measurement results. In Section 6 we fit fault-impact models to the measurement data, before concluding the paper in Section 7.

2   Motivating Examples

In this section we provide motivating examples for the fault-impact models studied in this work. Fault-impact models represent specific types of behaviour of the modelled system. We study two types of models: The first type of model represents the packet-loss characteristics of wireless networks, while the second type reflects delay characteristics.

Packet-loss characteristics are important for a wide range of applications in both public and private usage scenarios. For instance, wireless audio and video applications may adjust their transmission rate depending on packet loss or predict possible future error patterns [28].

Likewise, delay characteristics affect the service quality. Delays often depend not only on the conditions in the network, but on the traffic patterns as well. Our motivating example here is control traffic in the Smart Grid, where end nodes such as smart meters, smart plugs or power providers must be remotely controlled for e.g. billing, power production or power usage regulation. In the Smart Grid, the connection to the end nodes may in many cases be wireless on the last hop [10]. We consider the case where that hop uses IEEE 802.11 [14,15]. Control traffic on the last hop consists of data packets sent following different patterns with different requirements as necessary for the specific application. The traffic characteristics are given in Table 2. For comparison, it should be noted that Internet telephony/VOIP must satisfy a delay bound of 200 ms.

Table 1. Motivation for Smart Grid control traffic

Application	Latency	Remark
Telemonitoring	8- 10 ms	Short distances, often inside properties, utilities
Phasor Measurement Units (PMU)	10-20 ms (class A) 500 ms (class C)	NASPI (North American SynchroPhasor Initiative)
SCADA systems	200 ms	Short distances, often inside properties, utilities (power substations)
Smart Meter	200ms	Short distance data collection (concentrator)
Smart Meter	$\geq$1s	Cyclic data measurements

3   Related Work

There have been several studies dealing with wireless channel behaviour [1,4,11, 17]. Aguayo et al. [1] analyse the packet loss in a 38-node urban multi-hop IEEE 802.11b network. The paper makes the observations that the distribution of loss rates is relatively uniform, most links have relatively stable loss rates, only a small minority experience burst losses, and that signal-to-noise ratio and distance have little impact on the loss rate. Blywis et al. [4] made similar observations using the DES-testbed at Freie Universitt Berlin. They considered IEEE 802.11a and g networks. They found, however, that the distribution of loss rates follows a bathtub curve [24].

Markov models are used because of the simplicity of their characterisation and implementation. There are numerous publications especially on Gilbert-Elliot models [5–7, 11, 12, 17, 27–29]. Haßlinger and Hohlfeld [12] give a good overview of Gilbert-Elliot models in general. They use their own fitting approach to fit models to real internet traces. Yajnik et al. [28] examine the packet loss in multicast networks. They analyse the loss process of stationary traces, i.e. the error characteristics do not change over time. They introduce a way of fitting parameters based on the assumption of geometrically distributed loss runs.

There is also research combining both efforts, such as [5]. Carvalho et al. claim that the Gilbert model cannot capture certain characteristics of wireless errors in 802.11g networks. Wolter et al. [27] observed the same. That would suggest that links experience burst errors which contradicts some previous studies [1].

PH-distributions can be used to fit almost any empirical distribution and are recently studied extensively. There are some publications dealing with phase type fitting of empirical data, e.g. [13, 20, 26]. Wolter et al. [27] use PH-distributions for network loss modelling.

4   Experiment Setup and Methodology

We now present a measurement study conducted in the DES Testbed, an experimental network of 116 nodes distributed around the campus of Freie Universität Berlin [4,9]. The testbed spans three buildings and includes indoor and outdoor nodes. Approximate positions of the nodes on campus can be found in Figure 1. All nodes are equipped with three or more standard IEEE 802.11a/b/g wireless network adapters, and the nodes are additionally connected using the campus Ethernet.

In the first set of experiments we transmit large packets at the maximum rate between randomly-select pairs of nodes. The aim of these experiments is to observe end-to-end packet loss. In the second set of experiments we measure the end-to-end delay of Smart Grid control and data traffic.

4.1   Packet-Loss Measurements

We obtain packet-loss measurements by sending out probe packets, similar to [23], and monitoring the network, as follows: We send probe packets in broadcast mode on the primary network interface of a single node at a time. The second interface of all nodes (including the one sending) monitors the network, recording the header data of each packet it receives. We use the Tcpdump [25] tool to capture raw packet traces of all packets present on the same channel, including packets from other networks. Of course, for each node sending a packet, the majority of nodes in the testbed do not receive this packet, since they are out of the transmission range. In the experiment post-processing, all trace files that contain no packet of the original transmission are discarded, i.e. every remaining trace contains at least one packet sent out by the sending node.

Our experiment setup ensures that all relevant data is recorded. It is also used in a similar way by other authors [2, 11]. In particular, we are able to determine the sequence numbers of the probe packets that arrived successfully at the receivers.

We use a packet size of 1500 bytes and use all available bandwidth, by sending out packets as fast as possible. Nodes are configured to send at 54 Mbit/s fixed using an ad-hoc network on channel 13, which is not occupied by the campus network. However, channel 11 is used by campus access points, which overlaps with channel 13. In order to prevent a bias in the results, all mechanisms that

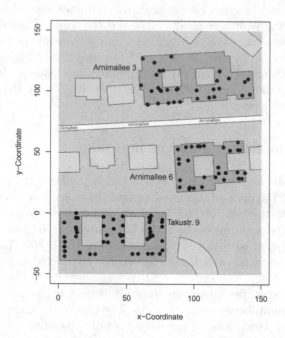

Fig. 1. Top view of the approximate positions of the nodes in the testbed with surroundings

could send additional packets as an overhead are switched off. Those include RTS/CTS (request to send / clear to send), MAC-retransmission and the address resolution protocol (ARP). Furthermore, no packet is allowed to be fragmented.

We obtained 3515 individual traces of 25000 packets each. Lost packets are detected based on a sequence number. All packets whose sequence numbers are not detected at the receiver are assumed to be lost.

In the data analysis we describe the packet error traces as a binary time series $\{X_i\}_{i=1}^n$, where x_i takes the value 1 if the ith packet is correctly transmitted and the value 0 if it was lost. In this paper, a sequence of consecutive 1's is defined to be a success run. In the same way, a sequence of consecutive 0's is called a loss run [17, 28].

We use the average packet error rate (PER) and its counterpart, the packet delivery ratio (PDR) [8, 14–16]. The PER is defined in this work as:

$$\text{PER} = \frac{\# \text{ of Packets lost}}{\# \text{ of Packets sent}} \tag{1}$$

Sometimes it is more intuitive to observe the packet delivery ratio (PDR) instead [4]:

$$\text{PDR} = 1 - \text{PER} = \frac{\# \text{ of Packets received}}{\# \text{ of Packets sent}} \tag{2}$$

4.2   Delay Measurements

In the absence of synchronised clocks on end nodes, measuring end-to-end delay is still a considerable challenge. Although end-to-end delay can be estimated from round-trip times observed at the sender without requiring synchronised clocks, this approach is not applicable in a shared medium, as the additional return packets compete for the medium and interfere with other packets, biasing other results.

We have therefore developed a new method that uses the Ethernet interface of the testbed nodes to measure end-to-end delay on the wireless medium with high accuracy. Our method assumes that the Ethernet backbone of the campus network has negligible jitter, meaning the round trip time over Ethernet is almost constant. Furthermore, we assume that the clock drift on the nodes is negligible.

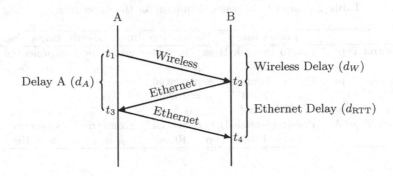

Fig. 2. Sequence diagram of packet exchange for delay measurement

The approach is illustrated in Figure 2: At time t_1 node A sends a packet over the wireless medium to B. Node B stores the time t_2 at which it received the packet, and immediately sends a response packet to A over the Ethernet connection. Node A receives this packet at time t_3 and computes the delay

$$d_A = t_3 - t_1. \tag{3}$$

Node A then sends d_A to node B, again using the Ethernet connection. Node B receives the packet at time t_4 and can thus compute the Ethernet round-trip time:

$$d_{\mathrm{RTT}} = t_4 - t_2 = t_4 - t_3 + t_3 - t_2, \tag{4}$$

Assuming that Ethernet round-trip times are symmetric,

$$d_{\mathrm{RTT}} = 2(t_3 - t_2). \tag{5}$$

Note that

$$d_A = t_2 - t_1 + t_3 - t_2, \tag{6}$$

and hence node B can compute the wireless one-way delay as

$$d_W = d_A - \frac{d_{\text{RTT}}}{2}. \tag{7}$$

We measure delays for Smart-Grid control traffic and Smart-Grid data traffic. Smart Grid control traffic is divided into three classes A, B, and C. The control-traffic classes represent small control messages (class A), large control messages (class B) and bursty traffic (class C) with real-time requirements. The data traffic represents billing information. The parameters of the traffic are shown in Table 2. The values are based on industry documentation and have been determined in discussion with industry experts.

Table 2. Smart-Grid traffic definition for the experiments

	Packet Size	Packet Gap	Burst Length	Burst Gap
Control Type A	Pareto (bounded)	exponential	geometric	exponential
	$\mu = 90B$	$\mu = 1s$	$\mu = 5$	$\mu = 5s$
Control Type B	Pareto (bounded)	exponential	-	-
	$\mu = 750B$	$\mu = 100ms$		
Control Type C	70 B fixed	20ms fixed	-	-
Data Type A	Pareto (bounded)	exponential	geometric	exponential
	$\mu = 490B$	$\mu = 100ms$	$\mu = 5$	$\mu = 10s$

5   Results

We now describe our measurement results. Although our focus is on deriving fault-impact models for these data (Section 6), we also provide a broader discussion of interesting properties.

5.1   Packet-Loss Measurements

We start with the packet-loss measurements. We first discuss general characteristics of the data set; in particular, we consider the distribution of packet-delivery ratios and the spatial distribution of link qualities throughout the testbed. We then focus on the behaviour of packet-loss runs, which will be modelled in Section 6.

We first study the effect of the distance on the packet-delivery ratio (PDR). Figure 4a shows the PDR as a function of the distance between the nodes. Although it is apparent that for longer distances the PDR has lower maximum values, i.e. the probability of a packet been transmitted successfully is smaller, there are also low PDR links with close distances. In particular, note that at distance 0, i.e. when the sender and receiver are the same node, the PDR is

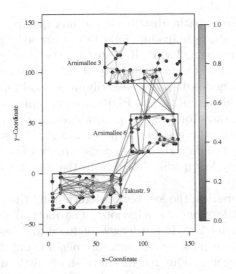

Fig. 3. Top-view map of the link quality in the testbed (1500 bytes packets)

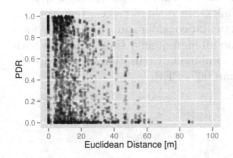

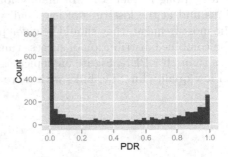

(a) Scatterplot of the PDR for each link over the distance of the two corresponding stations

(b) Histogram of the distribution of link qualities in the testbed measured using the PDR

Fig. 4. Packet loss measurement results

distributed over the entire range from 0 to 1. This is most likely an effect of over-saturation at the receiver. In general, distance seems to have a minor influence on packet losses.

In [27], we observed that there is a strong correlation between distance and average PDR. Our current results do not fully support this observation. However, our current measurement setup in the DES testbed is quite different from the one we used in [27]. In particular, the current setup involves a large set of nodes in separate buildings under real-life operating conditions, whereas in [27] we used an isolated setup of 2 nodes.

Figure 3 illustrates the spatial distribution of link qualities based on the average PER. The figure shows individual links in the testbed, with red indicating high PER and green indicating low PER. Note that the quality of links spanning buildings is, in general, lower than the quality of links within the same building; however, neighbouring nodes do not necessarily have good link quality.

Figure 4b shows a histogram of the PDRs over all links. Due to space constraints we omit histograms for different packet sizes; however, for every packet size under consideration, the general shape of the PDR distribution shows a similar bathtub curve. The data shows two dense groups of node-pairs that have either very high or very low quality links. Note that similar observations have been made in [4] and [1].

In most wireless scenarios, the key factor determining the link quality is not only the mean packet loss but its behaviour. The method commonly used to describe packet loss behaviour is the length of lossy and loss-free periods [5, 12, 17, 18, 28]. In [27], we took several measurements using different distances between sender and receiver. Our recent study shows little dependence of the PER on the distance between sender and receiver. Because of that, we investigate packet loss distributions in relation to their link qualities in this work.

Traces with long run-lengths show phenomena that do not show up in other traces. Long run-lengths are mainly found in the success-runs of high quality links and at the loss-runs of low quality links. Since these traces, however, are the most frequent, they should be given particular consideration. Figure 5a shows the combined distribution of success runs on all high quality links, i.e. links with less than 10% PER. Figure 5a depicts the combined distribution of loss runs on all low quality link, i.e. links with more than 90% PER.

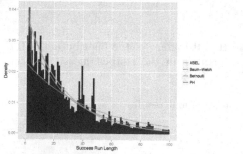 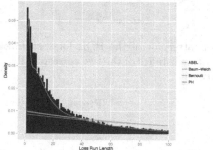

(a) Histogram of success runs on high quality links

(b) Histogram of loss runs on low quality links

Fig. 5. Results of the analysis of the length of consecutive packet loss or non-loss

While Figure 5b shows a monotonically decreasing curve, Figure 5a shows major fluctuations. Also, both traces show that there exist very long individual runs close to the maximum of total transmitted packets at 25.000.

5.2 Delay Measurements

In our analysis of the delay data we concentrate on the delay and on the medium access time.

The measured delay data is shown in Figure 6. Figure 6a is an example of the delays of one trace plotted over time. We have found that at the low data rate we used the medium access time is small compared to the actual transmission time of the packets.

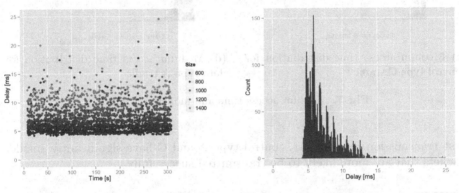

(a) Delay measurement between T9-K36a and T9-K46

(b) Histogram of delay between T9-K36a and T9-K46

Fig. 6. Delay measurement results for the node pair T9-K36a, T9-K46

The majority of the delay is composed of the transmission time of the packets. As expected, large packages take longer. In Figure 6b, which shows the distribution of the delay times, the Pareto distribution used for traffic generation is also apparent.

Nevertheless, the most important factor in the delay is the medium access time, since the actual transmission time of a packet of a certain size is deterministic and simple to compute. We therefore calculated the medium access time by subtracting the transmission time of a packet from the observed delay.

Figure 7 shows the observed medium access time distributions. The various traffic types show very different delay patterns. Notably, all traffic types show a local maximum at around 1 ms. Control type A and C show another local maximum close to 0.3 ms. Control type B and data type A show a smaller local maximum at around 0.6 ms.

A likely reason is the retransmission of packets. Control type B and data type A consist of larger packets, which are more likely to experience packet loss on a

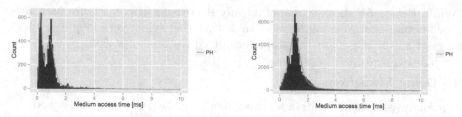

(a) Medium access time distribution for control type A traffic

(b) Medium access time distribution for control type B traffic

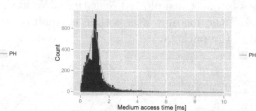

(c) Medium access time distribution for control type C traffic

(d) Medium access time distribution for data type A traffic

Fig. 7. Medium access time and modeling results

first transmission. In contrast, control type A and C have significantly smaller packets that are more likely to be transmitted successfully.

6 Fault-Impact Modelling

In the following section, we develop fault-impact models for IEEE 802.11. The remainder of this section is divided into two parts. First, we introduce loss models for the wireless medium for different link qualities based on PH distributions. For the actual usage of the models in IP packet loss injection [23], this is exceptionally useful, as it is easy to work with the PER as a parameter, rather than with distance as in [27]. We then present the modeling results of our delay measurements. This is also done using PH distributions. As fitting metrics, we use both moments and the mean squared error (MSE) between the distribution given by the models under consideration and the observed distributions.

6.1 Loss Models

For the modeling of wireless packet loss, we proposed the use of PH distributions in [27]. In the following section, we fitted a PH distribution to the samples. We used the G-FIT tool [26] to automatically select the best fit from a Hyper-Erlang distribution with 10 phases and 1–7 branches. The automatic selection process returned distributions with 4–5 branches, with maximum branch length of 5. We compare our phase-type modeling approach to traditional models, namely a

Bernoulli trial, a simple Gilbert model with two states and two parameters fitted to the average burst error length (ABEL) [3, 12, 28] and a Gilbert-Elliot model with two states and four parameters fitted with the Baum-Welch algorithm.

The resulting curves are shown in Figures 5. We observe that most of the measurements obtained were fitted well using a simple Gilbert-Model with two states and two parameters. However, we found that with increasing run-lengths, the models performed poorly. Because of that and because they are the most common link qualities, we concentrate on exceptionally high and low quality links in this work, namely the top and bottom 10% of the PDR range.

Especially Figure 5b, but also 5a show heavy-tailed behaviour, which the traditional models tend to fit rather than the maximum around 1. This applies both to the simple Gilbert model and to to the standard Gilbert-Elliot model in Figure 5b, although the fitting techniques are fundamentally different. This suggests that the number of states in the traditional models are not sufficient to reproduce the measured course of the distribution.

The PH distribution, on the other hand, approximates the distributions more accurately. Both the maximum at 1, as well as the tail are closely approximated. The distribution also follows characteristic bumps in the original distribution to some degree.

Table 3. Overview of performance metrics (% difference and MSE) for Bernoulli (BN), ABEL, Baum-Welch (BW) and PH-models (PH) of the loss-free run lengths of 1500 bytes packets between 0 and 0.1 PER

	BN	ABEL	BW	PH
Mean	38.44 %	0 %	40.01 %	0.01 %
Variance	90.79 %	75.33 %	91.26 %	26.29 %
MSE PMF	$8.47 \cdot 10^{-08}$	$1.20 \cdot 10^{-07}$	$8.66 \cdot 10^{-08}$	$9.05 \cdot 10^{-08}$

This also shows in the metrics, presented in Tables 3 and 4. For the distribution of success runs on high quality links, the mean run length is approximated accurately by the simple Gilbert model and the PH distribution. The PH distribution has the most accurate variance with a difference of 26.29% to the observed variance of the run length distribution and is by far the most accurate model under consideration. While the Bernoulli and Baum-Welch methods show a variance error of almost 100%, the ABEL approach captures the variance slightly better with around 70% error. Surprisingly, in terms of the PMF, the Bernoulli model has the lowest MSE. The Baum-Welch and PH models, however, only show a slightly higher MSE. Overall, the PH-distribution must be rated as the most suitable model for very high quality links.

For the distribution of loss runs on low quality links, the mean run length is approximated accurately by the simple Gilbert as well as the PH distribution. The PH distribution also shows the most accurate variance. In this particular case, the PH distribution also shows by far the lowest PMF MSE. The ABEL

Table 4. Overview of performance metrics (% difference and MSE) for Bernoulli (BN), ABEL, Baum-Welch (BW) and PH-models (PH) of the loss run lengths of 1500 bytes packets between 0.9 and 1 PER

	BN	ABEL	BW	PH
Mean	11.34 %	0 %	14.08 %	0.04 %
Variance	98.07 %	97.54 %	98.18 %	31.15 %
MSE PMF	$4.81 \cdot 10^{-07}$	$5.11 \cdot 10^{-07}$	$4.73 \cdot 10^{-07}$	$9.62 \cdot 10^{-08}$

method has the highest PMF MSE, although the traditional models are all similar with a magnitude of 10^{-07}.

6.2 Delay Models

For the modeling of wireless traffic delay, we also propose the use of PH distributions. In a delay model, it is also sufficient to approximate this medium access time, since the additional transmission time of the packet can be easily added. In the following section, we fit PH distributions with 20 phases to the measured data. Figure 7 shows the observed medium access time distributions with their respective PH distributions.

The PH distributions used generally capture the observed distributions well. Control type C traffic shows the poorest fit. The two distinct local maxima of the measured distribution are indistinguishable in the fitted model and in any case represent a significant simplification of the observation.

Table 5. Overview of fitting metrics for the medium access delay approximations

	First Moment	Second Moment	Third Moment	Variance
Control Type A	0.00%	7.23%	26.54%	14.90%
Control Type B	0.00%	4.01%	24.74%	5.16%
Control Type C	0.00%	12.27%	20.35%	22.11%
Data Type A	0.00%	0.02%	4.63%	0.03%

Table 5 shows the errors of the first three moments and the variance of the models in relation to the measured data. It confirms the assumptions that have been previously mentioned. The mean of all distributions is captured exactly at all traffic types. In general, the moments are well reproduced. Control type C is a challenging distribution, which would need additional phases to be represented accurately and shows the worst fitting. Outstanding is the exceptionally good fitting of data type A.

7   Conclusion

This work proposed fault-impact models based on delay and packet loss for IEEE 802.11g traffic. In a first step, the underlying wireless channel characteristics were presented in the form of the effect of load and packet size on the packet losses measured on a testbed. In a second step, the measured data was further analysed regarding characteristics of packet loss, like the impact of distance, PDR and RSSI distribution, and burst or loss-free run lengths. The main contribution developed throughout our research has been the development of new ways of characterising packet loss and delay, namely with PH-distributions, on the basis of the measured data.

In conclusion, the evaluation of the presented error models with appropriate metrics yield the following results:

- models based on PH-distributions should be considered for error modelling
- run distributions with long run lengths are better approximated with PH models
- models based on PH-distributions should be considered for delay modelling

A very important factor which is missing in the previously mentioned evaluation is the impact of the rate adaptation behaviour of the IEEE 802.11 interface. This would require that the measurements are repeated at different transmission rates. Additionally, different algorithms would have to be considered that determine when to change the transmission rate. Those algorithms would need additional study in themselves, as they are not publicly known, and their behaviour would greatly influence the performance of the system under study.

Rapid changes in technology provide further research opportunities in the area. In particular it would be interesting to study other technologies such as 802.11n, multiple-input and multiple-output (MIMO) technology and sensor networks. For all of these new technologies no models are available today.

Acknowledgments. We would like to thank Tilman Krauß for implementations of the Smart Grid control traffic patterns. We thank Detlef Hartmann from Bell Labs Berlin, Alcatel-Lucent, for many discussions of the requirements and characteristics of Smart Grid control traffic.

References

1. Aguayo, D., Bicket, J., Biswas, S., Judd, G., Morris, R.: Link-level measurements from an 802.11b mesh network. In: Proceedings of the 2004 Conference on Applications, Technologies, Architectures, and Protocols for Computer Communications, SIGCOMM 2004, pp. 121–132. ACM, New York (2004)
2. Arauz, J., Krishnamurthy, P.: Markov modeling of 802.11 channels. In: Vehicular Technology Conference, vol. 2, pp. 771–775. IEEE Computer Society (2003)
3. Billingsley, P.: Statistical Inference for Markov Processes. The University of Chicago Press (1961)

4. Blywis, B., Günes, M., Juraschek, F., Hahm, O.: Properties and Topology of the DES-Testbed. Technical Report TR-B-11-02, Freie Universität Berlin (2011), http://edocs.fu-berlin.de/docs/receive/FUDOCS_document_000000009836

5. Carvalho, L., Angeja, J., Navarro, A.: A new packet loss model of the IEEE 802.11g wireless network for multimedia communications. IEEE Transactions on Consumer Electronics 51(2), 809–814 (2005)

6. Elliott, E.O.: Estimates of Error Rates for Codes on Burst-Noise Channels. Bell System Technical Journal 42, 1977–1997 (1963)

7. Gilbert, E.N.: Capacity of a Burst-Noise Channel. Bell System Technical Journal 39, 1253–1266 (1960)

8. Goldsmith, A.: Wireless Communications. Cambridge University Press (2005)

9. Günes, M., Blywis, B., Juraschek, F.: Concept and Design of the Hybrid Distributed Embedded Systems Testbed. Technical Report TR-B-08-10, Freie Universität Berlin (2008), ftp://ftp.inf.fu-berlin.de/pub/reports/tr-b-08-10.pdf

10. Gungor, V.C., Sahin, D., Kocak, T., Ergut, S., Buccella, C., Cecati, C., Hancke, G.P.: Smart grid technologies: Communication technologies and standards. IEEE Transactions on Industrial Informatics 7(4), 529–539 (2011)

11. Hartwell, J.A., Fapojuwo, A.O.: Modeling and characterization of frame loss process in IEEE 802.11 wireless local area networks. In: Vehicular Technology Conference, vol. 6, pp. 4481–4485. IEEE Computer Society (2004)

12. Haßlinger, G., Hohlfeld, O.: The Gilbert-Elliott Model for Packet Loss in Real Time Services on the Internet. In: 14th GI/ITG Conference on Measurement, Modeling, and Evaluation of Computer and Communication Systems (MMB), pp. 269–286. VDE Verlag (2008)

13. Horváth, A., Telek, M.: PhFit: A General Phase-Type Fitting Tool. In: Field, T., Harrison, P.G., Bradley, J., Harder, U. (eds.) TOOLS 2002. LNCS, vol. 2324, pp. 82–91. Springer, Heidelberg (2002)

14. IEEE. IEEE 802.11g-2003: Further Higher Data Rate Extension in the 2.4 GHz Band. Institute of Electrical and Electronics Engineers, Inc. (2003)

15. IEEE. IEEE 802.11: Wireless LAN Medium Access Control (MAC) and Physical Layer (PHY) Specifications (2007 revision). Institute of Electrical and Electronics Engineers, Inc. (2007)

16. Khalili, R., Salamatian, K.: A new analytic approach to evaluation of packet error rate in wireless networks. In: Proceedings of the 3rd Annual Communication Networks and Services Research Conference, CNSR 2005, pp. 333–338. IEEE Computer Society (2005)

17. Konrad, A., Zhao, B.Y., Joseph, A.D., Ludwig, R.: A Markov-based channel model algorithm for wireless networks. Wireless Networks 9(3), 189–199 (2003)

18. McDougall, J., Miller, S.: Sensitivity of Wireless Network Simulations to a Two-State Markov Model Channel Approximation. In: Global Telecommunications Conference. GLOBACOM 2003, vol. 2, pp. 697–701. IEEE Computer Society (2003)

19. Neuts, M.F.: Matrix-Geometric Solutions in Stochastic Models: An Algorithmic Approach. Dover Publications Inc. (1981) (revised edition)

20. Reinecke, P., Krauß, T., Wolter, K.: Cluster-based fitting of phase-type distributions to empirical data. Computers & Mathematics with Applications (2012)

21. Reinecke, P., van Moorsel, A.P.A., Wolter, K.: The Fast and the Fair: A Fault-Injection-Driven Comparison of Restart Oracles for Reliable Web Services. In: Proc. 3rd International Conference on the Quantitative Evaluation of SysTems (QEST 2006), Riverside, CA, USA. IEEE (September 2006)

22. Reinecke, P., Wolter, K.: Phase-type approximations for message transmission times in web services reliable messaging. In: Kounev, S., Gorton, I., Sachs, K. (eds.) SIPEW 2008. LNCS, vol. 5119, pp. 191–207. Springer, Heidelberg (2008)
23. Reinecke, P., Wolter, K.: On Stochastic Fault-Injection for IP-Packet Loss Emulation. In: Thomas, N. (ed.) EPEW 2011. LNCS, vol. 6977, pp. 163–173. Springer, Heidelberg (2011)
24. Siewiorek, D.P., Swarz, R.S.: Reliable Computer Systems. A K Peters (1998)
25. Tcpdump/libpcap public repository
26. Thümmler, A., Buchholz, P., Telek, M.: A Novel Approach for Phase-Type Fitting with the EM Algorithm. IEEE Transactions on Dependable and Secure Computing 3, 245–258 (2006)
27. Wolter, K., Reinecke, P., Krauss, T., Happ, D., Eitel, F.: PH-distributed Fault Models for mobile Communication. In: Proceedings of the 2012 Winter Simulation Conference, WSC, Berlin, Germany (2012)
28. Yajnik, M., Moon, S.B., Kurose, J.F., Towsley, D.F.: Measurement and Modeling of the Temporal Dependence in Packet Loss. In: Eighteenth Annual Joint Conference of the IEEE Computer and Communications Societies. INFOCOM 1999, vol. 1, pp. 345–352. IEEE Computer Society (1999)
29. Zhang, Y., Duffield, N., Paxson, V., Shenker, S.: On the constancy of internet path properties. In: Proceedings of the 1st ACM SIGCOMM Workshop on Internet Measurement, IMW 2001, San Francisco, California, USA, pp. 197–211. ACM (2001)

VeriSiMPL: Verification via biSimulations of MPL Models*

Dieky Adzkiya[1] and Alessandro Abate[2]

[1] Delft Center for Systems and Control, TU Delft
[2] Department of Computer Science, University of Oxford

Abstract. VeriSiMPL ("very simple") is a software tool to obtain finite abstractions of Max-Plus-Linear (MPL) models. MPL models (Sect. 2), specified in MATLAB, are abstracted to Labeled Transition Systems (LTS). The LTS abstraction is formally put in relationship with the concrete MPL model via a (bi)simulation relation. The abstraction procedure (Sect. 3) runs in MATLAB and leverages sparse representations, fast manipulations based on vector calculus, and optimized data structures such as Difference-Bound Matrices. LTS abstractions can be exported to structures defined in the PROMELA. This enables the verification of MPL models against temporal specifications within the SPIN model checker (Sect. 4). The toolbox is available at
http://sourceforge.net/projects/verisimpl/

1 Motivations and Goals

Max-Plus-Linear (MPL) models are discrete-event systems [1] with continuous variables that express the timing of the underlying sequential events. MPL models are employed to describe the timing synchronization between interleaved processes, and as such are widely employed in the analysis and scheduling of infrastructure networks, such as communication and railway systems, production and manufacturing lines [1]. MPL models are classically analyzed by algebraic [1] or geometric techniques [2] over the max-plus algebra, which allows investigating properties such as transient and periodic regimes [1], or ultimate dynamical behavior. They can be simulated via the max-plus toolbox Scilab [3].

The recent work in [4,5] has explored a novel, alternative approach to analysis, which is based on finite-state abstractions of MPL models. The objective of this new approach is to allow a multitude of available tools that has been developed for finite-state models to be employed over MPL systems. We are in particular interested in the Linear Temporal Logic (LTL) model checking of MPL models via LTS abstractions.

This article presents VeriSiMPL, a software toolbox that implements and tests the abstraction technique in [4,5].

* This research is funded by the European Commission under the MoVeS project, FP7-ICT-2009-5 257005, by the European Commission under the NoE FP7-ICT-2009-5 257462, by the European Commission under Marie Curie grant MANTRAS PIRG-GA-2009-249295, and by NWO under VENI grant 016.103.020.

K. Joshi et al. (Eds.): QEST 2013, LNCS 8054, pp. 274–277, 2013.

2   Nuts and Bolts of Max-Plus-Linear Models

Define $\mathbb{R}_\varepsilon$ and ε respectively as $\mathbb{R} \cup \{\varepsilon\}$ and $-\infty$. For a pair $x, y \in \mathbb{R}_\varepsilon$, we define $x \oplus y = \max\{x, y\}$ and $x \otimes y = x + y$. Max-plus algebraic operations are extended to matrices as follows: if $A, B \in \mathbb{R}_\varepsilon^{m \times n}$ and $C \in \mathbb{R}_\varepsilon^{n \times p}$, then $[A \oplus B](i, j) = A(i, j) \oplus B(i, j)$ and $[A \otimes C](i, j) = \bigoplus_{k=1}^{n} A(i, k) \otimes C(k, j)$, for all i, j. An MPL model [1, Corollary 2.82] is defined as:

$$x(k) = A \otimes x(k - 1) \oplus B \otimes u(k) ,$$

where $A \in \mathbb{R}_\varepsilon^{n \times n}$, $B \in \mathbb{R}_\varepsilon^{n \times m}$, $x(k) \in \mathbb{R}_\varepsilon^n$, $u(k) \in \mathbb{R}_\varepsilon^m$, for $k \in \mathbb{N}$. In this work, the state and input spaces are taken to be $\mathbb{R}^n$ and $\mathbb{R}^m$, respectively: the independent variable k denotes an increasing discrete-event counter, whereas the n-dimensional state variable x defines the (continuous) timing of the discrete events and the m-dimensional input u characterizes external schedules. If the input matrix B contains at least a finite (not equal to ε) element, the MPL model is called *nonautonomous*, otherwise it is called *autonomous* since it evolves under no external schedule. Nonautonomous models embed nondeterminism in the form of a controller input.

Implementation: VeriSiMPL accepts MPL models written in MATLAB. For practical reasons, the state matrix A is assumed to be row-finite, namely characterized in each row with at least one element different from ε.

Example: Consider the following autonomous MPL model from [1, p. 4], representing the scheduling of train departures from two connected stations $i = 1, 2$ (event k denotes the time of the k-th departure at time $x_i(k)$ for station i)

$$x(k) = \begin{bmatrix} 3 & 7 \\ 2 & 4 \end{bmatrix} \otimes x(k - 1), \text{ i.e. } \begin{bmatrix} x_1(k) \\ x_2(k) \end{bmatrix} = \begin{bmatrix} \max\{3 + x_1(k - 1), 7 + x_2(k - 1)\} \\ \max\{2 + x_1(k - 1), 4 + x_2(k - 1)\} \end{bmatrix} .$$

3   From MPL Models to Labeled Transition Systems

We seek to construct a finite-state Labeled Transition System (LTS) as an abstraction of an (autonomous or nonautonomous) MPL model. An LTS comprises a set of finitely many states (Sect. 3.1), of a set of transitions relating pairs of states (Sect. 3.2), and is further decorated with labels on either states or transitions (Sect. 3.3).

3.1   LTS States. Partitioning of MPL Space

LTS states are obtained by partitioning the state space $\mathbb{R}^n$ based on the underlying dynamics, that is based on the state matrix A [4, Algorithms 1,2]. The partition can be further refined (in order to seek a bisimulation of the concrete model) or otherwise coarsened by merging adjacent regions (in order to reduce the cardinality of the set of abstract states).

Implementation: VeriSiMPL implements two alternative approaches [4, Algorithms 1,2]. In order to improve the performance of the procedure, standard pruning tricks are applied. Each generated region is shown to be a Difference-Bound Matrix (DBM) [6, Sect. 4.1]: this allows a computationally efficient representation based on the expression $x_i - x_j \bowtie \alpha_{i,j}$, $\bowtie \in \{<, \leq\}$. VeriSiMPL represents a DBM as a row cell with two elements: the first element is a real-valued matrix representing the upper bound $\alpha_{i,j}$, whereas the second is a Boolean matrix representing the value of $\bowtie$. A collection of DBM is also represented as a row with two elements, where the corresponding matrices are stacked along the third dimension. Quite importantly, DBM are closed under MPL operations.

Example: The partitioning regions generated for the MPL model in Sect. 2 are $R_1 = \{x \in \mathbb{R}^2 : x_1 - x_2 > 4\}$, $R_2 = \{x \in \mathbb{R}^2 : 2 < x_1 - x_2 \leq 4\}$, and $R_3 = \{x \in \mathbb{R}^2 : x_1 - x_2 \leq 2\}$.

3.2 LTS Transitions: Forward-Reachability Analysis

An LTS transition between any two abstract states R and R' is generated based on the relation between the two corresponding partitioning regions. At any given event counter k, there is a transition from R to R' if there exists an $x(k-1) \in R$ and possibly a $u(k) \in U \subseteq \mathbb{R}^m$ such that $x(k) \in R'$. Such a transition can be determined by a forward-reachability computation, i.e. checking the non-emptiness of $R' \cap \{x(k) : x(k-1) \in R, \ u(k) \in U\}$. We assume that the set of allowed inputs $U \subseteq \mathbb{R}^m$ is characterized via a DBM.

Implementation: VeriSiMPL performs forward reachability by mapping and manipulating DBM. It represents a transition in MATLAB as a sparse Boolean matrix. As in a precedence graph [1, Definition 2.8], the (i, j)-th element equals to 1 if there is a transition from j to i, else it is equal to 0.

Example: The transitions for the model in Sect. 2 are represented in Fig. 1. In a nonautonomous version of the model, the finite-state structure in Fig. 1 will simply present additional transitions.

3.3 LTS Labels: Fast Manipulation of DBM

LTS labels are quantities associated with states or transitions and characterize

1) the difference between the timing of a single event (k) for any two variables of the original MPL model, i.e. $x_i(k) - x_j(k)$, where $1 \leq i < j \leq n$; or

2) the time difference between consecutive events of the MPL model, i.e. $x_i(k) - x_i(k-1)$, for $1 \leq i \leq n$.

The first class of labels is determined by the representation of a partitioning region, whereas the second is derived from an outgoing partitioning region and its affine dynamics.

Implementation: Practically, in both cases VeriSiMPL stores the labels as (unions of) vectors of real-valued intervals in MATLAB. In the second case the labels are computed by fast DBM manipulations.

Example: The obtained LTS can be expressed as a simple text file and parsed by Graphviz for plotting, as displayed in Fig. 1.

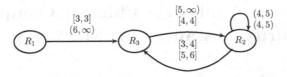

Fig. 1. LTS abstraction of the MPL model in Sect. 2, inclusive of abstract states, transitions, and labels

4   Computational Benchmark and Case Study

We have computed the runtime required to abstract an autonomous MPL system as a finite-state LTS, for increasing dimensions n of the MPL model, and kept track of the number of states and of transitions of the obtained LTS (memory requirement). Compared to partition-based abstraction procedures in the literature for other classes of dynamical systems [7], the present procedure comfortably manages MPL models with significant size (number of continuous variables).

Implementation: For any n, we have generated row-finite matrices A with 2 finite elements (random integers taking values between 1 and 100) placed randomly in each row. The algorithms have been implemented in MATLAB 7.13 (R2011b) and the experiments have been run on a 12-core Intel Xeon 3.47 GHz PC with 24 GB of memory. For $n = 15$, VeriSiMPL generates an LTS with about 10^4 states and 10^6 transitions, with a runtime limited within a few hours.

Example: The obtained LTS can be exported to PROMELA (a PROcess MEta LAnguage), to be later used by the SPIN model checker [8]. Consider the specification $\Psi : \forall k \in \mathbb{N}, \psi(k)$, where $\psi(k) = \{x_2(k+1) - x_2(k) \leq 6\}$. Notice that Ψ can be expressed as $\Box \psi$. We obtain the satisfiability set $\mathrm{Sat}(\Psi) = \{R_2, R_3\}$.

References

1. Baccelli, F., Cohen, G., Olsder, G., Quadrat, J.P.: Synchronization and Linearity, An Algebra for Discrete Event Systems. John Wiley and Sons (1992)
2. Katz, R.: Max-plus (A,B)-invariant spaces and control of timed discrete-event systems. IEEE Trans. Autom. Control 52(2), 229–241 (2007)
3. Plus, M.: Max-plus toolbox of Scilab (Online) (1998),
   http://www.cmap.polytechnique.fr/~gaubert/MaxplusToolbox.html
4. Adzkiya, D., De Schutter, B., Abate, A.: Abstraction and verification of autonomous max-plus-linear systems. In: Proc. 31st Amer. Control Conf., pp. 721–726 (2012)
5. Adzkiya, D., De Schutter, B., Abate, A.: Finite abstractions of nonautonomous max-plus-linear systems. In: Proc. 32nd Amer. Control Conf. (June 2013)
6. Dill, D.L.: Timing assumptions and verification of finite-state concurrent systems. In: Sifakis, J. (ed.) CAV 1989. LNCS, vol. 407, pp. 197–212. Springer, Heidelberg (1990)
7. Yordanov, B., Belta, C.: Formal analysis of discrete-time piecewise affine systems. IEEE Trans. Autom. Control 55(12), 2834–2840 (2010)
8. Holzmann, G.: The SPIN Model Checker: Primer and Reference Manual. Addison-Wesley (2003)

The BisimDist Library: Efficient Computation of Bisimilarity Distances for Markovian Models*

Giorgio Bacci, Giovanni Bacci, Kim Guldstrand Larsen, and Radu Mardare

Department of Computer Science, Aalborg University, Denmark
{grbacci,giovbacci,kgl,mardare}@cs.aau.dk

Abstract. This paper presents a library for exactly computing the bisimilarity Kantorovich-based pseudometrics between Markov chains and between Markov decision processes. These are distances that measure the behavioral discrepancies between non-bisimilar systems. They are computed by using an on-the-fly greedy strategy that prevents the exhaustive state space exploration and does not require a complete storage of the data structures. Tests performed on a consistent set of (pseudo)randomly generated instances show that our algorithm improves the efficiency of the previously proposed iterative algorithms, on average, with orders of magnitude. The tool is available as a Mathematica package library.

1 Introduction

Probabilistic bisimulation of Larsen and Skou [7] plays a central rôle in the verification of discrete-time Markov Chains (MCs), and this notion has been later extended to Markov Decision Processes with rewards (MDPs) [6]. Bisimulation equivalences may be used for comparing systems to a given model specification, or to make feasible the analysis of large systems by reducing their size by means of bisimilarity quotients. However, when the numerical values of probabilities are based on statistical samplings or subject to error estimates, any behavioral analysis based on a notion of equivalence is too fragile, as it only relates processes with identical behaviors. These problems motivated the study of *behavioral distances* (pseudometrics) for probabilistic systems, firstly developed for MCs [4,9,8] and later extended to MDPs [5]. The proposed pseudometrics are parametric in a *discount factor* $\lambda \in (0, 1]$ that controls the significance of the future in the measurement. These distances provide a way to measure the behavioral similarity between states and allow one to analyze models obtained as approximations of others, more accurate but less manageable, still ensuring that the obtained solution is close to the real one. These reasons motivate the development of algorithms for computing bisimilarity distances.

In [2] we proposed an efficient on-the-fly algorithm for computing the behavioral pseudometrics of Desharnais et al. [4] on MCs. Our method has been

* Work supported by the VKR Center of Excellence MT-LAB and the Sino-Danish Basic Research Center IDEA4CPS.

K. Joshi et al. (Eds.): QEST 2013, LNCS 8054, pp. 278–281, 2013.

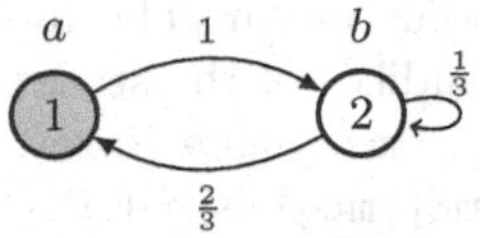

```
tm = MCtm[{{1,2}->1, {2,2}->1/3, {2,1}->2/3},2];
mc = MC[tm, {"a","b"}]
```

Fig. 1. Encoding of a Markov Chain as a data term in BISIMDIST

inspired by an alternative characterization of the pseudometric given in [3], that relates the pseudometric to the least solutions of a set of equation systems induced by a collection transportation schedules. The pseudometric is computed by successive refinements of over-approximations of the actual distance using a greedy strategy that always chooses a transportation schedule that better improves the current approximation. This strategy avoids the exhaustive exploration of the state space, and has the practical advantage that allows one to focus only on computing the distances between states that are of particular interest. Experimental results have shown that this technique performs, on average, orders of magnitude better then the corresponding iterative algorithms proposed in the literature, e.g., in [3]. The algorithm in [2] has been recently adapted in order to compute the bisimilarity pseudometric introduced by Ferns et al. in [5] for MDPs with rewards (see [1] for a detailed account on this extension).

In this paper, we present the BISIMDIST library, composed of two Mathematica packages which implement our on-the-fly algorithm for computing the bisimilarity distances for MCs and MDPs, respectively. BISIMDIST is available at http://people.cs.aau.dk/~giovbacci/tools.html together with simple tutorials presenting use case examples that show all the features of the library.

2   The BISIMDIST Library

The BISIMDIST library consists of two Mathematica packages: MCDIST and MDPDIST providing data structures and primitives for creating, manipulating, and computing bisimilarity distances for MCs and MDPs respectively. It also has methods to identify bisimilarity classes and to solve lumpability problems.

The MCDIST *Package:* An MC with n states is represented as a term of the form MC[<tm>, <lbl>], where <tm> is an $n \times n$ probability transition matrix (<tm> [[i,j]] denotes the probability of going from the state i to the state j) and <lbl> is a vector of strings of length n (<lbl> [[i]] is the label associated with the state i). Note that states are implicitly represented as indices $1 \leq i \leq n$.

The probability transition matrices can be defined explicitly as a matrix, or implicitly by listing only the transitions which have nonzero probability by means of the function MCtm (see Fig. 1). Given a list trRules of rules of the form $\{i,j\} \rightarrow p_{i,j}$, the function MCtm[trRules, n] returns an $n \times n$ matrix where each pair (i,j) is associated with the value $p_{i,j}$, otherwise 0. An MC mc is displayed by calling PlotMC[mc]. Given a sequence mc$_1$, ..., mc$_k$ of MCs,

`JoinMC[mc`$_1$`,...,mc`$_k$`]` yields an MC representing their disjoint union. The indices representing the set of states are obtained shifting the indices of the states of the arguments according to their order in the sequence (e.g. if `mc`$_1$ has n states, the index corresponding to the i-th state of `mc`$_2$ in `JoinMC[mc`$_1$`,mc`$_2$`]` is $n + i$).

Given an MC `mc` with n states, a list `Qpairs` of pairs of indices $1 \leq i, j \leq n$, and a rational discount factor $\lambda \in (0, 1]$, `BDistMC[mc, `λ`, Qpairs]` returns the list of all λ-discounted bisimilarity distances calculated between the pairs of states in `Qpairs` as list of rules of the form $\{i, j\} \rightarrow d_{i,j}$. The alias `All` is used for indicating the list of all pairs of states. `BDistMC` has the following options:

Verbose: (default `False`) displays all intermediate approximations steps;

ConsistencyCheck: (default `True`) checks that the term `mc` is a proper MC;

Estimates: (default `None`) takes a list of rules of the form $\{i, j\} \rightarrow d_{i,j}$ and computes the least over-approximation of the bisimilarity distance assuming $d_{i,j}$ to be the actual distance between the states i and j.

The package MCDIST provides also the functions `BisimClassesMC`, which calculates the bisimilarity classes of an MC, and `BisimQuotientMC` that, for a given an MC, yields its quotient w.r.t. probabilistic bisimilarity.

The MDPDIST Package: An MDP with n states and m action labels is represented as a term of the form `MPD[<tm>, <rw>, <act>]`, where `<tm>` is an $n \times m \times n$ labelled probability transition matrix (`<tm>`$[\![i,a,j]\!]$ is the probability of going from the state i to the state j, known that the action a as been chosen), `<rw>` is a $n \times m$ real-valued matrix representing a reward function, and `<act>` is a string-valued list of length m specifying the names of the action labels. States and action labels are implicitly encoded as indices.

Probability transition matrices of size $n \times m \times n$ can be defined by giving the nonzero transition probabilities as a list `trRules` of rules of the form $\{i, a, j\} \rightarrow p_{i,a,j}$ and calling `MDPtm[trRules, `n`, `m`]`. Analoguosly, $n \times m$ reward matrices can be defined by calling `MDPrm[<rwRules>, `n`, `m`]`, where `<rwRules>` is a list of rules of the form $\{i, a\} \rightarrow r_{i,a}$.

The MDPDIST package is provided with an interface similar to MCDIST with analogous semantics: `PlotMDP`, `JoinMDP`, `BDistMDP`, `BisimClassesMDP`, and `BisimQuotientMDP`.

3   Results and Conclusions

BISIMDIST is a research tool still undergoing development. While not yet mature enough to handle industrial case studies, the on-the-fly algorithm for computing the bisimilarity distance performs, on average, better than the iterative method proposed in [3]. Table 1 reports the average execution times of the on-the-fly algorithm run with discount factor $\lambda = 1/2$ on a collection of randomly generated MCs. We executed the iterative method on the same input instances, interrupting it as soon as it exceeded the running time of our method. The on-the-fly approach leads to a significant improvement in the performances: it yields the exact solution before the iterative method can under-approximate it with an

Table 1. Comparison between the on-the-fly and the iterative methods on MCs

# States	On-the-Fly (exact) Time (sec)	Iterative (approximated)		Approximation Error
		Time (sec)	# Iterations	
10	1.003	1.272	3.111	0.0946
12	4.642	5.522	4.042	0.0865
14	6.336	7.188	4.914	0.1189
20	34.379	38.205	7.538	0.1428

error of ≈ 0.1, which is a non-negligible error for a value in the interval $[0, 1]$. A more detailed analysis of the performances and scalability can be found in [2].

The BisimDist library provides primitives that aid the analysis on probabilistic systems by reasoning in terms of approximate behaviors. In [1], we further improved the efficiency of the implemented on-the-fly algorithm on MDPs, also in relation to the addition of primitives for handling algebraic operations over probabilistic systems, such as synchronous/asynchronous parallel composition. We plan to apply similar on-the-fly techniques for computing bisimilarity distances on continuous-time probabilistic systems and timed automata.

References

1. Bacci, G., Bacci, G., Larsen, K.G., Mardare, R.: Computing behavioral distances, compositionally. In: Chatterjee, K., Sgall, J. (eds.) MFCS 2013. LNCS, vol. 8087, pp. 74–85. Springer, Heidelberg (2013)
2. Bacci, G., Bacci, G., Larsen, K.G., Mardare, R.: On-the-Fly Exact Computation of Bisimilarity Distances. In: Piterman, N., Smolka, S.A. (eds.) TACAS 2013. LNCS, vol. 7795, pp. 1–15. Springer, Heidelberg (2013)
3. Chen, D., van Breugel, F., Worrell, J.: On the Complexity of Computing Probabilistic Bisimilarity. In: Birkedal, L. (ed.) FOSSACS 2012. LNCS, vol. 7213, pp. 437–451. Springer, Heidelberg (2012)
4. Desharnais, J., Gupta, V., Jagadeesan, R., Panangaden, P.: Metrics for labelled Markov processes. Theoretical Computer Science 318(3), 323–354 (2004)
5. Ferns, N., Panangaden, P., Precup, D.: Metrics for finite Markov Decision Processes. In: Proceedings of the 20th Conference on Uncertainty in Artificial Intelligence, UAI, pp. 162–169. AUAI Press (2004)
6. Givan, R., Dean, T., Greig, M.: Equivalence notions and model minimization in Markov decision processes. Artificial Intelligence 147(1-2), 163–223 (2003)
7. Larsen, K.G., Skou, A.: Bisimulation through probabilistic testing. Information and Computation 94(1), 1–28 (1991)
8. van Breugel, F., Sharma, B., Worrell, J.: Approximating a Behavioural Pseudo metric without Discount for Probabilistic Systems. Logical Methods in Computer Science 4(2), 1–23 (2008)
9. van Breugel, F., Worrell, J.: Approximating and computing behavioural distances in probabilistic transition systems. Theoretical Computer Science 360(1-3), 373–385 (2006)

Möbius Shell:
A Command-Line Interface for Möbius

Ken Keefe and William H. Sanders

University of Illinois, Urbana, IL 61801, USA
kjkeefe@illinois.edu whs@illinois.edu
https://www.mobius.illinois.edu

Abstract. The Möbius modeling environment is a mature, multi-formalism modeling and solution tool. Möbius provides a user-friendly graphical interface for creating discrete-event models, defining metrics, and solving for the metrics using a variety of solution techniques. For certain research needs, the graphical interface can become a limiting use pattern. This paper describes recent work that adds a comprehensive text-based interface for interacting with the Möbius tool, called the Möbius Shell. The Möbius Shell provides an interactive command shell and scriptable command language that can leverage all the existing and future features of Möbius.

Keywords: text-based interface, multi-formalism modeling, simulation, analytical solution, discrete-event systems.

1 The Möbius Modeling Environment

The Möbius Modeling Environment is an extensible modeling and solution tool. It offers a variety of existing modeling formalisms, including compositional modeling formalisms, a metric specification formalism that allows for time- and event-based rewards, global model parameterization with several means of defining experiments, and a set of analytical and simulation solution methods [1][2]. Until recently, Möbius has provided only a graphical user interface for working with each of its components. With the addition of the Möbius Shell, Möbius now offers a text-based user interface that enables an interactive or scriptable method of performing actions in the top-level tool or within one of the Möbius components.

2 Using the Möbius Shell

In the Möbius installation directory, the mobius executable launches the graphical version of the tool. The text-based, interactive version of Möbius can be launched using the mobius-shell executable:

```
$ mobius-shell
Welcome to Mobius 2.4.1!
  Enter "help" for a list of commands.
Mobius>
```

K. Joshi et al. (Eds.): QEST 2013, LNCS 8054, pp. 282–285, 2013.

Commands can then be executed at the Mobius> prompt. Each time a command is executed, text feedback is provided. Long-running jobs, such as the execution of a simulation, will provide continuous feedback and can be interrupted by hitting Ctrl+C.

Alternatively, the Möbius Shell can execute a script either by using a command pipe or by passing the script file path using a command switch:

```
$ cat myScript.txt | mobius-shell
$ mobius-shell -s myScript.txt
```

3 Key Möbius Shell Commands

The Möbius Shell is intended to be a full-fledged alternative to the traditional graphical user interface. Because of space limitations, here we detail only a few important commands. For a full treatment of the Möbius Shell command language, see the 2.4.1 (or later) version of the Möbius Manual [3].

3.1 Help

The Möbius Shell provides a comprehensive, integrated help system. Users can obtain a list of all available commands by executing the *help* command.

```
Mobius> help
Mobius Shell Command Help
Further help can be found for each command by executing:

help <command>
   archive    - Archive a project
   clean      - Clean a project or model component
   ...
```

Detailed help for each command can be accessed by including the command as an argument to the help command:

```
Mobius> help save
Generate and compile a model component command:

save <project name> (a|c|r|y|t|s) <component name>
   a   - Atomic model type
   c   - Composed model type
   r   - Reward model type
   y   - Study type
   t   - Transformer type
   s   - Solver type
```

3.2 Generate, Compile, and Save

When a user saves a model component in Möbius, the first step that Möbius performs is generation of a C++ representation of the model component. That typically consists of a set of classes that derive from base classes in the Möbius code library[1]. Next, Möbius compiles those classes and links them to code library archives that come with Möbius.

In the Möbius Shell, those steps can be performed individually or combined, as in the graphical tool. To generate the C++ representation of a model component, use the *generate* command. To compile the C++ representation, use the *compile* command. To do both, use the *save* command. For example, the below commands generate and compile the reward model called "perfEx" in the "satRelay" model.

```
Mobius> generate satRelay r perfEx
Generating code..............Done!
Mobius> compile satRelay r perfEx
make: Entering directory '/home/kjkeefe/MobiusProject/satRelay/Reward/perfEx'
make lib TARGET=libperfExPV_debug.a OBJS="perfExPVNodes.o perfExPVModel.o "
...
make: Leaving directory '/home/kjkeefe/MobiusProject/satRelay/Reward/perfEx'
Compile completed: SUCCESS
```

3.3   Run

The *run* command begins the execution of a transformer, analytical solver, or simulator. When a transformer or analytical solver is run, the feedback in the Möbius Shell is a summary, and the results are stored in a file in the component's directory. However, when a simulator runs, an aggregation of reward variable statistics is reported on the fly until all variables have converged within their defined confidence intervals, or until some other ending condition has been met (e.g., max number of iterations simulated). Those behaviors mirror those of the graphical version of the tool.

```
Mobius> run satRelay t AvNumSSG
Building State Space Generator for Linux architecture
Building for Linux systems on darboux
...
Generated: 8190 states
Computation Time (user + system): 2.160100e-01 seconds
State Generation of Experiment_3 on model AvNumSSG finished at Wed Mar 06 21:13:11
CST 2013.
```

3.4   Edit

The *edit* command allows the user to step into project components. For model components that have nested child elements (e.g., a Stochastic Activity Network (SAN) [4] model containing input gates and activities), the *edit* command can further step into those elements to make changes to their attributes (e.g., input predicate, firing distribution).

In the following example, we start by editing the "cpu_module" SAN model in the "Multi-Proc" project (which is included in the standard set of examples that come with Möbius). On line 3 we execute the *show* command to get a brief summary of this atomic model. Next, on line 10, we ask for further details on the activities in this SAN model. We could get a complete description of the "cpu_failure" activity by using the *show activity* command. On line 13 we begin editing the "cpu_failure" activity. We start by showing the details of the timing distribution. Next, on lines 19, 21, and 23, we alter the timing distribution type, mean, and variance, respectively. Having made the desired changes, we close that activity on line 25. Finally, we close the "cpu_module" SAN model. Möbius Shell then asks us if we would like to save our changes, which we do, and the model compiles successfully.

```
 1 Mobius> edit Multi-Proc a cpu_module
 2 Now editing the cpu_module SAN Atomic Model (enter the "close" command when finished)...
 3 Multi-Proc/Atomic/cpu_module> show
 4 Model contains 10 elements and 18 connections:
 5   1 activity (0 instantaneous, 1 timed)
 6   0 extended places
 7   1 input gate
 8   3 output gates
 9   5 places
10 Multi-Proc/Atomic/cpu_module> show activities
11 Model contains 1 activity:
12   cpu_failure (timed, incoming: Input_Gate1, outgoing: Case 1: OG1, Case 2: OG2, Case 3: OG3)
13 Multi-Proc/Atomic/cpu_module> edit cpu_failure
14 Now editing the cpu_failure Timed Activity (enter the "close" command when finished)...
15 Multi-Proc/Atomic/cpu_module/cpu_failure> show timing
16 Timing distribution: Exponential
17 Parameters:
18   Rate: 6.0 * failure_rate * cpus->Mark()
19 Multi-Proc/Atomic/cpu_module/cpu_failure> set timing distribution Normal
20 Timing distribution set to Normal (Mean: , Variance: )
21 Multi-Proc/Atomic/cpu_module/cpu_failure> set timing Mean "failure_rate * cpus->Mark()"
22 Timing distribution set to Normal (Mean: failure_rate * cpus->Mark(), Variance: )
23 Multi-Proc/Atomic/cpu_module/cpu_failure> set timing Variance "0.1"
24 Timing distribution set to Normal (Mean: failure_rate * cpus->Mark(), Variance: 0.1)
25 Multi-Proc/Atomic/cpu_module/cpu_failure> close
26 Closing the cpu_failure Timed Activity (changes will not be saved until atomic model is saved)...
27 Multi-Proc/Atomic/cpu_module> close
28 You have unsaved changes, would you like to save? (Y|n) y
29 Generating code..............Done!
30 make: Entering directory '/home/kjkeefe/MobiusProject/Multi-Proc/Atomic/cpu_module'
31 ...
32 Compile completed: SUCCESS
33 Mobius>
```

Acknowledgments. The authors would like to acknowledge the current and former members of the Möbius team and the outside contributors to the Möbius project. The authors would also like to thank Jenny Applequist for her editorial work.

References

1. Deavours, D.D., Clark, G., Courtney, T., Daly, D., Derisavi, S., Doyle, J.M., Sanders, W.H., Webster, P.G.: The Möbius framework and its implementation. IEEE Transactions on Software Engineering 28(10), 956–969 (2002)
2. Doyle, J.M.: Abstract model specification using the Möbius modeling tool. Master's thesis, University of Illinois at Urbana-Champaign, Urbana, Illinois (January 2000)
3. Möbius Team: The Möbius Manual. University of Illinois at Urbana-Champaign, Urbana, IL (2013), http://www.mobius.illinois.edu
4. Sanders, W.H., Meyer, J.F.: Stochastic activity networks: Formal definitions and concepts. In: Brinksma, E., Hermanns, H., Katoen, J.-P. (eds.) FMPA 2000. LNCS, vol. 2090, pp. 315–343. Springer, Heidelberg (2001)

A CTL Model Checker
for Stochastic Automata Networks*

Lucas Oleksinski, Claiton Correa, Fernando Luís Dotti, and Afonso Sales**

PUCRS - FACIN, Porto Alegre, Brazil
{lucas.oleksinski,claiton.correa}@acad.pucrs.br,
{fernando.dotti,afonso.sales}@pucrs.br

Abstract. Stochastic Automata Networks (SAN) is a Markovian formalism devoted to the quantitative evaluation of concurrent systems. Unlike other Markovian formalisms and despite its interesting features, SAN does not count with the support of model checking. This paper discusses the architecture, the main features and the initial results towards the construction of a symbolic CTL Model Checker for SAN. A parallel version of this model checker is also briefly discussed.

1 Introduction

Stochastic Automata Networks (SAN) was proposed by Plateau [12], being devoted to the quantitative evaluation of concurrent systems. It is a Markovian formalism that allows modeling a system into several subsystems which can interact with each other. Subsystems are represented by automata and interactions by synchronizing transitions of cooperating automata on same events. Dependencies among automata can also be defined, using functions. Functions evaluate on the global state of the automata network and can be used to specify the behavior of specific automata. The use of functions allows the description of complex behaviors in a very compact way [1]. Quantitative analysis of SAN models is possible using specialized software tools (*e.g.*, PEPS [13] or SAN Lite-Solver [14]), fundamentally allowing one to associate probabilities to the states of the model, using a steady state or transient analysis.

While developing models for involved situations it is highly desirable to reason about their computation histories and thus model checking becomes important. Indeed, many formalisms for quantitative analysis count with the support of specialized model checking tools. In the context of CTMC-based model checking, we can mention PRISM [8], SMART [4] and CASPA [7]. Such support however is lacking for SAN. In this paper we report our results towards the construction of the first SAN model checker. In this initial version, the tool is restricted to CTL model checking opposed to the stochastic verification as offered by the aforementioned tools.

* Paper partially sponsored by CNPq (560036/2010-8) and FAPERGS (PqG 1014867).
** Afonso Sales receives grant from PUCRS (Edital 01/2012 – Programa de Apoio à Atuação de Professores Horistas em Atividades de Pesquisa na PUCRS).

K. Joshi et al. (Eds.): QEST 2013, LNCS 8054, pp. 286–289, 2013.

2   Tool Overview

Fig. 1 illustrates the main processing steps of the SAN model checker. It has as input a model written in the SAN modeling language [13], a CTL (*Computation Tree Logic*) property, and an additional information if a witness or a counterexample to the property is desired. As output it offers the answer whether the property is true or false, and a witness or counterexample as chosen. The tool supports the standard CTL where atomic propositions are assertions about the global state of the automata network according to the SAN language [13].

The compilation of the SAN model generates a Markovian descriptor which is used as the system transition relation, *i.e.*, a set of tensors which operated by generalized Kronecker algebra allows the achievement of next states. The initial states of the model are those considered as reachable in the reachability declaration of the SAN model. Multi-valued Decision Diagrams (MDD) are used to encode the Reachable State Space (RSS) of the SAN model, which is calculated using an extension [15] of the saturation based approach [3]. The satisfaction sets calculation (SAT in Fig. 1) follows a breadth-first search algorithm. During this process, the RSS is labelled with all subformulas of the input formula.

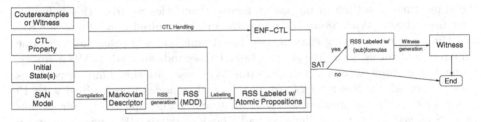

Fig. 1. The tool architecture

Whenever a counterexample is desired, the tool negates the input formula to generate a witness. The witness generator supports ENF-CTL operators and generates trace structured witnesses. To enrich witness information, whenever a branching is avoided the respective state of the trace is annotated with the subformula that holds from that state.

A parallel approach was proposed that replicates the entire RSS and assigns specific partitions of the state space to be computed by different nodes. Each node may locally compute successor states even these cross partition borders, without requiring communication. Communication is only required for fix-point calculation, which is executed as rounds of synchronization between nodes.

3   Experiments

We report CTL model checking results[1] on both sequential and parallel implementations of the model checker through set of experiments with two different models: the dining philosophers (DP) problem [15] and a model for an *ad hoc*

---

[1] As mentioned, our tool does not perform stochastic verification and thus numerical analysis is not carried out.

wireless network protocol (WN) [6]. For the DP model, starvation, mutual exclusion, deadlock presence and deadlock absence were checked for model variations with and without deadlock, respectively. Considering the DP model with 15 philosophers, corresponding to RSS of 470,832, all mentioned properties needed about 500 MB memory and 240 CPU seconds (using one core of a Intel Xeon Quad-Core E5520 2.27 GHz machine). For some properties, such as deadlock absence, the tool allows verification of a model with 20 philosophers which has 38,613,965 reachable states requiring around 600 MB memory and 1,650 CPU seconds. We experienced the parallel version in a cluster with 15 processors for the same DP model above with 15 philosophers. The worst speed-up took place with the "deadlock absence" property and the best speed-up took place with the "starvation" property, with speed-ups of 6 and 11, respectively. The verification of the deadlock absence property took a peak of 330 MB *per* node memory while the sequential solution took 500 MB. The starvation property took a peak of 140 MB *per* node memory while the sequential execution took 660 MB.

The WN model was build to obtain the end-to-end throughput traversing a route of *ad hoc* nodes, taking into consideration the interference range among nodes. Thus the model is build such that no two interferring nodes transmit at the same time, resulting in no packet losses. Properties assuring this behavior have been shown. With 28 *ad hoc* nodes the model resulted in a RSS of 568,518 and the property assuring that no two interfering nodes transmit at the same time required verification time around 130 CPU seconds and 361.46 MB memory. Using the parallel approach for the same WN case, for the same property as above reported, models with 24, 26 and 28 *ad hoc* nodes were verified with 15 processors, leading to speed-ups of 4.91; 4.40 and 7.40, respectively. For models with 28 *ad hoc* nodes, the verification had a peak of 289.08 MB *per* node memory while the sequential solution 361.46 MB.

To assess the tool correctness we have carried out experiments with the NuSMV tool. More specifically, a set of SAN models have been translated to the NuSMV, generating transition systems equivalent to the SAN model's underlying Markov Chains, and have been checked for the same CTL properties, leading to same results.

4   Conclusions and Future Works

In this paper we presented, at the authors' best knowledge, the first tool for model checking SAN models. We have discussed its key features, performance results and also initial results on a parallel version. As a first version of the SAN model checker, it has shown coherent results leading to a high confidence in its correctness, however with low performance. Even considering hardware differences, the results reported by PRISM [2], CASPA [7] and SMART [5] can be clearly considered much superior.

In this version we have adopted a Kronecker-based representation of the transition relation due to the usage of functional elements in the transition matrices which are necessary to represent SAN abstractions. The use of Kronecker representation has a considerable impact since the computation of next states implies that several tensors have to be operated, in a meaningful order, according to the number of submodels (automata) and synchronizing events. Moreover, whenever functions are used, they have to be evaluated in this process. In contrast, decision diagram based representation [11,10] would result in a more direct computation of transitions. This aspect is to be addressed in future works. Another aspect that we want to address is the use of saturation based model checking algorithms. A first step in this directed has been made in [15] where the reachable state space generation of SAN models is computed in a *saturated* way using both decision diagrams and Kronecker representations. Related works using decision diagrams, such as in [9], can contribute in relation to this aspect and must be more closely investigated.

References

1. Brenner, L., Fernandes, P., Sales, A.: The Need for and the Advantages of Generalized Tensor Algebra for Structured Kronecker Representations. Int. Journal of Simulation: Systems, Science & Technology (IJSIM) 6(3-4), 52–60 (2005)
2. PRISM (Probabilistic Model Checker), http://www.prismmodelchecker.org/
3. Ciardo, G., Lüttgen, G., Siminiceanu, R.I.: Saturation: An Efficient Iteration Strategy for Symbolic State-Space Generation. In: Margaria, T., Yi, W. (eds.) TACAS 2001. LNCS, vol. 2031, pp. 328–342. Springer, Heidelberg (2001)
4. Ciardo, G., Miner, A.S., Wan, M.: Advanced features in SMART: the stochastic model checking analyzer for reliability and timing. SIGMETRICS Performance Evaluation Review 36(4), 58–63 (2009)
5. Ciardo, G., Zhao, Y., Jin, X.: Ten Years of Saturation: A Petri Net Perspective. Transactions Petri Nets and Other Models of Concurrency 5, 51–95 (2012)
6. Dotti, F.L., Fernandes, P., Sales, A., Santos, O.M.: Modular Analytical Performance Models for Ad Hoc Wireless Networks. In: WiOpt 2005, pp. 164–173 (2005)
7. Kuntz, M., Siegle, M., Werner, E.: Symbolic Performance and Dependability Evaluation with the Tool CASPA. In: Núñez, M., Maamar, Z., Pelayo, F.L., Pousttchi, K., Rubio, F. (eds.) FORTE 2004. LNCS, vol. 3236, pp. 293–307. Springer, Heidelberg (2004)
8. Kwiatkowska, M., Norman, G., Parker, D.: PRISM 4.0: Verification of Probabilistic Real-time Systems. In: Gopalakrishnan, G., Qadeer, S. (eds.) CAV 2011. LNCS, vol. 6806, pp. 585–591. Springer, Heidelberg (2011)
9. Lampka, K., Siegle, M.: Activity-local symbolic state graph generation for high-level stochastic models. In: 13th MMB, pp. 245–264. VDE Verlag (2006)
10. Lampka, K., Siegle, M.: Analysis of Markov reward models using zero-suppressed multi-terminal BDDs. In: VALUETOOLS, p. 35 (2006)
11. Miner, A., Parker, D.: Symbolic Representations and Analysis of Large Probabilistic Systems. In: Baier, C., Haverkort, B.R., Hermanns, H., Katoen, J.-P., Siegle, M. (eds.) AUTONOMY 2003. LNCS, vol. 2925, pp. 296–338. Springer, Heidelberg (2004)
12. Plateau, B.: On the stochastic structure of parallelism and synchronization models for distributed algorithms. In: ACM SIGMETRICS Conf. on Measurements and Modeling of Computer Systems, Austin, USA, pp. 147–154. ACM Press (1985)
13. PEPS Project, http://www-id.imag.fr/Logiciels/peps/userguide.html
14. Sales, A.: SAN lite-solver: a user-friendly software tool to solve SAN models. In: SpringSim (TMS-DEVS), Orlando, FL, USA, vol. 44, pp. 9–16. SCS/ACM (2012)
15. Sales, A., Plateau, B.: Reachable state space generation for structured models which use functional transitions. In: QEST 2009, Budapest, Hungary, pp. 269–278 (2009)

The Steady-State Control Problem
for Markov Decision Processes

S. Akshay[1,2], Nathalie Bertrand[1], Serge Haddad[3], and Loïc Hélouët[1]

[1] Inria Rennes, France
[2] IIT Bombay, India
[3] LSV, ENS Cachan & CNRS & INRIA, France

Abstract. This paper addresses a control problem for probabilistic models in the setting of Markov decision processes (MDP). We are interested in the *steady-state control problem* which asks, given an ergodic MDP $\mathcal{M}$ and a distribution δ_{goal}, whether there exists a (history-dependent randomized) policy π ensuring that the steady-state distribution of $\mathcal{M}$ under π is exactly δ_{goal}. We first show that stationary randomized policies suffice to achieve a given steady-state distribution. Then we infer that the steady-state control problem is decidable for MDP, and can be represented as a linear program which is solvable in PTIME. This decidability result extends to labeled MDP (LMDP) where the objective is a steady-state distribution on labels carried by the states, and we provide a PSPACE algorithm. We also show that a related *steady-state language inclusion problem* is decidable in EXPTIME for LMDP. Finally, we prove that if we consider MDP under partial observation (POMDP), the steady-state control problem becomes undecidable.

1 Introduction

Probabilistic systems are frequently modeled as Markov chains, which are composed of a set of states and a probabilistic transition relation specifying the probability of moving from one state to another. When the system interacts with the environment, as is very often the case in real-life applications, in addition to the probabilistic moves, non-deterministic choices are possible. Such choices are captured by Markov Decision Processes (MDP), which extend Markov chains with non-determinism. Finally, in several applications, the system is not fully observable, and the information about the state of a system at a given instant is not precisely known. The presence of such uncertainty in observation can be captured by Partially Observable Markov Decision Processes (POMDP).

In all these settings, given a probabilistic system one is often interested in knowing whether, in the long run, it satisfies some property. For instance, one may want to make sure that the system does not, on an average, spend too much time in a faulty state. In the presence of non-deterministic choices (as in an MDP) or partial observation (as in a POMDP), a crucial question is whether we can always "control" these choices so that a long run property can be achieved.

In this paper, we are interested in control problems for Markov decision processes (MDP) and partially observable Markov decision processes (POMDP) with respect to long-run objectives. Given a Markov chain, it is well known [5,7] that one can compute

K. Joshi et al. (Eds.): QEST 2013, LNCS 8054, pp. 290–304, 2013.
© Springer-Verlag Berlin Heidelberg 2013

its set of steady-state distributions, depending on the initial distribution. In an open setting, *i.e.*, when considering MDP, computing steady-state distributions becomes more challenging. Controlling an MDP amounts to defining a policy, that is, a function that associates, with every history of the system, a distribution on non-deterministic choices.

We tackle the *steady-state control problem*: given an MDP with a fixed initial distribution, and a goal distribution over its state space, does there exist a policy realizing the goal distribution as its steady-state distribution? (1) We prove decidability of the steady-state control problem for the class of so-called ergodic MDP, and provide a PTIME algorithm using linear programming techniques. (2) We next lift the problem to the setting of LMDP, where we add labels to states and check if a goal distribution over these labels can be reached by the system under some policy. For LMDP we show decidability of the steady-state control problem and provide a PSPACE algorithm. (3) Finally, for POMDP, we establish that the steady-state control problem becomes undecidable.

We also consider the *steady-state language inclusion problem* for LMDP. Namely, given two LMDP the question is whether any steady-state distribution over labels realizable in one process can be realized in the other. Building on our techniques for the steady-state control problem, we show that the language inclusion problem for LMDP is decidable in EXPTIME.

As already mentionned, steady-state control can be useful to achieve a given error rate, and in general to enforce quantitative fairness in a system. Steady-state language inclusion is a way to guarantee that a refinement of a system does not affect its long term behaviors. The problem of controlling a system such that it reaches a steady-state has been vastly studied in control theory for continuous models, *e.g.* governed by differential equations and where reachability should occur in finite time. There is a large body of work which addresses control problems for Markov decision processes. However, the control objectives are usually defined in terms of an optimization of a cost function (see *e.g.* [8, 10]). On the contrary, in this work, the control objective is to achieve a given steady-state distribution. In a recent line of work [3, 6], the authors consider transient properties of MDP viewed as transformers of probability distributions. Compared to that setting, we are interested rather in long run properties. Finally, in [4], the authors consider the problem of language equivalence for labeled Markov chains (LMC) and LMDP. For LMC, this problem consists of checking if two given LMC have the same probability distribution on finite executions (over the set of labels) and is shown to be decidable in PTIME. The equivalence problem for LMDP is left open. As we are only interested in long run behaviors, we tackle a steady-state variant of this problem.

The paper is organized as follows. Section 2 introduces notations and definitions. Section 3 formalizes and studies the steady-state control problem: MDP are considered in Subsection 3.1; Subsection 3.2 extends the decidability results to LMDP and also deals with the steady-state language inclusion problem; and Subsection 3.3 establishes that partial observation entails undecidability of the steady-state control problem. We conclude with future directions in Section 4.

2 Preliminaries

In what follows, we introduce notations for matrices and vectors, assuming the matrix/vector size is understood from the context. We denote the identity matrix by $\mathbf{Id}$, the (row) vector with all entries equal to 1 by $\mathbb{1}$ and the (row) vector with only 0's by $\mathbf{0}$. The transpose of a matrix $\mathbf{M}$ (possibly a vector) is written $\mathbf{M}^t$. Given a square matrix $\mathbf{M}$, $\det(\mathbf{M})$ is its determinant.

2.1 Markov Chains

We recall some definitions and results about Markov chains. Given a countable set T, we let $\mathsf{Dist}(T)$ denote the set of distributions over T, that is, the set of functions $\delta : T \rightarrow [0, 1]$ such that $\sum_{t \in T} \delta(t) = 1$.

Definition 1. *A discrete time Markov chain (DTMC) is a tuple $\mathcal{A} = (S, \Delta, s_0)$ where:*

- *S is the finite or countable set of states.*
- *$\Delta : S \rightarrow \mathsf{Dist}(S)$ is the transition function describing the distribution over states reached in one step from a state.*
- *$s_0 \in \mathsf{Dist}(S)$ is the initial distribution.*

As usual the transition matrix $\mathbf{P}$ of the Markov chain $\mathcal{A}$ is the $|S| \times |S|$ row-stochastic matrix defined by $\mathbf{P}[s, s'] \stackrel{\text{def}}{=} \Delta(s)(s')$, i.e., the $(s, s')^{th}$ entry of the matrix $\mathbf{P}$ gives the value defined by Δ of the probability to reach s' from s in one step. When the DTMC $\mathcal{A}$ is finite, one defines an directed graph $G_\mathcal{A}$ whose vertices are states of $\mathcal{A}$ and such that there is an arc from s to s' if $\mathbf{P}[s, s'] > 0$. $\mathcal{A}$ is said to be *recurrent* if $G_\mathcal{A}$ is strongly connected. The periodicity of a graph p is the greatest integer such that there exists a partition of $S = \biguplus_{i=0}^{p-1} S_i$ such that for all $s \in S_i$ and $s' \in S$, there is an arc from s to s' only if $s' \in S_{(i+1 \mod p)}$. When the periodicity of $G_\mathcal{A}$ is 1, $\mathcal{A}$ is said to be *aperiodic*. Finally $\mathcal{A}$ is said to be *ergodic* if it is recurrent and aperiodic.

Now, consider the sequence of distributions $s_0, s_1, \ldots$ such that $s_i = s_0 \cdot \mathbf{P}^i$. This sequence does not necessarily converge (if the Markov chain is periodic)[1]. We write $\mathsf{sd}(\mathcal{A})$ when the limit exists and call it the *steady-state distribution* of $\mathcal{A}$. In case of an ergodic DTMC $\mathcal{A}$, (1) $\mathsf{sd}(\mathcal{A})$ exists, (2) it does not depend on s_0 and, (3) it is the unique distribution s which fulfills $s \cdot \mathbf{P} = s$. When $\mathcal{A}$ is only recurrent, there is still a single distribution, called the *invariant distribution*, that fulfills this equation, and it coincides with the Cesàro limit. However it is a steady-state distribution only for a subset of initial distributions.

Labeled Markov Chains. Let $L = \{l_1, l_2, \ldots\}$ be a finite set of labels. A *labeled Markov chain* is a tuple $(\mathcal{A}, \ell)$ where $\mathcal{A} = (S, \Delta, s_0)$ is a Markov chain and $\ell : S \rightarrow L$ is a function assigning a label to each state. Given $(\mathcal{A}, \ell)$ a labeled Markov chain, the *labeled steady-state distribution*, denoted by $\mathsf{lsd}(\mathcal{A}, \ell)$ or simply $\mathsf{lsd}(\mathcal{A})$ when ℓ is clear

---

[1] But it always admits a *Cesàro-limit*: the sequence $c_n = \frac{1}{n}(s_0 + \cdots + s_{n-1})$ converges (see e.g. [8, p.590]).

from the context, is defined when $\mathsf{sd}(\mathcal{A})$ exists and is its projection onto the labels in L, via ℓ. More formally, for every $l \in L$,

$$\mathsf{lsd}(\mathcal{A})(l) = \sum_{s \in S \mid \ell(s)=l} \mathsf{sd}(\mathcal{A})(s)$$

2.2 Markov Decision Processes

Definition 2. *A* Markov decision process *(MDP)* $\mathcal{M} = (S, \{A_s\}_{s \in S}, p, s_0)$ *is defined by:*

- *S, the finite set of states;*
- *For every state s, A_s, the finite set of actions enabled in s.*
- *$p : \{(s, a) \mid s \in S, a \in A_s\} \to \mathsf{Dist}(S)$ is the transition function. The conditional probability transition $p(s'|s, a)$ denotes the probability to go from s to s' if a is selected.*
- *$s_0 \in \mathsf{Dist}(S)$ is the initial distribution.*

To define the semantics of an MDP $\mathcal{M}$, we first define the notion of *history*: a possible finite or infinite execution of the MDP.

Definition 3. *Given an MDP $\mathcal{M}$, a* history *is a finite or infinite sequence alternating states and actions $\sigma = (s_0, a_0, \ldots, s_i, a_i, \ldots)$. The number of actions of σ is denoted $\lg(\sigma)$, and if σ is finite, we write $\mathsf{last}(\sigma)$ for this last state. One requires that for all $0 \le i < \lg(\sigma)$, $p(s_{i+1}|s_i, a_i) > 0$.*

Compared to Markov chains, MDP contain non-deterministic choices. From a state s, when an action $a \in A_s$ is chosen, the probability to reach state s' is $p(s'|s, a)$. In order to obtain a stochastic process, we need to fix the non-deterministic features of the MDP. This is done via (1) decision rules that select at some time instant the next action depending on the history of the execution, and (2) policies which specify which decision rules should be used at any time instant. Different classes of decision rules and policies are defined depending on two criteria: (1) the information used in the history and (2) the way the selection is performed (deterministically or randomly).

Definition 4. *Given an MDP $\mathcal{M}$ and $t \in \mathbb{N}$, a* decision rule *d_t associates with every history σ of length $t = \lg(\sigma) < \infty$ ending at a state s_t, a distribution $d_t(\sigma)$ over A_{s_t}.*

- *The set of all decision rules (also called* history-dependent randomized *decision rules) at time t is denoted D_t^{HR}.*
- *The subset of* history-dependent deterministic *decision rules at time t, denoted D_t^{HD}, consists of associating a single action (instead of a distribution) with each history σ of length $t < \infty$ ending at a state s_t. Thus, in this case $d_t(\sigma) \subset A_{s_t}$.*
- *The subset of* Markovian randomized *decision rules at time t, denoted D_t^{MR} only depends on the final state of the history. So one denotes $d_t(s)$ the distribution that depends on s.*
- *The subset of* Markovian deterministic *decision rules at time t, D_t^{MD} only depends on the final state of the history and selects a single action. So one denotes $d_t(s)$ this action belonging to A_s.*

When the time t is clear from context, we will omit the subscript and just write D^{HR}, D^{HD}, D^{MD} and D^{MR}.

Definition 5. *Given an* MDP $\mathcal{M}$, *a policy (also called a strategy)* π *is a finite or infinite sequence of decision rules* $\pi = (d_0, \ldots, d_t, \ldots)$ *such that* d_t *is a decision rule at time* t, *for every* $t \in \mathbb{N}$.

The set of policies such that for all t, $d_t \in D_t^K$ *is denoted* Π^K *for each* $K \in \{HR, HD, MR, MD\}$.

When decisions d_t *are Markovian and all equal to some rule* d, π *is said* stationary *and denoted* d^∞. *The set of stationary randomized (resp. deterministic) policies is denoted* Π^{SR} *(resp.* Π^{SD} *)*.

A Markovian policy only depends on the current state and the current time while a stationary policy only depends on the current state. Now, once a policy π is chosen, for each n, we can compute the probability distribution over the histories of length n of the MDP. That is, under the policy $\pi = d_0, d_1, \ldots d_n, \ldots$ and with initial distribution s_0, then, for any $n \in \mathbb{N}$, the probability of the history $\sigma_n = s_0 a_0 \ldots s_{n-1} a_{n-1} s_n$, is defined inductively by:

$$p^\pi(\sigma_n) = d_n(\sigma_{n-1})(a_{n-1}) \cdot p(s_n | s_{n-1}, a_{n-1}) \cdot p^\pi(\sigma_{n-1}) ,$$

and $p^\pi(\sigma_0) = s_0(s_0)$. Then, by summing over all histories of length n ending in the same state s, we obtain the probability of reaching state s after n steps. Formally, letting X_n denote the random variable corresponding to the state at time n, we have:

$$\mathbb{P}^\pi(X_n = s) = \sum_{\sigma | \lg(\sigma) = n \wedge \mathrm{last}(\sigma) = s} p^\pi(\sigma)$$

Observe that once a policy π is chosen, an MDP $\mathcal{M}$ can be seen as a discrete-time Markov chain (DTMC), written $\mathcal{M}_\pi$, whose states are histories. The Markov chain $\mathcal{M}_\pi$ has infinitely many states in general. When a stationary policy d^∞ is chosen, one can forget the history of states except for the last one, and thus consider the states of the DTMC $\mathcal{M}_\pi$ to be those of the MDP $\mathcal{M}$ and the transition matrix $\mathbf{P}_d$ is defined by:

$$\mathbf{P}_d[s, s'] \stackrel{\mathrm{def}}{=} \sum_{a \in A_s} d(s)(a) p(s' | s, a).$$

Thus, in this case the probability of being in state s at time n is just given by $\mathbb{P}(X_n = s) = (s_0 \cdot \mathbf{P}_d^n)(s)$.

Recurrence and Ergodicity. A Markov decision process $\mathcal{M}$ is called *recurrent* (resp. *ergodic*) if for every $\pi \in \Pi^{SD}$, $\mathcal{M}_\pi$ is recurrent (resp. ergodic). Recurrence and ergodicity of an MDP can be effectively checked, as the set of graphs $\{G_{\mathcal{M}_\pi} \mid \pi \in \Pi^{SD}\}$ is finite. Observe that when $\mathcal{M}$ is called recurrent (resp. ergodic) then for every $\pi \in \Pi^{SR}$, $\mathcal{M}_\pi$ is recurrent (resp. ergodic).

Steady-State Distributions. We fix a policy π of an MDP $\mathcal{M}$. Then, for any $n \in \mathbb{N}$, we define the distribution reached by $\mathcal{M}_\pi$ at the n-th stage, i.e., for any state $s \in S$ as: $\delta_n^\pi(s) = \mathbb{P}^\pi(X_n = s)$. Now when it exists, the steady-state distribution $\mathsf{sd}(\mathcal{M}_\pi)$ of the MDP $\mathcal{M}$ under policy π is defined as: $\mathsf{sd}(\mathcal{M}_\pi)(s) = \lim_{n \to \infty} \delta_n^\pi(s)$. Observe that when $\mathcal{M}$ is ergodic, for every decision rule d, $\mathcal{M}_{d^\infty}$ is ergodic and so $\mathsf{sd}(\mathcal{M}_{d^\infty})$ is defined.

Now, as we did for Markov chains, given a set of labels L, a labeled MDP, is a tuple $(\mathcal{M}, \ell)$ where $\mathcal{M}$ is an MDP and $\ell : S \to L$ is a labeling function. Then, for $\mathcal{M}$ an MDP, ℓ a labeling function, and π a strategy, we define $\mathsf{lsd}(\mathcal{M}_\pi, \ell)$ or simply $\mathsf{lsd}(\mathcal{M}_\pi)$ for the projection of $\mathsf{sd}(\mathcal{M}_\pi)$ (when it exists) onto the labels in L via ℓ.

3 The Steady-State Control Problem

3.1 Markov Decision Processes

Given a Markov decision process, the steady-state control problem asks whether one can come up with a policy to realize a given steady-state distribution. In this paper, we only consider ergodic MDP. Formally,

---

Steady-state control problem for MDP

Input: An ergodic MDP $\mathcal{M} = (S, \{A_s\}_{s \in S}, p, s_0)$, and a distribution $\delta_{\mathsf{goal}} \in \mathsf{Dist}(S)$.

Question: Does there exist a policy $\pi \in \Pi^{HR}$ s.t. $\mathsf{sd}(\mathcal{M}_\pi)$ exists and is equal to δ_{goal}?

The main contribution of this paper is to prove that, the above decision problem is decidable and belongs to PTIME for ergodic MDP. Furthermore it is effective: if the answer is positive, one can compute a witness policy. To establish this result we show that if there exists a witness policy, then there is a simple one, namely a stationary randomized policy $\pi \in \Pi^{SR}$. We then solve this simpler question by reformulating it as an equivalent linear programming problem, of size polynomial in the original MDP. More formally,

Theorem 1. *Let $\mathcal{M}$ be an ergodic* MDP. *Assume there exists $\pi \in \Pi^{HR}$ such that $\lim_{n \to \infty} \delta_n^\pi = \delta_{\mathsf{goal}}$. Then there exists $d^\infty \in \Pi^{SR}$ such that $\lim_{n \to \infty} \delta_n^{d^\infty} = \delta_{\mathsf{goal}}$.*

The following folk theorem states that Markovian policies (that is, policies based only on the history length and the current state) are as powerful as general history-dependent policies to achieve marginal distributions for $\{(X_n, Y_n)\}_{n \in \mathbb{N}}$ where $\{Y_n\}_{n \in \mathbb{N}}$ to denote the family of random variables corresponding to the chosen actions at time n. Observe that this is no more the case when considering joint distributions.

Theorem 2 ([8], Thm. 5.5.1). *Let $\pi \in \Pi^{HR}$ be a policy of an* MDP $\mathcal{M}$. *Then there exists a policy $\pi' \in \Pi^{MR}$ such that for all $n \in \mathbb{N}$, $s \in S$ and $a \in A_s$:*

$$\mathbb{P}^{\pi'}(X_n = s, Y_n = a) = \mathbb{P}^\pi(X_n = s, Y_n = a)$$

Hence for an history-dependent randomized policy, there exists a Markovian randomized one with the same transient distributions and so with the same steady-state transient distribution if the former exists. It thus suffices to prove Theorem 1 assuming that $\pi \in \Pi^{MR}$. To this aim, we establish several intermediate results.

Let $d \in D^{MR}$ be a Markovian randomized decision rule. d can be expressed as a convex combination of the finitely many Markovian deterministic decision rules: $d = \sum_{e \in D^{MD}} \lambda_e e$. We say that a sequence $d_n \in D^{MR}$ admits a limit d, denoted $d_n \to_{n \to \infty} d$, if, writing $d_n = \sum_{e \in D^{MD}} \lambda_{e,n} e$ and $d = \sum_{e \in D^{MD}} \lambda_e e$, then for all $e \in D^{MD}$, $\lim_{n \to \infty} \lambda_{e,n} = \lambda_e$.

Lemma 1. *Let $\mathcal{M}$ be an ergodic MDP and $(d_n)_{n \in \mathbb{N}} \in \Pi^{MR}$. If the sequence d_n has a limit d, then $\lim_{n \to \infty} \mathsf{sd}(\mathcal{M}_{d_n^\infty})$ exists and is equal to $\mathsf{sd}(\mathcal{M}_{d^\infty})$.*

In words, Lemma 1 states that the steady-state distribution under the limit policy d coincides with the limit of the steady-state distributions under the d_n's. The steady-state distribution operator is thus continuous over Markovian randomized decision rules.

Proof (of Lemma 1). Consider the following equation system with parameters $\{\lambda_e\}_{e \in D^{MD}}$, and a vector of variables X, obtained from

$$X \cdot (\mathbf{Id} - \sum_{e \in D^{MD}} \lambda_e \mathbf{P}_e) = 0$$

by removing one equation (any of them), and then adding $X \cdot \mathbb{1}^t = 1$. This system can be rewritten in the form $X \cdot \mathbf{M} = \mathbf{b}$. Using standard results of linear algebra, the determinant $\det(\mathbf{M})$ is a rational fraction in the λ_e's. Moreover due to ergodicity of $\mathcal{M}$, $\mathbf{M}$ is invertible for any tuple $(\lambda_e)_{e \in D^{MD}}$ with $\sum_{e \in D^{MD}} \lambda_e = 1$. Thus the denominator of this fraction does not cancel for such values. As a result, the function $f :$ $(\lambda_e)_{e \in D^{MD}} \mapsto \mathsf{sd}(\mathcal{M}_{(\sum \lambda_e e)^\infty})$, which is a vector of rational functions, is continuous which concludes the proof. $\square$

Note that Lemma 1 does not hold if we relax the assumption that $\mathcal{M}$ is ergodic. Indeed, consider an MDP with two states s_0, s_1 and two actions a, b, where action a loops with probability 1 on the current state, whereas action b moves from both states to state q_1, with probability 1. According to the terminology in [8, p. 348] this example models a multichain, not weakly communicating MDP. We assume the initial distribution to be the Dirac function in q_0. On this example, only the decision in state q_0 is relevant, since q_1 is a sink state. For every $n \in \mathbb{N}$, let $d_n \in D^{MR}$ be the Markovian randomized decision rule defined by $d_n(a) = 1 - \frac{1}{n+1}$ and $d_n(b) = \frac{1}{n+1}$. On one hand, the steady-state distribution in $\mathcal{M}$ under the stationary randomized policy d_n^∞ is $\mathsf{sd}(\mathcal{M}, d_n^\infty) = (0, 1)$. On the other hand, the sequence $(d_n)_{n \in \mathbb{N}}$ of decision rules admits a limit: $\lim_{n \to \infty} d_n = d$ with $d(a) = 1$ and $d(b) = 0$, and $\mathsf{sd}(\mathcal{M}, d^\infty) = (1, 0)$.

For the next lemma, we introduce further notations. For d a decision rule, we define its *greatest acceptable radius*, denoted r_d, as

$$r_d = \max\{r \in \mathbb{R} \mid \forall v \in \mathbb{R}^{|S|}, \ ||v - \mathsf{sd}(\mathcal{M}_{d^\infty})|| = r \implies \forall s \in S, v(s) \geq 0\},$$

where $||\cdot||$ is the Euclidean norm. Intuitively, r_d is the greatest radius of a neighborhood around $\mathcal{M}_{d^\infty}$ such that no element inside it has negative coordinates.

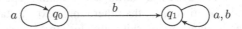

Clearly enough, for a fixed decision rule $d \in D^{MR}$, $r_d > 0$. Indeed, since $\mathcal{M}$ is ergodic, $\mathcal{M}$ equipped with the stationary policy d^∞ is a Markov chain consisting of a single recurrent class; hence, every state has a positive probability in the steady-state distribution $\mathsf{sd}(\mathcal{M}_{d^\infty})$. We also define the following set of distributions, that are r-away from a distribution w:

$$\mathcal{N}_{=r}(w) = \{v \mid v \in \mathsf{Dist}(S) \text{ and } ||v - w|| = r\}.$$

Lemma 2. *Let $\mathcal{M}$ be an ergodic* MDP. *Define*

$$\alpha \stackrel{def}{=} \inf_{d \in D^{MR}} \inf_{v \in \mathcal{N}_{=r_d}(\mathsf{sd}(\mathcal{M}_{d^\infty}))} \frac{||v \cdot (\mathbf{Id} - \mathbf{P}_d)||}{r_d}$$

Then, $\alpha > 0$.

Proof (of Lemma 2). First observe that for fixed $d \in D^{MR}$ and $v \in \mathcal{N}_{=r_d}(\mathsf{sd}(\mathcal{M}_{d^\infty}))$, $\frac{||v \cdot (\mathbf{Id} - \mathbf{P}_d)||}{r_d} > 0$. Indeed, if $v \in \mathsf{Dist}(S)$ and $||v - \mathsf{sd}(\mathcal{M}_{d^\infty})|| = r_d > 0$, then v is not the steady-state distribution under d^∞, so that $v \neq v \cdot \mathbf{P}_d$.

Towards a contradiction, let us assume that the infimum is 0:

$$\inf_{d \in D^{MR}} \inf_{v \in \mathcal{N}_{=r_d}(\mathsf{sd}(\mathcal{M}_{d^\infty}))} \frac{||v \cdot (\mathbf{Id} - \mathbf{P}_d)||}{r_d} = 0.$$

In this case, there exists a sequence of decisions $(d_n)_{n \in \mathbb{N}} \in D^{MR}$ and a sequence of distributions $(v_n)_{n \in \mathbb{N}}$, such that for each $n \in \mathbb{N}$, $v_n \in \mathcal{N}_{=r_{d_n}}(\mathsf{sd}(\mathcal{M}_{d_n^\infty}))$ and $\lim_{n \to \infty} \frac{||v_n \cdot (\mathbf{Id} - \mathbf{P}_{d_n})||}{r_{d_n}} = 0$. From these sequences (d_n) and (v_n), we can extract subsequences, for simplicity still indexed by $n \in \mathbb{N}$ such that:

(i) (d_n) converges, and we write d for its limit, and
(ii) (v_n) converges, and we write v for its limit.

Thanks to Lemma 1, $\lim_{n \to \infty} \mathsf{sd}(\mathcal{M}_{d_n^\infty}) = \mathsf{sd}(\mathcal{M}_{d^\infty})$. Moreover, using the continuity of the norm function $|| \cdot ||$, $\lim_{n \to \infty} r_{d_n} = r_d$, and $v \in \mathcal{N}_{=r_d}(\mathsf{sd}(\mathcal{M}_{d^\infty}))$. Still by continuity, we derive $\frac{||v \cdot (\mathbf{Id} - \mathbf{P}_d)||}{r_d} = 0$, a contradiction. □

$||v \cdot (\mathbf{Id} - \mathbf{P}_d)||$ is the distance between a distribution v and the resulting distribution after applying the decision rule d from v. Since we divide it by r_d, α roughly represents the minimum deviation "rate" (w.r.t. all d's) between v and its image when going away from $\mathsf{sd}(\mathcal{M}_{d^\infty})$.

Lemma 3. *Let $\mathcal{M}$ be an ergodic* MDP. *Assume there exists a policy $\pi \in \Pi^{MR}$ such that $\mathsf{sd}(\mathcal{M}_\pi)$ exists and is equal to δ_{goal}. Then for every $\varepsilon > 0$, there exists $d \in D^{MR}$ such that $||\mathsf{sd}(\mathcal{M}_{d^\infty}) - \delta_{\mathsf{goal}}|| < \varepsilon$.*

The above lemma states that if there is a Markovian randomized policy which at steady-state reaches the goal distribution δ_{goal}, then there must exist a stationary randomized policy d^∞ which at steady-state comes arbitrarily close to δ_{goal}.

Proof. Let us fix some arbitrary $\varepsilon > 0$. Since we assume that $\mathsf{sd}(\mathcal{M}_\pi) = \delta_{goal}$, for all $\gamma > 0$, there exists $n_0 \in \mathbb{N}$ such that for every $n \geq n_0$, $||\delta_n^\pi - \delta_{goal}|| < \gamma$. Let us choose $\gamma = \min\{\frac{\alpha\varepsilon}{4}, \frac{\varepsilon}{2}\}$.

Define $d \in D^{MR}$ as the decision made by π at the n_0-th step. That is, if $\pi = ((d_i)_{i \in \mathbb{N}})$, then $d_{n_0} = d$. Now, if $\delta_{n_0}^\pi = \mathsf{sd}(\mathcal{M}_{d^\infty})$ we are done since we will have $||\mathsf{sd}(\mathcal{M}_{d^\infty}) - \delta_{goal}|| < \varepsilon/2 < \varepsilon$. Otherwise, we let $\Theta = \frac{r_d}{||\delta_{n_0}^\pi - \mathsf{sd}(\mathcal{M}_{d^\infty})||}$ and $v = \Theta\delta_{n_0}^\pi + (1 - \Theta)\mathsf{sd}(\mathcal{M}_{d^\infty})$. Note that, under these definitions, $v \in \mathcal{N}_{=r_d}(\mathsf{sd}(\mathcal{M}_{d^\infty}))$. Observe also that $v \cdot (\mathbf{Id} - \mathbf{P}_d) = \Theta\delta_{n_0}^\pi \cdot (\mathbf{Id} - \mathbf{P}_d)$, by definition of v and since $\mathsf{sd}(\mathcal{M}_{d^\infty}) \cdot \mathbf{P}_d = \mathsf{sd}(\mathcal{M}_{d^\infty})$.

Thus we have

$$||v \cdot (\mathbf{Id} - \mathbf{P}_d)|| = ||\Theta\delta_{n_0}^\pi \cdot (\mathbf{Id} - \mathbf{P}_d)||$$
$$= \Theta||\delta_{n_0}^\pi - \delta_{n_0}^\pi \cdot \mathbf{P}_d|| = \Theta||\delta_{n_0+1}^\pi - \delta_{n_0}^\pi||$$
$$\leq \Theta(||\delta_{n_0+1}^\pi - \delta_{goal}|| + ||\delta_{n_0}^\pi - \delta_{goal}||) < \frac{\Theta\alpha\varepsilon}{2}.$$

By definition of α, we have $\alpha \leq \frac{||v \cdot (\mathbf{Id} - \mathbf{P}_d)||}{r_d}$. Then, combining this with the above equation and using the fact that $r_d > 0$, we obtain:

$$r_d \cdot \alpha \leq ||v \cdot (\mathbf{Id} - \mathbf{P}_d)|| < \frac{\Theta\alpha\varepsilon}{2}$$

By Lemma 2 we have $\alpha > 0$ which implies that $r_d < \frac{\Theta\varepsilon}{2}$. Substituting the definition of Θ we get after simplification:

$$||\delta_{n_0}^\pi - \mathsf{sd}(\mathcal{M}_{d^\infty})|| < \frac{\varepsilon}{2}$$

Thus, finally:

$$||\mathsf{sd}(\mathcal{M}_{d^\infty}) - \delta_{goal}|| \leq ||\mathsf{sd}(\mathcal{M}_{d^\infty}) - \delta_{n_0}^\pi|| + ||\delta_{n_0}^\pi - \delta_{goal}||$$
$$< \frac{\varepsilon}{2} + \frac{\varepsilon}{2} = \varepsilon$$

which proves the lemma. $\square$

Theorem 1 is a consequence of Lemma 3 because stationary randomized policies form a closed set due to Lemma 1. Thanks to Theorems 2 and 1 a naive algorithm to decide the steady-state control problem for MDP is the following: build a linear program whose non negative variables $\{\lambda_e\}$ are indexed by each $e \in D^{SD}$ and check whether $\delta_{goal} \cdot (\sum_e \lambda_e \mathbf{P}_e) = \delta_{goal}$ admits a solution with $\sum_e \lambda_e = 1$. This algorithm runs in exponential time w.r.t. the size of $\mathcal{M}$ since there are exponentially many stationary deterministic policies. Yet, a better complexity can be obtained as stated in the following theorem.

Theorem 3. *The steady-state control problem for ergodic* MDP *is effectively decidable in PTIME.*

Proof. According to Theorem 1, finding a policy with steady-state distribution δ_{goal} can be brought back to finding such a policy in Π^{SR}. Now, on the one hand, defining a randomized stationary policy for an MDP $\mathcal{M}$ consists in choosing decision rules $d_s \in D^{SR}$, or equivalently real numbers $\lambda_{s,a} \in [0, 1]$ for each pair (s, a) of states and action, and such that for every s, $\sum_{a \in A_s} \lambda_{s,a} = 1$. Intuitively, $\lambda_{s,a}$ represent the probability to choose action a when in state s. Note that the set $\Lambda = \{\lambda_{s,a} \mid s \in S, a \in A_s\}$ is of polynomial size (in the size of $\mathcal{M}$). Note also that once we have defined Λ, we have defined at the same time a policy $\pi_\Lambda \in \Pi^{SR}$, and that $\mathcal{M}_{\pi_\Lambda}$ is a Markov chain with state space S. The transition matrix $\mathbf{P}_\Lambda$ of $\mathcal{M}_{\pi_\Lambda}$ is such that $\mathbf{P}_\Lambda[s, s'] = \sum_{a \in A_s} \lambda_{s,a} \cdot p(s|s', a)$.

Due to ergodicity of $\mathcal{M}$, one only has to check whether $\delta_{goal} \cdot \mathbf{P}_\Lambda = \delta_{goal}$. Putting this altogether, we can derive a polynomial size linear programming specification of our problem: there exists a stationary randomized policy to achieve δ_{goal} if and only if we can find a set of non negative reals $\Lambda = \{\lambda_{s,a} \mid s \in S, a \in A_s\}$ such that

$$\forall s \in S, \sum_{s' \in S, a \in A_{s'}} \delta_{goal}(s').p(s|s', a).\lambda_{s',a} = \delta_{goal}(s) \text{ and } \sum_{a \in A_s} \lambda_{s,a} = 1$$

Solving this linear program can be done in polynomial time, using techniques such as the interior point methods (see for instance [9] for details). This proves the overall PTIME complexity. □

Discussion. Observe that Lemmas 1, 2 and 3 hold when $\mathcal{M}$ is only recurrent substituting the steady-state distribution of $M_{d\infty}$ by the (single) invariant distribution of this DTMC. Unfortunately, combining these lemmas in the recurrent case only provides a *necessary* condition namely: "If δ_{goal} is the steady-state distribution of some policy then it is the invariant distribution of some of $M_{d\infty}$".

3.2 Labeled Markov Decision Processes

The steady-state control problem for MDP describes random processes in which interactions with users or with the environment drive the system towards a desired distribution. However, the goal distribution is a very accurate description of the desired objective and, in particular, this assumes that the controller knows each state of the system. Labeling an MDP is a way to define higher-level objectives: labels can be seen as properties of states, and the goal distribution as a distribution over these properties. For instance, when resources are shared, a set of labels $L = \{l_1, \ldots, l_k\}$ may indicate which user (numbered from 1 to k) owns a particular resource. In such an example, taking δ_{goal} as the discrete uniform distribution over L encodes a guarantee of fairness.

In this section, we consider Markov decision processes in which states are labeled. Formally, let $\mathcal{M}$ be a Markov decision process, L be a finite set of labels and $\ell : S \to L$ a labeling function. We consider the following decision problem. Given $(\mathcal{M}, \ell)$ a labeled Markov decision process and $\delta_{goal} \in Dist(L)$ a distribution over labels, does there exist a policy π such that $lsd(\mathcal{M}_\pi) = \delta_{goal}$.

The steady-state control problem for LMDP is a generalization of the same problem for MDP: any goal distribution $\delta_{goal} \in Dist(L)$ represents the (possibly infinite) set of distributions in $Dist(S)$ that agree with δ_{goal} when projected onto the labels in L.

Theorem 4. *The steady-state control problem for ergodic* LMDP *is decidable in* PSPACE.

Proof. Given an ergodic MDP $\mathcal{M}$ labeled with a labeling function $\ell : S \to L$ and a distribution $\delta_{goal} \in Dist(L)$, the question is whether there exists a policy π for $\mathcal{M}$ such that the steady-state distribution $sd(\mathcal{M}_\pi)$ of $\mathcal{M}$ under π projected on labels equals δ_{goal}.

First, let us denote by $\Delta_{goal} = \ell^{-1}(\delta_{goal})$ the set of distributions in $Dist(S)$ that agree with δ_{goal}. If $\mathbf{x} = (x_s)_{s \in S} \in Dist(S)$ is a distribution, then $\mathbf{x} \in \Delta_{goal}$ can be characterized by the constraints:

$$\forall l \in L, \quad \sum_{s \in S | \ell(s) = l} x_s = \delta_{goal}(l) \ .$$

We rely on the proof idea from the unlabeled case: If there is a policy $\pi \in \Pi^{HR}$ with steady-state distribution in $Dist(S)$, then, there is a policy $\pi' \in \Pi^{SR}$ with the same steady-state distribution (thanks to Lemmas and Theorems from Subsection 3.1). This policy π' hence consists in repeatedly applying the same decision rule $d \in D^{SR}$: $\pi' = d^\infty$. As the MDP $\mathcal{M}$ is assumed to be ergodic, there is exactly one distribution $\mathbf{x} \in Dist(S)$ that is invariant under $\mathbf{P}_d$. The question then is: Is there a policy $d \in \Pi^{SR}$, and a distribution $\mathbf{x} = (x_s)_{s \in S} \in \Delta_{goal}$ such that $\mathbf{x} \cdot \mathbf{P}_d = \mathbf{x}$?

We thus derive the following system of equations, over non negative variables $\{x_s \mid s \in S\} \cup \{\lambda_{s,a} \mid s \in S, a \in A_s\}$:

$$\forall l \in L, \quad \sum_{s \in S | \ell(s) = l} x_s = \delta_{goal}(l)$$

$$\forall s \in S, \quad \sum_{s' \in S, a \in A_{s'}} x_{s'} . p(s \mid s', a) . \lambda_{s',a} = x_s \text{ and } \sum_{a \in A_s} \lambda_{s,a} = 1$$

The size of this system is still polynomial in the size of the LMDP. Note that the distribution $\mathbf{x}$ and the weights $\lambda_{s,a}$ in the decision rule d are variables. Therefore, contrary to the unlabeled case, the obtained system is composed of quadratic equations (*i.e.* equations containing products of two variables). We conclude by observing that quadratic equation systems as particular case of polynomial equation systems can be solved in PSPACE [2]. $\square$

Note that the above technique can be used to find policies enforcing *a set of distributions* Δ_{goal}. Indeed, if expected goals are defined through a set of constraints of the form $\delta_{goal}(s) \in [a, b]$ for every $s \in S$, then stationnary randomized policies achieving a steady-state distribution in Δ_{goal} are again the solutions for a polynomial equation system. Another natural steady-state control problem can be considered for LMDP. Here, the convergence to a steady-state distribution is assumed, and belongs to $\Delta_{\text{goal}} = \ell^{-1}(\delta_{\text{goal}})$. Alternatively, we could consider whether a goal distribution on labels can be realized by some policy π even when the convergence of the sequence $(\delta_n^\pi)_{n \in \mathbb{N}}$ is not guaranteed. This problem is more complex than the one we consider, and is left open.

Finally we study a "language inclusion problem". As mentioned in the introduction, one can define the *steady-state language inclusion problem* similar to the language equivalence problem for LMDP defined in [4]. Formally the steady-state language inclusion problem for LMDP asks whether given two LMDP $\mathcal{M}, \mathcal{M}'$, for every policy π of $\mathcal{M}$ such that $\text{lsd}(\mathcal{M}_\pi)$ is defined there exists a policy π' of $\mathcal{M}'$ such that $\text{lsd}(\mathcal{M}_\pi) = \text{lsd}(\mathcal{M}'_{\pi'})$. The following theorem establishes its decidability for ergodic LMDP.

Theorem 5. *The steady-state language inclusion problem for ergodic LMDP is decidable in EXPTIME.*

3.3 Partially Observable Markov Decision Processes

In the previous section, we introduced labels in MDP. As already mentioned, this allows us to talk about groups of states having some properties instead of states themselves. However, decisions are still taken according to the history of the system and, in particular, states are fully observable. In many applications, however, the exact state of a system is only partially known: for instance, in a network, an operator can only know the exact status of the nodes it controls, but has to rely on partial information for other nodes that it does not manage.

Thus, in a partially observable MDP, several states are considered as similar from the observer's point of view. As a consequence, decisions apply to a whole class of similar states, and have to be adequately chosen so that an objective is achieved regardless of which state of the class the system was in.

Definition 6. *A partially observable MDP (POMDP for short) is a tuple $\mathcal{M} = (S, \{A_s\}_{s \in S}, p, s_0, Part)$ where $(S, \{A_s\}_{s \in S}, p, s_0)$ is an MDP, referred to as the MDP underlying $\mathcal{M}$ and Part is a partition of S.*

The partition $Part$ of S induces an equivalence relation over states of S. For $s \in S$, we write $[s]$ for the equivalence class s belongs to, and elements of the set of equivalence classes will be denoted c, c_0, etc. We assume that for every $s, s' \in S$, $[s] = [s']$ implies $A_s = A_{s'}$, thus we write $A_{[s]}$ for this set of actions.

Definition 7. *Let $\mathcal{M} = (S, \{A_s\}_{s \in S}, p, s_0, Part)$ be a POMDP.*

- *A history in $\mathcal{M}$ is a finite or infinite sequence alternating state equivalence classes and actions $\sigma = (c_0, a_0, \cdots, c_i, a_i \cdots)$ such that there exists a history $(s_0, a_0, \cdots, s_i, a_i \cdots)$ in the underlying MDP with for all $0 \le i \le \lg(\sigma)$, $c_i = [s_i]$.*
- *A decision rule in $\mathcal{M}$ associates with every history of length $t < \infty$ a distribution $d_t(\sigma)$ over $A_{[s_t]}$.*
- *A policy of $\mathcal{M}$ is finite or infinite a sequence $\pi = (d_0, \cdots, d_t, \cdots)$ such that d_t is a decision rule at time t.*

Given a POMDP $\mathcal{M} = (S, \{A_s\}_{s \in S}, p, s_0, \textit{Part})$, any policy π for $\mathcal{M}$ induces a DTMC written $\mathcal{M}_\pi$. The notion of steady-state distributions extends from MDP to POMDP trivially, the steady-state distribution in the POMDP $\mathcal{M}$ under policy π is written $\mathsf{sd}(\mathcal{M}_\pi)$. Contrary to the fully observable case, the steady-state control problem cannot be decided for POMDP.

Theorem 6. *The steady-state control problem is undecidable for* POMDP.

Proof. The proof is by reduction from a variant of the emptiness problem for probabilistic finite automata. We start by recalling the definition of probabilistic finite automata (PFA). A PFA is a tuple $\mathcal{B} = (Q, \Sigma, \tau, F)$, where Q is the finite set of states, Σ the alphabet, $\tau : Q \times \Sigma \to \mathsf{Dist}(Q)$ defines the probabilistic transition function and $F \subseteq Q$ is the set of final states. The threshold language emptiness problem asks if there exists a finite word over Σ accepted by $\mathcal{B}$ with probability exactly $\frac{1}{2}$. This problem is known to be undecidable [1].

From a PFA $\mathcal{B} = (Q, \Sigma, \tau, F)$, we define a POMDP $\mathcal{M} = (S, \{A_s\}_{s \in S}, p, \textit{Part})$:

- $S = Q \cup \{good, bad\}$
- $A_s = \Sigma \cup \{\#\}$, for all $s \in S$
- $p(s'|s, a) = \tau(s, a)(s')$ if $a \in \Sigma$; $p(good|s, \#) = 1$ if $s \in F$; $p(bad|s, \#) = 1$ if $s \in Q \setminus F$; $p(good|good, a) = 1$ for any $a \in A_{good}$ and $p(bad|bad, a) = 1$ for any $a \in A_{bad}$.
- $\textit{Part} = S$, that is $\textit{Part}$ consists of a single equivalence class, S itself.

The construction is illustrated below.

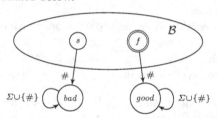

Assuming states in S are ordered such that $good$ and bad are the last states, we then let $\delta_{\text{goal}} = (0, \cdots, 0, 1/2, 1/2)$ be the goal distribution (thus assigning probability mass $1/2$ to both $good$ and bad). This construction ensures that the answer to the steady-state control problem on $\mathcal{M}$ with δ_{goal} is yes if and only if there exists a word $w \in \Sigma^*$ which is accepted in $\mathcal{B}$ with probability $1/2$.

Observe that in the POMDP $\mathcal{M}$ we built, all states are equivalent, so that a policy in $\mathcal{M}$ can only base its decision on the number of steps so far, and thus simply corresponds to a word on $A = \Sigma \cup \{\#\}$. Let us now prove the correctness of the reduction.

($\Longleftarrow$) Given a word w such that $\mathbb{P}_\mathcal{B}(w) = 1/2$, in $\mathcal{M}$ we define a policy π such that $\pi = w \#^\omega$. Then we can infer that $\mathsf{sd}(\mathcal{M}_\pi) = (0, \cdots, 0, 1/2, 1/2)$.

($\Longrightarrow$) First observe that π must contain a $\#$-action, otherwise $\mathsf{sd}(\mathcal{M}_\pi) = (\_, \cdots, \_, 0, 0)$. Thus we may write $\pi = w\#\rho$ with $w \in \Sigma^*$ and $\rho \in A^\omega$. So we obtain that $\mathsf{sd}(\mathcal{M}_\pi) = (0, \cdots, 0, \mathbb{P}_\mathcal{B}(w), 1 - \mathbb{P}_\mathcal{B}(w))$. From the assumption $\mathsf{sd}(\mathcal{M}_\pi) = (0, \cdots, 0, 1/2, 1/2)$, this implies that $\mathbb{P}_\mathcal{B}(w) = 1/2$.

This completes the undecidability proof.   □

Remark 1. In the above undecidability proof, the constructed POMDP is not ergodic. Further, to the best of our knowledge, the undecidability proofs (see for e.g., [1]) for the emptiness of PFA with threshold do not carry over to the ergodic setting. Thus, the status of the steady-state control problem for ergodic POMDP and ergodic PFA are left open in this paper.

4   Conclusion

In this paper, we have defined the steady-state control problem for MDP, and shown that this question is decidable for (ergodic) MDP in polynomial time, and for labeled MDP in polynomial space, but becomes undecidable when observation of states is restricted. It is an open question whether our algorithms are optimal and to establish matching lower-bounds or improve the complexities. Further, implementing our decision algorithm is an interesting next step to establish the feasibility of our approach on case studies. We would also like to extend the results to MDP that are not necessarily ergodic, and treat the case of ergodic POMDP. Another possible extension is to consider the control problem with a finite horizon: given an MDP $\mathcal{M}$, a goal distribution δ_{goal}, and a threshold ε, does there exist a strategy π and $k \in \mathbb{N}$ such that $||\delta_k^\pi - \delta_{goal}|| \leq \varepsilon$?

Finally, the results of this paper can have interesting potential applications in diagnosability of probabilistic systems [11]. Indeed, we would design strategies forcing the system to exhibit a steady-state distribution that depends on the occurrence of a fault.

Acknowledgments. We warmly thank the anonymous reviewers for their useful comments.

References

1. Bertoni, A.: The solution of problems relative to probabilistic automata in the frame of the formal languages theory. In: Siefkes, D. (ed.) GI 1974. LNCS, vol. 26, pp. 107–112. Springer, Heidelberg (1975)
2. Canny, J.F.: Some algebraic and geometric computations in PSPACE. In: 20th ACM Symp. on Theory of Computing, pp. 460–467 (1988)
3. Chadha, R., Korthikanti, V., Vishwanathan, M., Agha, G., Kwon, Y.: Model checking MDPs with a unique compact invariant set of distributions. In: QEST 2011 (2011)
4. Doyen, L., Henzinger, T.A., Raskin, J.-F.: Equivalence of labeled Markov chains. Int. J. Found. Comput. Sci. 19(3), 549–563 (2008)
5. Kemeny, J.G., Snell, J.L.: Finite Markov chains. Princeton University Press (1960)
6. Korthikanti, V.A., Viswanathan, M., Agha, G., Kwon, Y.: Reasoning about MDPs as transformers of probability distributions. In: QEST. IEEE Computer Society (2010)
7. Norris, J.R.: Markov chains. Cambridge series on statistical and probabilistic mathematics, vol. 2. Cambridge University Press (1997)
8. Puterman, M.L.: Markov decision processes: discrete stochastic dynamic programming. John Wiley & Sons (1994)

9. Roos, C., Terlaky, T., Vial, J.-P.: Theory and Algorithms for Linear Optimization: An interior point approach. John Wiley & Sons (1997)
10. Sigaud, O., Buffet, O. (eds.): Markov decision processes in artifical intelligence. John Wiley & Sons (2010)
11. Thorsley, D., Teneketzis, D.: Diagnosability of stochastic discrete-event systems. IEEE Trans. Automat. Contr. 50(4), 476–492 (2005)

Symbolic Control of Stochastic Switched Systems via Finite Abstractions[*]

Majid Zamani and Alessandro Abate[**]

Delft Center for Systems and Control
Delft University of Technology, The Netherlands
{m.zamani,a.abate}@tudelft.nl

Abstract. Stochastic switched systems are a class of continuous-time dynamical models with probabilistic evolution over a continuous domain and control-dependent discrete dynamics over a finite set of modes. As such, they represent a subclass of general stochastic hybrid systems. While the literature has witnessed recent progress in the dynamical analysis and controller synthesis for the stability of stochastic switched systems, more complex and challenging objectives related to the verification of and the synthesis for logic specifications (properties expressed as formulas in linear temporal logic or as automata on infinite strings) have not been formally investigated as of yet. This paper addresses these complex objectives by constructively deriving approximately equivalent (bisimilar) symbolic models of stochastic switched systems. More precisely, a finite symbolic model that is approximately bisimilar to a stochastic switched system is constructed under some dynamical stability assumptions on the concrete model. This allows to formally synthesize controllers (switching signals) over the finite symbolic model that are valid for the concrete system, by means of mature techniques in the literature.

1 Introduction

Stochastic hybrid systems are general dynamical systems comprising continuous and discrete dynamics interleaved with probabilistic noise and stochastic events [4]. Because of their versatility and generality they carry great promise in many safety critical applications [4], including power networks, automotive and financial engineering, air traffic control, biology, telecommunications, and embedded systems. Stochastic *switched* systems are a relevant class of stochastic hybrid systems: they consist of a finite set of modes of operation, each of which is associated to a probabilistic dynamical behavior; further, their discrete dynamics, in the form of mode changes, are governed by a deterministic control signal. However, unlike general stochastic hybrid systems,

---

[*] This work is supported by the European Commission STREP project MoVeS 257005, by the European Commission Marie Curie grant MANTRAS 249295, by the European Commission IAPP project AMBI 324432, by the European Commission NoE Hycon2 257462, and by the NWO VENI grant 016.103.020. A. Abate is also with the Department of Computer Science, University of Oxford.
[**] Corresponding author.

K. Joshi et al. (Eds.): QEST 2013, LNCS 8054, pp. 305–321, 2013.

they do not present probabilistic discrete dynamics (random switch of modes), nor continuous resets upon mode change.

It is known [12] that switched systems can be endowed with global dynamics that are not characteristic of the behavior of any of their single modes: for instance, global instability can arise by proper choice of the discrete switches between a set of stable dynamical modes. This global emergent behavior is one of the many features that makes switched systems theoretically interesting. With focus on *stochastic* switched systems, despite recent progress on basic dynamical analysis focused on stability properties [6], there are no notable results in terms of more complex objectives, such as those dealing with verification or (controller) synthesis for logical specifications. Specifications of interest are expressed as formulas in linear temporal logic or via automata on infinite strings, and as such they are not amenable to be handled by classical techniques for stochastic processes.

A promising direction to investigate these general properties is the use of *symbolic models*. Symbolic models are abstract descriptions of the original dynamics, where each abstract state (or symbol) corresponds to an aggregate of states in the concrete system. When a finite symbolic model is obtained and formally is in relationship with the original system, one can leverage mature techniques for controller synthesis over the discrete model [14] to automatically synthesize controllers for the original system. Towards this goal, a relevant approach is the construction of finite-state symbolic models that are *bisimilar* to the original system. Unfortunately, the class of continuous (time and space) dynamical systems admitting exactly bisimilar finite-state symbolic models is quite restrictive and in particular it covers mostly non-probabilistic models. The results in [5] provide a notion of exact stochastic bisimulation for a class of stochastic hybrid systems, however [5] does not provide any abstraction algorithm, nor looks at the synthesis problem. Therefore, rather than requiring exact equivalence, one can resort to *approximate bisimulation* relations [8], which introduce metrics between the trajectories of the abstract and the concrete models, and further require boundedness in time of these distances.

The construction of approximately bisimilar symbolic models has been recently studied for non-probabilistic continuous control systems, possibly endowed with non-determinism [13,18, and references therein], as well as for non-probabilistic switched systems [9]. However stochastic systems, particularly when endowed with switched dynamics, have only been partially explored. With focus on these models, only a few existing results deal with abstractions of discrete-time processes [2, and references therein]. Results for continuous-time models cover models with specific dynamics: probabilistic rectangular hybrid automata [20] and stochastic dynamical systems under contractivity assumptions [1]. Further, the results in [10] only *check* the (approximate) relationship between an uncountable abstraction and a class of stochastic hybrid systems via a notion of stochastic (bi)simulation function, however, these results do not provide any *construction* of the approximation, nor do they deal with *finite* abstractions, and appear to be computationally tractable only in the case of no inputs. In summary, to the best of our knowledge, there is no comprehensive work on the construction of finite bisimilar abstractions for continuous-time stochastic systems with control actions or with switched dynamics. A recent result [22] by the authors investigates this goal over stochastic control systems, however without any hybrid dynamics.

The main contribution of this work consists in showing the existence and the construction of approximate bisimilar symbolic models for incrementally stable stochastic switched systems. Incremental stability is a stability assumption applied to the stochastic switched systems under study: it can be described in terms of a so-called Lyapunov function (which can either be a single global function or correspond to a set of mode-dependent ones). It is an extension of a similar notion developed for non-probabilistic switched systems [9] in the sense that the results for non-probabilistic switched systems represent a special case of the results in this paper when the continuous dynamics are degenerate (they present no noise). The effectiveness of the results is illustrated with the synthesis of a controller (switching signal) for a room temperature regulation problem (admitting a global – or common – Lyapunov function), which is further subject to a constraint expressed by a finite automaton. More precisely, we display a switched controller synthesis for the purpose of temperature regulation toward a desired level, subject to the discrete constraint.

2 Stochastic Switched Systems

2.1 Notation

The identity map on a set A is denoted by 1_A. If A is a subset of B, we denote by $\imath_A : A \hookrightarrow B$ or simply by $\imath$ the natural inclusion map taking any $a \in A$ to $\imath(a) = a \in B$. The symbols $\mathbb{N}$, $\mathbb{N}_0$, $\mathbb{Z}$, $\mathbb{R}$, $\mathbb{R}^+$, and $\mathbb{R}_0^+$ denote the set of natural, nonnegative integer, integer, real, positive, and nonnegative real numbers, respectively. The symbols I_n, 0_n, and $0_{n \times m}$ denote the identity matrix, the zero vector, and the zero matrix in $\mathbb{R}^{n \times n}$, $\mathbb{R}^n$, and $\mathbb{R}^{n \times m}$, respectively. Given a vector $x \in \mathbb{R}^n$, we denote by x_i the i-th element of x, and by $\|x\|$ the infinity norm of x, namely, $\|x\| = \max\{|x_1|, |x_2|, ..., |x_n|\}$, where $|x_i|$ denotes the absolute value of x_i. Given a matrix $M = \{m_{ij}\} \in \mathbb{R}^{n \times m}$, we denote by $\|M\|$ the infinity norm of M, namely, $\|M\| = \max_{1 \leq i \leq n} \sum_{j=1}^m |m_{ij}|$, and by $\|M\|_F$ the Frobenius norm of M, namely, $\|M\|_F = \sqrt{\mathrm{Tr}(MM^T)}$, where $\mathrm{Tr}(P) = \sum_{i=1}^n p_{ii}$ for any $P = \{p_{ij}\} \in \mathbb{R}^{n \times n}$. The notations $\lambda_{\min}(A)$ and $\lambda_{\max}(A)$ stand for the minimum and maximum eigenvalues of matrix A, respectively.

The closed ball centered at $x \in \mathbb{R}^n$ with radius ε is defined by $\mathcal{B}_\varepsilon(x) = \{y \in \mathbb{R}^n \mid \|x - y\| \leq \varepsilon\}$. A set $B \subseteq \mathbb{R}^n$ is called a *box* if $B = \prod_{i=1}^n [c_i, d_i]$, where $c_i, d_i \in \mathbb{R}$ with $c_i < d_i$ for each $i \in \{1, ..., n\}$. The *span* of a box B is defined as $span(B) = \min\{|d_i - c_i| \mid i = 1, ..., n\}$. By defining $[\mathbb{R}^n]_\eta = \{a \in \mathbb{R}^n \mid a_i = k_i \eta, k_i \in \mathbb{Z}, \imath = 1, \cdots, n\}$, the set $\bigcup_{p \in [\mathbb{R}^n]_\eta} \mathcal{B}_\lambda(p)$ is a countable covering of $\mathbb{R}^n$ for any $\eta \in \mathbb{R}^+$ and $\lambda \geq \eta$. For a box B and $\eta \leq span(B)$, define the η-approximation $[B]_\eta = [\mathbb{R}^n]_\eta \cap B$. Note that $[B]_\eta \neq \varnothing$ for any $\eta \leq span(B)$. Geometrically, for any $\eta \in \mathbb{R}^+$ with $\eta \leq span(B)$ and $\lambda \geq \eta$, the collection of sets $\{\mathcal{B}_\lambda(p)\}_{p \in [B]_\eta}$ is a finite covering of B, i.e., $B \subseteq \bigcup_{p \in [B]_\eta} \mathcal{B}_\lambda(p)$. We extend the notions of *span* and approximation to finite unions of boxes as follows. Let $A = \bigcup_{j=1}^M A_j$, where each A_j is a box. Define $span(A) = \min\{span(A_j) \mid j = 1, \ldots, M\}$, and for any $\eta \leq span(A)$, define $[A]_\eta = \bigcup_{j=1}^{M}[A_j]_\eta$.

Given a set X, a function $\mathbf{d} : X \times X \to \mathbb{R}_0^+$ is a metric on X if for any $x, y, z \in X$, the following three conditions are satisfied: i) $\mathbf{d}(x, y) = 0$ if and only if $x = y$; ii) $\mathbf{d}(x, y) = \mathbf{d}(y, x)$; and iii) (triangle inequality) $\mathbf{d}(x, z) \leq \mathbf{d}(x, y) + \mathbf{d}(y, z)$. A continuous function $\gamma : \mathbb{R}_0^+ \to \mathbb{R}_0^+$, is said to belong to class $\mathcal{K}$ if it is strictly increasing and $\gamma(0) = 0$; γ is said to belong to class $\mathcal{K}_\infty$ if $\gamma \in \mathcal{K}$ and $\gamma(r) \to \infty$ as $r \to \infty$. A continuous function $\beta : \mathbb{R}_0^+ \times \mathbb{R}_0^+ \to \mathbb{R}_0^+$ is said to belong to class $\mathcal{KL}$ if, for each fixed s, the map $\beta(r, s)$ belongs to class $\mathcal{K}_\infty$ with respect to r and, for each fixed nonzero r, the map $\beta(r, s)$ is decreasing with respect to s and $\beta(r, s) \to 0$ as $s \to \infty$. We identify a relation $R \subseteq A \times B$ with the map $R : A \to 2^B$ defined by $b \in R(a)$ iff $(a, b) \in R$. Given a relation $R \subseteq A \times B$, R^{-1} denotes the inverse relation defined by $R^{-1} = \{(b, a) \in B \times A : (a, b) \in R\}$.

2.2 Stochastic Switched Systems

Let $(\Omega, \mathcal{F}, \mathbb{P})$ be a probability space endowed with a filtration $\mathbb{F} = (\mathcal{F}_s)_{s \geq 0}$ satisfying the usual conditions of completeness and right-continuity [11, p. 48]. Let $(W_s)_{s \geq 0}$ be a $\widehat{q}$-dimensional $\mathbb{F}$-Brownian motion [17].

Definition 1. *A stochastic switched system is a tuple* $\Sigma = (\mathbb{R}^n, \mathsf{P}, \mathcal{P}, F, G)$, *where*

- $\mathbb{R}^n$ *is the continuous state space;*
- $\mathsf{P} = \{1, \cdots, m\}$ *is a finite set of modes;*
- $\mathcal{P}$ *is a subset of* $\mathcal{S}(\mathbb{R}_0^+, \mathsf{P})$, *which denotes the set of piecewise constant functions (by convention continuous from the right) from* $\mathbb{R}_0^+$ *to* P, *and characterized by a finite number of discontinuities on every bounded interval in* $\mathbb{R}_0^+$;
- $F = \{f_1, \cdots, f_m\}$ *such that, for all* $p \in \mathsf{P}$, $f_p : \mathbb{R}^n \to \mathbb{R}^n$ *is a continuous function satisfying the following Lipschitz assumption: there exists a constant* $L \in \mathbb{R}^+$ *such that, for all* $x, x' \in \mathbb{R}^n$: $\|f_p(x) - f_p(x')\| \leq L\|x - x'\|$;
- $G = \{g_1, \cdots, g_m\}$, *such that for all* $p \in \mathsf{P}$, $g_p : \mathbb{R}^n \to \mathbb{R}^{n \times \widehat{q}}$ *is a continuous function satisfying the following Lipschitz assumption: there exists a constant* $Z \in \mathbb{R}^+$ *such that for all* $x, x' \in \mathbb{R}^n$: $\|g_p(x) - g_p(x')\| \leq Z\|x - x'\|$.

Let us discuss the semantics of model Σ. For any given $p \in \mathsf{P}$, we denote by Σ_p the subsystem of Σ defined by the stochastic differential equation

$$\mathrm{d}\xi = f_p(\xi)\,\mathrm{d}t + g_p(\xi)\,\mathrm{d}W_t, \tag{1}$$

where f_p is known as the drift, g_p as the diffusion, and again W_t is Brownian motion. A solution process of Σ_p exists and is uniquely determined owing to the assumptions on f_p and on g_p [17, Theorem 5.2.1, p. 68].

For the global model Σ, a continuous-time stochastic process $\xi : \Omega \times \mathbb{R}_0^+ \to \mathbb{R}^n$ is said to be a *solution process* of Σ if there exists a switching signal $\upsilon \in \mathcal{P}$ satisfying

$$\mathrm{d}\xi = f_\upsilon(\xi)\,\mathrm{d}t + g_\upsilon(\xi)\,\mathrm{d}W_t, \tag{2}$$

$\mathbb{P}$-almost surely ($\mathbb{P}$-a.s.) at each time $t \in \mathbb{R}_0^+$ when v is constant. Let us emphasize that v is a piecewise constant function defined over $\mathbb{R}_0^+$ and taking values in P, which simply dictates which mode the solution process ξ is in at any time $t \in \mathbb{R}_0^+$. Notice that the mode changes are non-probabilistic in that they are fully encompassed by a given function v in $\mathcal{P}$ and that, whenever a mode is changed (discontinuity in v), the value of the process ξ is not reset on $\mathbb{R}^n$ – thus ξ is a continuous function of time.

We further write $\xi_{av}(t)$ to denote the value of the solution process at time $t \in \mathbb{R}_0^+$ under the switching signal v from initial condition $\xi_{av}(0) = a$ $\mathbb{P}$-a.s., in which a is a random variable that is measurable in $\mathcal{F}_0$. Note that in general the stochastic switched system Σ may start from a random initial condition.

Finally, note that a solution process of Σ_p is also a solution process of Σ corresponding to the constant switching signal $v(t) = p$, for all $t \in \mathbb{R}_0^+$. We also use $\xi_{ap}(t)$ to denote the value of the solution process of Σ_p at time $t \in \mathbb{R}_0^+$ from the initial condition $\xi_{ap}(0) = a$ $\mathbb{P}$-a.s.

3   Notions of Incremental Stability

This section introduces some stability notions for stochastic switched systems, which generalize the concepts of incremental global asymptotic stability (δ-GAS) [3] for dynamical systems and of incremental global uniform asymptotic stability (δ-GUAS) [9] for non-probabilistic switched systems. The main results presented in this work rely on the stability assumptions discussed in this section.

Definition 2. *The stochastic subsystem Σ_p is incrementally globally asymptotically stable in the qth moment (δ-GAS-M_q), where $q \geq 1$, if there exists a $\mathcal{KL}$ function β_p such that for any $t \in \mathbb{R}_0^+$, and any $\mathbb{R}^n$-valued random variables a and a' that are measurable in $\mathcal{F}_0$, the following condition is satisfied:*

$$\mathbb{E}\left[\|\xi_{ap}(t) - \xi_{a'p}(t)\|^q\right] \leq \beta_p\left(\mathbb{E}\left[\|a - a'\|^q\right], t\right). \tag{3}$$

Intuitively, the notion requires (a higher moment of) the distance between trajectories to be bounded and decreasing in time. It can be easily checked that a δ-GAS-M_q stochastic subsystem Σ_p is δ-GAS [3] in the absence of any noise. Further, note that when $f_p(0_n) = 0_n$ and $g_p(0_n) = 0_{n \times \hat{q}}$ (drift and diffusion terms vanish at the origin), then δ-GAS-M_q implies global asymptotic stability in the qth moment (GAS-M_q) [6], which means that all the trajectories of Σ_p converge in the qth moment to the (constant) trajectory $\xi_{0_n p}(t) = 0_n$, for all $t \in \mathbb{R}_0^+$, (the equilibrium point). We extend the notion of δ-GAS-M_q to stochastic switched systems as follows.

Definition 3. *A stochastic switched system $\Sigma = (\mathbb{R}^n, \mathsf{P}, \mathcal{P}, F, G)$ is incrementally globally uniformly asymptotically stable in the qth moment (δ-GUAS-M_q), where $q \geq 1$, if there exists a $\mathcal{KL}$ function β such that for any $t \in \mathbb{R}_0^+$, any $\mathbb{R}^n$-valued random variables a and a' that are measurable in $\mathcal{F}_0$, and any switching signal $v \in \mathcal{P}$, the following condition is satisfied:*

$$\mathbb{E}\left[\|\xi_{av}(t) - \xi_{a'v}(t)\|^q\right] \leq \beta\left(\mathbb{E}\left[\|a - a'\|^q\right], t\right). \tag{4}$$

Essentially Definition 3 extends Definition 2 uniformly over any possible switching signal υ. As expected, the notion generalizes known ones in the literature: it can be easily seen that a δ-GUAS-M_q stochastic switched system Σ is δ-GUAS [9] in the absence of any noise and that, whenever $f_p(0_n) = 0_n$ and $g_p(0_n) = 0_{n \times \hat{q}}$ for all $p \in \mathsf{P}$, then δ-GUAS-M_q implies global uniform asymptotic stability in the qth moment (GUAS-M_q) [6].

For non-probabilistic systems the δ-GAS property can be characterized by scalar functions defined over the state space, known as Lyapunov functions [3]. Similarly, we describe δ-GAS-M_q in terms of the existence of *incremental Lyapunov functions*.

Definition 4. *Define the* diagonal set Δ *as:* $\Delta = \{(x, x) \mid x \in \mathbb{R}^n\}$. *Consider a stochastic subsystem Σ_p and a continuous function $V_p : \mathbb{R}^n \times \mathbb{R}^n \to \mathbb{R}_0^+$ that is twice continuously differentiable on $\{\mathbb{R}^n \times \mathbb{R}^n\} \backslash \Delta$. Function V_p is called an incremental global asymptotic stability in the qth moment (δ-GAS-M_q) Lyapunov function for Σ_p, where $q \geq 1$, if there exist $\mathcal{K}_\infty$ functions $\underline{\alpha}_p$, $\overline{\alpha}_p$, and a constant $\kappa_p \in \mathbb{R}^+$, such that*

(i) $\underline{\alpha}_p$ *(resp. $\overline{\alpha}_p$) is a convex (resp. concave) function;*
(ii) *for any $x, x' \in \mathbb{R}^n$, $\underline{\alpha}_p \left(\|x - x'\|^q \right) \leq V_p(x, x') \leq \overline{\alpha}_p \left(\|x - x'\|^q \right)$;*
(iii) *for any $x, x' \in \mathbb{R}^n$, such that $x \neq x'$,*

$$
\mathcal{L}V_p(x, x') := [\partial_x V_p \ \partial_{x'} V_p] \begin{bmatrix} f_p(x) \\ f_p(x') \end{bmatrix}
$$
$$
+ \frac{1}{2} Tr \left(\begin{bmatrix} g_p(x) \\ g_p(x') \end{bmatrix} [g_p^T(x) \ g_p^T(x')] \begin{bmatrix} \partial_{x,x} V_p & \partial_{x,x'} V_p \\ \partial_{x',x} V_p & \partial_{x',x'} V_p \end{bmatrix} \right) \leq -\kappa_p V_p(x, x').
$$

The operator $\mathcal{L}$ is the infinitesimal generator associated to the stochastic subsystem (1) [17, Section 7.3], which characterizes the derivative of the expected value of functions of the process with respect to time. For non-probabilistic systems, $\mathcal{L}$ allows computing the conventional functional derivative with respect to time. The symbols ∂_x and $\partial_{x,x'}$ denote first- and second-order partial derivatives with respect to x and x', respectively. Note that condition (i) is not required in the context of non-probabilistic systems [3].

The following theorem describes δ-GAS-M_q in terms of the existence of a δ-GAS-M_q Lyapunov function.

Theorem 1. *A stochastic subsystem Σ_p is δ-GAS-M_q if it admits a δ-GAS-M_q Lyapunov function.*

As qualitatively stated in the Introduction, it is known that a non-probabilistic switched system, whose subsystems are all δ-GAS, may exhibit some unstable behaviors under fast switching signals [9] and, hence, may not be δ-GUAS. The same occurrence can affect stochastic switched systems endowed with δ-GAS-M_q subsystems. The δ-GUAS property of non-probabilistic switched systems can be established by using a common (or global) Lyapunov function, or alternatively via multiple functions that are mode dependent [9]. This leads to the following extensions for δ-GUAS-M_q of stochastic switched systems.

Assume that for any $p \in \mathsf{P}$, the stochastic subsystem Σ_p admits a δ-GAS-M_q Lyapunov function V_p, satisfying conditions (i)-(iii) in Definition 4 with $\mathcal{K}_\infty$ functions $\underline{\alpha}_p$,

$\overline{\alpha}_p$, and a constant $\kappa_p \in \mathbb{R}^+$. Let us introduce functions $\underline{\alpha}$ and $\overline{\alpha}$ and constant κ for use in the rest of the paper. Let the $\mathcal{K}_\infty$ functions $\underline{\alpha}, \overline{\alpha}$, and the constant κ be defied as $\underline{\alpha} = \min\{\underline{\alpha}_1, \cdots, \underline{\alpha}_m\}$, $\overline{\alpha} = \max\{\overline{\alpha}_1, \cdots, \overline{\alpha}_m\}$, and $\kappa = \min\{\kappa_1, \cdots, \kappa_m\}$. First we show a result based on the existence of a common Lyapunov function, characterized by functions $\underline{\alpha} = \underline{\alpha}_1 = \cdots = \underline{\alpha}_m$ and $\overline{\alpha} = \overline{\alpha}_1 = \cdots = \overline{\alpha}_m$, and parameter κ.

Theorem 2. *Consider a stochastic switched system* $\Sigma = (\mathbb{R}^n, \mathsf{P}, \mathcal{P}, F, G)$. *If there exists a function V that is a common δ-GAS-M_q Lyapunov function for all the subsystems* $\{\Sigma_1, \cdots, \Sigma_m\}$, *then Σ is δ-GUAS-M_q.*

The condition conservatively requires the existence of a single function V that is valid for all the subsystems Σ_p, where $p \in \mathsf{P}$. When this common δ-GAS-M_q Lyapunov function V fails to exist, the δ-GUAS-M_q property of Σ can still be established by resorting to multiple δ-GAS-M_q Lyapunov functions (one per mode) over a restricted set of switching signals. More precisely, from Definition 1, let $\mathcal{S}_{\tau_d}(\mathbb{R}_0^+, \mathsf{P})$ denote the set of switching signals υ with *dwell time* $\tau_d \in \mathbb{R}_0^+$, meaning that $\upsilon \in \mathcal{S}(\mathbb{R}_0^+, \mathsf{P})$ has dwell time τ_d if the switching times $t_1, t_2, \ldots$ (occurring at the discontinuity points of υ) satisfy $t_1 > \tau_d$ and $t_i - t_{i-1} \geq \tau_d$, for all $i \geq 2$. We now show a result based on multiple Lyapunov functions.

Theorem 3. *Let $\tau_d \in \mathbb{R}_0^+$, and consider a stochastic switched system* $\Sigma_{\tau_d} = (\mathbb{R}^n, \mathsf{P}, \mathcal{P}_{\tau_d}, F, G)$ *with* $\mathcal{P}_{\tau_d} \subseteq \mathcal{S}_{\tau_d}(\mathbb{R}_0^+, \mathsf{P})$. *Assume that for any $p \in \mathsf{P}$, there exists a δ-GAS-M_q Lyapunov function V_p for subsystem $\Sigma_{\tau_d,p}$ and that in addition there exits a constant $\mu \geq 1$ such that*

$$\forall x, x' \in \mathbb{R}^n, \ \forall p, p' \in \mathsf{P}, \ V_p(x, x') \leq \mu V_{p'}(x, x'). \tag{5}$$

If $\tau_d > \log \mu / \kappa$, then Σ_{τ_d} is δ-GUAS-M_q.

The above result can be practically interpreted as the following fact: global stability is preserved under subsystem stability and enough time spent in each mode. Theorems 1, 2, and 3 provide sufficient conditions for certain stability properties, however they all hinge on finding proper Lyapunov functions.

For stochastic switched systems Σ (resp. Σ_{τ_d}) with f_p and g_p of the form of polynomials, for any $p \in \mathsf{P}$, one can resort to available software tools, such as **SOSTOOLS** [19], to search for appropriate δ-GAS-M_q Lyapunov functions.

We look next into special instances where these functions are known explicitly or can be easily computed based on the model dynamics. The first result provides a sufficient condition for a particular function V_p to be a δ-GAS-M_q Lyapunov function for a stochastic subsystem Σ_p, when $q = 1, 2$ (first or second moment).

Lemma 1. *Consider a stochastic subsystem Σ_p. Let $q \in \{1, 2\}$, $P_p \in \mathbb{R}^{n \times n}$ be a symmetric positive definite matrix, and the function $V_p : \mathbb{R}^n \times \mathbb{R}^n \to \mathbb{R}_0^+$ be defined as follows:*

$$V_p(x, x') := \left(\tilde{V}(x, x')\right)^{\frac{q}{2}} = \left(\frac{1}{q}(x - x')^T P_p (x - x')\right)^{\frac{q}{2}}, \tag{6}$$

and satisfies

$$(x - x')^T P_p(f_p(x) - f_p(x')) + \frac{1}{2} \left\| \sqrt{P_p} \left(g_p(x) - g_p(x') \right) \right\|_F^2 \leq -\kappa_p \left(V_p(x, x') \right)^{\frac{2}{q}} , \quad (7)$$

or, if f_p is differentiable, satisfies

$$(x - x')^T P_p \partial_x f_p(z)(x - x') + \frac{1}{2} \left\| \sqrt{P_p} \left(g_p(x) - g_p(x') \right) \right\|_F^2 \leq -\kappa_p \left(V_p(x, x') \right)^{\frac{2}{q}} , \quad (8)$$

for all x, x', z in $\mathbb{R}^n$, and for some constant $\kappa_p \in \mathbb{R}^+$. Then V_p is a δ-GAS-M_q Lyapunov function for Σ_p.

The next result provides a condition that is equivalent to (7) or to (8) for affine stochastic subsystems Σ_p (that is, for subsystems with affine drift and linear diffusion terms) in the form of a linear matrix inequality (LMI), which can be easily solved numerically.

Corollary 1. *Consider a stochastic subsystem Σ_p, where for any $x \in \mathbb{R}^n$, $f_p(x) := A_p x + b_p$ for some $A_p \in \mathbb{R}^{n \times n}, b_p \in \mathbb{R}^n$, and $g_p(x) := \begin{bmatrix} \sigma_{1,p} x & \sigma_{2,p} x & \dots & \sigma_{\widehat{q},p} x \end{bmatrix}$ for some $\sigma_{i,p} \in \mathbb{R}^{n \times n}$, where $i = 1, \dots, \widehat{q}$. Then, function V_p in (6) is a δ-GAS-M_q Lyapunov function for Σ_p if there exists a positive constant $\widehat{\kappa}_p \in \mathbb{R}^+$ satisfying the following LMI:*

$$P_p A_p + A_p^T P_p + \sum_{i=1}^{\widehat{q}} \sigma_{i,p}^T P_p \sigma_{i,p} \prec -\widehat{\kappa}_p P_p. \quad (9)$$

Notice that Corollary 1 allows obtaining tighter upper bounds for the inequalities (3) and (4) for any $p \in \mathsf{P}$, by selecting appropriate matrices P_p satisfying the LMI in (9).

4   Symbolic Models and Approximate Equivalence Relations

We employ the notion of *system* [21] to provide (in Sec. 5) an alternative description of stochastic switched systems that can be later directly related to their symbolic models.

Definition 5. *A system S is a tuple $S = (X, X_0, U, \longrightarrow, Y, H)$, where X is a set of states, $X_0 \subseteq X$ is a set of initial states, U is a set of inputs, $\longrightarrow \subseteq X \times U \times X$ is a transition relation, Y is a set of outputs, and $H : X \to Y$ is an output map.*

We write $x \xrightarrow{u} x'$ if $(x, u, x') \in \longrightarrow$. If $x \xrightarrow{u} x'$, we call state x' a u-successor, or simply a successor, of state x. For technical reasons, we assume that for each $x \in X$, there is some u-successor of x, for some $u \in U$ – let us remark that this is always the case for the considered systems later in this paper. A system S is said to be

- *metric*, if the output set Y is equipped with a metric $\mathbf{d} : Y \times Y \to \mathbb{R}_0^+$;
- *finite*, if X is a finite set;
- *deterministic*, if for any state $x \in X$ and any input u, there exists at most one u-successor.

For a system $S = (X, X_0, U, \longrightarrow, Y, H)$ and given any state $x_0 \in X_0$, a finite state run generated from x_0 is a finite sequence of transitions:

$$x_0 \xrightarrow{u_0} x_1 \xrightarrow{u_1} x_2 \xrightarrow{u_2} \cdots \xrightarrow{u_{n-2}} x_{n-1} \xrightarrow{u_{n-1}} x_n, \tag{10}$$

such that $x_i \xrightarrow{u_i} x_{i+1}$ for all $0 \leq i < n$. A finite state run can be trivially extended to an infinite state run as well. A finite output run is a sequence $\{y_0, y_1, \ldots, y_n\}$ such that there exists a finite state run of the form (10) with $y_i = H(x_i)$, for $i = 1, \ldots, n$. A finite output run can also be directly extended to an infinite output run as well.

Now, we recall the notion of approximate (bi)simulation relation, introduced in [8], which is useful when analyzing or synthesizing controllers for deterministic systems.

Definition 6. *Let* $S_a = (X_a, X_{a0}, U_a, \xrightarrow[a]{}, Y_a, H_a)$ *and* $S_b = (X_b, X_{b0}, U_b, \xrightarrow[b]{}, Y_b, H_b)$ *be metric systems with the same output sets* $Y_a = Y_b$ *and metric* d. *For* $\varepsilon \in \mathbb{R}_0^+$, *a relation* $R \subseteq X_a \times X_b$ *is said to be an* ε-approximate *simulation relation from* S_a *to* S_b *if the following three conditions are satisfied:*

(i) for every $x_{a0} \in X_{a0}$, *there exists* $x_{b0} \in X_{b0}$ *with* $(x_{a0}, x_{b0}) \in R$;
(ii) for every $(x_a, x_b) \in R$ *we have* $\mathbf{d}(H_a(x_a), H_b(x_b)) \leq \varepsilon$;
(iii) for every $(x_a, x_b) \in R$ *we have that* $x_a \xrightarrow[a]{u_a} x'_a$ *in* S_a *implies the existence of* $x_b \xrightarrow[b]{u_b} x'_b$ *in* S_b *satisfying* $(x'_a, x'_b) \in R$.

A relation $R \subseteq X_a \times X_b$ *is said to be an* ε-approximate bisimulation relation between S_a and S_b if R is an ε-approximate simulation relation from S_a to S_b and R^{-1} is an ε-approximate simulation relation from S_b to S_a.

System S_a *is* ε-approximately simulated by S_b, *or* S_b ε-approximately simulates S_a, *denoted by* $S_a \preceq_{\mathcal{S}}^{\varepsilon} S_b$, *if there exists an* ε-approximate simulation relation from S_a to S_b. *System* S_a *is* ε-approximate bisimilar to S_b, *denoted by* $S_a \cong_{\mathcal{S}}^{\varepsilon} S_b$, *if there exists an* ε-approximate bisimulation relation between S_a and S_b.

Note that when $\varepsilon = 0$, the condition (ii) in the above definition is changed to $(x_a, x_b) \in R$ if and only if $H_a(x_a) = H_b(x_b)$, and R becomes an exact simulation relation, as introduced in [16]. Similarly, when $\varepsilon = 0$ and whenever applicable, R translates into an exact bisimulation relation.

5 Symbolic Models for Stochastic Switched Systems

This section contains the main contributions of this work. We show that for any stochastic switched system Σ (resp. Σ_{τ_d} as in Theorem 3), admitting a common (resp. multiple) δ-GAS-M_q Lyapunov function(s), and for any precision level $\varepsilon \in \mathbb{R}^+$, we can construct a finite system that is ε-approximate bisimilar to Σ (resp. Σ_{τ_d}). In order to do so, we use systems as an abstract representation of stochastic switched systems, capturing all the information contained in them. More precisely, given a stochastic switched system $\Sigma = (\mathbb{R}^n, \mathsf{P}, \mathcal{P}, F, G)$, we define an associated metric system $S(\Sigma) = (X, X_0, U, \longrightarrow, Y, H)$, where:

- X is the set of all $\mathbb{R}^n$-valued random variables defined on the probability space $(\Omega, \mathcal{F}, \mathbb{P})$;
- X_0 is the set of all $\mathbb{R}^n$-valued random variables that are measurable over the trivial sigma-algebra $\mathcal{F}_0$, i.e., the system starts from a non-probabilistic initial condition, which is equivalently a random variable with a Dirac probability distribution;
- $U = \mathsf{P} \times \mathbb{R}^+$;
- $x \xrightarrow{p,\tau} x'$ if x and x' are measurable in $\mathcal{F}_t$ and $\mathcal{F}_{t+\tau}$, respectively, for some $t \in \mathbb{R}_0^+$, and there exists a solution process $\xi : \Omega \times \mathbb{R}_0^+ \to \mathbb{R}^n$ of Σ satisfying $\xi(t) = x$ and $\xi_{xp}(\tau) = x'$ $\mathbb{P}$-a.s.;
- Y is the set of all $\mathbb{R}^n$-valued random variables defined on the probability space $(\Omega, \mathcal{F}, \mathbb{P})$;
- $H = 1_X$.

We assume that the output set Y is equipped with the natural metric $\mathbf{d}(y, y') = \left(\mathbb{E}\left[\|y - y'\|^q \right] \right)^{\frac{1}{q}}$, for any $y, y' \in Y$ and some $q \geq 1$. Let us remark that the set of states of $S(\Sigma)$ is uncountable and that $S(\Sigma)$ is a deterministic system in the sense of Definition 5, since (cf. Subsection 2.2) its solution process is uniquely determined.

In subsequent developments, we will work with a sub-system of $S(\Sigma)$ obtained by selecting those transitions of $S(\Sigma)$ describing trajectories of duration τ, where τ is a given sampling time. This can be seen as a time discretization or a sampling of $S(\Sigma)$. This restriction is practically motivated by the fact that the switching in the original model Σ has to be controlled by a digital platform with a given clock period (τ). More precisely, given a stochastic switched system $\Sigma = (\mathbb{R}^n, \mathsf{P}, \mathcal{P}, F, G)$ and a sampling time $\tau \in \mathbb{R}^+$, we define the associated system $S_\tau(\Sigma) = \left(X_\tau, X_{\tau 0}, U_\tau, \xrightarrow{\ \ \tau\ \ }, Y_\tau, H_\tau \right)$, where $X_\tau = X$, $X_{\tau 0} = X_0$, $U_\tau = \mathsf{P}$, $Y_\tau = Y$, $H_\tau = H$, and

- $x_\tau \xrightarrow{p}{\tau} x'_\tau$ if x_τ and x'_τ are measurable, respectively, in $\mathcal{F}_{k\tau}$ and $\mathcal{F}_{(k+1)\tau}$ for some $k \in \mathbb{N}_0$, and there exists a solution process $\xi : \Omega \times \mathbb{R}_0^+ \to \mathbb{R}^n$ of Σ satisfying $\xi(k\tau) = x_\tau$ and $\xi_{x_\tau p}(\tau) = x'_\tau$ $\mathbb{P}$-a.s..

Note that a finite state run $x_0 \xrightarrow{u_0}{\tau} x_1 \xrightarrow{u_1}{\tau} \dots \xrightarrow{u_{N-1}}{\tau} x_N$ of $S_\tau(\Sigma)$, where $u_i \in \mathsf{P}$ and $x_i = \xi_{x_{i-1} u_{i-1}}(\tau)$ for $i = 1, \cdots, N$, captures the trajectory of the stochastic switched system Σ at times $t = 0, \tau, \cdots, N\tau$, started from the non-probabilistic initial condition x_0 and resulting from a switching signal υ obtained by the concatenation of the modes u_i (i.e. $\upsilon(t) = u_{i-1}$ for any $t \in [(i-1)\tau, i\tau[)$, for $i = 1, \cdots, N$.

Before introducing the symbolic model for the stochastic switched system, we proceed with the next lemma, borrowed from [22], which provides an upper bound on the distance (in the qth moment metric) between the solution processes of Σ_p and the corresponding non-probabilistic system obtained by disregarding the diffusion term (g_p).

Lemma 2. *Consider a stochastic subsystem Σ_p such that $f_p(0_n) = 0_n$ and $g_p(0_n) = 0_{n \times \hat{q}}$. Suppose there exists a δ-GAS-M_q Lyapunov function V_p for Σ_p such that its Hessian is a positive semidefinite matrix in $\mathbb{R}^{2n \times 2n}$ and $q \geq 2$. Then for any x in a compact set $\mathsf{D} \subset \mathbb{R}^n$ and any $p \in \mathsf{P}$, we have*

$$\mathbb{E}\left[\left\|\xi_{xp}(t) - \overline{\xi}_{xp}(t)\right\|^q\right] \leq h_p(g_p, t), \tag{11}$$

where $\overline{\xi}_{xp}$ is the solution of the ordinary differential equation (ODE) $\dot{\overline{\xi}}_{xp} = f_p\left(\overline{\xi}_{xp}\right)$ starting from the initial condition x, and the nonnegative valued function h_p tends to zero as $t \to 0$, $t \to +\infty$, or as $Z \to 0$, where Z is the Lipschitz constant, introduced in Definition 1.

Although the result in [22, Lemma 3.7] is based on the existence of δ-ISS-M_q Lyapunov functions, one can similarly show the result in Lemma 2 by using δ-GAS-M_q Lyapunov functions. In particular, one can compute explicitly function h_p using [22, Equation (9.4)] with slight modifications. Moreover, we refer the interested readers to [22, Lemma 3.9 and Corollary 3.10], providing explicit forms of the function h_p for (affine) stochastic subsystems Σ_p admitting a δ-GAS-M_q Lyapunov function V_p as in (6), where $q \in \{1, 2\}$. Note that one does not require the condition $f_p(0_n) = 0_n$ for affine subsystems Σ_p. For later use, we introduce function $h(G, t) = \max\{h_1(g_1, t), \cdots, h_m(g_m, t)\}$ for all $t \in \mathbb{R}_0^+$.

In order to show the main results, we raise the following supplementary assumption on the δ-GAS-M_q Lyapunov functions V_p: for all $p \in \mathsf{P}$, there exists a $\mathcal{K}_\infty$ and concave function $\widehat{\gamma}_p$ such that

$$|V_p(x, y) - V_p(x, z)| \leq \widehat{\gamma}_p\left(\|y - z\|\right), \tag{12}$$

for any $x, y, z \in \mathbb{R}^n$. This assumption is not restrictive, provided the function V_p is limited to a compact subset of $\mathbb{R}^n \times \mathbb{R}^n$. For all $x, y, z \in \mathsf{D}$, where D is a compact subset of $\mathbb{R}^n$, by applying the mean value theorem to the function $y \to V_p(x, y)$, one gets $|V_p(x, y) - V_p(x, z)| \leq \widehat{\gamma}_p\left(\|y - z\|\right)$, where $\widehat{\gamma}_p(r) = \left(\max_{(x,y) \in \mathsf{D} \setminus \Delta}\left\|\frac{\partial V_p(x,y)}{\partial y}\right\|\right) r$. In particular, for the δ-GAS-M_1 Lyapunov function V_p defined in (6), we obtain $\widehat{\gamma}_p(r) = \frac{\lambda_{\max}(P_p)}{\sqrt{\lambda_{\min}(P_p)}} r$ [21, Proposition 10.5]. For later use, let us define the $\mathcal{K}_\infty$ function $\widehat{\gamma}$ such that $\widehat{\gamma} = \max\{\widehat{\gamma}_1, \cdots, \widehat{\gamma}_m\}$. (Note that, for the case of a common Lyapunov function, we have: $\widehat{\gamma} = \widehat{\gamma}_1 = \cdots = \widehat{\gamma}_m$.) We proceed presenting the main results of this work.

5.1 Common Lyapunov Function

We first show a result based on the existence of a common δ-GAS-M_q Lyapunov function for subsystems $\Sigma_1, \cdots, \Sigma_m$. Consider a stochastic switched system $\Sigma = (\mathbb{R}^n, \mathsf{P}, \mathcal{P}, F, G)$ and a pair $\mathsf{q} = (\tau, \eta)$ of quantization parameters, where τ is the sampling time and η is the state space quantization. Given Σ and q, consider the following system: $S_{\mathsf{q}}(\Sigma) = (X_{\mathsf{q}}, X_{\mathsf{q}0}, U_{\mathsf{q}}, \xrightarrow[\mathsf{q}]{}, Y_{\mathsf{q}}, H_{\mathsf{q}})$, where $X_{\mathsf{q}} = [\mathbb{R}^n]_\eta$, $X_{\mathsf{q}0} = [\mathbb{R}^n]_\eta$, $U_{\mathsf{q}} = \mathsf{P}$, and

- $x_{\mathsf{q}} \xrightarrow[\mathsf{q}]{p} x_{\mathsf{q}}'$ if there exists a $x_{\mathsf{q}}' \in X_{\mathsf{q}}$ such that $\left\|\overline{\xi}_{x_{\mathsf{q}}p}(\tau) - x_{\mathsf{q}}'\right\| \leq \eta$, where $\dot{\overline{\xi}}_{x_{\mathsf{q}}p} = f_p\left(\overline{\xi}_{x_{\mathsf{q}}p}\right)$;

- Y_q is the set of all $\mathbb{R}^n$-valued random variables defined on the probability space $(\Omega, \mathcal{F}, \mathbb{P})$;
- $H_q = \imath : X_q \hookrightarrow Y_q$.

In order to relate models, the output set Y_q is taken to be that of the stochastic switched system $S_\tau(\Sigma)$. Therefore, in the definition of H_q, the inclusion map $\imath$ is meant, with a slight abuse of notation, as a mapping from a grid point to a random variable with a Dirac probability distribution centered at the grid point. There is no loss of generality to alternatively assume that $Y_q = X_q$ and $H_q = 1_{X_q}$.

The transition relation of $S_q(\Sigma)$ is well defined in the sense that for every $x_q \in [\mathbb{R}^n]_\eta$ and every $p \in \mathsf{P}$ there always exists $x'_q \in [\mathbb{R}^n]_\eta$ such that $x_q \xrightarrow[q]{p} x'_q$. This can be seen since by definition of $[\mathbb{R}^n]_\eta$, for any $\widehat{x} \in \mathbb{R}^n$ there always exists a state $\widehat{x}' \in [\mathbb{R}^n]_\eta$ such that $\|\widehat{x} - \widehat{x}'\| \leq \eta$. Hence, for $\overline{\xi}_{x_q p}(\tau)$ there always exists a state $x'_q \in [\mathbb{R}^n]_\eta$ satisfying $\left\|\overline{\xi}_{x_q p}(\tau) - x'_q\right\| \leq \eta$.

We can now present one of the main results of the paper, which relates the existence of a common δ-GAS-M_q Lyapunov function for the subsystems $\Sigma_1, \cdots, \Sigma_m$ to the construction of a finite symbolic model that is approximately bisimilar to the original system.

Theorem 4. *Let* $\Sigma = (\mathbb{R}^n, \mathsf{P}, \mathcal{P}, F, G)$ *be a stochastic switched system admitting a common δ-GAS-M_q Lyapunov function V, of the form of (6) or the one explained in Lemma 2, for subsystems* $\Sigma_1, \cdots, \Sigma_m$. *For any $\varepsilon \in \mathbb{R}^+$, and any double $\mathsf{q} = (\tau, \eta)$ of quantization parameters satisfying*

$$\overline{\alpha}\left(\eta^q\right) \leq \underline{\alpha}\left(\varepsilon^q\right), \tag{13}$$

$$e^{-\kappa\tau}\underline{\alpha}\left(\varepsilon^q\right) + \widehat{\gamma}\left((h(G, \tau))^{\frac{1}{q}} + \eta\right) \leq \underline{\alpha}\left(\varepsilon^q\right), \tag{14}$$

we have that $S_q(\Sigma) \cong_S^\varepsilon S_\tau(\Sigma)$.

It can be readily seen that when we are interested in the dynamics of Σ, initialized on a compact $\mathsf{D} \subset \mathbb{R}^n$ of the form of finite union of boxes and for a given precision ε, there always exist a sufficiently large value of τ and a small value of η such that $\eta \leq span(\mathsf{D})$ and the conditions in (13) and (14) are satisfied. For a given fixed sampling time τ, the precision ε is lower bounded by:

$$\varepsilon > \left(\underline{\alpha}^{-1}\left(\frac{\widehat{\gamma}\left((h(G, \tau))^{\frac{1}{q}}\right)}{1 - e^{-\kappa\tau}}\right)\right)^{\frac{1}{q}}. \tag{15}$$

One can easily verify that the lower bound on ε in (15) goes to zero as τ goes to infinity or as $Z \to 0$, where Z is the Lipschitz constant, introduced in Definition 1. Furthermore, one can try to minimize the lower bound on ε in (15) by appropriately choosing a common δ-GAS-M_q Lyapunov function V.

Note that the results in [9, Theorem 4.1] for non-probabilistic models are fully recovered by the statement in Theorem 4 if the stochastic switched system Σ is not affected by any noise, implying that $h_p(g_p, t)$ is identically zero for all $p \in \mathsf{P}$, and that the δ-GAS-M_q common Lyapunov function simply reduces to being the δ-GAS one.

5.2 Multiple Lyapunov Functions

If a common δ-GAS-M$_q$ Lyapunov function does not exist, one can still attempt computing approximately bisimilar symbolic models by seeking mode-dependent Lyapunov functions and by restricting the set of switching signals using a dwell time τ_d. For simplicity and without loss of generality, we assume that τ_d is an integer multiple of τ, i.e. there exists $N \in \mathbb{N}$ such that $\tau_d = N\tau$.

Given a stochastic switched system $\Sigma_{\tau_d} = (\mathbb{R}^n, \mathsf{P}, \mathcal{P}_{\tau_d}, F, G)$ and a sampling time $\tau \in \mathbb{R}^+$, we define the system $S_\tau(\Sigma_{\tau_d}) = (X_\tau, X_{\tau 0}, U_\tau, \xrightarrow[\tau]{}, Y_\tau, H_\tau)$, where:

- $X_\tau = \mathcal{X} \times \mathsf{P} \times \{1, \ldots, N-1\}$, where $\mathcal{X}$ is the set of all $\mathbb{R}^n$-valued random variables defined on the probability space $(\Omega, \mathcal{F}, \mathbb{P})$;
- $X_{\tau 0} = \mathcal{X}_0 \times \mathsf{P} \times \{0\}$, where $\mathcal{X}_0$ is the set of all $\mathbb{R}^n$-valued random variables that are measurable with respect to the trivial sigma-algebra $\mathcal{F}_0$, i.e., the stochastic switched system starts from a non-probabilistic initial condition;
- $U_\tau = \mathsf{P}$;
- $(x_\tau, p, i) \xrightarrow[\tau]{p} (x'_\tau, p', i')$ if x_τ and x'_τ are measurable, respectively, in $\mathcal{F}_{k\tau}$ and $\mathcal{F}_{(k+1)\tau}$ for some $k \in \mathbb{N}_0$, and there exists a solution process $\xi : \Omega \times \mathbb{R}_0^+ \to \mathbb{R}^n$ of Σ satisfying $\xi(k\tau) = x_\tau$ and $\xi_{x_\tau p}(\tau) = x'_\tau$ $\mathbb{P}$-a.s. and one of the following holds:
  - $i < N-1$, $p' = p$, and $i' = i+1$: switching is not allowed because the time elapsed since the latest switch is strictly smaller than the dwell time;
  - $i = N-1$, $p' = p$, and $i' = N-1$: switching is allowed but no mode switch occurs;
  - $i = N-1$, $p' \neq p$, and $i' = 0$: switching is allowed and a mode switch occurs.
- $Y_\tau = \mathcal{X}$ is the set of all $\mathbb{R}^n$-valued random variables defined on the probability space $(\Omega, \mathcal{F}, \mathbb{P})$;
- H_τ is the map taking $(x_\tau, p, i) \in \mathcal{X} \times \mathsf{P} \times \{1, \cdots, N-1\}$ to $x_\tau \in \mathcal{X}$.

We assume that the output set Y_τ is equipped with the natural metric $\mathbf{d}(y, y') = \left(\mathbb{E}\left[\|y - y'\|^q \right] \right)^{\frac{1}{q}}$, for any $y, y' \in Y_\tau$ and some $q \geq 1$. One can readily verify that the (in)finite output runs of $S_\tau(\Sigma_{\tau_d})$ are the (in)finite output runs of $S_\tau(\Sigma)$ corresponding to switching signals with dwell time $\tau_d = N\tau$.

Consider a stochastic switched system $\Sigma_{\tau_d} = (\mathbb{R}^n, \mathsf{P}, \mathcal{P}_{\tau_d}, F, G)$ and a pair $\mathsf{q} = (\tau, \eta)$ of quantization parameters, where τ is the sampling time and η is the state space quantization. Given Σ_{τ_d} and q, consider the following system: $S_{\mathsf{q}}(\Sigma_{\tau_d}) = (X_{\mathsf{q}}, X_{\mathsf{q}0}, U_{\mathsf{q}}, \xrightarrow[\mathsf{q}]{}, Y_{\mathsf{q}}, H_{\mathsf{q}})$, where $X_{\mathsf{q}} = [\mathbb{R}^n]_\eta \times \mathsf{P} \times \{0, \cdots, N-1\}$, $X_{\mathsf{q}0} = [\mathbb{R}^n]_\eta \times \mathsf{P} \times \{0\}$, $U_{\mathsf{q}} = \mathsf{P}$, and

- $(x_{\mathsf{q}}, p, i) \xrightarrow[\mathsf{q}]{p} (x'_{\mathsf{q}}, p', i')$ if there exists a $x'_{\mathsf{q}} \in X_{\mathsf{q}}$ such that $\left\| \bar{\xi}_{x_{\mathsf{q}} p}(\tau) - x'_{\mathsf{q}} \right\| \leq \eta$, where $\bar{\xi}_{x_{\mathsf{q}} p} = f_p\left(\bar{\xi}_{x_{\mathsf{q}} p} \right)$ and one of the following holds:
  - $i < N-1$, $p' = p$, and $i' = i+1$;
  - $i = N-1$, $p' = p$, and $i' = N-1$;
  - $i = N-1$, $p' \neq p$, and $i' = 0$.
- $Y_{\mathsf{q}} = \mathcal{X}$ is the set of all $\mathbb{R}^n$-valued random variables defined on the probability space $(\Omega, \mathcal{F}, \mathbb{P})$;

- H_q is the map taking $(x_q, p, i) \in [\mathbb{R}^n]_\eta \times \mathsf{P} \times \{1, \cdots, N-1\}$ to a random variable with a Dirac probability distribution centered at x_q.

Similar to what we showed in the case of a common Lyapunov function, the transition relation of $S_q(\Sigma_{\tau_d})$ is well defined in the sense that for every $(x_q, p, i) \in [\mathbb{R}^n]_\eta \times \mathsf{P} \times \{0, \cdots, N-1\}$ there always exists $(x'_q, p', i') \in [\mathbb{R}^n]_\eta \times \mathsf{P} \times \{0, \cdots, N-1\}$ such that $(x_q, p, i) \xrightarrow[q]{p} (x'_q, p', i')$.

We present the second main result of the paper, which relates the existence of multiple Lyapunov functions for a stochastic switched system to that of a symbolic model.

Theorem 5. *Consider* $\tau_d \in \mathbb{R}_0^+$, *and a stochastic switched system* $\Sigma_{\tau_d} = (\mathbb{R}^n, \mathsf{P}, \mathcal{P}_{\tau_d}, F, G)$ *such that* $\tau_d = N\tau$, *for some* $N \in \mathbb{N}$. *Let us assume that for any* $p \in \mathsf{P}$, *there exists a* δ-GAS-M_q *Lyapunov function* V_p, *of the form in* (6) *or as the one in Lemma 2, for subsystem* $\Sigma_{\tau_d, p}$. *Moreover, assume that* (5) *holds for some* $\mu \geq 1$. *If* $\tau_d > \log \mu / \kappa$, *for any* $\varepsilon \in \mathbb{R}^+$, *and any pair* $q = (\tau, \eta)$ *of quantization parameters satisfying*

$$\overline{\alpha}(\eta^q) \leq \underline{\alpha}(\varepsilon^q), \tag{16}$$

$$\widehat{\gamma}\left((h(G, \tau))^{\frac{1}{q}} + \eta\right) \leq \frac{\frac{1}{\mu} - e^{-\kappa\tau_d}}{1 - e^{-\kappa\tau_d}} (1 - e^{-\kappa\tau}) \underline{\alpha}(\varepsilon^q), \tag{17}$$

we have that $S_q(\Sigma_{\tau_d}) \cong_{\mathcal{S}}^\varepsilon S_\tau(\Sigma_{\tau_d})$.

It can be readily seen that when we are interested in the dynamics of Σ_{τ_d}, initialized on a compact $\mathsf{D} \subset \mathbb{R}^n$ of the form of finite union of boxes, and for a precision ε, there always exist sufficiently large value of τ and small value of η such that $\eta \leq span(\mathsf{D})$ and the conditions in (16) and (17) are satisfied. For a given fixed sampling time τ, the precision ε is lower bounded by:

$$\varepsilon \geq \left(\underline{\alpha}^{-1}\left(\frac{\widehat{\gamma}\left((h(G, \tau))^{\frac{1}{q}}\right)}{1 - e^{-\kappa\tau}} \cdot \frac{1 - e^{-\kappa\tau_d}}{\frac{1}{\mu} - e^{-\kappa\tau_d}}\right)\right)^{\frac{1}{q}}. \tag{18}$$

The properties of the bound in (18) are analogous to those of the case of a common Lyapunov function. Similarly, Theorem 5 subsumes [9, Theorem 4.2] over non-probabilistic models.

6   Case Study

We experimentally demonstrate the effectiveness of the results. In the example below, the computation of the abstraction $S_q(\Sigma)$ has been performed via the software tool **Pessoa** [15] on a laptop with CPU 2GHz Intel Core i7. Controller enforcing the specification was found by using standard algorithms from game theory [14], as implemented in **Pessoa**. The terms $W_t^i, i = 1, 2$, denote the standard Brownian motion.

The stochastic switched system Σ is a simple thermal model of a two-room building, borrowed from [7], affected by noise and described by the following stochastic differential equations:

$$\begin{cases} d\xi_1 = (\alpha_{21}(\xi_2 - \xi_1) + \alpha_{e1}(T_e - \xi_1) + \alpha_f(T_f - \xi_1)(p-1)) \, dt + \sigma_1\xi_1 \, dW_t^1, \\ d\xi_2 = (\alpha_{12}(\xi_1 - \xi_2) + \alpha_{e2}(T_e - \xi_2)) \, dt + \sigma_2\xi_2 \, dW_t^2, \end{cases} \quad (19)$$

where ξ_1 and ξ_2 denote the temperature in each room, $T_e = 10$ (degrees Celsius) is the external temperature and $T_f = 50$ is the temperature of a heater that can be switched off ($p = 1$) or on ($p = 2$): these two operations correspond to the modes P of the model, whereas the state space is $\mathbb{R}^2$. The drifts f_p and diffusion terms g_p, $p = 1, 2$, can be simply written out of (19) and are affine. The parameters of the drifts are chosen based on the ones in [7] as follows: $\alpha_{21} = \alpha_{12} = 5 \times 10^{-2}, \alpha_{e1} = 5 \times 10^{-3}, \alpha_{e2} = 3.3 \times 10^{-3}$, and $\alpha_f = 8.3 \times 10^{-3}$. We work on the subset D $= [20, 22] \times [20, 22] \subset \mathbb{R}^2$ of the state space of Σ. Within D one can conservatively overapproximate the multiplicative noises in (19) as additive noises with variance between 0.02 and 0.022.

It can be readily verified that the function $V(x_1, x_2) = \sqrt{(x_1 - x_2)^T(x_1 - x_2)}$ is a common δ-GAS-M$_1$ Lyapunov function for Σ, satisfying the LMI condition (9) with $P_p = I_2$, and $\widehat{\kappa}_p = 0.0083$, for $p \in \{1, 2\}$.

For a given sampling time $\tau = 20$ time units, using inequality (15), the precision ε is lower bounded by the quantity 1.09. While one can reduce this lower bound by increasing the sampling time, as discussed later the empirical bound computed in the experiments is significantly lower than the theoretical bound $\varepsilon = 1.09$. For a selected precision $\varepsilon = 1.1$, the discretization parameter η of $S_q(\Sigma)$, computed from Theorem 4, equals to 0.003. This has lead to a symbolic system $S_q(\Sigma)$ with a resulting number of states equal to 895122. The CPU time employed to compute the abstraction amounted to 506.32 seconds.

Consider the objective to design a controller (switching policy) forcing the first moment of the trajectories of Σ to stay within D. This objective can be encoded via the LTL specification $\Box$D. Furthermore, to add an additional discrete component to the problem, we assume that the heater has to stay in the off mode ($p = 1$) at most one time slot in every two slots. A time slot is an interval of the form $[k\tau, (k+1)\tau[$, with $k \in \mathbb{N}$ and where τ is the sampling time. Possible switching policies are for instance:

$$|12|12|12|12|12|12|12| \cdots, \quad |21|21|21|21|21|21|21| \cdots, \quad |12|21|22|12|12|21|22| \cdots,$$

where 2 denotes a slot where the heater is on ($p = 2$) and 1 denotes a slot where heater is off ($p = 1$). This constraint on the switching policies can be represented by the finite system (labeled automaton) in Figure 1, where the allowed initial states are distinguished as targets of a sourceless arrow. The CPU time for synthesizing the controller amounted to 21.14 seconds. In Figure 2, we show several realizations of closed-loop trajectory ξ_{x_0v} stemming from initial condition $x_0 = (21, 21)$ (left panel), as well as the corresponding evolution of switching signal v (right panel), where the finite system is initialized from state q_1. Furthermore, in Figure 2 (middle panels), we show the average value over 100 experiments of the distance in time of the solution process ξ_{x_0v} to the set D, namely $\|\xi_{x_0v}(t)\|_D$, where the point-to-set distance is defined as $\|x\|_D = \inf_{d \in D} \|x - d\|$. Notice that the average distance is significantly lower than the precision $\varepsilon = 1.1$, as expected since the conditions based on Lyapunov functions

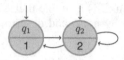

Fig. 1. Finite system describing the constraint over the switching policies. The lower part of the states are labeled with the outputs (2 and 1) denoting whether heater is on ($p = 2$) or off ($p = 1$).

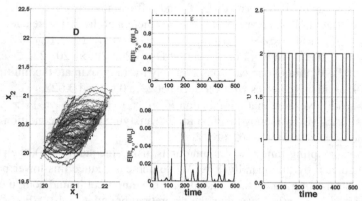

Fig. 2. Several realizations of the closed-loop trajectory $\xi_{x_0 \upsilon}$ with initial condition $x_0 = (21, 21)$ (left panel). Average values (over 100 experiments) of the distance of the solution process $\xi_{x_0 \upsilon}$ to the set D, in different vertical scales (middle panels). Evolution of the synthesized switching signal υ (right panel), where the finite system initialized from state q_1.

can lead to conservative bounds. (As discussed in Corollary 1, bounds can be improved by seeking optimized Lyapunov functions.)

References

1. Abate, A.: A contractivity approach for probabilistic bisimulations of diffusion processes. In: Proceedings of 48th IEEE Conference on Decision and Control, pp. 2230–2235 (December 2009)
2. Abate, A., D'Innocenzo, A., Di Benedetto, M.D.: Approximate abstractions of stochastic hybrid systems. IEEE Transactions on Automatic Control 56(11), 2688–2694 (2011)
3. Angeli, D.: A Lyapunov approach to incremental stability properties. IEEE Transactions on Automatic Control 47(3), 410–421 (2002)
4. Blom, H.A.P., Lygeros, J.: Stochastic Hybrid Systems: Theory and Safety Critical Applications. LNCIS, vol. 337. Springer, Heidelberg (2006)
5. Bujorianu, M.L., Lygeros, J., Bujorianu, M.C.: Bisimulation for General Stochastic Hybrid Systems. In: Morari, M., Thiele, L. (eds.) HSCC 2005. LNCS, vol. 3414, pp. 198–214. Springer, Heidelberg (2005)
6. Chatterjee, D., Liberzon, D.: Stability analysis of deterministic and stochastic switched systems via a comparison principle and multiple Lyapunov functions. SIAM Journal on Control and Optimization 45(1), 174–206 (2006)

7. Girard, A.: Low-complexity switching controllers for safety using symbolic models. In: Proceedings of 4th IFAC Conference on Analysis and Design of Hybrid Systems, pp. 82–87 (2012)
8. Girard, A., Pappas, G.J.: Approximation metrics for discrete and continuous systems. IEEE Transactions on Automatic Control 25(5), 782–798 (2007)
9. Girard, A., Pola, G., Tabuada, P.: Approximately bisimilar symbolic models for incrementally stable switched systems. IEEE Transactions on Automatic Control 55(1), 116–126 (2010)
10. Julius, A.A., Pappas, G.J.: Approximations of stochastic hybrid systems. IEEE Transaction on Automatic Control 54(6), 1193–1203 (2009)
11. Karatzas, I., Shreve, S.E.: Brownian Motion and Stochastic Calculus, 2nd edn. Graduate Texts in Mathematics, vol. 113. Springer, New York (1991)
12. Liberzon, D.: Switching in Systems and Control. Systems & Control: Foundations & Applications. Birkhäuser (2003)
13. Majumdar, R., Zamani, M.: Approximately bisimilar symbolic models for digital control systems. In: Madhusudan, P., Seshia, S.A. (eds.) CAV 2012. LNCS, vol. 7358, pp. 362–377. Springer, Heidelberg (2012)
14. Maler, O., Pnueli, A., Sifakis, J.: On the synthesis of discrete controllers for timed systems. In: Mayr, E.W., Puech, C. (eds.) STACS 1995. LNCS, vol. 900, pp. 229–242. Springer, Heidelberg (1995)
15. Mazo Jr., M., Davitian, A., Tabuada, P.: PESSOA: A tool for embedded controller synthesis. In: Touili, T., Cook, B., Jackson, P. (eds.) CAV 2010. LNCS, vol. 6174, pp. 566–569. Springer, Heidelberg (2010)
16. Milner, R.: Communication and Concurrency. Prentice-Hall, Inc. (1989)
17. Oksendal, B.K.: Stochastic differential equations: An introduction with applications, 5th edn. Springer (November 2002)
18. Pola, G., Tabuada, P.: Symbolic models for nonlinear control systems: Alternating approximate bisimulations. SIAM Journal on Control and Optimization 48(2), 719–733 (2009)
19. Prajna, S., Papachristodoulou, A., Seiler, P., Parrilo, P.A.: SOSTOOLS: Control applications and new developments. In: Proceedings of IEEE International Symposium on Computer Aided Control Systems Design, pp. 315–320 (2004)
20. Sproston, J.: Discrete-time verification and control for probabilistic rectangular hybrid automata. In: Proceedings of 8th International Conference on Quantitative Evaluation of Systems, pp. 79–88 (2011)
21. Tabuada, P.: Verification and Control of Hybrid Systems, A symbolic approach, 1st edn. Springer (June 2009)
22. Zamani, M., Mohajerin Esfahani, P., Majumdar, R., Abate, A., Lygeros, J.: Symbolic control of stochastic systems via approximately bisimilar finite abstractions. arXiv: 1302.3868 (2013)

Synthesis for Multi-objective Stochastic Games: An Application to Autonomous Urban Driving

Taolue Chen, Marta Kwiatkowska, Aistis Simaitis, and Clemens Wiltsche

Department of Computer Science, University of Oxford, United Kingdom

Abstract. We study strategy synthesis for stochastic two-player games with multiple objectives expressed as a conjunction of LTL and expected total reward goals. For stopping games, the strategies are constructed from the Pareto frontiers that we compute via value iteration. Since, in general, infinite memory is required for deterministic winning strategies in such games, our construction takes advantage of randomised memory updates in order to provide compact strategies. We implement our methods in PRISM-games, a model checker for stochastic multi-player games, and present a case study motivated by the DARPA Urban Challenge, illustrating how our methods can be used to synthesise strategies for high-level control of autonomous vehicles.

1 Introduction

The increasing reliance on sensor-enabled smart devices in a variety of applications, for example, autonomous parking and driving, medical devices, and communication technology, has called for software quality assurance technologies for the development of their embedded software controllers. To that end, techniques such as formal verification, validation and synthesis from specifications have been advocated, see e.g. [17,3]. One advantage of synthesis is that it guarantees correctness of the controllers by construction. This technique has been successfully demonstrated for LTL (linear temporal logic) specifications in robotic control and urban driving [19,21], where it can be used to synthesise controllers that maximise the probability of the system behaving according to its specification, which is expressed as an LTL formula. As a natural continuation of the aforementioned studies, one may wish to synthesise controllers that satisfy several objectives simultaneously. These objectives might be LTL formulae with certain probabilities, or a broader range of quantitative, reward-based specifications. Such *multi-objective* specifications enable the designers to choose the best controller for the given application by exploiting the Pareto trade-offs between objectives, such as increasing the probability of reaching a goal, while at the same time reducing the probability of entering an unsafe state or keeping the expected value of some reward function above a given value.

The majority of research in (discrete) controller synthesis from LTL specifications has centred on modelling in the setting of Markov chains or Markov decision processes (MDPs). In this paper, we focus on stochastic two-player games, which

K. Joshi et al. (Eds.): QEST 2013, LNCS 8054, pp. 322–337, 2013.

generalise MDPs, and which provide a natural view of the system (represented as Player 1) playing a game against an adversarial environment (Player 2). In the setting of multi-objective stochastic games, several challenges for controller synthesis need to be overcome that we address in this paper. Firstly, the games are *not* determined, and thus determinacy cannot be leveraged by the synthesis algorithms. Secondly, the winning strategies for such games require both memory and randomisation; previous work has shown that the number of different probability distributions required by the winning strategy may be exponential or even infinite, and so one needs to adopt a representation of strategies which allows us to encode such distributions using finitely many memory elements.

Modelling via multi-objective stochastic games is well suited to navigation problems, such as urban driving, where the environment makes choices to which the system has to react by selecting appropriate responses that, for example, avoid obstacles or minimise the likelihood of accidents. The choices of the environment can be nondeterministic, but can also be modelled probabilistically, e.g. where statistical observations about certain hazards are available. In addition to probabilities, one can also annotate the model with rewards, to evaluate various quantities by means of expectations. For instance, we can model a trade-off between the probability p of the car reaching a certain position without an accident, and the expected quality r of the roads it drives over to get there. While both $(p, r) = (0.9, 4)$ and $(p', r') = (0.8, 5)$ might be achievable, the combination $(p'', r'') = (0.9, 5)$ could be unattainable by any controller, since (p, r) and (p', r') are already Pareto optimal.

In this paper we extend the results of [9], where verification techniques for multi-objective stochastic games were proposed. We focus here on quantitative multi-objective conjunctions for stochastic two-player games, where each property in the conjunction can be either an LTL formula or a reward function. We formulate a value iteration method to compute an approximation of the Pareto frontiers for exploring the trade-offs between different controllers and show how to construct control strategies. We also develop a prototype implementation as an extension of PRISM-games [7] using the Parma Polyhedra Library [1], and apply it to a case study of urban driving inspired by the 2007 DARPA Urban Challenge [11]. We construct and evaluate strategies for autonomous driving based on OpenStreetMap [16] data for a number of English villages.

Contributions. The paper makes the following contributions:

- We show how to solve stopping games with conjunctions of LTL objectives by reduction to reachability reward objectives.
- We formulate and implement algorithms for computing Pareto frontiers for reachability games using value iteration.
- We construct strategies from the results of the value iteration algorithm and evaluate their respective trade-offs.
- We present a case study of urban driving demonstrating all of the above.

Related Work. Multi-objective optimisation has been applied in the context of formal verification, mainly for MDPs and (non-stochastic) games, where it is usually referred to as multidimensional optimisation. For MDPs, [5,2] introduced multiple discounted objectives and multiple long-run objectives, respectively, whereas [13] studied the problem of Boolean combinations of quantitative ω-regular objectives, which were extended in [15] with total rewards. In the setting of non-stochastic games, multidimensional energy games were introduced, including their complexity and the strategy synthesis problem [4,14,6,18]. In [8] stochastic games, where the objective of Player 1 is to achieve the expectation of an objective function "precisely p", were introduced. This problem is a special case of the problem that we study.

Our case study is inspired by the DARPA Urban Challenge [11], whose guidelines and technical evaluation criteria we use, and particularly the report by the winning team, which encourages the use of formal verification of urban driving systems [17]. Team Caltech suggested the use of temporal logic synthesis for high level planning [3]. This was later formulated for MDPs and LTL goals for autonomous vehicles in adversarial environments (pedestrians crossing a street) [19]. Receding horizon control [20] and incremental approaches [21] have been suggested to alleviate the computational complexity by leveraging structure specific to the autonomous control problems.

2   Preliminaries

Given a vector $\boldsymbol{x} \in \mathbb{R}^n_{\geq 0}$ and a scalar $z \in \mathbb{R}$, we write x_i for the i-th component of $\boldsymbol{x}$ where $1 \leq i \leq n$, and $\boldsymbol{x} + z$ for the vector $(x_1 + z, \ldots, x_n + z)$. Moreover, the *dot product* of vectors $\boldsymbol{x}$ and $\boldsymbol{y}$ is defined by $\boldsymbol{x} \cdot \boldsymbol{y} \stackrel{\text{def}}{=} \sum_{i=1}^{n} x_i \cdot y_i$ and the sum of two sets of vectors $X, Y \subseteq \mathbb{R}^n_{\geq 0}$ is defined by $X + Y \stackrel{\text{def}}{=} \{\boldsymbol{x} + \boldsymbol{y} \mid \boldsymbol{x} \in X, \boldsymbol{y} \in Y\}$. Given a set $X \subseteq \mathbb{R}^n_{\geq 0}$, we define the *downward closure* of X as $\mathsf{dwc}(X) \stackrel{\text{def}}{=} \{\boldsymbol{y} \mid \exists \boldsymbol{x} . \boldsymbol{x} \in X \text{ and } \boldsymbol{y} \leq \boldsymbol{x}\}$, and its convex hull as $\mathsf{conv}(X) \stackrel{\text{def}}{=} \{\boldsymbol{x} \mid \exists \boldsymbol{x}_1, \boldsymbol{x}_2 \in X, \alpha \in [0,1] . \boldsymbol{x} = \alpha \boldsymbol{x}_1 + (1 - \alpha)\boldsymbol{x}_2\}$.

A *discrete probability distribution* over a (countable) set S is a function $\mu : S \to [0,1]$ such that $\sum_{s \in S} \mu(s) = 1$. We write $\mathcal{D}(S)$ for the set of all discrete distributions over S. Let $\mathsf{supp}(\mu) \stackrel{\text{def}}{=} \{s \in S \mid \mu(s) > 0\}$ be the *support* of $\mu \in \mathcal{D}(S)$. A distribution $\mu \in \mathcal{D}(S)$ is a *Dirac distribution* if $\mu(s) = 1$ for some $s \in S$. Sometimes we identify a Dirac distribution μ with the unique element in $\mathsf{supp}(\mu)$. We represent a distribution $\mu \in \mathcal{D}(S)$ on a set $S = \{s_1, \ldots, s_n\}$ as a map $[s_1 \mapsto \mu(s_1), \ldots, s_n \mapsto \mu(s_n)] \in \mathcal{D}(S)$, and we usually omit the elements of S outside $\mathsf{supp}(\mu)$ to simplify the presentation.

Definition 1 (Stochastic two-player game). *A stochastic two-player game (called* game *henceforth) is a tuple* $\mathcal{G} = \langle S, (S_\square, S_\Diamond, S_\bigcirc), \Delta \rangle$, *where*

- S *is a countable set of states partitioned into sets* $S_\square$, $S_\Diamond$, *and* $S_\bigcirc$; *and*
- $\Delta : S \times S \to [0,1]$ *is a transition function such that* $\Delta(\langle s, t \rangle) \in \{0, 1\}$ *if* $s \in S_\square \cup S_\Diamond$, *and* $\sum_{t \in S} \Delta(\langle s, t \rangle) = 1$ *if* $s \in S_\bigcirc$.

$S_\square$ and $S_\Diamond$ represent the sets of states controlled by players Player 1 and Player 2, respectively, while $S_\bigcirc$ is the set of stochastic states. For a state $s \in S$, the set of successor states is denoted by $\Delta(s) \stackrel{\text{def}}{=} \{t \in S \mid \Delta(\langle s, t \rangle) > 0\}$. We assume that $\Delta(s) \neq \emptyset$ for all $s \in S$. Moreover, we denote a set of *terminal states* Term $\stackrel{\text{def}}{=} \{s \in S \mid \Delta(\langle s, t \rangle) = 1$ iff $s = t\}$.

An infinite *path* λ of a stochastic game $\mathcal{G}$ is an infinite sequence $s_0 s_1 \ldots$ of states such that $s_{i+1} \in \Delta(s_i)$ for all $i \geq 0$. A finite path is a finite such sequence. For a finite or infinite path λ we write $\text{len}(\lambda)$ for the number of states in the path. For $i < \text{len}(\lambda)$ we write λ_i to refer to the i-th state s_i of λ, and we denote the suffix of the path λ starting at position i by $\lambda^i \stackrel{\text{def}}{=} s_i s_{i+1} \ldots$. For a finite path $\lambda = s_0 s_1 \ldots s_n$ we write $\text{last}(\lambda)$ for the last state of the path, i.e., $\text{last}(\lambda) = s_n$. We write $\Omega_{\mathcal{G}, s}$ for the set of infinite paths starting in state s.

Strategy. In this paper we use an alternative formulation of strategies [2] that generalises the concept of strategy automata [12].

Definition 2. *A strategy of Player 1 in a game $\mathcal{G} = \langle S, (S_\square, S_\Diamond, S_\bigcirc), \Delta \rangle$ is a tuple $\pi = \langle \mathcal{M}, \pi_u, \pi_n, \alpha \rangle$, where:*

- *$\mathcal{M}$ is a countable set of memory elements,*
- *$\pi_u \colon \mathcal{M} \times S \to \mathcal{D}(\mathcal{M})$ is a memory update function,*
- *$\pi_n \colon S_\square \times \mathcal{M} \to \mathcal{D}(S)$ is a next move function s.t. $\pi_n(s, m)[s'] > 0$ only if $s' \in \Delta(s)$,*
- *$\alpha \colon S \to \mathcal{D}(\mathcal{M})$ defines for each state of $\mathcal{G}$ an initial memory distribution*

A strategy σ for Player 2 is defined in an analogous manner. We denote the set of all strategies for Player 1 and Player 2 by Π and Σ, respectively.

A strategy is *memoryless* if $|\mathcal{M}| = 1$. We say that a strategy requires finite memory if $|\mathcal{M}| < \infty$ and infinite memory if $|\mathcal{M}| = \infty$. We also classify the strategies based on the use of randomisation. A strategy $\pi = \langle \mathcal{M}, \pi_u, \pi_n, \alpha \rangle$ is *pure* if π_u, π_n, and α map to Dirac distributions; *deterministic update* if π_u and α map to Dirac distributions, while π_n maps to an arbitrary distributions; and *stochastic update* where π_u, π_n, and α can map to arbitrary distributions.

Markov Chain Induced by Strategy Pairs. Given a game $\mathcal{G}$ with initial state distribution $\varsigma \in \mathcal{D}(S)$, a Player 1 strategy $\pi = \langle \mathcal{M}_1, \pi_u, \pi_n, \alpha_1 \rangle$ and a Player 2 strategy $\sigma = \langle \mathcal{M}_2, \sigma_u, \sigma_n, \alpha_2 \rangle$ induce a countable Markov chain $\mathcal{G}(\varsigma, \pi, \sigma) = \langle S', (\emptyset, \emptyset, S'), \Delta' \rangle$ with initial state distribution $\varsigma(\pi, \sigma) \in \mathcal{D}(S')$, where

- $S' = S \times \mathcal{M}_1 \times \mathcal{M}_2$,
- $\Delta' \colon S' \times S' \to [0, 1]$ is such that for all $(s, m_1, m_2), (s', m_1', m_2') \in S'$ we have

$$(\langle (s, m_1, m_2), (s', m_1', m_2') \rangle) \mapsto$$

$$\begin{cases} \pi_n(s, m_1)[s'] \cdot \pi_u(m_1, s')[m_1'] \cdot \sigma_u(m_2, s')[m_2'] & \text{if } s \in S_\square \\ \sigma_n(s, m_2)[s'] \cdot \pi_u(m_1, s')[m_1'] \cdot \sigma_u(m_2, s')[m_2'] & \text{if } s \in S_\Diamond \\ \Delta(\langle s, s' \rangle) \cdot \pi_u(m_1, s')[m_1'] \cdot \sigma_u(m_2, s')[m_2'] & \text{if } s \in S_\bigcirc, \end{cases}$$

- $\varsigma(\pi, \sigma) \colon S' \to [0, 1]$ is defined such that for all $(s, m_1, m_2) \in S'$ we have that $\varsigma(\pi, \sigma)[s, m_1, m_2] = \varsigma[s] \cdot \alpha_1(s)[m_1] \cdot \alpha_2(s)[m_2]$.

Probability Measure. A stochastic game $\mathcal{G}$ together with a strategy pair $(\pi, \sigma) \in \Pi \times \Sigma$ and a starting state s induces a (possibly infinite) Markov chain on the game. We define the probability measure over the set of paths $\Omega_{\mathcal{G},s}$ and a strategy pair (π, σ) in the following way. The basic open sets of $\Omega_{\mathcal{G},s}$ are the *cylinder sets* $\mathrm{Cyl}(\lambda) \stackrel{\text{def}}{=} \lambda \cdot S^\omega$ for every finite path $\lambda = s_0 s_1 \ldots s_k$ of $\mathcal{G}$, and the probability assigned to $\mathrm{Cyl}(\lambda)$ equals $\prod_{i=0}^{k} \Delta(\langle s_i, s_{i+1} \rangle) \cdot p_i(s_0, \ldots, s_i)$, where $p_i(\lambda) = \pi(\lambda)$ if $\mathrm{last}(\lambda) \in S_\square$, $p_i = \sigma(\lambda)$ if $\mathrm{last}(\lambda) \in S_\Diamond$ and 1 otherwise. This definition induces a probability measure on the algebra of cylinder sets, which can be extended to a unique probability measure $\mathrm{Pr}_{\mathcal{G},s}^{\pi,\sigma}$ on the σ-algebra generated by these sets. The expected value of a measurable function $f : S^\omega \to \mathbb{R} \cup \{\infty\}$ under a strategy pair $(\pi, \sigma) \in \Pi \times \Sigma$ is defined as $\mathbb{E}_{\mathcal{G},s}^{\pi,\sigma}[f] \stackrel{\text{def}}{=} \int f \, d\mathrm{Pr}_{\mathcal{G},s}^{\pi,\sigma}$. We say that a game $\mathcal{G}$ is a *stopping game* if for every pair of strategies π and σ a terminal state is reached with probability 1.

Winning Objectives. In this paper we study objectives which are conjunctions of LTL and expected total reward goals. This section provides definitions of these concepts, as well as that of Pareto frontiers representing the possible trade-offs.

LTL. To specify the LTL goals, we use the following standard notation:

$$\varXi ::= T \mid \neg \varXi \mid \varXi_1 \wedge \varXi_2 \mid \mathsf{X} \varXi \mid \varXi_1 \mathsf{U} \varXi_2,$$

where $T \subseteq S$. Given a path λ and a LTL formula $\varXi$, we define $\lambda \models \varXi$ as:

$$
\begin{aligned}
\lambda &\models T & &\Leftrightarrow \lambda_0 \in T \\
\lambda &\models \neg \varXi & &\Leftrightarrow \lambda \not\models \varXi \\
\lambda &\models \varXi_1 \wedge \varXi_2 & &\Leftrightarrow \lambda \models \varXi_1 \text{ and } \lambda \models \varXi_2 \\
\lambda &\models \mathsf{X} \varXi & &\Leftrightarrow \lambda^1 \models \varXi \\
\lambda &\models \varXi_1 \mathsf{U} \varXi_2 & &\Leftrightarrow \lambda^i \models \varXi_2 \text{ for some } i \in \mathbb{N}_0 \\
& & &\quad\text{and } \lambda^j \models \varXi_1 \text{ for } 0 \le j < i.
\end{aligned}
$$

Operators $\mathsf{F} \varXi \stackrel{\text{def}}{=} S \mathsf{U} \varXi$ and $\mathsf{G} \varXi \stackrel{\text{def}}{=} \neg \mathsf{F} \neg \varXi$ have their usual meaning. We use a formula and the set of paths satisfying the formula interchangeably, e.g. the set of paths reaching a state in $T \subseteq S$ is denoted by $\mathsf{F} T = \{\omega \in \Omega_{\mathcal{G}} \mid \exists i . \omega_i \in T\}$.

Expected total reward. To specify the reward goals, we define a (k-dimensional) reward function $r : S \to \mathbb{R}_{\ge 0}^k$, which for each state s of the game $\mathcal{G}$ assigns a reward vector $r(s) \in \mathbb{R}_{\ge 0}^k$. We define a vector of *total reward* random variables $rew(r)$ as $rew(r)(\lambda) \stackrel{\text{def}}{=} \sum_{j \ge 0} r(\lambda_j)$ for any path λ. Under a fixed strategy pair (π, σ) for both players, the expected total reward is the expected value of the total reward random variable, i.e.,

$$\mathbb{E}_{\mathcal{G},s}^{\pi,\sigma}[rew(r)] = \int_{\Omega_{\mathcal{G},s}} rew(r) \, d\mathrm{Pr}_{\mathcal{G},s}^{\pi,\sigma}.$$

We require the expected total rewards to be bounded. Hence, due to the self-loops in terminal states, in particular we require $r(s) = 0$ for all $s \in \mathsf{Term}$.

Conjunctive queries. A *conjunctive query* (CQ) is a tuple of LTL formulae, reward functions, and their respective *lower* bounds,

$$\varphi = (\boldsymbol{\varXi}, \boldsymbol{r}, \boldsymbol{v}) = ((\varXi_1, \ldots, \varXi_m), (r_1, \ldots, r_{n-m}), (v_1, \ldots, v_m, v_{m+1}, \ldots, v_n)),$$

where m is the number of LTL objectives, $n-m$ is the number of expected total reward objectives, $v_i \in [0,1]$ for $1 \le i \le m$ and $v_i \in \mathbb{R}$ for $m+1 \le i \le n$, Ξ_i is an LTL formula, and r_i is a reward function. We call $\boldsymbol{v}$ the *target vector* of the CQ. Additionally, we define a *reward conjunctive query* (rCQ) to be a CQ with $m = 0$. Player 1 *achieves* a CQ φ at state s if there exists a strategy $\pi \in \Pi$ such that, for any strategy $\sigma \in \Sigma$ of Player 2, it holds that

$$\bigwedge_{i=1}^{m} (\mathrm{Pr}_{\mathcal{G},s}^{\pi,\sigma}(\Xi_i) \ge v_i) \wedge \bigwedge_{i=m+1}^{n} (\mathbb{E}_{\mathcal{G},s}^{\pi,\sigma}[rew(r)_{i-m}] \ge v_i).$$

Pareto Frontiers. A *Pareto frontier* for objectives $\boldsymbol{\Xi}$ and $\boldsymbol{r}$ is a set of points $P \subseteq \mathbb{R}^n$ such that for any $\boldsymbol{p} \in P$ the following hold:

(a) for all $\varepsilon > 0$ the query $(\boldsymbol{\Xi}, \boldsymbol{r}, \boldsymbol{p} - \varepsilon)$ is achievable, and
(b) for all $\varepsilon > 0$ the query $(\boldsymbol{\Xi}, \boldsymbol{r}, \boldsymbol{p} + \varepsilon)$ is not achievable.

Given $\varepsilon > 0$, an ε-*approximation* of the Pareto frontier P is a set of points Q satisfying that for any $\boldsymbol{q} \in Q$ there is a point $\boldsymbol{p} \in P$ such that $\|\boldsymbol{p} - \boldsymbol{q}\| \le \varepsilon$, and for every $\boldsymbol{p} \in P$ there is a vector $\boldsymbol{q} \in Q$ such that $\|\boldsymbol{p} - \boldsymbol{q}\| \le \varepsilon$, where $\| \cdot \|$ is the Manhattan distance of two points in $\mathbb{R}$.

3 Computing the Pareto Frontiers

We now turn our attention to the computation of Pareto frontiers for the states of the game, which will be required for the strategy construction. First, we recall from our work in [9] how to compute them for rCQs, and then extend this by showing how a stopping game $\mathcal{G}$ with a general CQ φ can be reduced to a stopping game $\mathcal{G}'$ with a rCQ φ', such that there exists a strategy for Player 1 in $\mathcal{G}$ to satisfy φ if and only if there is a strategy for Player 1 in $\mathcal{G}'$ to satisfy φ'.

Value iteration. To compute the successive approximations of the Pareto frontiers we iteratively apply the functional from the theorem below. For stopping games this is guaranteed to compute an ε-approximation of the Pareto frontiers.

Theorem 1 (Pareto frontier approximation). *For a stopping game $\mathcal{G}$ and a rCQ $\varphi = (\boldsymbol{r}, \boldsymbol{v})$, an ε-approximation of the Pareto frontiers for all states can be computed in $k = |S| + \lceil |S| \cdot \frac{\ln(\varepsilon \cdot (n \cdot M)^{-1})}{\ln(1-\delta)} \rceil$ iterations of the operator $F : (S \to \mathcal{P}(\mathbb{R}_{\ge 0}^n)) \to (S \to \mathcal{P}(\mathbb{R}_{\ge 0}^n))$ defined by*

$$F(X)(s) \stackrel{\text{def}}{=} \begin{cases} dwc(conv(\bigcup_{t \in \Delta(s)} X_t) + \boldsymbol{r}(s)) & \text{if } s \in S_{\sqcup} \\ dwc(\bigcap_{t \in \Delta(s)} X_t + \boldsymbol{r}(s)) & \text{if } s \in S_{\Diamond} \\ dwc(\sum_{t \in \Delta(s)} \Delta(\langle s, t \rangle) \times X_t + \boldsymbol{r}(s)) & \text{if } s \in S_{\bigcirc}, \end{cases}$$

where initially $X_s^0 \stackrel{\text{def}}{=} \{\boldsymbol{x} \in \mathbb{R}_{\ge 0}^n \mid \boldsymbol{x} \le \boldsymbol{r}(s)\}$ for all $s \in S$, $M = |S| \cdot \frac{\max_{s \in S, i} r_i(s)}{\delta}$ for $\delta = p_{\min}^{|S|}$, and $p_{\min}$ is the smallest positive probability in $\mathcal{G}$.

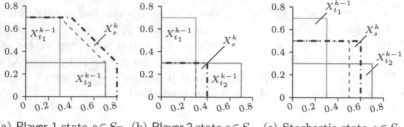

(a) Player 1 state $s \in S_\square$. (b) Player 2 state $s \in S_\diamond$. (c) Stochastic state $s \in S_\bigcirc$.
$$\Delta(\langle s, t_1 \rangle) = \Delta(\langle s, t_2 \rangle) = \tfrac{1}{2}.$$

Fig. 1. Illustration of value iteration operations for rCQs. We plot the polytopes of s and its successors t_1 and t_2 on the same graph. The reward at s is $r(s) = (0, 0.1)$ and $\Delta(s) = \{t_1, t_2\}$. The dashed (green) polytope is the result before adding the reward, which is shown in the dash-dotted (red) polytope.

The value iteration computes a sequence of polytopes X_s^k for each state s that converges to the Pareto frontiers. It starts with the bottom element $\bot : S \to \mathcal{P}(\mathbb{R}^n_{\geq 0})$ defined by $\bot(s) \stackrel{\text{def}}{=} X_s^0$. The operations in the operator F are illustrated in Figure 1. We assume that the expected reward for the game is bounded under any pair of strategies, which in particular implies that, for any terminal state $t \in \mathsf{Term}$, $r(t) = \mathbf{0}$.

Strategy Iteration. We use the sets X^k computed using the above iteration to obtain the memory elements of the stochastic update strategy (see Section 4). Thus, the iteration performed above can be regarded as a strategy iteration algorithm: it generates strategies that are optimal for achieving the goal in k steps (initially $k = 1$). After performing another iteration, we obtain the optimal strategy for $k + 1$ steps, etc. Since all rewards are non-negative, F is monotone, i.e. $X_s^k \subseteq X_s^{k+1}$ for all states s, and thus in every step the strategy either improves or it has converged to the optimal one. The strategy iteration can be performed either until the objective vector v is in X_s^k, or until it is within the required accuracy determined by the stopping criterion.

LTL to Expected Reward. We present a reduction from LTL objectives to expected total reward objectives for stopping games. The reduction is based on a similar reduction for Markov decision processes ([10,13]) which relies on constructing a deterministic Rabin automaton for each LTL formula and then building their product with the game $\mathcal{G}$. As the game $\mathcal{G}$ is stopping, almost all runs reach some terminal state with a positive probability. Hence, it is sufficient to analyse the outcomes of the formulae in the *terminal* states of the product game in order to establish whether the runs ending in them satisfy the formula.

Definition 3 (Rabin automaton). *A deterministic Rabin automaton is a tuple* $\langle Q, \Sigma, \tau, q^0, ((L^1, R^1), \ldots, (L^j, R^j)) \rangle$, *where Q are the states of the automaton with initial state $q^0 \in Q$, $\Sigma = S$ is the alphabet, $\tau : Q \times \Sigma \to Q$ is a transition function and $L^l, R^l \subseteq Q$ are Rabin pairs.*

Theorem 2 (LTL to expected reward). *Given a stopping game $\mathcal{G}$, a state s and a CQ $\varphi = ((\Xi_1, \ldots, \Xi_m), (r_1 \ldots r_{n-m}), v)$, there exists a game $\mathcal{G}'$, a state*

s', and a rCQ $\varphi' = ((r'_1 \ldots r'_n), v)$ such that there is a **Player 1** strategy to satisfy φ in s of $\mathcal{G}$ if and only if there is a **Player 1** strategy to satisfy φ' in s' of $\mathcal{G}'$.

For each LTL formula Ξ_i, we construct a *deterministic* Rabin automaton $\mathcal{A}_i = \langle Q_i, \Sigma_i, \tau_i, q_i^0, ((L_i^1, R_i^1), \ldots, (L_i^j, R_i^j)) \rangle$ with $\Sigma_i = S$, such that any path λ satisfies Ξ_i iff λ is accepted by $\mathcal{A}_i$. W.l.o.g. we assume each DRA is *complete*.

We index elements of tuples by subscripts, e.g., for a tuple $s' = (s, q^0)$, $s'_0 = s$ and $s'_1 = q^0$. Given a stopping game $\mathcal{G} = \langle S, (S_\square, S_\lozenge, S_\bigcirc), \Delta \rangle$ and DRAs $\mathcal{A}_i$ for $1 \leq i \leq m$, we construct a stopping game $\mathcal{G}' = \langle S', (S'_\square, S'_\lozenge, S'_\bigcirc), \Delta' \rangle$, where:

- $S' = S \times Q_1 \times \cdots \times Q_m \cup \{term\}$, $S'_\# = \{s' \in S' \mid s'_0 \in S_\#\}$ for $\# \in \{\square, \lozenge, \circ\}$;
- for states $s', t' \in S'$, using $T' \stackrel{\text{def}}{=} \{t' \in S' \mid t'_0 \in \mathsf{Term}\}$,

$$\Delta'(s', t') = \begin{cases} \Delta(s'_0, t'_0) & \text{if } s' \notin T' \text{ and } \forall . 1 \leq i \leq m, \tau_i(s'_i, t'_0) = t'_i \\ 1 & \text{if } s' \in T' \cup \{term\} \text{ and } t' = term \\ 0 & \text{otherwise;} \end{cases}$$

- the initial state is $s' \in S'$ s.t. $s'_0 = s$ and, for all $1 \leq i \leq m$, $\tau_i(q_i^0, s) = s'_i$.

The new multi-objective query is $\varphi' = ((r'_1, \ldots, r'_n), (v_1, \ldots, v_n))$, where each r'_i for $1 \leq i \leq m$ is defined as

$$r'_i(s') = \begin{cases} 1 & \text{if } s' \in T' \text{ and } {s'_0}^\omega \text{ is accepted by } \mathcal{A}_i \text{ with initial state } s'_i \\ 0 & \text{otherwise,} \end{cases}$$

and for $m + 1 \leq i \leq n$, $r'_i(s') = r_{i-m}(s'_0)$ for $s' \neq term$ and $r'_i(term) = 0$, i.e., the reward functions of φ for $m + 1 \leq i \leq n$ stay unchanged with respect to S. Correctness follows by the standard argument of one-to-one correspondence between strategies in $\mathcal{G}'$ and $\mathcal{G}$ for both players, see e.g. [15].

4 Strategy Synthesis

In this section we present a construction of the strategy for **Player 1**, which, given a stopping game $\mathcal{G}$, the sets X^k computed using the iteration from Section 3, and a state s, achieves the rCQ $\varphi = (r, v)$, where $v \in X_s^k$. The resulting strategy uses random updates on its memory elements to ensure that at every step the expected total reward is kept above the target vector v. The assumption that the game $\mathcal{G}$ is stopping implies that $\mathcal{G}$ terminates with probability 1, and thus the target vector is achieved. Note that the stochastic memory update is crucial to provide a compact representation of the strategy, as our previous work has demonstrated the need for exponential (Proposition 3 in [8]), or even infinite (Theorem 2 in [9]) memory if the strategy is only allowed to use randomisation in **Player 1** states.

We denote the set of vertices (corner points) of a polytope X as $Cnr(X)$. The strategy $\pi = \langle \mathcal{M}, \pi_u, \pi_n, \alpha \rangle$ is defined as follows.

- $\mathcal{M} = \bigcup_{t' \in S'} \{(t', \boldsymbol{p}) \mid \boldsymbol{p} \in Cnr(X_{t'}^k)\}$.
- $\pi_u((t', \boldsymbol{p}), u) = [(u', \boldsymbol{q}_0^{u'}) \mapsto \beta_0^u, \ldots, (u', \boldsymbol{q}_l^{u'}) \mapsto \beta_l^u]$, where $t', u' \in S'$ such that $t_0' = t, u_0' = u$ for $t, u \in S$, and such that $\Delta'(t', u') > 0$ (note that because DRAs are deterministic, such u' is unique), and where for all $0 \leq i \leq l$, $\boldsymbol{q}_i^{u'} \in Cnr(X_{u'}^k)$, $\beta_i^u \in [0, 1]$, and $\sum_i \beta_i^u = 1$, such that
  - for $t' \in S'_\square \cup S'_\lozenge$ we have $\sum_i \beta_i^u \cdot \boldsymbol{q}_i^{u'} \geq \boldsymbol{p} - \boldsymbol{r}'(t)$,
  - for $t' \in S'_\bigcirc$, $\boldsymbol{q}_i^{u'}$ and β_i^u have to be chosen together with the respective values $\boldsymbol{q}_i^{v'}$, and β_i^v assigned by $\pi_u((t', \boldsymbol{p}), v)$ for the remaining successors $v \in S \setminus \{u\}$ of t, so that they satisfy

$$\Delta(t, u) \cdot \sum_i \beta_i^u \cdot \boldsymbol{q}_i^{u'} + \sum_{v \in S \setminus \{u\}} \Delta(t, v) \cdot \sum_i \beta_i^v \cdot \boldsymbol{q}_i^{v'} \geq \boldsymbol{p} - \boldsymbol{r}'(t'),$$

    which ensures that the expected total reward is kept larger than the current memory element.
- $\pi_n(t, (t', \boldsymbol{p})) = [u \mapsto 1]$ for some $u \in S$ such that $\Delta'(t', u') > 0$ (where $u' \in S'$, $u_0' = u$), and for all $0 \leq i \leq l$ there exist $\boldsymbol{q}_i^{u'} \in Cnr(X_{u'}^k)$, $\beta_i^u \in [0, 1]$, such that $\sum_i \beta_i^u = 1$ and $\sum_i \beta_i^u \cdot \boldsymbol{q}_i^{u'} \geq \boldsymbol{p} - \boldsymbol{r}'(t')$.
- $\alpha(s) = [(s', \boldsymbol{q}_0^{s'}) \mapsto \beta_0^s, \ldots, (s', \boldsymbol{q}_l^{s'}) \mapsto \beta_l^s]$, where s' is the respective initial state of $\mathcal{G}'$, and $\boldsymbol{q}_i^{s'} \in Cnr(X_{s'}^k)$, $\beta_i^s \in [0, 1]$ (for all $0 \leq i \leq l$), and $\sum_i \beta_i^s = 1$ such that $\sum_i \beta_i^s \cdot \boldsymbol{q}_i^{s'} \geq \boldsymbol{v}'$.

Note that, for all $u' \in S'$, it is always possible to choose $l \leq n$, i.e. the number of points $\boldsymbol{q}_i^{u'}$ and respective coefficients β_i^u may be less than the number of objectives. Also, the points $\boldsymbol{q}_i^{t'}$ can indeed be picked from $X_{t'}^k$ because they exist both in $X_{t'}^{k-1}$, and $X_{t'}^k \supseteq X_{t'}^{k-1}$ due to the monotonicity of F.

Theorem 3. *The strategy constructed above achieves the expectation for the total reward functions which is greater than or equal to $\boldsymbol{v}$ for the rCQ φ.*

The proof follows the structure from [8], establishing, by induction on the length of finite prefixes, that the expectation of the vector held in memory is always between the target vector $\boldsymbol{v}$ and the expected total reward.

5   Case Study

We implement a prototype multi-objective strategy synthesis engine in PRISM-games, a model checking and synthesis tool for stochastic games [7], and present a case study applying our value iteration and strategy synthesis methods to perform control tasks. Our case study is motivated by the DARPA Urban Challenge 2007 (henceforth referred to as the Challenge), a competition for autonomous cars to navigate safely and effectively in an urban setting [11]. In the Challenge, cars were exposed to a set of traffic situations, and had to plan routes, avoid hazards, and cope with the presence of other vehicles.

Table 1. Model parameters for our prototype

Hazard	Abbreviation	λ	Reaction	Accident Probability
Pedestrian	p	0.05	Brake	0.01
			Honk	0.04
			Change Lane	0.03
Jam	j	0.1	Honk	0.01
			U-Turn	0.02
Obstacle	o	0.02	Change Lane	0.02
			U-Turn	0.02

5.1   Problem Setting

We identify desired functions of vehicle controllers from the Challenge, and model a scenario of a car driving through a map imported from OpenStreetMap [16].

We model the problem as a two-player stochastic game, where Player 1 represents the car navigating through the map and Player 2 represents the environment. The game setting allows us to model the nondeterministic, adversarial nature of hazards selected by the environment, and the car's reaction to each hazard is represented by the Player 1 strategy. Probabilities are used to model the relative likelihood of events in a given road segment and between different road segments, and are understood to be the parameters obtained from statistical observations; for example, certain road types are more prone to accidents. Finally, considering multiple objectives enables the exploration of trade-offs when constructing controllers.

Each road segment is modelled as part of a stochastic game between the environment (Player 2) that selects from a set of available hazards, and the car (Player 1) that selects reactions to the hazards, as well as making route selection (steering) decisions. Hazards occur with certain probabilities depending on the properties of the road segment the car is on, and each reaction the car takes is only successful with a given probability. We also model other parameters of the road, e.g. we use rewards for road quality. In our autonomous driving scenario, we consider three types of hazards with the corresponding reactions the car can take (see Table 1). These hazards, as well as the reactions, are chosen according to the Challenge event guidelines and technical evaluation criteria [11].

5.2   Model

We model the road network as a directed graph $G = (V, E)$, where each edge $e \in E$ represents a road segment, see Figure 2(a). To each edge $e \subset E$, we associate a subgame $\mathcal{G}_e$, parameterised by the properties of the corresponding road segment (e.g., length, quality, number of lanes).

An example illustrating a subgame $\mathcal{G}_e$ is shown in Figure 3. The states have labels of the form $\langle s, e \rangle$; when the context is clear, we simply use s. In state s_0, a set of at most two hazards is selected probabilistically, from which Player 2 can then choose. To each hazard $h \in \{p, j, o\}$ we associate a tuning parameter λ, and

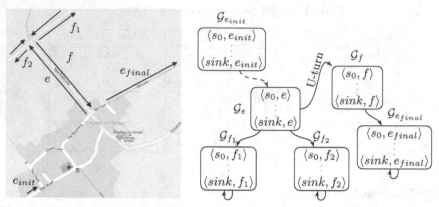

(a) Map of Charlton-on-Otmoor, (b) Connection of subgames. All roads that are
UK, with overlaid graph (only par- not one ways have U-turn connections, but only
tially shown).   one is shown for the sake of clarity. The dashed
                                    arrow abstracts several intermediate subgames.

Fig. 2. Illustrating the graph $G = (V, E)$ and the corresponding subgame connections

let the probability of a set of hazards $\{h_1, h_2\}$ be $p_{h_1 h_2} \overset{\text{def}}{=} \tanh(\lambda_1 \lambda_2 \mathsf{len}(e))/k$,
where $\mathsf{len}(e)$ is the length of the road corresponding to e in meters, and $k = 6$
is the number of sets of hazards.[1] For a single hazard h, p_h is defined similarly,
and the empty set of hazards is chosen with the residual probability p_{none}. The
parameters for our prototype model are given in Table 1.

Once a set of possible hazards is chosen in s_0, and Player 2 has selected a
specific one in s_1, s_2 and s_3, Player 1 must select an appropriate reaction in s_4,
s_5 and s_6. Then the game enters either the terminal accident state "acc", or a
Player 1 state "sink", where the next edge can be chosen. If the reaction is not
appropriate in the given road segment (e.g., changing lane in a single lane road),
a "violation" terminal is entered with probability 1 (not shown in Figure 3).

From the subgames $\mathcal{G}_e$ and the graph G a game $\mathcal{G}$ is constructed that connects
the games $\mathcal{G}_e$ as shown in Figure 2(b). In a local sink, e.g. $\langle sink, e \rangle$, Player 1 makes
the decision as to which edge to go to next and enters the corresponding local
initial state, e.g. $\langle s_0, f_1 \rangle$. Also, in a U-turn, the subgame $\mathcal{G}_f$ of the reverse edge f
of e is entered at its initial state, e.g. $\langle s_0, f \rangle$. If a road segment does not have any
successors, or is the goal, the local sink for the corresponding game is made into
a terminal state. Note that the above construction results in a stopping game.

5.3 Objectives

We study three objectives for the autonomous driving scenario. Formally, we
consider the CQ $\varphi = ((\mathsf{F}\, T_1, \mathsf{G}\, \neg T_2), (r), (v_1, v_2, v_3))$, where $T_1 = \{\langle s_0, e_{final} \rangle\}$,
$T_2 = \{\langle acc, e \rangle \mid e \in E\}$, and r is a reward function explained below.

---

[1] The use of the sigmoid tanh achieves a valid probability distribution independently of
the road length, while the weights λ tune the relative frequency of hazard occurrence.

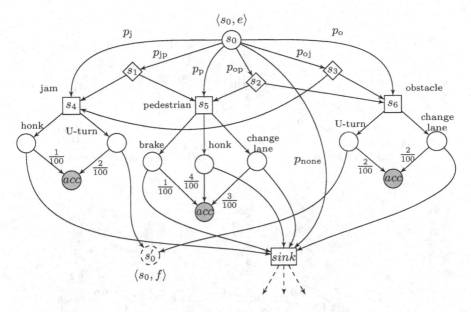

Fig. 3. Subgame $\mathcal{G}_e$ with reverse edge f of e. The reward in *sink* is rval(e)len(e). Hazards and reactions are indicated as annotations.

Target location reachability. From the initial location, reach a target at a particular orientation with probability v_1, i.e. achieve $\Pr^{\pi,\sigma}_{\mathcal{G},\langle s_0,e_{init}\rangle}(\mathsf{F}\,T_1) \geq v_1$. Note that the orientation of the car is implicit, as two-way streets are modelled as separate edges. On a high level, reachability at a correct orientation is a primary goal also in the Challenge.

Accident avoidance. Achieve a probability v_2 to never make an accident, that is, achieve $\Pr^{\pi,\sigma}_{\mathcal{G},\langle s_0,e_{init}\rangle}(\mathsf{G}\,\neg T_2) \geq v_2$. Note that a traffic rule violation, represented by the "violation" state is not considered an accident. This safety goal represents the other primary goal of the Challenge.

Road quality. Achieve a certain road quality v_3 over the duration of driving, i.e. achieve $\mathbb{E}^{\pi,\sigma}_{\mathcal{G},\langle s_0,e_{init}\rangle}[rew((r))] \geq v_3$. The road quality is determined according to the road type and length extracted from the map data. Hence, each edge e is assigned a value rval(e), and the reward function r is defined by $r(\langle e, sink\rangle) \overset{\text{def}}{=}$ rval(e) $\cdot$ len(e). In the Challenge, cars must be able to navigate over different road types, and select adequate roads.

5.4   Implementation

We now give the details of our prototype implementation, focusing on how to make the computation of strategies achieving CQs more efficient. Since the sets $Cnr(X_s^k)$ can be represented as convex polytopes, we use the Parma Polyhedra Library [1] to perform the value iteration operations.

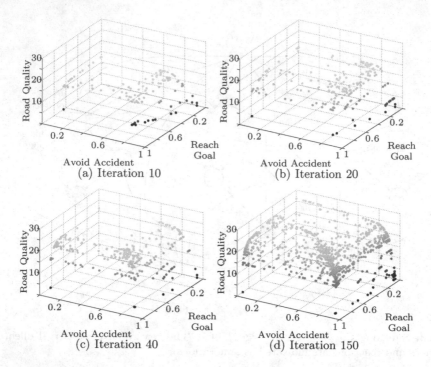

Fig. 4. Successive (under-)approximations of the Pareto frontier in $\langle s_0, e_{init} \rangle$ of $\mathcal{G}$ for Charlton-on-Otmoor, UK

Gauss-Seidel update. Optionally, in-place updates can be used when computing X^{k+1} from X^k. That is, when X_s^{k+1} is computed from X^k, the result is stored in the same memory location as X_s^k. Subsequently, if X_t^{k+1} is computed for $t \neq t$, and $s \in \Delta(t)$, then X_s^{k+1} is used instead of X_s^k. Correctness of this method, also called Gauss-Seidel update, follows from the monotonicity of the functional F from Theorem 1.

Dynamic Accuracy Adaptation. During value iteration, $|Cnr(X_s^k)|$ may increase exponentially with k. To mitigate this, for each step k, we fix a baseline accuracy $a_k \gg 1$, and round down each coordinate i of each corner of X_s^k to a multiple of $\frac{M_i}{a_k}$, where M_i is of the same order of magnitude as the maximum reward in dimension i. The resulting polytopes are denoted by $\tilde{X}_s^k$. Note that we must round *down* in order to maintain safe under-approximations of the Pareto frontiers.[2]

Starting from a_0, we dynamically increase the accuracy a_k by some factor $\lambda > 1$ after N_k steps, while at the same time increasing the number N_k of steps until the next increase by the same factor λ. In the long run, this yields an additive increase in the accuracy of $\frac{a_0(\lambda-1)}{N_0}$ per step. With this approach, we obtain under-approximations of the Pareto frontiers with a small number of points that are gradually refined by allowing more points in the polytopes.

---

[2] C.f. the induction hypothesis in the proof of Theorem 1, see [9].

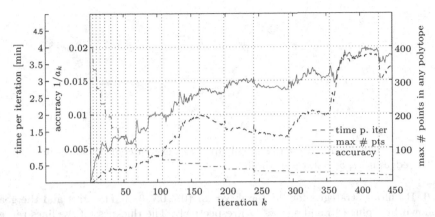

Fig. 5. Performance indicators for Charlton-on-Otmoor, cf. Figure 4. Note the changes in the number of points and time to complete an iteration as the accuracy changes.

Note that, while $X_s^k \subseteq F(X^k)(s)$, it is no longer true that $X_s^k \subseteq \tilde{X}'$, where $X' = F(X^k)(s)$. Therefore we use $\tilde{X}_s^{k+1} \cup \tilde{X}_s^k$ to preserve monotonicity.

Stopping criterion. Given a rCQ $\varphi = (r, v)$, a strategy which achieves v is sufficient, even if the Pareto frontier contains a point $w > v$. It is therefore possible to terminate value iteration prematurely after iteration k, and yet apply the strategy construction in Section 4 to achieve any points in the polytopes X_s^k.

5.5 Results

In this section we present the experimental results of the case study, which are computed using the CQ φ from Section 5.3.

Value Iteration. We visualise the results of the value iteration, providing the intuition behind the trade-offs involved. In Figure 4 we show the polytopes computed for the initial state of the game for Charlton-on-Otmoor for several values of k. Rounding decreases the number of corner points, and Gauss-Seidel updates increase the convergence speed, albeit not the time per iteration. In Figure 5 we show performance indicators of the value iteration of Figure 4.

Strategy Evaluation. For $v = (0.7, 0.7, 6.0)$, we evaluate the constructed strategy π for an adversary σ that picks hazards uniformly at random, build the induced Markov chain $\mathcal{G}(\pi, \sigma, [\langle s_0, e_{init}\rangle \mapsto 1])$, and illustrate the resulting strategies for two villages in the UK in Figure 6. In Figure 6(b), one can observe that roads are picked that do not lead towards the goal. This is due to the strategy achieving a point on (or below) the Pareto frontier representing a trade-off between the three objectives, as opposed to maintaining a hard constraint of having to reach the

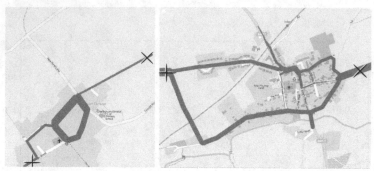

(a) Charlton-on-Otmoor: 43 (b) Islip: 125 edges in G, 1527 states in edges in G, 501 states in $\mathcal{G}$. $\mathcal{G}$.

Fig. 6. Resulting strategies for the target vector $(0.7, 0.7, 6.0)$. The start and the goal are shown by a plus $(+)$ and a cross $(\times)$ respectively. The thickness of the lines represents the expected proportion of trip time spent on the respective road by the car.

goal. Moreover, since maximising road quality is traded off against other goals, it may be suboptimal to take roads several times to improve the expectation while possibly incurring accidents and violations.

6   Conclusion

In this paper we have provided the first application of multi-objective stochastic two-player games. We have proposed algorithms for strategy synthesis and extended the approach to support important classes of LTL objectives. To evaluate the applicability of our techniques, we have developed a prototype implementation of the algorithms in the PRISM-games model checker and conducted a case study, synthesising and evaluating strategies for autonomous urban driving using real map data. There are many directions for future work, including extending the approach to support minimisation of the reward functions, application to assume-guarantee synthesis, and handling more complex multi-objective queries (e.g., combinations of conjunctions and disjunctions of objectives).

Acknowledgments. The authors thank Vojtěch Forejt, Mateusz Ujma and Klaus Dräger for helpful discussions and comments. The authors are partially supported by ERC Advanced Grant VERIWARE, the Institute for the Future of Computing at the Oxford Martin School, EPSRC grant EP/F001096, and the German Academic Exchange Service (DAAD).

References

1. Bagnara, R., Hill, P.M., Zaffanella, E.: The Parma Polyhedra Library: Toward a complete set of numerical abstractions for the analysis and verification of hardware and software systems. Sci. Comput. Program. 72(1-2), 3–21 (2008)

2. Brázdil, T., Brozek, V., Chatterjee, K., Forejt, V., Kučera, A.: Two views on multiple mean-payoff objectives in Markov decision processes. In: LICS, pp. 33–42 (2011)
3. Campbell, M., Egerstedt, M., How, J.P., Murray, R.M.: Autonomous driving in urban environments: approaches, lessons and challenges. Phil. Trans. R. Soc. A 368(1928), 4649–4672 (2010)
4. Chatterjee, K., Doyen, L., Henzinger, T.A., Raskin, J.F.: Generalized mean-payoff and energy games. In: FSTTCS. LIPIcs, vol. 8, pp. 505–516 (2010)
5. Chatterjee, K., Majumdar, R., Henzinger, T.A.: Markov decision processes with multiple objectives. In: Durand, B., Thomas, W. (eds.) STACS 2006. LNCS, vol. 3884, pp. 325–336. Springer, Heidelberg (2006)
6. Chatterjee, K., Randour, M., Raskin, J.-F.: Strategy synthesis for multi-dimensional quantitative objectives. In: Koutny, M., Ulidowski, I. (eds.) CONCUR 2012. LNCS, vol. 7454, pp. 115–131. Springer, Heidelberg (2012)
7. Chen, T., Forejt, V., Kwiatkowska, M., Parker, D., Simaitis, A.: PRISM-games: A model checker for stochastic multi-player games. In: Piterman, N., Smolka, S.A. (eds.) TACAS 2013. LNCS, vol. 7795, pp. 185–191. Springer, Heidelberg (2013)
8. Chen, T., Forejt, V., Kwiatkowska, M., Simaitis, A., Trivedi, A., Ummels, M.: Playing stochastic games precisely. In: Koutny, M., Ulidowski, I. (eds.) CONCUR 2012. LNCS, vol. 7454, pp. 348–363. Springer, Heidelberg (2012)
9. Chen, T., Forejt, V., Kwiatkowska, M., Simaitis, A., Wiltsche, C.: On stochastic games with multiple objectives. In: MFCS (accepted, 2013)
10. Courcoubetis, C., Yannakakis, M.: Markov decision processes and regular events. IEEE Trans. Autom. Control 43(10), 1399–1418 (1998)
11. DARPA. Urban Challenge (2007) (online accessed March 8, 2013)
12. Dziembowski, S., Jurdzinski, M., Walukiewicz, I.: How much memory is needed to win infinite games? In: LICS, pp. 99–110 (1997)
13. Etessami, K., Kwiatkowska, M.Z., Vardi, M.Y., Yannakakis, M.: Multi-objective model checking of Markov decision processes. LMCS 4(4) (2008)
14. Fahrenberg, U., Juhl, L., Larsen, K.G., Srba, J.: Energy games in multiweighted automata. In: Cerone, A., Pihlajasaari, P. (eds.) ICTAC 2011. LNCS, vol. 6916, pp. 95–115. Springer, Heidelberg (2011)
15. Forejt, V., Kwiatkowska, M., Norman, G., Parker, D., Qu, H.: Quantitative multi-objective verification for probabilistic systems. In: Abdulla, P.A., Leino, K.R.M. (eds.) TACAS 2011. LNCS, vol. 6605, pp. 112–127. Springer, Heidelberg (2011)
16. OpenStreetMap (2013) (online; accessed March 8, 2013)
17. Urmson, C., Anhalt, J., Bagnell, D., Baker, C., Bittner, R., Clark, M.N., Dolan, J., Duggins, D., Galatali, T., Geyer, C., et al.: Autonomous driving in urban environments: Boss and the urban challenge. J. Field Robot. 25(8), 425–466 (2008)
18. Velner, Y., Chatterjee, K., Doyen, L., Henzinger, T.A., Rabinovich, A., Raskin, J.F.: The complexity of multi-mean-payoff and multi-energy games. CoRR, abs/1209.3234 (2012)
19. Wongpiromsarn, T., Frazzoli, E.: Control of probabilistic systems under dynamic, partially known environments with temporal logic specifications. In: CDC, pp. 7644–7651 (2012)
20. Wongpiromsarn, T., Topcu, U., Murray, R.M.: Receding horizon temporal logic planning. IEEE Trans. Automat. Contr. 57(11), 2817–2830 (2012)
21. Wongpiromsarn, T., Ulusoy, A., Belta, C., Frazzoli, E., Rus, D.: Incremental synthesis of control policies for heterogeneous multi-agent systems with linear temporal logic specification. In: ICRA (accepted, 2013)

Stochastic Parity Games on Lossy Channel Systems

Parosh Aziz Abdulla[1], Lorenzo Clemente[2], Richard Mayr[3], and Sven Sandberg[1]

[1] Uppsala University
[2] LaBRI, University of Bordeaux I
[3] University of Edinburgh

Abstract. We give an algorithm for solving stochastic parity games with almost-sure winning conditions on *lossy channel systems*, for the case where the players are restricted to finite-memory strategies. First, we describe a general framework, where we consider the class of $2\frac{1}{2}$-player games with almost-sure parity winning conditions on possibly infinite game graphs, assuming that the game contains a *finite attractor*. An attractor is a set of states (not necessarily absorbing) that is almost surely re-visited regardless of the players' decisions. We present a scheme that characterizes the set of winning states for each player. Then, we instantiate this scheme to obtain an algorithm for *stochastic game lossy channel systems*.

1 Introduction

Background. 2-player games can be used to model the interaction of a controller (player 0) who makes choices in a reactive system, and a malicious adversary (player 1) who represents an attacker. To model randomness in the system (e.g., unreliability; randomized algorithms), a third player 'random' is defined who makes choices according to a predefined probability distribution. The resulting stochastic game is called a $2\frac{1}{2}$-player game in the terminology of [15]. The choices of the players induce a run of the system, and the winning conditions of the game are expressed in terms of predicates on runs.

Most classic work on algorithms for stochastic games has focused on finite-state systems (e.g., [24,17,19,15]), but more recently several classes of infinite-state systems have been considered as well. Stochastic games on infinite-state probabilistic recursive systems (i.e., probabilistic pushdown automata with unbounded stacks) were studied in [21,22,20]. A different (and incomparable) class of infinite-state systems are channel systems, which use unbounded communication buffers instead of unbounded recursion.

Channel Systems consist of finite-state machines that communicate by asynchronous message passing via unbounded FIFO communication channels. They are also known as communicating finite-state machines (CFSM) [13].

A *Lossy Channel System (LCS)* [7] consists of finite-state machines that communicate by asynchronous message passing via unbounded unreliable (i.e., lossy) FIFO communication channels, i.e., messages can spontaneously disappear from channels.

A *Probabilistic Lossy Channel System (PLCS)* [10,8] is a probabilistic variant of LCS where, in each computation step, messages are lost from the channels with a given probability. In [5], a *game extension* of PLCS was introduced where the players control transitions in the control graph and message losses are probabilistic.

The original motivation for LCS and PLCS was to capture the behavior of communication protocols; such protocols are designed to operate correctly even if the communication medium is unreliable (i.e., if messages can be lost). However, Channel Systems

K. Joshi et al. (Eds.): QEST 2013, LNCS 8054, pp. 338–354, 2013.

(aka CFSM) are a very expressive model that can encode the behavior of Turing machines, by storing the content of a Turing tape in a channel [13]. The only reason why certain questions are decidable for LCS/PLCS is that the message loss induces a quasi-order on the configurations, which has the properties of a simulation. Similarly to Turing machines and CFSM, one can encode many classes of infinite-state probabilistic transition systems into a PLCS. The only requirement is that the system re-visits a certain finite core region (we call this an attractor; see below) with probability one, e.g.,

- Queuing systems where waiting customers in a queue drop out with a certain probability in every time interval. This is similar to the well-studied class of queuing systems with impatient customers which practice *reneging*, i.e., drop out of a queue after a given maximal waiting time; see [26] section II.B. Like in some works cited in [26], the maximal waiting time in our model is exponentially distributed. In basic PLCS, unlike in [26], this exponential distribution does not depend on the current number of waiting customers. However, an extension of PLCS with this feature would still be analyzable in our framework (except in the pathological case where a high number of waiting customers increases the customers patience exponentially, because such a system would not necessarily have a finite attractor).
- Probabilistic resource trading games with probabilistically fluctuating prices. The given stores of resources are encoded by counters (i.e., channels), which exhibit a probabilistic decline (due to storage costs, decay, corrosion, obsolescence).
- Systems modeling operation cost/reward, which is stored in counters/channels, but probabilistically discounted/decaying over time.
- Systems which are periodically restarted (though not necessarily by a deterministic schedule), due to, e.g., energy depletion or maintenance work.

Due to this wide applicability of PLCS, we focus on this model in this paper. However, our main results are formulated in more general terms referring to infinite Markov chains with a finite attractor; see below.

Previous work. Several algorithms for symbolic model checking of PLCS have been presented [1,23]. Markov decision processes (i.e., $1\frac{1}{2}$-player games) on infinite graphs induced by PLCS were studied in [9], which shows that $1\frac{1}{2}$-player games with almost-sure Büchi objectives are pure memoryless determined and decidable. This result was later generalized to $2\frac{1}{2}$-player games [5], and further extended to generalized Büchi objectives [11]. On the other hand, $1\frac{1}{2}$-player games on PLCS with positive probability Büchi objectives (i.e., almost-sure co-Büchi objectives from the (here passive) opponent's point of view) can require infinite memory to win and are also undecidable [9]. (Undecidability and infinite memory requirement are separate results, since decidability does not imply the existence of finite-memory strategies in infinite-state games). If players are restricted to finite-memory strategies, the $1\frac{1}{2}$ player game with positive probability parity objectives (even the more general *Streett objectives*) becomes decidable [9]. Note that the finite-memory case and the infinite-memory one are a priori incomparable problems, and neither subsumes the other. Cf. Section 7.

Non-stochastic (2-player) parity games on infinite graphs were studied in [27], where it is shown that such games are determined, and that both players possess winning memoryless strategies in their respective winning sets. Furthermore, a scheme for computing the winning sets and winning strategies is given. Stochastic games ($2\frac{1}{2}$-player games) with parity conditions on *finite* graphs are known to be memoryless determined and effectively solvable [18,15,14].

Our contribution. We give an algorithm to decide almost-sure parity games for probabilistic lossy channel systems in the case where the players are restricted to finite memory strategies. We do that in two steps. First, we give our result in general terms (Section 4): We consider the class of $2\frac{1}{2}$-player games with almost-sure *parity* winning conditions on possibly infinite game graphs, under the assumption that the game contains a *finite attractor*. An attractor is a set A of states such that, regardless of the strategies used by the players, the probability measure of the runs which visit A infinitely often is one.[1] Note that this means neither that A is absorbing, nor that every run must visit A. We present a general scheme characterizing the set of winning states for each player. The scheme is a non-trivial generalization of the well-known scheme for non-stochastic games in [27] (see the remark in Section 4). In fact, the constructions are equivalent in the case that no probabilistic states are present. We show correctness of the scheme for games where each player is restricted to a finite-memory strategy. The correctness proof here is more involved than in the non-stochastic case of [27]; we rely on the existence of a finite attractor and the restriction of the players to use finite-memory strategies. Furthermore, we show that if a player is winning against all finite-memory strategies of the other player then he can win using a *memoryless* strategy. In the second step (Section 6), we show that the scheme can be instantiated for lossy channel systems. The instantiation requires the use of a much more involved framework than the classical one for well quasi-ordered transition systems [3] (see the remark in Section 6). The above two steps yield an algorithm to decide parity games in the case when the players are restricted to finite memory strategies. If the players are allowed infinite memory, then the problem is undecidable already for $1\frac{1}{2}$-player games with co-Büchi objectives (a special case of 2-color parity objectives) [9]. Note that even if the players are restricted to finite memory strategies, such a strategy (even a memoryless one) on an infinite game graph is still an infinite object. Thus, unlike for finite game graphs, one cannot solve a game by just guessing strategies and then checking if they are winning. Instead, we show how to effectively compute a finite, symbolic representation of the (possibly infinite) set of winning states for each player as a regular language.

Full proofs are available in the technical report [4].

2   Preliminaries

Notation. Let $\mathbb{O}$ and $\mathbb{N}$ denote the set of ordinal resp. natural numbers. We use $f : X \to Y$ to denote that f is a total function from X to Y, and use $f : X \rightharpoonup Y$ to denote that f is a partial function from X to Y. We write $f(x) = \bot$ to denote that f is undefined on x, and define $dom(f) := \{x \mid f(x) \neq \bot\}$. We say that f is an *extension* of g if $g(x) = f(x)$ whenever $g(x) \neq \bot$. For $X' \subseteq X$, we use $f|X'$ to denote the restriction of f to X'. We

---

[1] In the game community (e.g., [27]) the word *attractor* is used to denote what we call a *force set* in Section 3. In the infinite-state systems community (e.g., [1,6]), the word is used in the same way as we use it in this paper.

will sometimes need to pick an arbitrary element from a set. To simplify the exposition, we let $select(X)$ denote an arbitrary but fixed element of the nonempty set X.

A *probability distribution* on a countable set X is a function $f : X \to [0,1]$ such that $\sum_{x \in X} f(x) = 1$. For a set X, we use X^* and X^ω to denote the sets of finite and infinite words over X, respectively. The empty word is denoted by ε.

Games. A *game* (of *rank n*) is a tuple $G = (S, S^0, S^1, S^R, \longrightarrow, P, \mathrm{Col})$ defined as follows. S is a set of *states*, partitioned into the pairwise disjoint sets of *random states* S^R, states S^0 of Player 0, and states S^1 of Player 1. $\longrightarrow \subseteq S \times S$ is the *transition relation*. We write $s \longrightarrow s'$ to denote that $(s, s') \in \longrightarrow$. We assume that for each s there is at least one and at most countably many s' with $s \longrightarrow s'$. The *probability function* $P : S^R \times S \to [0,1]$ satisfies both $\forall s \in S^R. \forall s' \in S.(P(s, s') > 0 \iff s \longrightarrow s')$ and $\forall s \in S^R. \sum_{s' \in S} P(s, s') = 1$. (The sum is well-defined since we assumed that the number of successors of any state is at most countable.) $\mathrm{Col} : S \to \{0, \ldots, n\}$, where $\mathrm{Col}(s)$ is called the *color* of state s. Let $Q \subseteq S$ be a set of states. We use $\overset{G}{\_} Q := S - Q$ to denote the *complement* of Q. Define $[Q]^0 := Q \cap S^0$, $[Q]^1 := Q \cap S^1$, $[Q]^{0,1} := [Q]^0 \cup [Q]^1$, and $[Q]^R := Q \cap S^R$. For $n \in \mathbb{N}$ and $\sim \in \{=, \leq\}$, let $[Q]^{\mathrm{Col} \sim n} := \{s \in Q | \mathrm{Col}(s) \sim n\}$ denote the sets of states in Q with color $\sim n$. A *run* ρ in G is an infinite sequence $s_0 s_1 \cdots$ of states s.t. $s_i \longrightarrow s_{i+1}$ for all $i \geq 0$; $\rho(i)$ denotes s_i. A *path* π is a finite sequence $s_0 \cdots s_n$ of states s.t. $s_i \longrightarrow s_{i+1}$ for all $i : 0 \leq i < n$. We say that ρ (or π) *visits* s if $s = s_i$ for some i. For any $Q \subseteq S$, we use Π_Q to denote the set of paths that end in some state in Q. Intuitively, the choices of the players and the resolution of randomness induce a run $s_0 s_1 \cdots$, starting in some initial state $s_0 \in S$; state s_{i+1} is chosen as a successor of s_i, and this choice is made by Player 0 if $s_i \in S^0$, by Player 1 if $s_i \in S^1$, and it is chosen randomly according to the probability distribution $P(s_i, \cdot)$ if $s_i \in S^R$.

Strategies. For $x \in \{0, 1\}$, a *strategy* of Player x is a partial function $f^x : \Pi_{S^x} \rightharpoonup S$ s.t. $s_n \longrightarrow f^x(s_0 \cdots s_n)$ if $f^x(s_0 \cdots s_n)$ is defined. The strategy f^x prescribes for Player x the next move, given the current prefix of the run. A run $\rho = s_0 s_1 \cdots$ is said to be *consistent* with a strategy f^x of Player x if $s_{i+1} = f^x(s_0 s_1 \cdots s_i)$ whenever $f^x(s_0 s_1 \cdots s_i) \neq \bot$. We say that ρ is *induced* by (s, f^x, f^{1-x}) if $s_0 = s$ and ρ is consistent with both f^x and f^{1-x}. We use $Runs(G, s, f^x, f^{1-x})$ to denote the set of runs in G induced by (s, f^x, f^{1-x}). We say that f^x is *total* if it is defined for every $\pi \in \Pi_{S^x}$. A strategy f^x of Player x is *memoryless* if the next state only depends on the current state and not on the previous history of the run, i.e., for any path $s_0 \cdots s_n \in \Pi_{S^x}$, we have $f^x(s_0 \cdots s_n) = f^x(s_n)$.

A *finite-memory strategy* updates a finite memory each time a transition is taken, and the next state depends only on the current state and memory. Formally, we define a *memory structure* for Player x as a quadruple $\mathcal{M} = (M, m_0, \tau, \mu)$ satisfying the following properties. The nonempty set M is called the *memory* and $m_0 \in M$ is the *initial memory configuration*. For a current memory configuration m and a current state s, the next state is given by $\tau : S^x \times M \to S$, where $s \longrightarrow \tau(s, m)$. The next memory configuration is given by $\mu : S \times M \to M$. We extend μ to paths by $\mu(\varepsilon, m) = m$ and $\mu(s_0 \cdots s_n, m) = \mu(s_n, \mu(s_0 \cdots s_{n-1}, m))$. The total strategy $strat_{\mathcal{M}} : \Pi_{S^x} \to S$ induced by $\mathcal{M}$ is given by $strat_{\mathcal{M}}(s_0 \cdots s_n) := \tau(s_n, \mu(s_0 \cdots s_{n-1}, m_0))$. A total strategy f^x is said to have *finite memory* if there is a memory structure $\mathcal{M} = (M, m_0, \tau, \mu)$ where M is finite and $f^x = strat_{\mathcal{M}}$. Consider a run $\rho = s_0 s_1 \cdots \in Runs(G, s, f^x, f^{1-x})$ where f^{1-x} is induced by $\mathcal{M}$. We say that ρ visits the configuration (s, m) if there is an i such that $s_i = s$

and $\mu(s_0s_1\cdots s_{i-1},m_0) = m$. We use $F_{all}^x(G)$, $F_{finite}^x(G)$, and $F_0^x(G)$ to denote the set of *all, finite-memory*, and *memoryless* strategies respectively of Player x in G. Note that memoryless strategies and strategies in general can be partial, whereas for simplicity we only define total finite-memory strategies.

Probability Measures. We use the standard definition of probability measures for a set of runs [12]. First, we define the measure for total strategies, and then we extend it to general (partial) strategies. Let $\Omega^s = sS^\omega$ denote the set of all infinite sequences of states starting from s. Consider a game $G = (S, S^0, S^1, S^R, \longrightarrow, P, \mathrm{Col})$, an initial state s, and total strategies f^x and f^{1-x} of Players x and $1-x$. For a measurable set $\mathfrak{R} \subseteq \Omega^s$, we define $\mathcal{P}_{G,s,f^x,f^{1-x}}(\mathfrak{R})$ to be the probability measure of $\mathfrak{R}$ under the strategies f^x, f^{1-x}. This measure is well-defined [12]. For (partial) strategies f^x and f^{1-x} of Players x and $1-x$, $\sim \in \{<,\leq,=,\geq,>\}$, a real number $c \in [0,1]$, and any measurable set $\mathfrak{R} \subseteq \Omega^s$, we define $\mathcal{P}_{G,s,f^x,f^{1-x}}(\mathfrak{R}) \sim c$ iff $\mathcal{P}_{G,s,g^x,g^{1-x}}(\mathfrak{R}) \sim c$ for all total strategies g^x and g^{1-x} that are extensions of f^x resp. f^{1-x}.

Winning Conditions. The winner of the game is determined by a predicate on infinite runs. We assume familiarity with the syntax and semantics of the temporal logic CTL^* (see, e.g., [16]). Formulas are interpreted on the structure $(S, \longrightarrow)$. We use $[\![\varphi]\!]^s$ to denote the set of runs starting from s that satisfy the CTL^* path-formula φ. This set is measurable [25], and we just write $\mathcal{P}_{G,s,f^x,f^{1-x}}(\varphi) \sim c$ instead of $\mathcal{P}_{G,s,f^x,f^{1-x}}([\![\varphi]\!]^s) \sim c$.

We will consider games with *parity* winning conditions, whereby Player 1 wins if the largest color that occurs infinitely often in the infinite run is odd, and Player 0 wins if it is even. Thus, the winning condition for Player x can be expressed in CTL^* as $x\text{-}Parity := \bigvee_{i\in\{0,\dots,n\}\wedge(i \bmod 2)=x}(\Box\Diamond[S]^{\mathrm{Col}=i} \wedge \Diamond\Box[S]^{\mathrm{Col}\leq i})$.

Winning Sets. For a strategy f^x of Player x, and a set F^{1-x} of strategies of Player $1-x$, we define $W^x(f^x, F^{1-x})(G, \varphi^{\sim c}) := \{s \mid \forall f^{1-x} \in F^{1-x}.\mathcal{P}_{G,s,f^x,f^{1-x}}(\varphi) \sim c\}$. If there is a strategy f^x such that $s \in W^x(f^x, F^{1-x})(G, \varphi^{\sim c})$, then we say that s is a *winning state* for Player x in G wrt. $\varphi^{\sim c}$ (and f^x is *winning at* s), provided that Player $1-x$ is restricted to strategies in F^{1-x}. Sometimes, when the parameters G, s, F^{1-x}, φ, and $\sim c$ are known, we will not mention them and may simply say that "s is a winning state" or that "f^x is a winning strategy", etc. If $s \in W^x(f^x, F^{1-x})(G, \varphi^{=1})$, then we say that Player x *almost surely (a.s.)* wins from s. If $s \in W^x(f^x, F^{1-x})(G, \varphi^{>0})$, then we say that Player x wins *with positive probability (w.p.p.)*. We define $V^x(f^x, F^{1-x})(G, \varphi) := \{s \mid \forall f^{1-x} \in F^{1-x}. Runs(G, s, f^x, f^{1-x}) \subseteq [\![\varphi]\!]^s\}$. If $s \in V^x(f^x, F^{1-x})(G, \varphi)$, then we say that Player x *surely* wins from s. Notice that any strategy that is surely winning from a state s is also winning from s a.s., i.e., $V^x(f^x, F^{1-x})(G, \varphi) \subseteq W^x(f^x, F^{1-x})(G, \varphi^{=1})$.

Determinacy and Solvability. A game is called *determined*, wrt. a winning condition and two sets F^0, F^1 of strategies of Player 0, resp. Player 1, if, from every state, one of the players x has a strategy $f^x \in F^x$ that wins against all strategies $f^{1-x} \in F^{1-x}$ of the opponent. By *solving* a determined game, we mean giving an algorithm to compute symbolic representations of the sets of states which are winning for either player.

Attractors. A set $A \subseteq S$ is said to be an *attractor* if, for each state $s \in S$ and strategies f^0, f^1 of Player 0 resp. Player 1, it is the case that $\mathcal{P}_{G,s,f^0,f^1}(\Diamond A) = 1$. In other words, regardless of where we start a run and regardless of the strategies used by the players, we will reach a state inside the attractor a.s.. It is straightforward to see that this also implies that $\mathcal{P}_{G,s,f^0,f^1}(\Box\Diamond A) = 1$, i.e., the attractor will be visited infinitely often a.s..

Transition Systems. Consider strategies $f^x \in F_0^x$ and $f^{1-x} \in F_{finite}^{1-x}$ of Player x resp. Player $1-x$, where f^x is memoryless and f^{1-x} is finite-memory. Suppose that f^{1-x} is induced by memory structure $\mathcal{M} = (M, m_0, \tau, \mu)$. We define the *transition system* $\mathcal{T}$ induced by G, f^{1-x}, f^x to be the pair $(S_M, \rightsquigarrow)$ where $S_M = S \times M$, and $\rightsquigarrow \subseteq S_M \times S_M$ such that $(s_1, m_1) \rightsquigarrow (s_2, m_2)$ if $m_2 = \mu(s_1, m_1)$, and one of the following three conditions is satisfied: (i) $s_1 \in S^x$ and either $s_2 = f^x(s_1)$ or $f^x(s_1) = \bot$, (ii) $s_1 \in S^{1-x}$ and $s_2 = \tau(s_1, m_1)$, or (iii) $s_1 \in S^R$ and $P(s_1, s_2) > 0$. Consider the directed acyclic graph (DAG) of maximal strongly connected components (SCCs) of the transition system $\mathcal{T}$. An SCC is called a *bottom SCC (BSCC)* if no other SCC is reachable from it. Observe that the existence of BSCCs is not guaranteed in an infinite transition system. However, if G contains a finite attractor A and M is finite then $\mathcal{T}$ contains at least one BSCC, and in fact each BSCC contains at least one element (s_A, m) with $s_A \in A$. In particular, for any state $s \in S$, any run $\rho \in Runs(G, s, f^x, f^{1-x})$ will visit a configuration (s_A, m) infinitely often a.s. where $s_A \in A$ and $(s_A, m) \in B$ for some BSCC B.

3 Reachability

In this section we present some concepts related to checking reachability objectives in games. First, we define basic notions. Then we recall a standard scheme (described e.g. in [27]) for checking reachability winning conditions, and state some of its properties that we use in the later sections. Below, fix a game $G = (S, S^0, S^1, S^R, \longrightarrow, P, \text{Col})$.

Reachability Properties. Fix a state $s \in S$ and sets of states $Q, Q' \subseteq S$. Let $Post_G(s) := \{s' : s \longrightarrow s'\}$ denote the set of *successors* of s. Extend it to sets of states by $Post_G(Q) := \bigcup_{s \in Q} Post_G(s)$. Note that for any given state $s \in S^R$, $P(s, \cdot)$ is a probability distribution over $Post_G(s)$. Let $Pre_G(s) := \{s' : s' \longrightarrow s\}$ denote the set of *predecessors* of s, and extend it to sets of states as above. We define $\widetilde{Pre}_G(Q) := \stackrel{G}{\sim} Pre_G\left(\stackrel{G}{\sim} Q\right)$, i.e., it denotes the set of states whose successors *all* belong to Q. We say that Q is *sink-free* if $Post_G(s) \cap Q \neq \emptyset$ for all $s \in Q$, and *closable* if it is sink-free and $Post_G(s) \subseteq Q$ for all $s \in [Q]^R$. If Q is closable then each state in $[Q]^{0,1}$ has at least one successor in Q, and all the successors of states in $[Q]^R$ are in Q.

If $\stackrel{G}{\sim} Q$ is closable, we define the *subgame* $G \ominus Q := (Q', [Q']^0, [Q']^1, [Q']^R, \longrightarrow', P', \text{Col}')$, where $Q' := \stackrel{G}{\sim} Q$ is the new set of states, $\longrightarrow' := \longrightarrow \cap (Q' \times Q')$, $P' := P|([Q']^R \times Q')$, $\text{Col}' := \text{Col}|Q'$. Notice that $P'(s)$ is a probability distribution for any $s \in S^R$ since $\stackrel{G}{\sim} Q$ is closable. We use $G \ominus Q_1 \ominus Q_2$ to denote $(G \ominus Q_1) \ominus Q_2$.

For $x \in \{0, 1\}$, we say that Q is an *x-trap* if it is closable and $Post_G(s) \subseteq Q$ for all $s \in [Q]^x$. Notice that S is both a 0-trap and a 1-trap, and in particular it is both sink-free and closable. The following lemma (adapted from [27]) states that, starting from a state inside a set of states Q that is a trap for one player, the other player can surely keep the run inside Q.

Lemma 1. *If Q is a $(1-x)$-trap, then there exists a memoryless strategy $f^x \in F_0^x(G)$ for Player x such that $Q \subseteq V^x(f^x, F_{all}^{1-x}(G))(G, \square Q)$.*

Scheme. Given a set $\texttt{Target} \subseteq S$, we give a scheme for computing a partitioning of S into two sets $Force^x(G, \texttt{Target})$ and $Avoid^{1-x}(G, \texttt{Target})$ that are winning for Players x and $1-x$. More precisely, we define a memoryless strategy that allows Player x to force the game to $\texttt{Target}$ w.p.p.; and define a memoryless strategy that allows Player $1-x$ to surely avoid $\texttt{Target}$.

First, we characterize the states that are winning for Player x, by defining an increasing set of states each of which consists of winning states for Player x, as follows:

$$\mathcal{R}_0 := \texttt{Target};$$

$$\mathcal{R}_{i+1} := \mathcal{R}_i \cup [Pre_G(\mathcal{R}_i)]^R \cup [Pre_G(\mathcal{R}_i)]^x \cup [\widetilde{Pre_G}(\mathcal{R}_i)]^{1-x} \quad \text{if } i+1 \text{ is a successor ordinal;}$$

$$\mathcal{R}_i := \bigcup_{j<i} \mathcal{R}_j \quad \text{if } i > 0 \text{ is a limit ordinal;}$$

$$Force^x(G, \texttt{Target}) := \bigcup_{i \in \mathbb{O}} \mathcal{R}_i; \quad Avoid^{1-x}(G, \texttt{Target}) := \overset{\mathcal{G}}{\setminus} Force^x(G, \texttt{Target}).$$

First, we show that the iteration above converges (possibly in infinitely many steps). To this end, we observe that $\mathcal{R}_i \subseteq \mathcal{R}_{i+1}$ if $i+1$ is a successor ordinal and $\mathcal{R}_j \subseteq \mathcal{R}_i$ if $j < i$ and i is a limit ordinal. Therefore $\mathcal{R}_0 \subseteq \mathcal{R}_1 \subseteq \cdots$. Since the sequence is non-decreasing and since the sequence is bounded by S, it will eventually converge. Define α to be the smallest ordinal such that $\mathcal{R}_\alpha = \mathcal{R}_i$ for all $i \geq \alpha$. This gives the following lemma, which also implies that the $Avoid^{1-x}$ set is a trap for Player x. (Lemmas 2 and 3 are adapted from [27], where they are stated in a non-probabilistic setting.)

Lemma 2. *There is an $\alpha \in \mathbb{O}$ such that $\mathcal{R}_\alpha = \bigcup_{i \in \mathbb{O}} \mathcal{R}_i$.*

Lemma 3. *$Avoid^{1-x}(G, \texttt{Target})$ is an x-trap.*

The following lemma shows correctness of the construction. In fact, it shows that a winning player also has a memoryless winning strategy.

Lemma 4. *There is a memoryless strategy $force^x(G, \texttt{Target}) \in F_0^x(G)$ such that $Force^x(G, \texttt{Target}) \subseteq W^x(force^x(G, \texttt{Target}), F_{all}^{1-x}(G))(G, \Diamond \texttt{Target}^{>0})$; and a memoryless strategy $avoid^{1-x}(G, \texttt{Target}) \in F_0^{1-x}(G)$ such that $Avoid^x(G, \texttt{Target}) \subseteq V^{1-x}(avoid^{1-x}(G, \texttt{Target}), F_{all}^x(G))(G, \square(\overset{\mathcal{G}}{\setminus} \texttt{Target}))$.*

The first claim of the lemma can be proven using transfinite induction on i to show that it holds for each state $s \in \mathcal{R}_i$. The second claim follows from Lemma 3 and Lemma 1.

4   Parity Conditions

We describe a scheme for solving stochastic parity games with almost-sure winning conditions on infinite graphs, under the conditions that the game has a finite attractor (as defined in Section 2), and that the players are restricted to finite-memory strategies.

By induction on n, we define two sequences of functions $C_0, C_1, \ldots$ and $\mathcal{D}_0, \mathcal{D}_1, \ldots$ s.t., for each $n \geq 0$ and game G of rank at most n, $C_n(G)$ characterizes the states from which Player x is winning a.s., where $x = n \bmod 2$, and $\mathcal{D}_n(G)$ characterizes the set of states from which Player x is winning w.p.p.. The scheme for C_n is related to [27]; cf. the remark at the end of this section. In both cases, we provide a memoryless strategy that is winning for Player x; Player $1 - x$ is always restricted to finite-memory.

For the base case, let $C_0(G) := S$ and $\mathcal{D}_0(G) := S$ for any game G of rank 0. Indeed, from any configuration Player 0 trivially wins a.s./w.p.p. because there is only color 0.

For $n \geq 1$, let G be a game of rank n. $C_n(G)$ is defined with the help of two auxiliary transfinite sequences $\{X_i\}_{i \in \mathbb{O}}$ and $\{\mathcal{Y}_i\}_{i \in \mathbb{O}}$. The construction ensures that $X_0 \subseteq \mathcal{Y}_0 \subseteq X_1 \subseteq \mathcal{Y}_1 \subseteq \cdots$, and that the elements of $X_i, \mathcal{Y}_i$ are winning w.p.p. for Player $1 - x$. The construction alternates as follows. In the inductive step, we have already constructed X_j and $\mathcal{Y}_j$ for all $j < i$. Our construction of X_j and $\mathcal{Y}_j$ is in three steps:

1. X_i is the set of states where Player $1 - x$ can force the run to visit $\bigcup_{j<i} \mathcal{Y}_j$ w.p.p..
2. Find a set of states where Player $1 - x$ wins w.p.p. in $G \ominus X_i$.
3. Take $\mathcal{Y}_i$ to be the union of X_j and the set constructed in step 2.

We next show how to find the winning states in $G \ominus X_i$ in step 2. We first compute the set of states where Player x can force the play in $G \ominus X_i$ to reach a state with color n w.p.p.. We call this set Z_i. The subgame $G \ominus X_i \ominus Z_i$ does not contain any states of color n. Therefore, this game can be completely solved, using the already constructed function $\mathcal{D}_{n-1}(G \ominus X_i \ominus Z_i)$. We will prove that the states where Player $1 - x$ wins w.p.p. in $G \ominus X_i \ominus Z_i$ are winning w.p.p. also in G. We thus take $\mathcal{Y}_i$ as the union of X_i and $\mathcal{D}_{n-1}(G \ominus X_i \ominus Z_i)$. We define the sequences formally:

$$X_i := Force^{1-x}(G, \bigcup_{j<i} \mathcal{Y}_j),$$
$$Z_i := Force^x(G \ominus X_i, [_\sqsubseteq^G X_i]^{\mathrm{Col}=n}), \qquad C_n(G) := _\sqsubseteq^G (\bigcup_{i \in \mathbb{O}} X_i).$$
$$\mathcal{Y}_i := X_i \cup \mathcal{D}_{n-1}(G \ominus X_i \ominus Z_i),$$

Notice that the subgames $G \ominus X_i$ and $G \ominus X_i \ominus Z_i$ are well-defined since (by Lemma 3) $_\sqsubseteq^G X_i$ is closable in G, and $_\sqsubseteq^{G \ominus X_i} Z_i$ is closable in $G \ominus X_i$.

We now construct $\mathcal{D}_n(G)$. Assume that we can construct $C_n(G)$. We will define the transfinite sequence $\{\mathcal{U}_i\}_{i \in \mathbb{O}}$ and the auxiliary transfinite sequence $\{\mathcal{V}_i\}_{i \in \mathbb{O}}$. We again precede the formal definition with an informal explanation of the idea. The construction ensures that $\mathcal{U}_0 \subseteq \mathcal{V}_0 \subseteq \mathcal{U}_1 \subseteq \mathcal{V}_1 \subseteq \cdots$, and that all $\mathcal{U}_i, \mathcal{V}_i$ are winning w.p.p. for Player x in G. The construction alternates in a similar manner to the construction of C_n. In the inductive step, we have already constructed $\mathcal{V}_j$ for all $j < i$. We first compute the set of states where Player x can force the play to reach $\mathcal{V}_j$ w.p.p. for some $j < i$. We call this set $\mathcal{U}_i$. It is clear that $\mathcal{U}_i$ is winning w.p.p. for Player x in G, given the induction hypothesis that all $\mathcal{V}_j$ are winning. Then, we find a set of states where Player x wins w.p.p. in $G \ominus \mathcal{U}_i$. It is clear that $C_n(G \ominus \mathcal{U}_i)$ is such a set. This set is winning w.p.p. for Player x, because a play starting in $C_n(G \ominus \mathcal{U}_i)$ either stays in this set and Player x wins with probability 1, or the play leaves $C_n(G \ominus \mathcal{U}_i)$ and enters $\mathcal{U}_i$ which, as we already know, is winning w.p.p.. We thus take $\mathcal{V}_i$ as the union of $\mathcal{U}_i$ and $C_n(G \ominus \mathcal{U}_i)$. We define the sequences formally by

$$\mathcal{U}_i := Force^x(G, \bigcup_{j<i} \mathcal{V}_j), \qquad \mathcal{D}_n(G) := \bigcup_{i \in \mathbb{O}} \mathcal{U}_i.$$
$$\mathcal{V}_i := \mathcal{U}_i \cup C_n(G \ominus \mathcal{U}_i),$$

By the definitions, for $j < i$ we get $\mathcal{Y}_j \subseteq X_i \subseteq \mathcal{Y}_i$ and $\mathcal{V}_j \subseteq \mathcal{U}_i \subseteq \mathcal{V}_i$. As in Lemma 2, we can prove that these sequences converge.

Lemma 5. *There are* $\alpha, \beta \in \mathbb{O}$ *such that (i)* $X_\alpha = \mathcal{Y}_\alpha = \bigcup_{i \in \mathbb{O}} \mathcal{Y}_i$, *(ii)* $C_n(G) =^G X_\alpha$, *(iii)* $\mathcal{U}_\beta = \mathcal{V}_\beta = \bigcup_{i \in \mathbb{O}} \mathcal{V}_i$, *and (iv)* $\mathcal{D}_n(G) = \mathcal{U}_\beta$.

The following lemma shows the correctness of the construction. Recall that we assume that G is of rank n and that it contains a finite attractor. Let $x = n \bmod 2$.

Lemma 6. *There are memoryless strategies* $f_c^x, f_d^x, \in F_\emptyset^x(G)$ *and* $f_c^{1-x}, f_d^{1-x} \in F_\emptyset^{1-x}(G)$ *such that the following properties hold:*
(i) $C_n(G) \subseteq W^x(f_c^x, F_{finite}^{1-x}(G))(G, x\text{-}Parity^{=1})$.
(ii) $\stackrel{G}{=} C_n(G) \subseteq W^{1-x}(f_c^{1-x}, F_{finite}^x(G))(G, (1-x)\text{-}Parity^{>0})$.
(iii) $\mathcal{D}_n(G) \subseteq W^x(f_d^x, F_{finite}^{1-x}(G))(G, x\text{-}Parity^{>0})$.
(iv) $\stackrel{G}{=} \mathcal{D}_n(G) \subseteq W^{1-x}(f_d^{1-x}, F_{finite}^x(G))(G, (1-x)\text{-}Parity^{=1})$.

Proof. Using induction on n, we define the strategies $f_c^x, f_d^x, f_c^{1-x}, f_d^{1-x}$, and prove that the strategies are indeed winning.

f_c^x. For $n \geq 1$, let α be as defined in Lemma 5. Let $\overline{X_\alpha} :=^G X_\alpha$ and $\overline{Z_\alpha} :=^G Z_\alpha$. We know that $C_n(G) = \overline{X_\alpha}$. For a state $s \in C_n(G)$, we define $f_c^x(s)$ depending on the membership of s in one of the following three partitions of $C_n(G)$: (1) $\overline{X_\alpha} \cap \overline{Z_\alpha}$, (2) $\overline{X_\alpha} \cap [Z_\alpha]^{Col<n}$, and (3) $\overline{X_\alpha} \cap [Z_\alpha]^{Col=n}$.

1. $s \in \overline{X_\alpha} \cap \overline{Z_\alpha}$. Define $G' := G \ominus X_\alpha \ominus Z_\alpha$. From Lemma 5, we have that $X_{\alpha+1} - X_\alpha = \emptyset$. By the construction of $\mathcal{Y}_i$ we have, for arbitrary i, that $\mathcal{D}^{n-1}(G \ominus X_i \ominus Z_i) = \mathcal{Y}_i - X_i$, and by the construction of X_{i+1}, we have that $\mathcal{Y}_i - X_i \subseteq X_{i+1} - X_i$. By combining these facts we obtain $\mathcal{D}^{n-1}(G') \subseteq X_{\alpha+1} - X_\alpha = \emptyset$. Since $G \ominus X_i \ominus Z_i$ does not contain any states of color n (or higher), it follows by the induction hypothesis that there is a memoryless strategy $f_1 \in F_\emptyset^x(G')$ such that $\stackrel{G'}{=} \mathcal{D}_{n-1}(G') \subseteq W^x(f_1, F_{finite}^{1-x}(G'))(G', x\text{-}Parity^{=1})$. We define $f_c^x(s) := f_1(s)$.
2. $s \in \overline{X_\alpha} \cap [Z_\alpha]^{Col<n}$. Define $f_c^x(s) := force^x(G \ominus X_\alpha, [Z_\alpha]^{Col=n})(s)$.
3. $s \in \overline{X_\alpha} \cap [Z_\alpha]^{Col=n}$. By Lemma 3 we know that $Post_G(s) \cap \overline{X_\alpha} \neq \emptyset$. Define $f_c^x(s) := select(Post_G(s) \cap \overline{X_\alpha})$.

Let $f^{1-x} \in F_{finite}^{1-x}(G)$ be a finite-memory strategy for Player $1-x$. We show that $\mathcal{P}_{G,s,f_c^x,f^{1-x}}(x\text{-}Parity) = 1$ for any state $s \in C_n(G)$. First, we show that, any run $s_0 s_1 \cdots \in Runs(G, s, f_c^x, f^{1-x})$ will always stay inside $\overline{X_\alpha}$, i.e., $s_i \in \overline{X_\alpha}$ for all $i \geq 0$. We use induction on i. The base case follows from $s_0 = s \in \overline{X_\alpha}$. For the induction step, we assume that $s_i \in \overline{X_\alpha}$, and show that $s_{i+1} \in \overline{X_\alpha}$. We consider the following cases:

- $s_i \in [\overline{X_\alpha}]^{1-x} \cup [\overline{X_\alpha}]^R$. The result follows since $\overline{X_\alpha}$ is a $(1-x)$-trap in G (by Lemma 3).
- $s_i \in [\overline{X_\alpha} \cap \overline{Z_\alpha}]^x$. We know that $s_{i+1} = f_1(s_i)$. Since $f_1 \in F_\emptyset^x(G \ominus X_\alpha \ominus Z_\alpha)$ it follows that $s_{i+1} \in \overline{X_\alpha} \cap \overline{Z_\alpha}$ and in particular $s_{i+1} \in \overline{X_\alpha}$.
- $s_i \in [\overline{X_\alpha} \cap [Z_\alpha]^{Col<n}]^x$. We know that $s_{i+1} = force^x(G \ominus X_\alpha, [Z_\alpha]^{Col=n})(s_i)$. The result follows by the fact that $force^x(G \ominus X_\alpha, [Z_\alpha]^{Col=n})$ is a strategy in $G \ominus X_\alpha$.
- $s_i \in [\overline{X_\alpha} \cap [Z_\alpha]^{Col=n}]^x$. We have $s_{i+1} \in Post_G(s_i) \cap \overline{X_\alpha}$ and in particular $s_{i+1} \in \overline{X_\alpha}$.

Let us again consider a run $\rho \in Runs(G, s, f^x, f^{1-x})$. We show that ρ is a.s. winning for Player x with respect to x-Parity in G. Let f^{1-x} be induced by a memory structure $\mathcal{M} = (M, m_0, \tau, \mu)$. Let $\mathcal{T}$ be the transition system induced by G, f^x, and f^{1-x}. As explained in Section 2, ρ will a.s. visit a configuration $(s_A, m) \in B$ for some BSCC B in $\mathcal{T}$. This implies that each state that occurs in B will a.s. be visited infinitely often by ρ. There are two possible cases: (i) There is a configuration $(s_B, m) \in B$ with $\mathrm{Col}(s_B) = n$. Since each state in G has color at most n, Player x will a.s. win. (ii) There is no configuration $(s_B, m) \in B$ with $\mathrm{Col}(s_B) = n$. This implies that $\{s_B | (s_B, m) \in B\} \subseteq \overline{Z}$, and hence Player x uses the strategy f_1 to win the game.

f_c^{1-x}. We define a strategy f_c^{1-x} such that $X_i \subseteq \mathcal{Y}_i \subseteq W^{1-x}(f_c^{1-x}, F_{finite}^x(G))(G, (1-x)\text{-}Parity^{>0})$ for all i. The result follows then from the definition of $C_n(G)$. The inclusion $X_i \subseteq \mathcal{Y}_i$ holds by the definition of $\mathcal{Y}_i$. For any state $s \in \overline{C_n(G)}$, we define $f_c^{1-x}(s)$ as follows. Let β be the smallest ordinal such that $s \in \mathcal{Y}_\beta$. Such a β exists by the well-ordering of ordinals and since $\overline{C_n(G)} = \bigcup_{i \in \mathbb{O}} X_i = \bigcup_{i \in \mathbb{O}} \mathcal{Y}_i$. Now there are two cases:

- $s \in X_\beta - \bigcup_{j < \beta} \mathcal{Y}_j$. Define $f_c^{1-x}(s) := f_1(s) := force^{1-x}(G, \bigcup_{j<\beta} \mathcal{Y}_j)(s)$.
- $s \in \mathcal{D}_{n-1}(G \ominus X_\beta \ominus Z_\beta)$. By the induction hypothesis (on n), there is a memoryless strategy $f_2 \in F_0^{1-x}(G)$ of Player $1-x$ such that $s \in W^{1-x}(f_2, F_{finite}^x(G \ominus X_\beta \ominus Z_\beta))(G \ominus X_\beta \ominus Z_\beta, (1-x)\text{-}Parity^{>0})$. Define $f_c^{1-x}(s) := f_2(s)$.

Let $f^x \in F_{finite}^x(G)$ be an arbitrary finite-memory strategy for Player x. We now use induction on i to show that $\mathcal{P}_{G,s,f_c^{1-x},f^x}((1-x)\text{-}Parity) > 0$ for any state $s \in \mathcal{Y}_i$. There are three cases:

1. If $s \in \bigcup_{j<i} \mathcal{Y}_j$ then the result follows by the induction hypothesis (on i).
2. If $s \in X_i - \bigcup_{j<i} \mathcal{Y}_j$ then we know that Player $1-x$, can use f_1 to force the game to $\bigcup_{j<i} \mathcal{Y}_j$ from which she wins w.p.p..
3. If $s \in \mathcal{D}_{n-1}(G \ominus X_i \ominus Z_i)$ then Player $1-x$ uses f_2. There are now two sub-cases: either (i) there is a run from s consistent with f^x and f_c^{1-x} that reaches X_i; or (ii) there is no such run. In sub-case (i), the run reaches X_i w.p.p. and then by cases 1 and 2 Player $1-x$ wins w.p.p.. In sub-case (ii), any run stays forever outside X_i. So the game is in effect played on $G \ominus X_i$. Notice then that any run from s that is consistent with f^x and f_c^{1-x} stays forever in $G \ominus X_i \ominus Z_i$. The reason is that (by Lemma 3) $\overset{G \ominus X_i}{\longrightarrow} Z_i$ is an x-trap in $G \ominus X_i$. Since any run remains inside $G \ominus X_i \ominus Z_i$, Player $1-x$ wins w.p.p. wrt. $(1-x)\text{-}Parity$ using f_2.

f_d^x. For any state s, let β be the smallest ordinal such that $s \in \mathcal{Y}_\beta$. We define $f_d^x(s)$ by two cases:

- $s \in \mathcal{U}_\beta - \bigcup_{j<\beta} \mathcal{V}_j$. Define $f_d^x(s) := f_1(s) := force^x(G, \bigcup_{j<\beta} \mathcal{V}_j)(s)$.
- $s \in C_n(G \ominus \mathcal{U}_\beta)$. By the induction hypothesis (on n), Player x has a winning memoryless strategy f_2 inside $G \ominus \mathcal{U}_i$. Define $f_d^x(s) := f_2(s)$.

Let $f^{1-x} \in F_{all}^{1-x}(G)$ be an arbitrary strategy for Player $1-x$. We now use induction on i to show that $\mathcal{P}_{G,s,f_d^x,f^{1-x}}(x\text{-}Parity) > 0$ for any state $s \in \mathcal{V}_i$. There are three cases:

1. If $s \in \bigcup_{j<i} \mathcal{V}_j$ then the result follows by the induction hypothesis (on i).
2. If $s \in \mathcal{U}_i - \bigcup_{j<i} \mathcal{V}_j$ then we know that Player x can use f_1 to force the game to $\bigcup_{j<i} \mathcal{V}_j$ from which she wins w.p.p. by the previous case.
3. If $s \in C_n(\mathcal{G} \ominus \mathcal{U}_i)$ then Player x uses f_2. There are now two sub-cases: either (i) there is a run from s consistent with f_d^x and f^{1-x} that reaches $\mathcal{U}_i$; or (ii) there is no such run. In sub-case (i), the run reaches $\mathcal{U}_i$ w.p.p. and then by cases 1 and 2 Player x wins w.p.p.. In sub-case (ii), any run stays forever outside $\mathcal{U}_i$. Hence, Player x wins a.s. wrt. x-Parity using f_2.

f_d^{1-x}. By the definition of $\mathcal{U}_i$ we know that $\bigcup_{j<i} \mathcal{V}_j \subseteq \mathcal{U}_i$, and by the definition of $\mathcal{V}_i$ we know that $\mathcal{U}_i \subseteq \mathcal{V}_i$. Thus, $\mathcal{U}_0 \subseteq \mathcal{V}_0 \subseteq \mathcal{U}_1 \subseteq \mathcal{V}_1 \subseteq \cdots$, and hence there is an $\alpha \in \mathbb{O}$ such that $\mathcal{U}_i = \mathcal{V}_i = \mathcal{U}_\alpha$ for all $i \geq \alpha$. This means that $\mathcal{D}_n(\mathcal{G}) = \mathcal{U}_\alpha$ and hence by Lemma 3 we know that $\stackrel{\mathcal{G}}{\rightharpoonup} \mathcal{D}_n(\mathcal{G})$ is an x-trap. Furthermore, since $\mathcal{V}_\alpha = \mathcal{U}_\alpha \cup C_n(\mathcal{G} \ominus \mathcal{U}_\alpha)$, where the union is disjoint, it follows that $C_n(\mathcal{G} \ominus \mathcal{U}_\alpha) = \emptyset$ and hence, by the induction hypothesis, Player $1 - x$ has a memoryless strategy $f \in F_\emptyset^{1-x}(\mathcal{G})$ that is winning w.p.p. against all finite memory strategies $f^x \in F_{finite}^x(\mathcal{G})$ on all states in $\stackrel{\mathcal{G}}{\rightharpoonup} \mathcal{U}_\alpha = \stackrel{\mathcal{G}}{\rightharpoonup} \mathcal{D}_n(\mathcal{G})$. Below, we show that f indeed allows Player $1 - x$ to win almost surely.

Fix a finite-memory strategy $f^x \in F_{finite}^x(\mathcal{G})$. Let f^x be induced by a memory structure $\mathcal{M} = (M, m_0, \tau, \mu)$. Consider a run $\rho \in Runs(\mathcal{G}, s, f, f^x)$. Then, ρ will surely stay inside $\mathcal{G} \ominus \mathcal{U}_\alpha$. The reason is that $\stackrel{\mathcal{G}}{\rightharpoonup} \mathcal{U}_\alpha$ is a trap for Player x by Lemma 3, and that f is a strategy defined inside $\mathcal{G} \ominus \mathcal{U}_\alpha$. Let $\mathcal{T}$ be the transition system induced by $\mathcal{G}$, f^x, and f. As explained in Section 2, ρ will a.s. visit a configuration $(s_A, m) \in B$ for some BSCC B in $\mathcal{T}$. This implies that each configuration in B will a.s. be visited infinitely often by ρ. Let n be the maximal color occurring among the states of B. Then, either (i) $n \bmod 2 = x$ in which case all states inside B are almost sure losing for Player $1 - x$; or (ii) $n \bmod 2 = 1 - x$ in which case all states inside B are almost sure winning for Player $1 - x$. The result follows from the fact that case (i) gives a contradiction since all states in $\stackrel{\mathcal{G}}{\rightharpoonup} \mathcal{U}_\alpha = \stackrel{\mathcal{G}}{\rightharpoonup} \mathcal{D}_n(\mathcal{G})$ (including those in B) are winning for Player $1 - x$ w.p.p.. Define $f_d^{1-x}(s) := f(s)$.

The following theorem follows immediately from the previous lemmas.

Theorem 1. *Stochastic parity games with almost sure winning conditions on infinite graphs are memoryless determined, provided there exists a finite attractor and the players are restricted to finite-memory strategies.*

Remark. The scheme for C_n is adapted from the well-known scheme for non-stochastic games in [27]; in fact, the constructions are equivalent in the case that no probabilistic states are present. Our contribution *to the scheme* is: (1) C_n is a non-trivial extension of the scheme in [27] to handle probabilistic states; (2) we introduce the alternation between C_n and $\mathcal{D}_n$; (3) the construction of $\mathcal{D}_n$ is new and has no counterpart in the non-stochastic case of [27].

5   Lossy Channel Systems

A *lossy channel system (LCS)* [7] is a finite-state machine equipped with a finite number of unbounded fifo channels (queues). The system is *lossy* in the sense that, before and after a transition, an arbitrary number of messages may be lost from the channels. We consider *stochastic game-LCS (SG-LCS)*: each individual message is lost independently with probability λ in every step, where $\lambda > 0$ is a parameter of the system. The set of states is partitioned into states belonging to Player 0 and 1. The player who owns the current control-state chooses an enabled outgoing transition. Formally, a SG-LCS of rank n is a tuple $\mathcal{L} = (\mathsf{S}, \mathsf{S}^0, \mathsf{S}^1, \mathsf{C}, \mathsf{M}, \mathsf{T}, \lambda, \mathsf{Col})$ where S is a finite set of *control-states* partitioned into states $\mathsf{S}^0, \mathsf{S}^1$ of Player 0 and 1; C is a finite set of *channels*, M is a finite set called the *message alphabet*, T is a set of *transitions*, $0 < \lambda < 1$ is the *loss rate*, and $\mathsf{Col} : S \to \{0, \ldots, n\}$ is the *coloring* function. Each transition $\mathsf{t} \in \mathsf{T}$ is of the form $\mathsf{s} \xrightarrow{\mathrm{op}} \mathsf{s}'$, where $\mathsf{s}, \mathsf{s}' \in \mathsf{S}$ and op is one of the following three forms: $\mathsf{c}!\mathsf{m}$ (send message $\mathsf{m} \in \mathsf{M}$ in channel $\mathsf{c} \in \mathsf{C}$), $\mathsf{c}?\mathsf{m}$ (receive message m from channel c), or nop (do not modify the channels). The SG-LCS $\mathcal{L}$ induces a game $\mathcal{G} = (S, S^0, S^1, S^R, \longrightarrow, P, \mathsf{Col})$, where $S = \mathsf{S} \times (\mathsf{M}^*)^{\mathsf{C}} \times \{0, 1\}$. That is, each state in the game consists of a control-state, a function that assigns a finite word over the message alphabet to each channel, and one of the symbols 0 or 1. States where the last symbol is 0 are random: $S^R = \mathsf{S} \times (\mathsf{M}^*)^{\mathsf{C}} \times \{0\}$. The other states belong to a player according to the control-state: $S^x = \mathsf{S}^x \times (\mathsf{M}^*)^{\mathsf{C}} \times \{1\}$. Transitions out of states of the form $s = (\mathsf{s}, \mathsf{x}, 1)$ model transitions in T leaving state s. On the other hand, transitions leaving states of the form $s = (\mathsf{s}, \mathsf{x}, 0)$ model message losses. If $s = (\mathsf{s}, \mathsf{x}, 1), s' = (\mathsf{s}', \mathsf{x}', 0) \in S$, then there is a transition $s \longrightarrow s'$ in the game iff one of the following holds: (i) $\mathsf{s} \xrightarrow{\mathrm{nop}} \mathsf{s}'$ and $\mathsf{x} = \mathsf{x}'$; (ii) $\mathsf{s} \xrightarrow{\mathsf{c}!\mathsf{m}} \mathsf{s}'$, $\mathsf{x}'(\mathsf{c}) = \mathsf{x}(\mathsf{c})\mathsf{m}$, and for all $\mathsf{c}' \in \mathsf{C} - \{\mathsf{c}\}$, $\mathsf{x}'(\mathsf{c}') = \mathsf{x}(\mathsf{c}')$; and (iii) $\mathsf{s} \xrightarrow{\mathsf{c}?\mathsf{m}} \mathsf{s}'$, $\mathsf{x}(\mathsf{c}) = \mathsf{m}\mathsf{x}'(\mathsf{c})$, and for all $\mathsf{c}' \in \mathsf{C} - \{\mathsf{c}\}$, $\mathsf{x}'(\mathsf{c}') = \mathsf{x}(\mathsf{c}')$. Every state of the form $(\mathsf{s}, \mathsf{x}, 0)$ has at least one successor, namely $(\mathsf{s}, \mathsf{x}, 1)$. If a state $(\mathsf{s}, \mathsf{x}, 1)$ does not have successors according to the rules above, then we add a transition $(\mathsf{s}, \mathsf{x}, 1) \longrightarrow (\mathsf{s}, \mathsf{x}, 0)$, to ensure that the induced game is sink-free. To model message losses, we introduce the subword ordering $\preceq$ on words: $x \preceq y$ iff x is a word obtained by removing zero or more messages from arbitrary positions of y. This is extended to channel states $\mathsf{x}, \mathsf{x}' : \mathsf{C} \to \mathsf{M}^*$ by $\mathsf{x} \preceq \mathsf{x}'$ iff $\mathsf{x}(\mathsf{c}) \preceq \mathsf{x}'(\mathsf{c})$ for all channels $\mathsf{c} \in \mathsf{C}$, and to game states $s = (\mathsf{s}, \mathsf{x}, i), s' = (\mathsf{s}', \mathsf{x}', i') \in S$ by $s \preceq s'$ iff $\mathsf{s} = \mathsf{s}'$, $\mathsf{x} \preceq \mathsf{x}'$, and $i = i'$. For any $s = (\mathsf{s}, \mathsf{x}, 0)$ and any x' such that $\mathsf{x}' \preceq \mathsf{x}$, there is a transition $s \longrightarrow (\mathsf{s}, \mathsf{x}', 1)$. The probability of random transitions is given by $P((\mathsf{s}, \mathsf{x}, 0), (\mathsf{s}, \mathsf{x}', 1)) = a \cdot \lambda^b \cdot (1 - \lambda)^c$, where a is the number of ways to obtain x' by losing messages in x, b is the total number of messages needed to be lost in all channels in order to obtain x' from x, and c is the total number of messages in all channels of x' (see [1] for details). Finally, for a state $s = (\mathsf{s}, \mathsf{x}, i)$, we define $\mathsf{Col}(s) := \mathsf{Col}(\mathsf{s})$. Notice that the graph of the game is bipartite, in the sense that a state in S^R has only transitions to states in $[S]^{0,1}$, and vice versa.

In the qualitative *parity game problem* for SG-LCS, we want to characterize the sets of configurations where Player x can force the x-*Parity* condition to hold a.s., for both players.

6   From Scheme to Algorithm

We transform the scheme of Section 4 into an algorithm for deciding the a.s. parity game problem for SG-LCS. Consider an SG-LCS $\mathcal{L} = (\mathsf{S}, \mathsf{S}^0, \mathsf{S}^1, \mathsf{C}, \mathsf{M}, \mathsf{T}, \lambda, \mathsf{Col})$ and the induced game $\mathcal{G} = (S, S^0, S^1, S^R, \longrightarrow, P, \mathsf{Col})$ of some rank n. Furthermore, assume that the players are restricted to finite-memory strategies. We show the following.

Theorem 2. *The sets of winning states for Players 0 and 1 are effectively computable as regular languages. Furthermore, from each state, memoryless strategies suffice for the winning player.*

We give the proof in several steps. First, we show that the game induced by an SG-LCS contains a finite attractor (Lemma 7). Then, we show that the scheme in Section 3 for computing winning states wrt. reachability objectives is guaranteed to terminate (Lemma 9). Furthermore, we show that the scheme in Section 4 for computing winning states wrt. a.s. parity objectives is guaranteed to terminate (Lemma 15). Notice that Lemmas 9 and 15 imply that for SG-LCS our transfinite constructions stabilize below ω (the first infinite ordinal). Finally, we show that each step in the above two schemes can be performed using standard operations on regular languages (Lemmas 16 and 17).

Finite attractor. In [1] it was shown that any Markov chain induced by a Probabilistic LCS contains a finite attractor. The proof can be carried over in a straightforward manner to the current setting. More precisely, the finite attractor is given by $A = (S \times \boldsymbol{\varepsilon} \times \{0,1\})$ where $\boldsymbol{\varepsilon}(c) = \varepsilon$ for each $c \in C$. In other words, A is given by the set of states in which all channels are empty. The proof relies on the observation that if the number of messages in some channel is sufficiently large, it is more likely that the number of messages decreases than that it increases in the next step. This gives the following.

Lemma 7. G *contains a finite attractor.*

Termination of Reachability Scheme. For a set of states $Q \subseteq S$, we define the *upward closure* of Q by $Q\!\uparrow := \{s|\ \exists s' \in Q. s' \preceq s\}$. A set $U \subseteq Q \subseteq S$ is said to be *Q-upward-closed* (or *Q-u.c.* for short) if $(U\!\uparrow) \cap Q = U$. We say that U is *upward closed* if it is S-u.c.

Lemma 8. *If $Q_0 \subseteq Q_1 \subseteq \cdots$, and for all i it holds that $Q_i \subseteq Q$ and Q_i is Q-u.c., then there is an $\alpha \in \mathbb{N}$ such that $Q_i = Q_\alpha$ for all $i \geq \alpha$.*

Now, we can show termination of the reachability scheme.

Lemma 9. *There exists an $\alpha \in \mathbb{N}$ such that $\mathcal{R}_i = \mathcal{R}_\alpha$ for all $i \geq \alpha$.*

Proof. First, we show that $[\mathcal{R}_i - \texttt{Target}]^R$ is $(\overset{G}{\leftharpoondown}\texttt{Target})$-u.c. for all $i \in \mathbb{N}$. We use induction on i. For $i = 0$ the result is trivial since $\mathcal{R}_i - \texttt{Target} = \emptyset$. For $i > 0$, suppose that $s = (\mathsf{s}, \mathsf{x}, 0) \in [\mathcal{R}_i]^R - \texttt{Target}$. This means that $s \longrightarrow (\mathsf{s}, \mathsf{x}', 1) \in \mathcal{R}_{i-1}$ for some $\mathsf{x}' \preceq \mathsf{x}$, and hence $s' \longrightarrow (\mathsf{s}, \mathsf{x}', 1)$ for all $s \preceq s'$.

By Lemma 8, there is an $\alpha' \in \mathbb{N}$ such that $[\mathcal{R}_i]^R - \texttt{Target} = [\mathcal{R}_{\alpha'}]^R - \texttt{Target}$ for all $i \geq \alpha'$. Since $\mathcal{R}_i \supseteq \texttt{Target}$ for all $i \geq 0$ it follows that $[\mathcal{R}_i]^R = [\mathcal{R}_{\alpha'}]^R$ for all $i \geq \alpha'$.

Since the graph of G is bipartite (as explained in Section 5), we have $[Pre_G(\mathcal{R}_i)]^x = [Pre_G([\mathcal{R}_i]^R)]^x$ and $[\widetilde{Pre_G}(\mathcal{R}_i)]^{1-x} = [\widetilde{Pre_G}([\mathcal{R}_i]^R)]^{1-x}$. Since $[\mathcal{R}_i]^R = [\mathcal{R}_{\alpha'}]^R$ for all $i \geq \alpha'$, we thus have $[Pre_G(\mathcal{R}_i)]^x = [Pre_G([\mathcal{R}_{\alpha'}]^R)]^x \subseteq \mathcal{R}_{\alpha'+1}$ and $[\widetilde{Pre_G}(\mathcal{R}_i)]^{1-x} = [\widetilde{Pre_G}([\mathcal{R}_{\alpha'}]^R)]^{1-x} \subseteq \mathcal{R}_{\alpha'+1}$. It then follows that $\mathcal{R}_i = \mathcal{R}_\alpha$ for all $i \geq \alpha := \alpha' + 1$.

Termination of Parity Scheme. We use several auxiliary lemmas. The following lemma states that sink-freeness is preserved by the reachability scheme.

Lemma 10. *If* $\texttt{Target}$ *is sink-free then* $Force^x(G, \texttt{Target})$ *is sink-free.*

Lemma 11. *If* Target *is sink-free then* $[Force^x(G, \text{Target})]^R$ *is upward closed.*

Lemma 12. *Let* $\{Q_i\}_{i \in \mathbb{O}}$ *and* $\{Q'_i\}_{i \in \mathbb{O}}$ *be sequences of sets of states such that (i) Each* Q'_i *is sink-free; (ii)* $Q_i = Q'_i \cup Force^x(G, \bigcup_{j<i} Q_j)$; *(iii)* Q'_i *and* $Force^x(G, \bigcup_{j<i} Q_j)$ *are disjoint for all i. Then, there is an* $\alpha \in \mathbb{N}$ *such that* $Q_i = Q_\alpha$ *for all* $i \geq \alpha$.

To apply Lemma 12, we prove the following two lemmas.

Lemma 13. $C_n(G)$ *is a* $(1-x)$-*trap.*

Proof. $C_0(G)$ is trivially a $(1-x)$-trap. For $i \geq 1$, the result follows immediately from Lemma 5 and Lemma 3.

Lemma 14. *For any game of rank n both* $C_n(G)$ *and* $D_n(G)$ *are sink-free.*

Proof. If $n = 0$, then by definition $C_n(G) = D_n(G) = S$, which is sink-free by assumption. Next, assume $n \geq 1$. By Lemma 13 we know that $C_n(G)$ is a $(1-x)$-trap and hence also sink-free. To prove the claim for $D_n(G)$, we use induction on i and prove that both U_i and V_i are sink-free. Assume U_j and V_j are sink-free for all $j < i$. Then $\bigcup_{j<i} V_j$ is sink-free, and hence U_i is sink-free by Lemma 10. Since $C_n(G \ominus U_i)$ and U_i are sink-free, it follows that V_i is sink-free.

Now, we apply Lemma 12 to prove that the sequences $\{X_i\}_{i \in \mathbb{O}}$ and $\{U_i\}_{i \in \mathbb{O}}$ terminate. First, by Lemma 14 we know that $D_{n-1}(G \ominus X_i \ominus Z_i)$ is sink-free. We know that X_i and $D_{n-1}(G \ominus X_i \ominus Z_i)$ are disjoint since $D_{n-1}(G \ominus X_i \ominus Z_i) \subseteq^G (X_i \cup Z_i)$. Hence, we can apply Lemma 12 with $Q_i = Y_i$, $Q'_i = D_{n-1}(G \ominus X_i \ominus Z_i)$, and conclude that $\{Y_i\}_{i \in \mathbb{O}}$ terminates, and hence $\{X_i\}_{i \in \mathbb{O}}$ terminates. Second, by Lemma 14 we know that $C_n(G \ominus U_i)$ is sink-free. Since $C_n(G \ominus U_i) \subseteq^G U_i$, we know that U_i and $C_n(G \ominus U_i)$ are disjoint. Hence, we can apply Lemma 12 with $Q_i = V_i$, $Q'_i = C_{n-1}(G \ominus U_i)$, and conclude that $\{V_i\}_{i \in \mathbb{O}}$ terminates, and hence $\{U_i\}_{i \in \mathbb{O}}$ terminates. This gives the following lemma.

Lemma 15. *There is an* $\alpha \in \mathbb{N}$ *such that* $X_i = X_\alpha$ *for all* $i \geq \alpha$. *There is a* $\beta \in \mathbb{N}$ *such that* $U_i = U_\beta$ *for all* $i \geq \beta$.

Computability. For a given regular set R, the set $Pre_G(R)$ is effectively regular [2], i.e., computable as a regular language. The following lemma then follows from the fact that the other operations used in computing $Force^x(G, \text{Target})$ are those of set complement and union, which are effective for regular languages.

Lemma 16. *If* Target *is regular then* $Force^x(G, \text{Target})$ *is effectively regular.*

Lemma 17. *For each n, both* $C_n(G)$ *and* $D_n(G)$ *are effectively regular.*

Proof. The set S is regular, and hence $C_0(G) = D_0(G) = S$ is effectively regular. The result for $n > 0$ follows from Lemma 16 and from the fact that the rest of the operations used to build $C_n(G)$ and $D_n(G)$ are those of set complement and union.

Remark. Although we use Higman's lemma for showing termination of our fixpoint computations, our proof differs significantly from the standard ones for well quasi-ordered transition systems [3]. For instance, the generated sets are in general not upward closed wrt. the underlying ordering $\preceq$. Therefore, we need to use the notion of Q-upward closedness for a set of states Q. More importantly, we need to define new (and much more involved) sufficient conditions for the termination of the computations (Lemma 12), and to show that these conditions are satisfied (Lemma 14).

7   Conclusions and Discussion

We have presented a scheme for solving stochastic games with a.s. parity winning conditions under the two requirements that (i) the game contains a finite attractor and (ii) both players are restricted to finite-memory strategies. We have shown that this class of games is memoryless determined. The method is instantiated to prove decidability of a.s. parity games induced by lossy channel systems. The two above requirements are both necessary for our method. To see why our scheme fails if the game lacks a **finite attractor**, consider the game in Figure 1 (a) (a variant of the Gambler's ruin problem). All states are random, i.e., $S^0 = S^1 = \emptyset$, and $\mathtt{Col}(s_0) = 1$ and $\mathtt{Col}(s_i) = 0$ when $i > 0$.

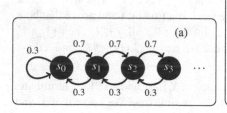 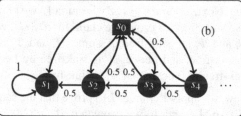

Fig. 1. (a) Finite attractor requirement. (b) Finite strategy requirement.

The probability to go right from any state is 0.7 and the probability to go left (or to make a self-loop in s_0) is 0.3. This game does not have any finite attractor. It can be shown that the probability to reach s_0 infinitely often is 0 for all initial states. However, our construction will classify all states as winning for player 1. More precisely, the construction of $C_1(\mathcal{G})$ converges after one iteration with $Z_i = S, X_i = \emptyset$ for all i and $C_1(\mathcal{G}) = S$. Intuitively, the problem is that even if the force-set of $\{s_0\}$ (which is the entire set of states) is visited infinitely many times, the probability of visiting $\{s_0\}$ infinitely often is still zero, since the probability of returning to $\{s_0\}$ gets smaller and smaller. Such behavior is impossible in a game graph that contains a finite attractor.

We restrict both players to **finite-memory strategies**. This is a different problem from when arbitrary strategies are allowed (not a sub-problem). In fact, it was shown in [9] that for arbitrary strategies, the problem is undecidable. Figure 1 (b) gives an example of a game graph where the two problems yield different results (see also [9]). Player 1 controls s_0, whereas $s_1, s_2, \ldots$ are random; $\mathtt{Col}(s_0) = 0, \mathtt{Col}(s_1) = 2, \mathtt{Col}(s_i) = 1$ if $i \geq 2$. The transition probabilities are $P(s_1, s_1) = 1$ and $P(s_n, s_{n-1}) = P(s_n, s_0) = \frac{1}{2}$ when $n \geq 2$. Player 1 wants to ensure that the highest color that is seen infinitely often is odd, and thus wants to avoid state s_1 (which has color 2). If the players can use arbitrary strategies, then although player 1 cannot win with probability 1, he can win with a probability arbitrarily close to 1 using an infinite-memory strategy: player 1 goes from s_0 to s_{k+i} when the play visits s_0 for the i'th time. Then player 1 wins with probability $\prod_{i=1}^{\infty}(1 - 2^{-k-i+1})$, which can be made arbitrarily close to 1 for sufficiently large k. In particular, player 0 does not win a.s. in this case. On the other hand, if the players are limited to finite-memory strategies, then no matter what strategy player 1 uses, the play visits s_1 infinitely often with probability 1, so player 0 wins almost surely; this is also what our algorithm computes.

As future work, we will consider extending our framework to (fragments of) probabilistic extensions of other models such as Petri nets and noisy Turing machines [6].

References

1. Abdulla, P.A., Bertrand, N., Rabinovich, A., Schnoebelen, P.: Verification of probabilistic systems with faulty communication. Information and Computation 202(2), 105–228 (2005)
2. Abdulla, P.A., Bouajjani, A., d'Orso, J.: Deciding monotonic games. In: Baaz, M., Makowsky, J.A. (eds.) CSL 2003. LNCS, vol. 2803, pp. 1–14. Springer, Heidelberg (2003)
3. Abdulla, P.A., Čerāns, K., Jonsson, B., Tsay, Y.-K.: Algorithmic analysis of programs with well quasi-ordered domains. Information and Computation 160, 109–127 (2000)
4. Abdulla, P.A., Clemente, L., Mayr, R., Sandberg, S.: Stochastic parity games on lossy channel systems. Technical Report EDI-INF-RR-1416, University of Edinburgh (2013), http://arxiv.org/abs/1305.5228, http://www.inf.ed.ac.uk/publications/report/1416.html
5. Abdulla, P.A., Henda, N.B., de Alfaro, L., Mayr, R., Sandberg, S.: Stochastic games with lossy channels. In: Amadio, R.M. (ed.) FOSSACS 2008. LNCS, vol. 4962, pp. 35–49. Springer, Heidelberg (2008)
6. Abdulla, P.A., Henda, N.B., Mayr, R.: Decisive Markov chains. Logical Methods in Computer Science 3 (2007)
7. Abdulla, P.A., Jonsson, B.: Verifying programs with unreliable channels. In: LICS, pp. 160–170 (1993)
8. Abdulla, P.A., Rabinovich, A.: Verification of probabilistic systems with faulty communication. In: Gordon, A.D. (ed.) FOSSACS 2003. LNCS, vol. 2620, pp. 39–53. Springer, Heidelberg (2003)
9. Baier, C., Bertrand, N., Schnoebelen, P.: Verifying nondeterministic probabilistic channel systems against ω-regular linear-time properties. ACM Trans. on Comp. Logic 9 (2007)
10. Bertr, N., Schnoebelen, P.: Model checking lossy channels systems is probably decidable. In: Gordon, A.D. (ed.) FOSSACS 2003. LNCS, vol. 2620, pp. 120–135. Springer, Heidelberg (2003)
11. Bertrand, N., Schnoebelen, P.: Solving stochastic büchi games on infinite arenas with a finite attractor. In: Proceedings of the 11th International Workshop on Quantitative Aspects of Programming Languages (QAPl 2013), Roma, Italy (to appear March 2013)
12. Billingsley, P.: Probability and Measure, 2nd edn. Wiley, New York (1986)
13. Brand, D., Zafiropulo, P.: On communicating finite-state machines. Journal of the ACM 2(5), 323–342 (1983)
14. Chatterjee, K., de Alfaro, L., Henzinger, T.: Strategy improvement for concurrent reachability games. In: QEST, pp. 291–300. IEEE Computer Society Press (2006)
15. Chatterjee, K., Jurdziński, M., Henzinger, T.: Simple stochastic parity games. In: Baaz, M., Makowsky, J.A. (eds.) CSL 2003. LNCS, vol. 2803, pp. 100–113. Springer, Heidelberg (2003)
16. Clarke, E.M., Grumberg, O., Peled, D.: Model Checking. MIT Press (December 1999)
17. Condon, A.: The complexity of stochastic games. Information and Computation 96(2), 203–224 (1992)
18. de Alfaro, L., Henzinger, T.: Concurrent omega-regular games. In: LICS, Washington - Brussels - Tokyo, pp. 141–156. IEEE (2000)

19. de Alfaro, L., Henzinger, T., Kupferman, O.: Concurrent reachability games. In: FOCS, pp. 564–575. IEEE Computer Society Press (1998)
20. Etessami, K., Wojtczak, D., Yannakakis, M.: Recursive stochastic games with positive rewards. In: Aceto, L., Damgård, I., Goldberg, L.A., Halldórsson, M.M., Ingólfsdóttir, A., Walukiewicz, I. (eds.) ICALP 2008, Part I. LNCS, vol. 5125, pp. 711–723. Springer, Heidelberg (2008)
21. Etessami, K., Yannakakis, M.: Recursive markov decision processes and recursive stochastic games. In: Caires, L., Italiano, G.F., Monteiro, L., Palamidessi, C., Yung, M. (eds.) ICALP 2005. LNCS, vol. 3580, pp. 891–903. Springer, Heidelberg (2005)
22. Etessami, K., Yannakakis, M.: Recursive concurrent stochastic games. LMCS 4 (2008)
23. Rabinovich, A.: Quantitative analysis of probabilistic lossy channel systems. In: Baeten, J.C.M., Lenstra, J.K., Parrow, J., Woeginger, G.J. (eds.) ICALP 2003. LNCS, vol. 2719, pp. 1008–1021. Springer, Heidelberg (2003)
24. Shapley, L.S.: Stochastic games. Proceedings of the National Academy of Sciences 39(10), 1095–1100 (1953)
25. Vardi, M.Y.: Automatic verification of probabilistic concurrent finite-state programs. In: FOCS, pp. 327–338 (1985)
26. Wang, K., Li, N., Jiang, Z.: Queueing system with impatient customers: A review. In: IEEE International Conference on Service Operations and Logistics and Informatics (SOLI), pp. 82–87. IEEE (2010)
27. Zielonka, W.: Infinite games on finitely coloured graphs with applications to automata on infinite trees. TCS 200, 135–183 (1998)

Transient Analysis of Networks of Stochastic Timed Automata Using Stochastic State Classes

Paolo Ballarini[1], Nathalie Bertrand[2], András Horváth[3],
Marco Paolieri[4], and Enrico Vicario[4]

[1] École Centrale Paris, France
[2] Inria Rennes, France
[3] Università di Torino, Italy
[4] Università di Firenze, Italy

Abstract. Stochastic Timed Automata (STA) associate logical locations with continuous, generally distributed sojourn times. In this paper, we introduce Networks of Stochastic Timed Automata (NSTA), where the components interact with each other by message broadcasts. This results in an underlying stochastic process whose state is made of the vector of logical locations, the remaining sojourn times, and the value of clocks. We characterize this general state space Markov process through transient stochastic state classes that sample the state and the absolute age after each event. This provides an algorithmic approach to transient analysis of NSTA models, with fairly general termination conditions which we characterize with respect to structural properties of individual components that can be checked through straightforward algorithms.

1 Introduction

Timed Automata (TA) extend standard automata by adding real-time clocks to states and clock constraints to transitions [4]. While this produces a continuous, infinite state-space, various finite abstractions based on regions [4], zones [17], or clock difference diagrams [7] were developed to allow the solution of verification problems in a qualitative perspective, i.e., with reference to possible or necessary behaviors. Various probabilistic extensions were then proposed to enable quantitative evaluation of the probability of feasible behaviors, or to restrain qualitative verification to behaviors with non-null probability.

In Probabilistic Timed Automata (PTA) [24], non-deterministic continuous-time delays are mixed with discrete distributions over actions, and models are checked against PTCTL [22]. In real-time probabilistic processes [2,3], also event durations are randomized. The underlying stochastic process becomes continuous time and may fall in the class of Generalized Semi-Markov Processes (GSMPs). Model-checking can be performed with respect to PTCTL by relying on finite-state abstraction. Continuous Probabilistic Timed Automata (CPTA) [23] extend PTA with randomized clock updates and enable approximate checking against PTCTL.

K. Joshi et al. (Eds.): QEST 2013, LNCS 8054, pp. 355–371, 2013.

Stochastic Timed Automata (STA) [5] were proposed with the aim of relaxing the idealized aspects of TA through a semantics that distinguishes null-probability behaviors. To this end, STA associate locations with sojourn time distributions and transition edges with weights for the probabilistic choice among multiple enabled transitions. In the most general formulation, both of these quantities may depend on clock valuations. For single-clock STA, almost-sure verification of LTL specifications was shown decidable [6] and an approximated technique for quantitative model-checking was proposed [8]. Decidability of almost sure verification was also shown for *reactive* timed automata [10], even with multiple-clocks but under restrictions on sojourn times.

Combination of qualitative real-time constraints with quantitative probabilistic information was also largely addressed on the ground of various classes of Stochastic Petri Nets (SPNs) with generally distributed transitions. In general, the underlying stochastic process of such models belongs to the class of GSMPs [18], for which simulation or statistical model checking are the only general viable approaches to quantitative evaluation. Analytic treatment becomes possible under the so-called enabling restriction that basically requires that no more than one generally distributed transition be enabled at the same time [15,9].

More recently, the method of stochastic state classes addressed models with multiple generally distributed transitions possibly supported over bounded domains, through the symbolic characterization of supports and distributions of remaining times to fire after each transition firing. This was first proposed for steady state analysis [12,26], and then extended to transient analysis [20] and probabilistic model checking [19]. A similar approach was developed for the analysis of Duration Probabilistic Automata (DPA) [25], which compose a set of acyclic semi-Markov processes under control of a non-deterministic scheduler. Symbolic derivation of probability density functions over equivalence classes was proposed also in [1] with a calculus similar to that of [12] but leveraging finite state-space abstractions based on regions rather than zones.

In this paper, we extend the STA formalism by introducing the so-called Networks of Stochastic Timed Automata (NSTA), where multiple STA may synchronize through message passing over a broadcast channel (Sect. 2). We then propose an analytic approach to transient analysis of NSTA models based on the method of stochastic state classes. To this end, we describe the construction of stochastic state classes that sample the state after each transition, we show how these classes provide transient probabilities, and we characterize conditions for termination of the analysis (Sect. 3). An example is then discussed to highlight modeling patterns of NSTA, and analysis results are provided through a preliminary tool-chain implemented on top of the ORIS tool [11,14] (Sect. 4).

2   Model Definition and Semantics

2.1   Timed Automata

Timed automata were introduced in the 90's as a model for real-time systems [4]. Given a finite set of clocks X, we write $\mathcal{G}(X)$ for the set of *guards*, i.e., conjunctions of atomic constraints of the form $x \sim c$ where $x \in X$, $\sim \in \{<, \leq, =, \geq, >\}$ and $c \in \mathbb{N}$. For a *clock valuation* $v \in \mathbb{R}_{\geq 0}^X$ and a guard $g \in \mathcal{G}(X)$, we write $v \models g$ when v satisfies g.

Definition 1 (Timed automaton). *A* timed automaton *is a tuple* $\langle L, \ell_0, \Sigma, X, E \rangle$ *where L is a finite set of locations, $\ell_0 \in L$ is the initial location, Σ is the action alphabet, X is a set of clocks and $E \subseteq L \times \Sigma \times \mathcal{G}(X) \times 2^X \times L$ is a set of edges.*

The semantics of a timed automaton is a transition system where the states are pairs (ℓ, v) of a location $\ell \in L$ and a valuation $v \in \mathbb{R}_{\geq 0}^X$ for the clocks. From any state (ℓ, v) and for any delay $\tau \in \mathbb{R}_{\geq 0}$, there is a delay transition leading to the state $(\ell, v + \tau)$, where $v + \tau$ is a notation for the valuation defined by $(v+\tau)(x) = v(x) + \tau$ for all $x \in X$. Also, for every edge $e = (\ell, a, g, r, \ell') \in E$ and from every state (ℓ, v) such that $v \models g$, there exists a discrete transition leading to $(\ell', v_{[r \leftarrow 0]})$ where $v_{[r \leftarrow 0]}$ denotes the valuation defined by $v_{[r \leftarrow 0]}(x) = 0$ if $x \in r$, and $v_{[r \leftarrow 0]}(x) = x$ otherwise. In this case, we say that edge e is enabled in (ℓ, v) and we write $(\ell, v) \xrightarrow{e} (\ell', v')$ or $e(\ell, v) = (\ell', v')$, with $v' = v_{[r \leftarrow 0]}$.

2.2 Stochastic Timed Automata

We consider stochastic timed automata, a continuous-time probabilistic model associating locations with sojourn time probability density functions (PDFs) and edges with probabilistic choices based on weights [6].

Definition 2 (Stochastic timed automaton). *A stochastic timed automaton is a tuple* $\mathcal{A} = \langle L, \ell_0, \Sigma, X, E, \mu, w \rangle$ *consisting of a timed automaton* $\langle L, \ell_0, \Sigma, X, E \rangle$ *equipped with sojourn time probability density functions* $\mu = (\mu_\ell)_{\ell \in L}$ *and natural weights* $w = (w_e)_{e \in E}$.

Transitions of a STA are determined as follows. When a location l is entered with clock valuation v, then (1) the sojourn time T is chosen according to the PDF μ_ℓ, (2) after the delay T has elapsed, denoting by $E(\ell, v + T)$ the set of edges enabled in state $(\ell, v+T)$, edge $e \in E(\ell, v+T)$ is selected with probability $w_e / \sum_{f \in E(\ell, v+T)} w_f$, (3) assuming e was selected, and if $(\ell', v') = e(\ell, v + T)$, then a transition to ℓ' with clock valuation v' occurs.

The underlying stochastic process of STA ranges from CTMCs to GSMPs, and it will be discussed in Sect. 2.4. In general, the underlying process of a STA is a general state space Markov chain whose state is composed of a discrete component (the location) and a continuous one (the clock valuation and the remaining sojourn times). In this Markov process, if the current state is (ℓ, v, T), the probability to fire an edge $e \in E(\ell, v+T)$ with $(\ell, v+T) \xrightarrow{e} (\ell', v')$ and sample a new sojourn time $T' \leq t$ is given by $(w_e / \sum_{f \in E(\ell, v+T)} w_f) \cdot \int_0^t \mu_{\ell'}(\tau) \, d\tau$.

Remark 1 (Comparison with the model from [6]). Note that we consider a restricted class of stochastic timed automata where the sojourn time probability density functions only depend on the current location, and not on the clock valuation when entering that location. This has consequences on the PDFs that are possible in a location, but also on the structure of the underlying timed automaton itself. However, natural large classes of stochastic timed automata such as reactive STA [10] are covered in our framework.

Also, differently from the traditional approach, we choose a slightly alternative view of STA by considering states (ℓ, v, T) where the sojourn time has already been sampled, rather than states (ℓ, v). This choice is motivated by the extension to networks that we propose in the following section.

2.3   Networks of Stochastic Timed Automata

We now introduce a model where several stochastic timed automata form a network and interact by broadcasting messages. Therefore, in each component, the set of actions is partitioned into sending and receiving actions.

Definition 3 (Network of STA). *A* network of STA *is a tuple* $\langle \mathcal{A}_1, \ldots, \mathcal{A}_n \rangle$ *of n stochastic timed automata associated with global weights* $w(\mathcal{A}_1), \ldots, w(\mathcal{A}_n)$ *and sharing an alphabet* $\Sigma = \Sigma_b \cup \Sigma_r$ *partitioned into* broadcasts *(Σ_b) and* receptions *(Σ_r). Broadcasts are of the form* $!m$ *and receptions of the form* $?m$, *for some message m from a fixed alphabet* M, *i.e.,* $\Sigma_b = !$M *and* $\Sigma_r = ?$M.

The intuitive semantics of a network is as follows. The network starts in a configuration where each STA is in its initial location and samples an initial sojourn time. When the minimum sampled sojourn time elapses, say for STA $\mathcal{A}_i$, the component $\mathcal{A}_i$ performs an action selected according to the weights of its enabled edges, and broadcasts the associated message (races among equal, deterministic times to fire are solved by the global weights $w(\mathcal{A}_i)$). When $\mathcal{A}_i$ performs a broadcast (sending action), then all other components for which the corresponding receiving action is enabled must synchronize and perform the corresponding reception. On occurrence of such inter-process communication, all components involved in the exchange (the sender and the receivers) update their locations and sample new sojourn times, and then the execution in the network proceeds.

Formally, states of the network $\langle \mathcal{A}_1, \ldots, \mathcal{A}_n \rangle$ are n-tuples of triplets (ℓ_i, v_i, T_i), one for each component, consisting of current location (ℓ_i), clock valuation (v_i), and remaining sojourn time (T_i). Given the current state of the network $\mathbf{s} = \langle (\ell_1, v_1, T_1), \ldots, (\ell_n, v_n, T_n) \rangle$, the next transition is determined by selecting the component $\mathcal{A}_i$ with lowest remaining sojourn time T_i; deterministically, the delay T_i is elapsed from $\mathbf{s}$ and a broadcast edge e_i enabled in $(\ell_i, v_i + T_i)$ is selected according to weights (as in STA), resulting in the local transition $(\ell_i', v_i') = e_i(\ell_i, v_i + T_i)$. For every other component $\mathcal{A}_j$ with $j \neq i$, an enabled matching receiving edge e_j (if any) is selected according to the weights, and the state of components with a selected (synchronizing) action is updated as $(\ell_j', v_j') = e_j(\ell_j, v_j + T_i)$. New sojourn times are sampled for components that performed an action: T_i' according to $\mu_{\ell_i'}$, and possibly T_j' according to $\mu_{\ell_j'}$; the resulting state is then $\mathbf{s}' = \langle (\ell_1', v_1', T_1'), \cdots (\ell_n', v_n', T_n') \rangle$ where $(\ell_k', v_k', T_k') = (\ell_k, v_k + T_k, T_k' - T_i)$ for components $\mathcal{A}_k$ that did not take an action.

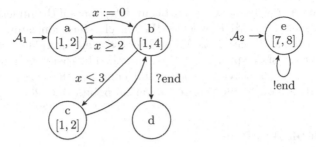

Fig. 1. Network of STA

Example. Consider the NSTA depicted in Fig. 1, where all sojourn times of A_1 and A_2 are uniformly distributed over the depicted supports. Automaton A_2 broadcasts an END message at each termination of its computation: if A_1 is in b at the moment of the broadcast, its current computation is interrupted by the receiving edge (b, d) and the network reaches the absorbing state $\vec{\ell} = (d, e)$. The clock x in A_1 tracks the time elapsed since the last transition from a to b.

2.4 Underlying Stochastic Process of STA and NSTA

The underlying stochastic process of a STA is a CTMC only under severe restrictions: (1) the (non-null) sojourn times of locations must be exponentially distributed and (2) the choice of the next location cannot depend on the value of clocks, so that clocks do not have an impact on the behavior (i.e., the control-flow) of the STA. Under these conditions, the STA is *memoryless* in every location and every time instant is a regeneration point, i.e., the current location alone determines the future of the process in a stochastic sense.

If clocks are reset at every transition, or those that are not reset cannot affect future choices of the next location (i.e., they do not appear in guards), then every transition constitutes a regeneration point and the underlying stochastic process is a semi-Markov process (SMP). When clocks can carry memory from a location to another and influence the behavior of the STA, then there can be transitions that do not result in a regeneration point. If regeneration points are guaranteed to appear infinitely often, the process is Markov regenerative (MRP); otherwise, we have a generalized semi-Markov process (GSMP). Regeneration points can be exploited in the analysis of the process [21]; if regeneration is not guaranteed, two general analysis techniques can be adopted: the so-called supplementary variable approach [16] and the method of stochastic state classes [20].

In the case of a NSTA, the underlying process is a CTMC if all STA are memoryless in every location, i.e., their underlying process is also a CTMC. If this condition does not hold, we can still have regeneration points when (1) all automata are memoryless in the current location, or (2) some automata are memoryless in the current location and the others "lost memory" at the same time due to one or more receiving transitions, i.e., they performed a transition that reset all clocks except those that cannot have an impact on the future behavior. The way regeneration points occur determines if the underlying process of the NSTA is a SMP, a MRP or a GSMP. The classification cannot be based on the analysis of the automata in isolation because their interplay can be decisive. The transient analysis we propose in this paper can be easily modified to provide information on the underlying process and compute the kernels of the MRP required in calculations exploiting regeneration points (see [20]).

3   Transient Analysis

3.1   Stochastic State Classes for NSTA

Stochastic state classes characterize the underlying stochastic process by representing explicitly, for each state transition, the resulting logical state and joint probability density function of the continuous random variables that govern the evolution of the system (clocks and remaining sojourn times). To support transient analysis, we include an additional clock x_{age} to encode the absolute time of the last transition [20]. Let us formalize the concept of stochastic state class.

Definition 4 (Stochastic state class). *A stochastic state class for the STA network* $\langle \mathcal{A}_1, \ldots, \mathcal{A}_n \rangle$ *is a tuple* $\langle \ell, D, f \rangle$ *where* $\ell = (\ell_1, \ldots, \ell_n)$ *specifies the current location of each automaton and* $f: D \to [0, 1]$ *is the probability density function of the random remaining sojourn times* $\vec{\tau} = (\tau_1, \ldots, \tau_n)$ *and clocks* $\langle x_{age}, \vec{x} \rangle$, *with* $\vec{x} = (x_1, \ldots, x_m)$, *on the support* $D \subseteq \mathbb{R}_{\geq 0}^n \times \mathbb{R}_{\leq 0}^{m+1}$.

Note that the support of remaining sojourn times τ_i is $\mathbb{R}_{\geq 0}$, while that of clock random variables x_i is $\mathbb{R}_{\leq 0}$; this is required to allow an efficient representation of the joint support D. In fact, stochastic state classes were originally developed in the context of Stochastic Time Petri Nets (STPNs), where times to fire of transitions decrease with unitary rate and the support of joint PDFs can be represented as a Difference Bounds Matrix zone (DBM zone), i.e., the set of solutions of a system of linear inequalities $\tau_i - \tau_j \leq b_{ij}$ for all $i \neq j \in \{*, 1, \ldots, n\}$ with $b_{ij} \in \mathbb{R} \cup \{+\infty\}$ and $\tau_* = 0$. This form allows a compact representation of the state space, and it is preserved by all the operations required in the computation of successor state classes through the firing of a transition; these operations include: reducing all the variables of a stochastic class by one of them, marginalizing variables, and adding new variables in product form. In STA, sojourn time random variables can be managed with the same operations: they can be reduced by the minimum one, or marginalized and added in product form for the automata performing the transition. In contrast, clocks should be *increased* by the minimum sojourn time; in order to preserve the DBM form for the support of PDFs of random variables, we thus encode clocks as negative variables that are initially set to zero and *decreased* at each transition. In so doing, all the random variables of the stochastic state class (sojourn times and clocks) are simply decreased by one of them, as in STPNs.

   We now define the stochastic state classes of a NSTA, starting from the initial class, and then describing the derivation of successor classes.

Initial class. In the initial class Σ_0, all the clocks are set to zero and each automaton $\mathcal{A}_i$ is in its initial location ℓ_i with a sojourn time independently distributed according to the probability density function μ_{ℓ_i}. Hence, we have $\Sigma_0 = \langle \vec{\ell_0}, D_0, f_0 \rangle$ with $\vec{\ell_0} = (\ell_1, \ldots, \ell_n)$ and

$$D_0 = ([a_0, b_0] \times \cdots \times [a_n, b_n]) \times [0, 0] \times [0, 0]^m$$

$$f_0(\vec{\tau}, x_{age}, \vec{x}) = \prod_{i=0}^{n} \mu_{\ell_i}(\tau_i) \cdot \delta(x_{age}) \cdot \prod_{i=0}^{m} \delta(x_i)$$

where δ is the Dirac delta function, μ_{ℓ_i} is the PDF associated with location ℓ_i, and $[a_i, b_i]$ its support (we indicate the Cartesian product of supports by $\times$).

Computation of successor classes. The computation of successor classes characterizes the set of states (locations, remaining sojourn times and clock valuations) that can be reached after a state transition in the STA network, their probability density function, and the probability of the state transition itself.

In general, a transition is identified by the automaton $\mathcal{A}_i$ with minimum remaining sojourn time, an edge e in $\mathcal{A}_i$, and a possible set E of receiving edges for the message broadcast by e. In addition, a transition also depends on the clock valuation, which may restrict the set of enabled edges because of the guards (in $\mathcal{A}_i$ or in the rest of the network). This partitions the space of clock valuations in *decision domains*, such that any two valuations in the same domain satisfy the same guards and thus result in a probabilistic choice within the same set of enabled edges. Since guards are expressed by conjunctions of simple inequalities, each guard is satisfied within a hyper-rectangular domain, but a decision domain may be the difference among different hyper-rectangles. When guards on the outgoing edges of a location involve multiple clocks, decision domains are not hyper-rectangular, but they can be anyway partitioned into a set of hyper-rectangular subdomains, which we call *decision zones*. We denote with $\mathcal{R}(\vec{\ell})$ the set of decision zones r associated with location $\vec{\ell}$ of the network.

Example. For the NSTA depicted in Fig. 1, $\mathcal{A}_1$ has outgoing edges in location b with guards on x that are both satisfied for $x \in [2, 3]$, and exclusively satisfied for $x \notin [2, 3]$. The decision zones in b are thus $r_{b,1} = [0, 2)$, $r_{b,2} = [2, 3]$, $r_{b,3} = [3, +\infty)$, while any other location is associated with a single decision zone $[0, +\infty)$. If $x \in r_{b,2}$, assuming equally weighted edges $w(b, a) = w(b, c)$, either enabled edge (b, a) or (b, c) is selected with probability $1/2$.

Definition 5 (Succession relation). *We say that $\Sigma' = \langle \vec{\ell'}, D', f' \rangle$ is the successor of $\Sigma - \langle \vec{\ell}, D, f \rangle$ through the edge e of $\mathcal{A}_i$, for a decision zone $r \subset \mathcal{R}(\vec{\ell})$ and a set of receiving edges E, with probability p (and we write $\Sigma \overset{\xi, p}{\Longrightarrow} \Sigma'$ with $\xi = (\mathcal{A}_i, r, e, E)$), if, given that the location of the NSTA is $\vec{\ell}$ and the sojourn times and clocks are random variables distributed over D according to f, then:*

(i) with non-null probability p, $\mathcal{A}_i$ is the automaton with minimum remaining sojourn time in Σ, the random clock valuation $\vec{x}$ belongs to the decision zone

$r \in \mathcal{R}(\vec{\ell})$, the sender edge e and the receiving edges E are enabled by r, and they are selected as outgoing event;

(ii) conditioned to (i), the state transition yields the location $\vec{\ell}'$ with sojourn times and clock random variables distributed over D' according to f'.

Given this definition, the successors of a stochastic state class can be derived through the following steps.

1. *Conditioning on the minimum sojourn time.* For each automaton of the network, we compute the probability that its remaining time to fire is the minimum, and we condition the sojourn time and clock random variables by this event. Up to a renaming of the components, we assume that the automaton $\mathcal{A}_1$ has minimum remaining time to fire with probability

$$p_{\tau_1} = \int_{\{\langle \vec{\tau}, x_{age}, \vec{x} \rangle \in D \,|\, \tau_1 \leq \tau_j \;\forall j\}} f(\vec{\tau}, x_{age}, \vec{x}) \, d\vec{\tau} \, dx_{age} \, d\vec{x}.$$

By conditioning on this event, we obtain the random vector of sojourn times and clocks $\vec{v}_a = \langle \vec{\tau}, x_{age}, \vec{x} \,|\, \{\tau_1 \leq \tau_j \;\forall j\}\rangle$ distributed over $D_a = \{\langle \vec{\tau}, x_{age}, \vec{x} \rangle \in D \,|\, \tau_1 \leq \tau_j \;\forall j\}$ according to $f_a = f/p_{\tau_1}$. Note that races among equal deterministic times to fire are resolved by the global weights $w(\mathcal{A}_1), \ldots, w(\mathcal{A}_n)$ associated with the automata.

2. *Shifting all the variables by the minimum sojourn time and marginalizing the minimum sojourn time.* In order to account for the time elapsed in the previous state, all the variables are decreased by the minimum sojourn time τ_1, that is in turn marginalized. This leads to a random vector of sojourn times and clocks $\vec{v}_b = \langle \tau_2 - \tau_1, \ldots, \tau_n - \tau_1, x_{age} - \tau_1, \vec{x} - \tau_1 \rangle$ distributed over

$$D_b = \{(\tau_2, \ldots, \tau_n, x_{age}, \vec{x}) \in \mathbb{R}_{\geq 0}^{n-1} \times \mathbb{R}_{\leq 0}^{m+1} \,|$$
$$\exists \tau_1 \in \mathbb{R}_{\geq 0} : (\tau_1, \tau_2 + \tau_1, \ldots, \tau_n + \tau_1, x_{age} + \tau_1, \vec{x} + \tau_1) \in D_a\}$$

according to

$$f_b(\tau_2, \ldots, \tau_n, x_{age}, \vec{x}) = \int_{L_1}^{U_1} f_a(\tau_1, \tau_2 + \tau_1, \ldots, \tau_n + \tau_1, x_{age} + \tau_1, \vec{x} + \tau_1) \, d\tau_1$$

where

$$L_1(\tau_2, \ldots, \tau_n, x_{age}, \vec{x}) = \min_{j \neq \tau_1}\{b_{\tau_1, j} + j\}$$
$$U_1(\tau_2, \ldots, \tau_n, x_{age}, \vec{x}) = \max_{i \neq \tau_1}\{-b_{i, \tau_1} + i\}$$

are the piecewise linear functions of the minimum and maximum constraints on the variable τ_1 within the DBM zone D_a of coefficients b_{ij} and variables $i, j \in \{\tau_1, \ldots, \tau_n, x_{age}, x_1, \ldots, x_m, *\}$. Because of this piecewise dependency of integration bounds on different expressions of the form $b_{\tau_1, j} + j$ or $-b_{i, \tau_1} + i$, the result of the symbolic integration is in general a piecewise continuous function on a partitioning of D_b in DBM subzones [13]. In this case, all of the following steps have to be performed individually on each subzone.

3. *Conditioning on a decision zone.* In order to analyze fixed sets of enabled edges, each decision zone $r \in \mathcal{R}(\vec{\ell})$ is taken into account separately by imposing that the clock variables $\vec{x}$ belong to $r \subseteq \mathbb{R}^m_{\leq 0}$. This event has probability

$$p_r = \int_{\{\langle \vec{\tau}, x_{age}, \vec{x} \rangle \in D_b \,|\, \vec{x} \in r\}} f_b(\vec{\tau}, x_{age}, \vec{x}) \, d\vec{\tau} \, dx_{age} \, d\vec{x}$$

and, by conditioning on it, we obtain the vector of sojourn times and clocks $\vec{v}_c = \langle \vec{v}_b \,|\, \{\vec{x} \in r\} \rangle$ distributed over $D_c = \{\langle \vec{\tau}, x_{age}, \vec{x} \rangle \in D_b \,|\, \vec{x} \in r\}$ according to $f_c = f_b/p_r$.

4. *Selection of a sender edge.* Since the set of enabled edges W is fixed within the decision zone r, the edge $e \in W$ is selected by automaton $\mathcal{A}_1$ in r with probability $p_e = w(e)/\sum_{e' \in W} w(e')$.

5. *Selection of receiving edges.* For each automaton with more than one receiving edge for the symbol broadcast by the edge e and enabled in r, the choice is resolved with weights, so that the probability p_E of each distinct set E of receiving edges is determined.

6. *Locations update.* The locations are updated according to the transitions performed by the automaton with minimum sojourn time and by those with receiving edges in E. A new locations vector $\vec{\ell'}$ is computed.

7. *Variables removal.* For automata that updated their locations, remaining sojourn time variables and reset clocks are marginalized. As an example, marginalization of a variable τ_2 is performed as

$$f_d(\tau_3, \ldots, \tau_n, x_{age}, \vec{x}) = \int_{E_2}^{L_2} f_c(\tau_2, \ldots, \tau_n, x_{age}, \vec{x}) \, d\tau_2.$$

Similarly to *shift and project* operations, subzones can be introduced by the piecewise integration bounds L_2 and E_2.

8. *Variables addition.* Similarly to the definition of the initial stochastic state class, new Dirac deltas are added in product form for all reset clocks, and sojourn time PDFs are added in product form to f_d for automata that updated their locations. This results in the final domain D' and PDF f'.

Given a stochastic class $\Sigma = \langle \vec{\ell}, D, f \rangle$, the probability associated with the successor resulting from the transition given by the sender edge e of $\mathcal{A}_1$ and receiving edges E within the decision zone $r \in \mathcal{R}(\vec{\ell})$ is thus $p = p_{\tau_1} p_r p_e p_E$. Note that, in general, the automaton with minimum time to fire can select different edges, and same edge e can result in several stochastic successors with distinct sets of receiving edges or distinct decision zones.

Moreover, if no edge is enabled in the current location of $\mathcal{A}_1$ when the clocks belong to the decision zone $r \in \mathcal{R}(\vec{\ell})$, the remaining sojourn time variable τ_1 is marginalized (step 1) but not reintroduced in product form (step 8). According to the semantics of NSTA, $\mathcal{A}_1$ reaches (with probability $p = p_{\tau_1} p_r$) a state from which it can perform a transition only by receiving a broadcast symbol; if all of the automata are in such receive-only condition, a deadlock has occurred.

3.2 Transient Tree Enumeration and Transient Measures

The transient evolution of the logical state $\vec{\ell}$ in a NSTA can be analyzed through the probability and time distribution of discrete events representing the end of sojourn times sampled according to the PDFs μ_ℓ. We consider as discrete event abstraction the tuple $(\mathcal{A}_i, r, e, E)$: given the current locations $\vec{\ell}$, the event $(\mathcal{A}_i, r, e, E)$ is the next event of the network if

- the sojourn time of $\mathcal{A}_i$ is the minimum;
- the clock valuation $\vec{v}$ belongs to the decision zone $r \in \mathcal{R}(\vec{\ell})$;
- the sender and receiving edges e and E (respectively) are randomly selected ($e = \mathrm{NIL}$ and $E = \emptyset$ if no edge is enabled for $\mathcal{A}_i$ in ℓ_i when $\vec{v} \in r$).

Since decision zones represent a partition of the space of clock valuations, the events $(\mathcal{A}_i, r, e, E)$ for each automaton, decision zone, and distinct enabled sender and receiving edges, are mutually exclusive and collectively exhaustive. Moreover, given a PDF and support for clocks and remaining sojourn times in the current locations, the computation of successor classes presented in Sect. 3.1 allows to compute the probability of an event $(\mathcal{A}_i, r, e, E)$ and the PDF and support conditioned to it. Successive events can then be evaluated independently after conditioning, since a stochastic state class is a full characterization of the future evolution of the probabilistic model.

This construction leads to the enumeration of a tree in which each node is labeled with a stochastic state class, and where each edge carries an event $\xi = (\mathcal{A}_i, r, e, E)$ and a probability p.

Definition 6 (Transient tree). *The* transient tree *from an initial stochastic state class Σ_0 is a tuple* TRANSIENT-TREE$(\Sigma_0) = \langle N, A, n_0, \Sigma, p, \xi \rangle$ *where*

- *N is a set of nodes and $n_0 \in N$ is the root of the tree;*
- *the labeling function Σ associates each node $n \in N$ with a stochastic state class $\Sigma(n)$, with $\Sigma(n_0) = \Sigma_0$;*
- *A is the smallest set of edges (n, n') with $n, n' \in N$ such that $\Sigma(n')$ is a successor of $\Sigma(n)$, i.e., $\Sigma(n) \xrightarrow{\xi, p} \Sigma(n')$; in such a case, the edge is labeled with the probability $p(n, n')$ and the event $\xi(n, n')$ that it bears.*

The transient tree from an initial stochastic state class Σ_0 can be enumerated by repeatedly computing the successor classes of a leaf node until some stopping criterion is satisfied. A node n_k in the transient stochastic tree can thus be associated with a sequence of events $\xi_1, \xi_2, \ldots, \xi_k$ such that

$$\Sigma(n_0) \xrightarrow{\xi_1, p_1} \Sigma(n_1) \xrightarrow{\xi_2, p_2} \ldots \xrightarrow{\xi_k, p_k} \Sigma(n_k).$$

The probability that the sequence of events $\xi_1, \xi_2, \ldots, \xi_k$ happens, leading from the initial node n_0 to the node n_k, is given by the *reaching probability*

$\eta(n_k) = \prod_{i=1}^{k} p_i$. The stochastic state class $\Sigma(n_k)$ defines the PDF of remaining sojourn times and clocks after the sequence of events $\xi_1, \xi_2, \ldots, \xi_k$. This information allows to impose additional constraints on the execution time and compute more specific measures. Notably, if $\Sigma(n_k) = \langle \vec{\ell}, D, f \rangle$, the probability that the system has performed, at time t, all and only the events $\xi_1, \xi_2, \ldots, \xi_k$ is given by

$$\pi(n_k, t) = \eta(n_k) \cdot \int_{D(t)} f(\vec{\tau}, x_{age}, \vec{x}) \, d\vec{\tau} \, dx_{age} \, d\vec{x}$$

with $D(t) = \{\langle \vec{\tau}, x_{age}, \vec{x} \rangle \in D \mid -x_{age} \leq t \text{ and } -x_{age} + \tau_i > t \text{ for } i = 1, \ldots, n\}$. The reaching probability $\eta(n)$ accounts for the probability of performing the sequence of events, while the restricted domain $D(t)$ imposes that the last event happened at a time $-x_{age} \leq t$ and the next one will happen at a time $\min_i\{-x_{age} + \tau_i\}$ greater than t. By definition of the PDF f, integrating over this restricted set of remaining sojourn time and age values results in the required measure.

Transient probabilities for a specific vector of locations $\vec{\ell}^*$ can then be defined as the sum of measures $\pi(n, t)$ for all nodes associated with a stochastic state class $\Sigma(n) = \langle \vec{\ell}, D, f \rangle$ with $\vec{\ell} = \vec{\ell}^*$:

$$\pi(\vec{\ell}^*, t) = \sum_{n \in N : \Sigma(n) = \langle \vec{\ell}^*, D, f \rangle} \pi(n, t).$$

Events associated with the measures $\pi(n, t)$ are in fact mutually exclusive for distinct n: at any given time instant t, a node n uniquely identifies a sequence of transitions performed by the system and $\pi(n, t)$ is the probability that all and only the transitions in the sequence have been performed.

Note that, in the enumeration of the transient tree, each decision zone has to be taken into account separately and, moreover, the piecewise partition of PDFs in DBM subzones requires that the derivation of density functions be repeated on each subdomain. In the worst case, the number of successors and subdomains within each successor grows exponentially with the depth of the transient tree, comprising the dominating factor of complexity in practical implementations.

3.3   Termination

The *exact* evaluation of transient probabilities up to a given time bound T requires the enumeration of all and only the stochastic state classes that can be reached within T, i.e., all classes where the support of x_{age} includes values greater or equal to $-T$. If an *approximation* bound $\epsilon \geq 0$ is allowed, construction of the tree can be halted as soon as the total probability of reaching any leaf node before T is lower than ϵ; such probability, indicated as $\eta_T(n)$, can be computed for any node n associated with $\Sigma(n) = \langle \vec{\ell}, D, f \rangle$ as

$$\eta_T(n) = \eta(n) \cdot \int_{\{\langle \vec{\tau}, x_{age}, \vec{x} \rangle \in D \mid -x_{age} \leq T\}} f(\vec{\tau}, x_{age}, \vec{x}) \, d\vec{\tau} \, dx_{age} \, d\vec{x}.$$

For both exact and approximate evaluation, termination involves two orthogonal aspects pertaining to: (1) the cyclic behaviors arising in the composition of multiple automata; (2) the guarantee of time progression in each cyclic execution, in certainty (i.e., surely) or in probability (i.e., almost surely). We characterize termination through sufficient conditions that can be checked on the structure of individual STA, relaxing on aspect (1) so as to obtain conditions that can be easily checked by the designer in the construction of a model without resorting to state space analysis. Necessary and sufficient conditions require instead a non-deterministic analysis of the state class graph of the NSTA, with concepts analogous to those applied in [19] and without affecting the conditions given in the following for the advancement of time along cyclic behaviors.

Termination in exact analysis depends on *necessarily tangible* and *possibly vanishing* locations, i.e., locations where the *minimum* sojourn time is strictly greater than zero or equal to zero, respectively.

Theorem 1 (Exact time-bounded termination). *Given an initial stochastic state class Σ_0, the enumeration of the transient tree* TRANSIENT-TREE(Σ_0) *within any time bound T terminates in a finite number of steps provided that every cycle of each automaton visits at least one necessarily tangible location that cannot be preempted by messages broadcast on exit from possibly vanishing locations.*

Proof. Since the NSTA has a finite number of locations and the branching factor of TRANSIENT-TREE(Σ_0) is bounded, an infinite path would traverse some cycle within some automaton infinitely often. In so doing, at least one necessarily tangible location is visited infinitely often; at each visit, the minimum time to complete the path would increase by at least the minimum sojourn time of the location or the minimum sojourn time of the location in some other automaton that broadcasts a preempting message, thus exceeding T in a finite number of steps.   □

When a cycle can be traversed in a time arbitrarily close to zero, termination can still be guaranteed if we exclude cycles that are *bound* to complete in zero time. In this case, termination depends on *possibly tangible* and *necessarily vanishing* locations, i.e., locations where the *maximum* sojourn time is strictly greater than zero or equal to zero, respectively.

Theorem 2 (Time-bounded termination with ϵ-error). *Given an initial stochastic state class Σ_0, a time bound T and an allowed error ϵ, if the transient tree is enumerated in breadth first order, there exists a number $N(\Sigma_0, T, \epsilon)$ such that, after $N(\Sigma_0, T, \epsilon)$ classes have been traversed, the probability of reaching any leaf node before T is lower than ϵ, provided that every cycle within each automaton visits at least one possibly tangible location that cannot be preempted by a message broadcast on exit from a necessarily vanishing location.*

Proof. Since each cycle of the NSTA visits at least some possibly tangible location of a STA, there exists a finite number $m \in \mathbb{N}$ such that the sojourn time of a possibly tangible location elapses every m state transitions of the NSTA. For every number n, after $n \cdot m$ state transitions, at least n tangible sojourn times have been completed and the total elapsed time $T_{n \cdot m}$ is at least $\sum_{i=1}^{n} \tau_i$, where τ_i denotes the i-th sojourn time completed in a possibly tangible location. Since the random variables τ_i are independent and distributed according to a finite number of PDFs with supports $[a_i, b_i]$ with $b_i > 0$, for all $\epsilon > 0$ there exists $k(\epsilon, T) \in \mathbb{N}$ such that $Prob\{T_{k(\epsilon,T)} < T\} < \epsilon$. According to the construction of the transient tree, $Prob\{T_{k(\epsilon,T)} < T\}$ is the total probability obtained by summing up the quantity $\eta_T(\cdot)$ over all nodes at depth $k(\epsilon, T)$. Since the tree is constructed in breadth first order, for any allowed approximation ϵ, there exists $k(\epsilon, T)$ such that, when the tree reaches depth $k(\epsilon, T)$, the sum of probabilities $\eta_T(\cdot)$ over all the nodes of the frontier of the tree is lower than ϵ.   □

4   Example

Fig. 2 shows the NSTA model of a flexible manufacturing system where one consumer $\mathcal{C}$ alternates between two producers $\mathcal{A}$ and $\mathcal{B}$, scheduling optional activities when time laxity occurs. In so doing, the consumer uses time observations to reduce the probability of performing an optional task while the next producer is waiting for the consumption of the previously produced item: two clocks (a and b) keep memory of the time elapsed since the last task accepted from the two producers. An optional activity is scheduled when the time elapsed since the last synchronization with the next producer is lower than a given threshold. For simplicity of the model, sojourn times are assumed to be uniformly distributed on the supports (but any expolynomial probability density function could be managed as well).

The synchronization between the STA occurs through symmetric handshakes: after the end of a production, producer $\mathcal{A}$ broadcasts a message A and waits for the echo AA; on reception of the echo, it immediately broadcasts a new A message. With this protocol, the first A message (indicating the end of a production phase of $\mathcal{A}$) can be lost if $\mathcal{C}$ is still consuming the item produced by producer $\mathcal{B}$, forcing producer $\mathcal{A}$ to wait for the end of consumption signaled by $\mathcal{C}$ with message AA; the availability of a produced unit is then communicated again with the second A message (the protocol for producer $\mathcal{B}$ is analogous).

The example is intended to point out some basic modeling patterns of the NSTA formalism: (1) clock values do not affect the distribution of location sojourn times, but they can be used to restrict the choice of the next location; (2) reception of a broadcast preempts the sojourn time in the location; (3) broadcast is non-blocking for the sender, and it is relevant only for receivers in an accepting location; (4) on completion of the sojourn time in any location, if no outgoing edge is enabled, an automaton remains blocked until some accepted action is received or clocks reach a value that satisfy some guard on an outgoing edge.

Structural complexities with relevant impact on the analysis are also illustrated by this example: (1) the overall state is composed of the vector of (discrete) locations in the three automata, and of the (continuous) values of the two consumer clocks and three remaining sojourn times; (2) the two consumer clocks

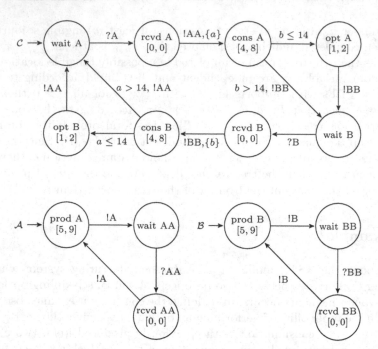

Fig. 2. An alternating producer/consumer scheme

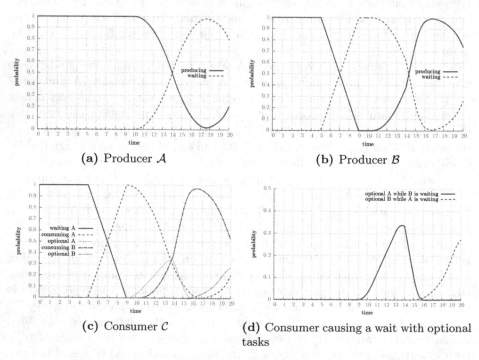

(a) Producer $\mathcal{A}$

(b) Producer $\mathcal{B}$

(c) Consumer $\mathcal{C}$

(d) Consumer causing a wait with optional tasks

Fig. 3. Transient analysis of the producer/consumer example

are independent and overlap their activity cycles between reset points and guard usage; (3) automata include both tangible and vanishing states, and broadcast actions can also be issued on exit from vanishing locations.

For any given time bound T, termination of transient analysis is guaranteed by Theorem 1: every cycle in every automaton visits at least one location with minimum sojourn time greater than zero that cannot be preempted by any broadcast issued on exit from some vanishing state. If sojourn times of all the locations included a minimum sojourn equal to zero, then termination would be guaranteed by Theorem 2 for any allowed error $\epsilon > 0$.

Fig. 3 reports, for each automaton, the transient state probabilities derived through a preliminary implementation of the analysis algorithm of Sect. 3 based on the symbolic calculus library of the ORIS tool [11,14]. As an example of relevant measure, we consider also the probability (for each time instant) that the consumer is performing an optional task while the next producer is waiting.

5   Conclusions

In this paper, we introduced the model of networks of stochastic timed automata, an extension of timed automata with stochastic semantics and message broadcasts between its components. We provided a characterization of the underlying stochastic process of STA and NSTA, identifying the conditions corresponding to each family of stochastic processes (i.e., CTMCs, SMPs, MRPs and GSMPs).

A technique for the transient analysis of NSTA based on stochastic state classes was proposed. We defined an iterative procedure for the construction of the transient tree of stochastic state classes, and identified sufficient conditions for its termination. Modeling examples were presented in order to illustrate common design patterns of the formalism, providing transient state probabilities computed analytically with the ORIS tool (and validated through simulation). Notably, this result constitutes the basis for the application of time-bounded probabilistic model checking techniques based on stochastic state classes to networks of STA.

References

1. Alur, R., Bernadsky, M.: Bounded model checking for GSMP models of stochastic real-time systems. In: Hespanha, J.P., Tiwari, A. (eds.) HSCC 2006. LNCS, vol. 3927, pp. 19–33. Springer, Heidelberg (2006)
2. Alur, R., Courcoubetis, C., Dill, D.: Model-checking for Probabilistic Real-time Systems. In: Leach Albert, J., Monien, B., Rodríguez-Artalejo, M. (eds.) ICALP 1991. LNCS, vol. 510, pp. 115–126. Springer, Heidelberg (1991)
3. Alur, R., Courcoubetis, C., Dill, D.L.: Verifying Automata Specifications of Probabilistic Real-time Systems. In: Huizing, C., de Bakker, J.W., Rozenberg, G., de Roever, W.-P. (eds.) REX 1991. LNCS, vol. 600, pp. 28–44. Springer, Heidelberg (1992)
4. Alur, R., Dill, D.L.: A theory of timed automata. Theor. Comput. Sci. 126(2), 183–235 (1994)

5. Baier, C., Bertrand, N., Bouyer, P., Brihaye, T., Größer, M.: Probabilistic and topological semantics for timed automata. In: Arvind, V., Prasad, S. (eds.) FSTTCS 2007. LNCS, vol. 4855, pp. 179–191. Springer, Heidelberg (2007)
6. Baier, C., Bertrand, N., Bouyer, P., Brihaye, T., Größer, M.: Almost-sure model checking of infinite paths in one-clock timed automata. In: LICS 2008, pp. 217–226. IEEE CS (2008)
7. Behrmann, G., Larsen, K.G., Pearson, J., Weise, C., Yi, W.: Efficient timed reachability analysis using clock difference diagrams. In: Halbwachs, N., Peled, D.A. (eds.) CAV 1999. LNCS, vol. 1633, pp. 341–353. Springer, Heidelberg (1999)
8. Bertrand, N., Bouyer, P., Brihaye, T., Markey, N.: Quantitative model-checking of one-clock timed automata under probabilistic semantics. In: QEST 2008, pp. 55–64. IEEE CS (2008)
9. Bobbio, A., Telek, M.: Markov regenerative SPN with non-overlapping activity cycles. In: IPDS 1995, pp. 124–133. IEEE CS (1995)
10. Bouyer, P., Brihaye, T., Jurdzinski, M., Menet, Q.: Almost-sure model-checking of reactive timed automata. In: QEST 2012, pp. 138–147. IEEE CS (2012)
11. Bucci, G., Carnevali, L., Ridi, L., Vicario, E.: Oris: a tool for modeling, verification and evaluation of real-time systems. Int. J. on Softw. Tools for Techn. Transfer 12(5), 391–403 (2010)
12. Bucci, G., Piovosi, R., Sassoli, L., Vicario, E.: Introducing probability within state class analysis of dense time dependent systems. In: QEST 2005, pp. 13–22. IEEE CS (2005)
13. Carnevali, L., Grassi, L., Vicario, E.: State-density functions over DBM domains in the analysis of non-Markovian models. IEEE Trans. Softw. Eng. 35(2), 178–194 (2009)
14. Carnevali, L., Ridi, L., Vicario, E.: A framework for simulation and symbolic state space analysis of non-Markovian models. In: Flammini, F., Bologna, S., Vittorini, V. (eds.) SAFECOMP 2011. LNCS, vol. 6894, pp. 409–422. Springer, Heidelberg (2011)
15. Ciardo, G., German, R., Lindemann, C.: A characterization of the stochastic process underlying a stochastic Petri net. IEEE Trans. Softw. Eng. 20(7), 506–515 (1994)
16. Cox, D.R.: The analysis of non-Markovian stochastic processes by the inclusion of supplementary variables. Math. Proc. Cambridge 51, 433–441 (1955)
17. Dill, D.L.: Timing assumptions and verification of finite-state concurrent systems. In: Sifakis, J. (ed.) CAV 1989. LNCS, vol. 407, pp. 197–212. Springer, Heidelberg (1990)
18. Haas, P.J.: Stochastic Petri Nets: Modelling, Stability, Simulation. Springer (2002)
19. Horváth, A., Paolieri, M., Ridi, L., Vicario, E.: Probabilistic model checking of non-Markovian models with concurrent generally distributed timers. In: QEST 2011, pp. 131–140. IEEE CS (2011)
20. Horváth, A., Paolieri, M., Ridi, L., Vicario, E.: Transient analysis of non-Markovian models using stochastic state classes. Perform. Eval. 69(7-8), 315–335 (2012)
21. Kulkarni, V.: Modeling and analysis of stochastic systems. Chapman & Hall (1995)
22. Kwiatkowska, M., Norman, G., Parker, D.: PRISM 4.0: Verification of probabilistic real-time systems. In: Gopalakrishnan, G., Qadeer, S. (eds.) CAV 2011. LNCS, vol. 6806, pp. 585–591. Springer, Heidelberg (2011)
23. Kwiatkowska, M., Norman, G., Segala, R., Sproston, J.: Verifying quantitative properties of continuous probabilistic timed automata. In: Palamidessi, C. (ed.) CONCUR 2000. LNCS, vol. 1877, pp. 123–137. Springer, Heidelberg (2000)

24. Kwiatkowska, M., Norman, G., Segala, R., Sproston, J.: Automatic verification of real-time systems with discrete probability distributions. Theor. Comput. Sci. 282(1), 101–150 (2002)
25. Maler, O., Larsen, K.G., Krogh, B.H.: On zone-based analysis of duration probabilistic automata. In: INFINITY, pp. 33–46 (2010)
26. Vicario, E., Sassoli, L., Carnevali, L.: Using stochastic state classes in quantitative evaluation of dense-time reactive systems. IEEE Trans. Softw. Eng. 35(5), 703–719 (2009)

Automated Rare Event Simulation
for Stochastic Petri Nets

Daniël Reijsbergen, Pieter-Tjerk de Boer,
Werner Scheinhardt, and Boudewijn Haverkort

Center for Telematics & Information Technology,
University of Twente, Enschede, The Netherlands

Abstract. We introduce an automated approach for applying rare event simulation to stochastic Petri net (SPN) models of highly reliable systems. Rare event simulation can be much faster than standard simulation because it is able to exploit information about the typical behaviour of the system. Previously, such information came from heuristics, human insight, or analysis on the full state space. We present a formal algorithm that obtains the required information from the high-level SPN-description, without generating the full state space. Essentially, our algorithm reduces the state space of the model into a (much smaller) graph in which each node represents a set of states for which the most likely path to failure has the same form. We empirically demonstrate the efficiency of the method with two case studies.

1 Introduction

The first step towards the analysis of a highly dependable system is its specification as a state transition system. When the behaviour of the system is stochastic, a common model is the (discrete- or continuous-time) Markov chain. The state space of the Markov chain can be very large (even infinite), but the chain often has enough structure to allow for implicit specification using a high-level description language. Classical examples of such languages are stochastic Petri nets (SPNs) [1], and stochastic activity networks [24].

Given an SPN, one specifies a measure for the performance of the highly dependable system in terms of its stochastic properties. The measure that we focus on in this paper is the probability that one reaches a certain uncommon set of states (the *goal* set) before reaching a more typical set (the *taboo* set). This probability can be interesting by itself, but is particularly interesting as it appears in expressions for, e.g., the Mean Time To Failure, the time-bounded unreliability and the steady-state unavailability. Numerical methods for computing this probability are well-established, but since they operate mostly on the complete state space, which is often very large, they can be computationally infeasible (an issue commonly referred to as the *state space explosion problem*).

A remedy is then to use stochastic (discrete-event) simulation [16], i.e., repeatedly generating random executions of the system model and using the average behaviour observed in the executions to obtain an estimate of the probability of

K. Joshi et al. (Eds.): QEST 2013, LNCS 8054, pp. 372–388, 2013.

interest. Discrete-event simulation can be carried out on the level of the SPN and only requires that, overall, the current state in the system is stored instead of the entire state space. A common problem is that when the goal set is rare (like failure states in a highly reliable system) one needs an infeasibly large number of executions to obtain an accurate estimate.

In order to reduce the number of executions needed, several *efficient simulation methods* have been proposed in the past few decades. They can be largely divided into two main categories: *importance sampling* methods [10], and *RESTART* and *multilevel splitting* [8,27] methods. Both can use knowledge of the typical *paths toward* or the *distance to* the goal set to their advantage. Several techniques have been implemented in the past two decades [4,12,21,26,28], but all of these rely on user input or the adequacy of heuristics in order to perform well.

In this paper we show that the required information can be obtained in an automated way from the SPN and the description of the goal and taboo sets. As such, we present a formal algorithm that achieves this. It uses the structure of the SPN to divide the implied state space into zones, in each of which the distance to the goal set can be expressed using the same distance function. In this way we can find the overall distance function, which can then be used in an efficient simulation procedure. We demonstrate the potential gain of the method, both for a simple example (which is also used as a running example throughout the paper), and a more demanding model of a multicomponent system with interdependent component types.

The structure of the rest of this paper is as follows: in Section 2, we explain the position of this paper in the context of the earlier scientific literature. In Section 3 we discuss the exact definition of an SPN that we will use throughout this paper, and explain the foundations of (rare event) simulation. The core algorithm that determines the distance function in an automated way is the topic of Section 4. Section 5 contains a simulation study involving the simple model and a more realistic model. In Section 6, we discuss a few challenges associated with the new method and ways to overcome them, before we conclude the paper.

2   Context within the Literature

One way to obtain knowledge about the way the system progresses toward the goal set is to divide the transitions in the SPN into failure and repair transitions that respectively take the system towards or away from the goal set. One can then apply *failure biasing* [25]. This has been implemented in, among others, SAVE (see [4]) and in UltraSAN [21], the predecessor to the tool Möbius [6].

One variation of failure biasing that is especially noteworthy in the context of this paper is distance failure biasing [5]. It is based on a notion of distance similar to the one we introduce in Section 3. However, the technique presented in [5] can only be applied to a very narrow class of models (namely models with independent component types) and the gains compared to failure biasing may not justify the numerical effort of the minimal cut algorithm that is used (see also the discussion in [20]).

Another technique is to split the simulation effort into two different stages: one to obtain information about the typical behaviour to the rare set and one to use this knowledge in an importance sampling scheme. This idea forms the basis of the *cross-entropy method* for importance sampling [23] [11] and Kelling's framework for RESTART in SPN [14]. The cross entropy method has recently been implemented in the PLASMA-platform [12].

For RESTART and splitting, one implicitly divides the state space of the model in several *level sets*. Some examples of how to determine these level sets are to let the user specify them by hand [18,26], or to use a two-step approach similar to the one underlying the cross-entropy method [14]. The splitting framework has been implemented in the Stochastic Petri Net Package [26] and the tool TimeNet [28]. The methods based on this principle are largely heuristic in nature.

3   Model and Preliminaries

The outline of this section is as follows. In Section 3.1, we describe the type of Petri nets we consider throughout the paper. In Section 3.2, we illustrate this with an example that we use throughout this paper. In Section 3.3, we discuss the performance property of interest, and we discuss simulation in Section 3.4.

3.1   Discrete-Time Stochastic Petri Nets

We assume that the reader is familiar with the general concept of a Petri net (if not, see e.g. [19]). We use Multi-Guarded Petri Nets as in [13], although we extend the net with marking-dependent firing rates for the transitions. We define a Petri net to be $(P, T, Pre, Post, G)$, where

- $P = \{1, 2, \ldots, |P|\}$ denotes the set of *places*,
- $T = \{t_1, \ldots, t_{|T|}\}$ denotes the set of *transitions*,
- $Pre : P \times T \to \mathbb{N}$ and $Post : P \times T \to \mathbb{N}$ are the *pre-* and *post- incidence functions*.[1]
- G denotes the set of guards (more details are given below).

We are interested in the (embedded) *discrete-time behaviour* of the Petri net; let $X_i(n)$ be the number of tokens in place i after the n-th time a transition is fired, $n \in \mathbb{N}$. Let $\boldsymbol{X}(n) = (X_1(n), \ldots, X_{|P|}(n))^{\mathrm{T}}$ be the *marking* (or *state*) of the net at time n. Let $\mathcal{X} = \mathbb{N}^{|P|}$ be the set of all possible markings; then we let transition t_i have exponential rate $\lambda_i(\boldsymbol{x})$ with $\boldsymbol{x} \in \mathcal{X}$. Importantly, although we allow the rates λ_i to depend on the marking $\boldsymbol{x}$ we assume that these rates are functions of ϵ (see below) and that numbers r_i exist such that for all $\boldsymbol{x} \in \mathcal{X}$, $\lambda_i(\boldsymbol{x}) = \Theta(\epsilon^{r_i})$, i.e.,

$$0 < \lim_{\epsilon \downarrow 0} \frac{\lambda_i(\boldsymbol{x})}{\epsilon^{r_i}} < \infty$$

---

[1] We use $\mathbb{N} = \{0, 1, 2, \ldots\}$.

The rate $\lambda_i(\boldsymbol{x}(n))$ determines the relative likelihood of the transition to *fire* at step n. The number ϵ is the so-called *rarity parameter*, which is typically a small number that signifies how rare the event of interest is.

When transition t fires, the marking changes as follows: $Pre(p,t)$ tokens are removed from place p while $Post(p',t)$ tokens are added to place p'. A transition cannot fire if this would result in a negative number of tokens in a place, nor can it fire when one of its guards is not enabled (as discussed below). The guards can be described in terms of *constraints*, a concept that we will use often in Section 4. A constraint $c = (\boldsymbol{\alpha}, \beta, \bowtie)$ is an element of $\mathbb{Z}^{|P|} \times \mathbb{Z} \times \{\leq, \geq\}$, and we say that marking $\boldsymbol{x}$ satisfies constraint c if $\boldsymbol{\alpha}^\mathrm{T} \boldsymbol{x} \bowtie \beta$. A *guard* g is then a 4-tuple $(p, t, \beta, \bowtie)$ that imposes upon a transition t the necessary condition that it can only fire in $\boldsymbol{x}$ if the number of tokens in place p satisfies the inequality $x_p(\cdot) \bowtie \beta$. Let

$$\mathbf{1}_i(\boldsymbol{x}) = \begin{cases} 1 & \text{if } \forall (p, t_i, \beta, \bowtie) \in G : x_p(\cdot) \bowtie \beta, \\ 0 & \text{otherwise,} \end{cases} \tag{1}$$

If $\mathbf{1}_i(\boldsymbol{x}(n)) = 1$, we say that transition t_i is enabled at time n. If there are no guards $g \in G$ such that $g = (\cdot, t, \cdot, \cdot)$ then the transition t is *always* enabled. Let the total incidence vector $\boldsymbol{u}_i = (u_{i1}, \ldots, u_{i|P|})$ of transition t_i be the vector that describes the effect of firing t_i on the marking. It is defined by $u_{ij} = Post(j, i) - Pre(j, i)$. Then the probability measure governing the marking process $\boldsymbol{X}(n)$ is uniquely characterised by

$$\mathbb{P}(\boldsymbol{x}(n) \to \boldsymbol{x}(n+1)) = \mathbb{P}\left(\boldsymbol{X}(n+1) = \boldsymbol{x}(n+1) \mid \boldsymbol{X}(n) = \boldsymbol{x}(n)\right)$$
$$= \frac{\sum_{i \in \mathcal{I}} \lambda_i(\boldsymbol{x}(n)) \mathbf{1}_i(\boldsymbol{x}(n))}{\sum_{j=1}^{|T|} \lambda_j(\boldsymbol{x}(n)) \mathbf{1}_j(\boldsymbol{x}(n))}, \tag{2}$$

where $\mathcal{I} = \{i \in \mathbb{N} : t_i \in T, \boldsymbol{x}(n+1) = \boldsymbol{x}(n) + \boldsymbol{u}_i\}$.

3.2 Running Example

The running example that we use throughout this paper is a reliability model equivalent to a two-node $M/M/1$ tandem queue. It can be seen as a single component with an infinite number of hot spares; when a component or a spare breaks down, two repair phases have to be completed consecutively. Component and spares fail according to a Poisson process with rate $\lambda = \Theta(\epsilon^2)$. The times between first phase repairs are exponentially distributed with rate $\mu = \Theta(\epsilon)$. The times between second phase repairs are exponentially distributed with rate $\nu = \Theta(1)$. We assume that none of the rates depend on the marking, and that both queues will be empty most of the time. This system can be modelled using an SPN as depicted in Figure 1. The typical rare event that we are interested in is having n or more components awaiting the second phase of repair before all components have been repaired, starting from the first break-down of the main component. This rare event can be cast in the more general framework outlined in Section 3.3.

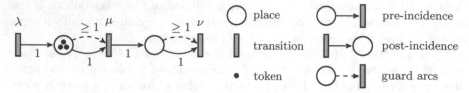

Fig. 1. Tandem queue, depicted in the form of a stochastic Petri net

3.3 Problem Setting

From now on, we will call elements of the taboo set a-markings and elements of the goal set b-markings. We then seek to estimate the probability of reaching a b-marking before reaching an a-marking starting from an initial marking x_0. Let $G_a = \{g_a^1, \ldots, g_a^{|G_a|}\} \subset P \times (\mathbb{N} \cup 0) \times \{\leq, \geq\}$ be the set of a-constraints, and let

$$1_g(x) = \begin{cases} 1 & \text{if } g = (p, c, \bowtie) \text{ and } x_p \bowtie c, \\ 0 & \text{otherwise.} \end{cases}$$

for all $x \in \mathcal{X}$. Let $X \subset \mathcal{X}$ be a a-based hyperrectangle if $\forall g \in G_a\colon \forall x \in X\colon$ $1_g(x) \equiv c(g, X)$, where $c(g, X) \in \{0, 1\} \forall g \in G_a$. Then let the taboo set $\mathcal{X}_a$ be any union of a-based hyperrectangles. The goal set $\mathcal{X}_b$ are defined similarly for G_b. If a marking is both an a- and b-marking, we will consider it to be a b-marking only. In LTL-notation [3], the event of interest can be written as $\neg a \,\mathrm{U}\, b$; in this paper we will denote the event of interest by $\Psi_{x_0} = \{(\omega_0, \ldots, \omega_m) : m \in \mathbb{N} : \omega_0 = x_0, \omega_m \in \mathcal{X}_b, \omega_k \notin \mathcal{X}_a \,\forall k = 0, \ldots, m - 1\}$ in order to emphasise the dependence on the initial state, and denote its probability of interest as $\mathbb{P}(\Psi_{x_0})$.

3.4 Efficient Simulation

We will estimate the probability $\mathbb{P}(\Psi_{x_0})$ using a series of N simulation runs, for some constant $N \in \mathbb{N}$. In each run, we initialise the marking to be x_0. We then iteratively fire transitions using the probability measure $\mathbb{P}$ as defined in (2) until we reach an a- or b-marking. When we terminate, we can set $w_i = 1$ if the event Ψ_{x_0} occurred on run i (i.e. if we ended in $\mathcal{X}_b$) and to $w_i = 0$ otherwise, and then obtain the standard Monte Carlo (MC) estimator $\widehat{p}$ for $\mathbb{P}(\Psi_{x_0})$ as

$$\widehat{p}_\mathbb{P} = \frac{1}{N} \sum_{i=1}^{N} w_i.$$

A confidence interval for $\widehat{p}$ can be constructed for large N using the Central Limit Theorem [16].

Our focus will be the case where $\mathbb{P}(\Psi_{x_0})$ is small, as this is typically the case in a highly reliable system setting. In this situation, N needs to be very large to obtain a reasonable estimate for $\widehat{p}$. To remedy this, we apply importance sampling.[2] Instead of sampling directly from $\mathbb{P}$, we use a different probability

---

[2] We note here that the distance function d could also be used to construct level sets for RESTART/splitting.

measure $\mathbb{Q}$; after sampling the runs $(\boldsymbol{x}_i(0), \boldsymbol{x}_i(1), \ldots, \boldsymbol{x}_i(n_i))$, $i = 1, \ldots, N$, we use the importance sampling (IS) estimator

$$\widehat{p}_{\mathbb{Q}} = \frac{1}{N} \sum_{i=1}^{N} w_i \prod_{j=0}^{n_i-1} \frac{\mathbb{P}(\boldsymbol{x}_i(j) \to \boldsymbol{x}_i(j+1))}{\mathbb{Q}(\boldsymbol{x}_i(j) \to \boldsymbol{x}_i(j+1))}. \tag{3}$$

If a suitable new measure $\mathbb{Q}$ is chosen, the number of runs required to obtain a reasonable estimate can be reduced dramatically. The choice of the new measure $\mathbb{Q}$ is non-trivial, however. Typically, good simulation measures $\mathbb{Q}$ increase the likelihood of $\Psi_{\boldsymbol{x}_0}$ occurring, albeit not too strongly. In order to make $\Psi_{\boldsymbol{x}_0}$ more likely, $\mathbb{Q}$ must in some way push the marking in the direction of the goal set and away from the taboo set. The first challenge that arises is then to determine how *far* a marking is from the goal set, so that the simulation $\mathbb{Q}$ can increase the likelihood of moving to a marking with lower distance. For this paper, we define the distance function $d(\boldsymbol{x})$ as

$$d(\boldsymbol{x}) = \min\{r : \exists \omega \in \Psi_{\boldsymbol{x}} \text{ s.t. } \mathbb{P}(\omega) = \Theta(\epsilon^r)\}, \tag{4}$$

where we use the fact that, in essence, the event $\Psi_{\boldsymbol{x}}$ is simply a set of sequences of markings. In words, $d(\boldsymbol{x})$ is the *minimal distance* or *cost* in terms of the ϵ-order of the path from $\boldsymbol{x}$ to the goal set. If the set over which the minimum is taken is empty, we let $d(\boldsymbol{x}) = \infty$. Given $d(\boldsymbol{x})$, we use the following measure $\mathbb{Q}$:

$$\mathbb{Q}(\boldsymbol{x}(j) \to \boldsymbol{x}(j+1)) = \frac{\mathbb{P}(\boldsymbol{x}(j) \to \boldsymbol{x}(j+1))\epsilon^{d(\boldsymbol{x}(j+1))}}{\sum_{\boldsymbol{x}'} \mathbb{P}(\boldsymbol{x}(j) \to \boldsymbol{x}')\epsilon^{d(\boldsymbol{x}')}}. \tag{5}$$

This estimator can be proven to have so-called bounded relative error under some assumptions (more on this in Section 6.2). The remaining problem is then to find $d(\boldsymbol{x})$ for each possible marking $\boldsymbol{x}$. This will be the topic of Section 4.

4 An Algorithm for Determining the Distance Function

In this section we discuss an automated algorithm for finding the function d as defined in (4). The algorithm is executed during a pre-processing phase, before the actual simulation phase starts. Since d is the solution to a shortest path problem in a weighted graph,[3] we could apply Dijkstra's algorithm to find d explicitly for each state (i.e. marking) in the state space $\mathcal{X}$. However, since Dijkstra's algorithm uses the complete state space, it is not better than standard numerical algorithms. Hence, our aim will be to partition $\mathcal{X}$ into *zones* such that for each zone, all the states in this zone have a *similar* cost function d in a sense to be detailed below. Formally, let a zone z be a set of *constraints* $\{c_1^z, \ldots, c_{|z|}^z\}$, as defined in Section 3.1. Let the *zone set* $\mathcal{X}_z$ be the set of states that satisfy

---

[3] Namely one which corresponds to the underlying Markov chain and with the costs of the transitions in terms of ϵ-orders as weights. For another application of Dijkstra's algorithm to finding the most likely paths in a Markov chain, see [9].

all constraints in z. The idea is then to find a set of zones Z such that the sets $\mathcal{X}_z, z \in Z$, form a partition of $\mathcal{X}$ and that we can find functions $d_z(\boldsymbol{x})$ that give an easy expression for the distance to $\mathcal{X}_b$ of all states $\boldsymbol{x} \in z$.

Particularly, we aim to construct a *zone graph*; a graph where the nodes correspond to the zones of Z and in which there is an arc from zones z to z' if for each state $\boldsymbol{x} \in \mathcal{X}_z$ we can reach some state in z' through repeated firing of *a single transition*. We will call such a repeated firing a *stutter step*, as in, e.g., [2]. Furthermore, we want the shortest path from any state in z to $\mathcal{X}_b$ to correspond to the same path through the zone graph. Finally, we want the cost in terms of ϵ-orders of firing the transition of the stutter step to be the same in all states in the same zone set. If all these conditions hold, then for each zone z we can find a function d_z that is the *same* affine function for all $\boldsymbol{x} \in \mathcal{X}_z$ (a function f is affine if $f(\boldsymbol{x}) = \boldsymbol{\alpha}^{\mathrm{T}}\boldsymbol{x} + \beta$ for some $\boldsymbol{\alpha} \in \mathbb{R}^{|P|}$ and $\beta \in \mathbb{R}$). In this section, we will clarify how this can be done.

To make the preceding concrete, consider the running example. The first two zone sets that we create are $\mathcal{X}_a$ and $\mathcal{X}_b$; in particular, $\mathcal{X}_b$ consists of the states in which $x_2 \geq n$; we assume $n \geq 3$. In the state $(1, n-1)$, t_2 needs to fire once to reach $\mathcal{X}_b$, and the distance of this step is 1 (because t_2 needs to 'win the race' from t_3, which fires ϵ^{-1} times faster). In $(2, n-2)$, we need to fire t_2 twice, giving a total distance of 2. The same holds for all states $(x, n-x)$, $x \geq 1$; we fire t_2 x times and the total distance is x. It then makes sense to group all these states together in a zone set. However, for $(n, 0)$, we need to fire t_2 n times, but the total cost is $n-1$ as t_2 does not need to compete against t_3 in the first step. Hence, $(n, 0)$ and $(n-1, 1)$ will not be in the same zone. The complete set of zones, with their distance functions and shortest paths to the taboo set, is illustrated in Figure 2.

Algorithm 1. Main loop.

```
1: initZoneGraph()
2: while S ≠ ∅ do
3:     s = (z°, tᵢ, zᵈ) := some element from S
4:     possibilitySplit(s)
5:     if d_{z°} ≠ unassigned then costSplit(s)
6:     update(s); S := S\s
7: end while
```

In Section 4.1, we will outline the main algorithm. As in the previous example, an initial partitioning is always necessary, as we will discuss in Section 4.2. However, this initialisation alone is not sufficient. It may be that it is not possible for all markings in an initial zone to reach another zone by the same stutter step; this is the topic of Section 4.3. Also, there may exist markings within a single zone for which the shortest path follows a different sequence of stutter steps; more on that in Section 4.4.

4.1   Main Loop

Let a *stutter step* s be a triple (z^o, t_i, z^d), where z^o is the source/origin zone, z^d is the destination zone and t_i is the transition that is repeatedly fired. The algorithm works as follows: we keep a list S of stutter steps that could be part of shortest paths. After initialising the list, we repeatedly take stutter steps s out of S and check whether for all markings in the origin zone of s it holds that

1) we can indeed reach the destination zone of s using only the given stutter step, and
2) the new distance function indeed gives shorter distance than what was known before.

If not, we split up the source zone and (potentially) add new stutter steps to S. Finally, we discard s, pick a new stutter step, and repeat until S is empty. The precise way in which this is done is given by Algorithm 1.

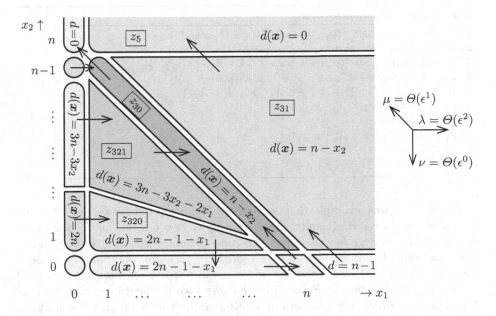

Fig. 2. The final result of a call to the algorithm, excluding lines 2 and 3 of initZoneGraph()

4.2   Initialisation Phase (initZoneGraph())

During the initialisation phase, the state space is divided into zones such that

a) from all states in the same zone set the same transitions are enabled, and
b) all states in a zone set are either in $\mathcal{X}_a$, all in $\mathcal{X}_b$ or all in neither.

Condition a) implies that the cost of firing a transition is always the same in a zone (because the cost depends on which other transitions can be fired). During the initialisation we can already assign distance ∞ to the states in $\mathcal{X}_a$ and 0 to the states in $\mathcal{X}_b$. Furthermore, we initialise the stutter step list S during this phase; its initial elements will be those stutter steps that directly lead into $\mathcal{X}_b$. The precise way in which all this is done is given in Algorithm 2.

Lines 2 and 3 deal with a technical obstacle; when for a stutter step (z^o, t_i, z^d) it holds that $z^o = z^d$, line 1 of possibilitySplit() will fail. However, we cannot exclude these 'self-loops' in the zone graph; there exist cases in which the shortest path moves to the edge of an initial zone without crossing it. To remedy this, we also create 'edges' around the initial zones of line 1.

In line 4 of Algorithm 2, we use the negation $\neg c$ of a constraint c. If $c = (\alpha, \beta, \leq)$, then its negation is given by $\neg c = (\alpha, \beta + 1, \geq)$, and if $c = (\alpha, \beta, \geq)$ then $\neg c = (\alpha, \beta - 1, \leq)$. If all elements of α are at least 1, then the resulting zone sets $\mathcal{X}_{\{c\}}$ and $\mathcal{X}_{\{\neg c\}}$ are each other's complements with respect to $\mathcal{X}$.

Algorithm 2. initZoneGraph()

1: $C' := \{c = (p, \beta, \bowtie) : (p, \cdot, \beta, \bowtie) \in G \vee c \in G_a \cup G_b\}$
2: $u^{\max} := \max_{i=1,\ldots,|T|} \max_{k=1,\ldots,|P|} |u_{ik}|$
3: $C := \{c = (p, \beta, \bowtie) : (p, \beta + k, \bowtie) \in C', k \in \mathbb{Z}, |k| \leq u^{\max}\}$
4: $Z := \{z \in \mathcal{Z} : \forall c \in C : c \in z \vee \neg c \in z, \mathcal{X}_z \neq \emptyset\}$ $\triangleright$ $\mathcal{Z}$ = set of all zones
5: $V := \{(z^o, t_i, z^d) : \exists x \in \mathcal{X} : x \in \mathcal{X}_{z^o}, x + u_i \in \mathcal{X}_{z^d}\}$
6: $Z_a := \{z \in Z : \forall x \in \mathcal{X}_z : x \in \mathcal{X}_a\}$
7: $\forall z \in Z_a : d_z := \infty$
8: $Z_b := \{z \in Z : \forall x \in \mathcal{X}_z : x \in \mathcal{X}_b\}$
9: $\forall z \in Z_b : d_z = 0$
10: $S := \{v \in V : v = (z, \cdot, z'), z \notin Z_a, z' \in Z_b\}$

For the running example as displayed in Figure 1, the transition structure first gives us four initial zones: z_0 where only t_1 can fire, z_1 for t_1 and t_2, z_2 for t_1 and t_3, and z_3 for all three. The zone structure resulting from a call to initZoneGraph() is displayed in Figure 3(a). In fact, for the running example the algorithm would also work well if we would not include margins, i.e. omit lines 2 and 3, resulting in Figure 3(b). For the sake of clarity, we will continue based on the latter, even though our implementation does include the margins. We get two additional zones, z_4 and z_5, to distinguish $\mathcal{X}_b$. S is initialised with all stutter steps leading into these two zones; the only stutter steps satisfying this requirement are the two t_2-stutter steps going from z_3 into z_4 and z_5.

4.3   Divide Zones According to Possibility of Firing (possibilitySplit())

To determine the cost of a stutter step $s = (z^o, t_i, z^d)$, we need to determine the number of times y that t_i must fire to take a marking in z^o to z^d. This is done by findNumberOfTransitions(). The main idea is to find a function $y(x)$ (written as y for brevity) such that after firing t_i $y - 1$ times, the marking is still in z^o, and after firing one more time the marking is in z^d. In order to find this number, we choose any constraint c_1 from z^o and c_2 from z^d that exclude each other, i.e., $\mathcal{X}_{z^o} \cap \mathcal{X}_{z^d} = \emptyset$, and chooses y to be the smallest number of firings to enable c_2. Since all constraints are non-strict inequalities, y is chosen such that $x + y u_i$ exactly satisfies the constraint. The remaining constraints in z^o and z^d then impose restrictions on x that must be satisfied in order for this stutter step to be carried out.

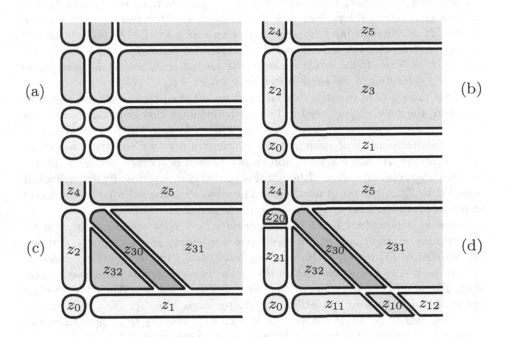

Fig. 3. Figure (a) illustrates the result of a call to `initZoneGraph()` when lines 2 and 3 are included (we only show the margins around the axes). Figures (b-d) depict the zones after several iterations of the algorithm, without lines 2 and 3 of `initZoneGraph()`.

Algorithm 3. possibilitySplit().

Require: stutter step s
1: $(c_1, c_2) :=$ some two constraints such that
        1) $c_1 \in z^o$,   2) $c_2 \in z^d$ and   3) $\mathcal{X}_{\{c_1\}} \cap \mathcal{X}_{\{c_2\}} = \emptyset$
2: $y :=$ findNumberOfTransitions($c_2, \boldsymbol{u}_i$)
3: $C_1 := \{c : c = \boldsymbol{a}(\boldsymbol{x} + (y - 1)\boldsymbol{u}_i) \bowtie b \wedge \boldsymbol{ax} \bowtie b \in z^o \backslash c_1\}$
4: $C_2 := \{c : c = \boldsymbol{a}(\boldsymbol{x} + y\boldsymbol{u}_i) \bowtie b \wedge \boldsymbol{ax} \bowtie b \in z^d \backslash c_2\}$
5: $C := C_1 \cup C_2$
6: $Z_{\text{new}} := \{z : \forall c \in C : c \in z \vee \neg c \in z \wedge \forall c \in z^o : c \in z \wedge \exists \boldsymbol{x} \in \mathcal{X} : \boldsymbol{x} \in \mathcal{X}_z\}$
7: $z^n := z \in Z_{\text{new}} : \forall c \in C : c \in z$
8: $d_{z^n}(\boldsymbol{x}) := d_{z^d}(\boldsymbol{x} + y\boldsymbol{u}_i) + y\kappa_i(\boldsymbol{x})$ $\triangleright$ where $\kappa_i(\boldsymbol{x}) = \frac{1_i(\boldsymbol{x})r_i}{\sum_{j=1}^{|T|} 1_j(\boldsymbol{x})r_j}$

Assume that we happen to first consider the μ-stutter step from z_3 to z_4. After the initialisation phase, there are two pairs of constraints from z_3 and z_4 that exclude each other; the pair $x_1 \geq 1$ and $x_1 \leq 0$, and the pair $x_2 \leq n - 1$ and $x_2 \geq n$. If we consider the first pair, we end up with $y = x_1$. The two constraints that we end up through lines 4 and 5 of Algorithm 3 are $x_1 + x_2 - 1 \leq n - 1$ and $x_1 + x_2 \geq n$. If we would consider the second pair, we would have found $y = n - x_2$, leading to the same restrictions on $x_1 + x_2$.

Given the set C of constraints that must be satisfied for the stutter step s to be taken, the zone z^o may need to be subdivided such that one zone remains in which the stutter step s is always possible. This is done in line 6 of Algorithm 3; all zones that consist of combinations of constraints in C or their negations are considered. If such a zone is non-empty (which is checked using an Integer Linear Programming-solver, although this can be computationally expensive), it is added to Z_{new}, the set of new zones. The zone z^n is the subzone (i.e. a subset in terms of constraints) of z^o for which s was possible.

Since we obtained the additional constraints $x_1 + x_2 \leq n$ and $x_1 + x_2 \geq n$ for the running example, we obtain three new non-empty zones; z_{30}, z_{31} and z_{32}, all depicted in Figure 3(c). Of those, z_{30} has cost $d_{z_{30}}(\boldsymbol{x}) = x_1$ or, equivalently, $d_{z_{30}}(\boldsymbol{x}) = n - x_2$, depending on which of the two constraint pairs was considered. The other two zones do not have any cost assigned yet. When the function update() *in Algorithm 1 is called, the stutter steps from z_1, z_2, z_{31} and z_{32} to z_{30} are added to S. Furthermore, the stutter step from z_3 to z_5 is removed, as z_3 no longer exists. It is replaced by the μ-stutter step from z_{31} to z_5.*

Upon further calls to possibilitySplit()*, the zone z_1 is subdivided into three new zones and z_2 into two new zones, and distance functions are assigned to all. This is displayed in Figure 3(d). Furthermore, zones z_{31} and z_{32} have distance assigned to them. In particular, we mention the distance function of z_{32}: $d_{z_{32}}(\boldsymbol{x}) = 3n - 2x_1 - 3x_2$. In the next section, z_{32} is split into two zones, only one of which retains this distance function.*

4.4   Divide Zones According to Costs (costSplit())

Algorithm 4. costSplit()

Require: step s
1: $c_n := d_{z^n}(\boldsymbol{x}) - d_{z^d}(\boldsymbol{x}) < 0$
2: $z' := z^n \cup \{c_n\}$
3: $z'' := z^n \cup \{\neg c_n\}$
4: $d_{z'}(\boldsymbol{x}) := d_{z^n}(\boldsymbol{x})$
5: $d_{z''}(\boldsymbol{x}) := d_{z^d}(\boldsymbol{x})$
6: **if** $\exists \boldsymbol{x} \in \mathcal{X} : \boldsymbol{x} \in \mathcal{X}_{z'}$ **then**
7:     **if** $\nexists \boldsymbol{x} \in \mathcal{X} : \boldsymbol{x} \in \mathcal{X}_{z''}$ **then**
8:         $Z_{\text{new}} := Z_{\text{new}} \backslash z^n \cup z'$
9:     **else**
10:         $Z_{\text{new}} := Z_{\text{new}} \backslash z^n \cup z' \cup z''$
11:     **end if**
12: **end if**

When the algorithm as described so far is executed, it will consecutively consider zones to which no distance function has yet been assigned yet, split them and assign costs to them. However, when a zone is considered that *already has* a distance function assigned to it, the new path may be the shortest only for a subset of the zone. We need costSplit() for these situations.

Say that, after running Algorithm 3, one has found a subzone z^n of z^o for which the stutter step under consideration can be applied, and for which

d_{z^n} is the distance function. If d_{z^o} has already been assigned, then the stutter step under consideration is only interesting for those markings x for which $d_{z^n}(x) < d_{z^o}(x)$. This constraint is exactly the one constructed in line 1. The zone z^n is then divided into two new zones: z', for which this constraint holds, and z'', for which it does not. If z' is empty, the stutter step under consideration has been irrelevant, and the list S should not be updated. If only z'' is empty, then z' fully replaces z^n. However, if both z' and z'' are non-empty, the two of them are added to Z_{new} instead of z^n.

For the running example, the distance function $d_{z_{32}}(x) = 3n - 2x_0 - 3x_1$ had already been assigned to the zone z_{32} as depicted in Figure 3(d). Assume that the next stutter step to be considered is the t_3-stutter step from z_{32} to z_{11}. Since the distance function in z_{11} is $2n - 1 - x_1$, and the cost of firing t_3 in z_{32} is zero, the new zones z_{320} and z_{321} are separated by the line $3x_2 \leq n - x_1$. In the next and final iteration z_{21} is further divided into the zones z_{210} and z_{211} by possibilitySplit()*.*

5   Empirical Results

We present numerical results obtained using the algorithm to find d in Section 5.1, while in Section 5.2 we use d to apply simulation.

Case Description. We use two case studies. The first is the running example from Section 3.2, where the system is failed if $x_2 > n$, $n \in \mathbb{N}$. The second is a more realistic multicomponent system with interdependent component types, taken from [22]. For the latter we have six component types, with n_i components of type i and $(n_1, \ldots, n_6) = (n+2, n+1, n+3, n, n+4, n+2)$. In the benchmark setting, $n = 3$. If k components of type i have failed, the rate at which the next component of type i fails is $(n_i - k)\lambda_i\epsilon$, where $(\lambda_1, \ldots, \lambda_6) = (2.5, 1, 5, 3, 1, 5)$. There is a single repairman who repairs components following a preemptive priority repair strategy, where components of type i have priority over components of type j if $i < j$. The repair rate for type i is always μ_i, $(\mu_1, \ldots, \mu_6) = (1, 1.5, 1, 2, 1, 1.5)$. The system is said to have failed when all components of any type are down. We estimate the probability that, after the first component failure (drawn randomly), the system fails before all components are repaired.

5.1   Results of the Distance Finding Algorithm

A summary of the results of our algorithm is displayed in Table 1. The number of initial constraints is the main factor that determines the runtime of the algorithm. For the initial zones, we distinguish between the $(a \cup b)$- and $\neg(a \cup b)$-zones because only the latter have an impact on the runtime of the rest of the algorithm. A few things to mention: the number of zones may depend on n because for small n some zones will be empty, which are discarded. Also, the final number of zones may depend on the way stutter steps are chosen from S in the main loop, because if a zone is split by a stutter step that later turns out to be insignificant,

Table 1. Results of the numerical analysis for the running example

	Running Example		Multicomponent System	
n	3	10	3	5
# initial constraints	5	5	18	18
# initial zones	15	18	38880	46656
# initial $\neg(a \cup b)$-zones	8	11	3071	4095
# final $\neg(a \cup b)$-zones	14	27	3557	5477
# iterations in main loop	57	114	26421	42189
# markings in $\mathcal{X}$	∞	∞	40320	241920
time to construct (sec)	1.77	1.78	41.38	194.79

these zones are not recombined by our implementation, so both the number of zones and the number of iterations are implementation-dependent. For the multicomponent system, for small n the number of zones is almost equal to the size of the state space. This is a (for this case study unnecessary) consequence of the margins defined in lines 2 and 3 of `initZoneGraph()`.

5.2   Simulation Results

The simulation results are summarised in Tables 2 and 3. In both tables, we display the results for three simulation methods: standard Monte Carlo (MC), importance sampling (IS) using Balanced Failure Biasing (BFB) and IS based on our distance finding algorithm (Zone-IS). Under BFB, the total probability of firing a failure transition is set to $\frac{1}{2}$, uniformly distributed over the individual failure transitions (and similarly for the repairs — for more information, see [25]). In our implementation, we only consider the ν-transition t_3 to be a repair transition. Next to the simulation results, we display numerical approximations obtained using the model checking tool PRISM [15].

For the efficiency of the methods we look at the *relative error* (r. error) of the estimates, defined as the ratio of the estimator's standard deviation to the estimate. A lower value generally means a better estimate; however, if a change of measure is poorly suited for the system, IS may suffer from underestimation [7]. An example of this are the results for BFB for $n = 10$ and $\epsilon = 0.01$ in Table 2. For the sake of consistency with [22], we used 200 000 000 runs per MC-estimate and 10 000 000 runs per IS-estimate. In all cases Zone-IS outperforms BFB, except for $n = 5$ in Table 3. The reason is that BFB needs a clear distinction between failures and repairs to work well.

6   Discussion and Conclusions

6.1   Conclusions

We have presented a novel method to automatically construct a change of measure for speeding up the simulation of rare events in stochastic Petri nets. Our

Table 2. Results of the simulation analysis for the running example

		MC		BFB		Zone-IS		PRISM
n	ϵ	$\widehat{p}$	r. error	$\widehat{p}$	r. error	$\widehat{p}$	r. error	$\widehat{p}$
3	10^{-1}	$1.11 \cdot 10^{-4}$	0.007	$1.096 \cdot 10^{-4}$	0.007	$1.100 \cdot 10^{-4}$	$6.31 \cdot 10^{-4}$	$1.100 \cdot 10^{-4}$
	10^{-2}	$1.50 \cdot 10^{-8}$	0.577	$1.007 \cdot 10^{-8}$	0.011	$1.010 \cdot 10^{-8}$	$2.21 \cdot 10^{-4}$	$1.010 \cdot 10^{-8}$
	10^{-3}	—	—	$1.026 \cdot 10^{-12}$	0.011	$1.001 \cdot 10^{-12}$	$7.16 \cdot 10^{-5}$	$1.001 \cdot 10^{-12}$
	10^{-4}	—	—	$1.003 \cdot 10^{-16}$	0.011	$1.000 \cdot 10^{-16}$	$2.39 \cdot 10^{-5}$	$1.000 \cdot 10^{-16}$
5	10^{-1}	$1.00 \cdot 10^{-8}$	0.707	$1.140 \cdot 10^{-8}$	0.040	$1.098 \cdot 10^{-8}$	0.001	$1.100 \cdot 10^{-8}$
	10^{-2}	—	—	$9.843 \cdot 10^{-17}$	0.083	$1.010 \cdot 10^{-16}$	$5.09 \cdot 10^{-4}$	$1.010 \cdot 10^{-16}$
10	10^{-1}	—	—	$1.638 \cdot 10^{-18}$	0.970	$1.109 \cdot 10^{-18}$	0.006	$1.100 \cdot 10^{-18}$
	10^{-2}	—	—	$3.144 \cdot 10^{-42}$	0.865	$1.017 \cdot 10^{-36}$	0.003	$1.010 \cdot 10^{-36}$

Table 3. Results of the simulation analysis for the multicomponent system

		MC		BFB		Zone-IS	
n	ϵ	$\widehat{p}$	r. error	$\widehat{p}$	r. error	$\widehat{p}$	r. error
3	10^{-3}	$7.25 \cdot 10^{-7}$	0.083	$7.535 \cdot 10^{-7}$	0.019	$7.283 \cdot 10^{-7}$	0.007
	10^{-4}	$1.0 \cdot 10^{-8}$	0.707	$4.815 \cdot 10^{-9}$	0.027	$4.861 \cdot 10^{-8}$	0.002
5	10^{-3}	—	—	$1.155 \cdot 10^{-10}$	0.123	$4.368 \cdot 10^{-11}$	0.288
	10^{-4}	—	—	$1.901 \cdot 10^{-15}$	0.288	$1.381 \cdot 10^{-15}$	0.351

approach uniquely combines two characteristics: it uses a *high-level description of the model* with much flexibility and expressivity (a Petri net) and it works *without generating the entire state-space*.

The heart of our method is an algorithm which automatically partitions the state-space into a collection of zones. Each zone comprises states in which the same so-called change of measure is needed in the rare-event simulation scheme. The zones are demarcated by a set of affine inequalities, thus avoiding enumeration of all states. The number of zones in typical models does not need to increase as the model's size increases.

We have demonstrated that our algorithm works well in two examples. More experimentation will be needed to fully understand its possibilities and limitations and to optimise the implementation, and some extensions of the algorithm may be needed to handle certain classes of models (see below).

6.2   Discussion

In order to mathematically prove that the method always performs well, it remains to deal with three issues. The first is the correctness of the algorithm; i.e., whether the returned distance function really satisfies the definition in (4). The second is *termination* of the algorithm within finite time. The third is *efficiency* of the resulting importance sampling estimator. The first issue can be dealt with using a suitable invariant statement. For the latter two, we give a short discussion.

Termination. If the state space is infinite, it is possible that the (current) algorithm will not terminate. For example, if transition t_1 takes the system closer to the goal states and enables a transition t_2 with a very high firing rate, but firing the t_2 disables itself and does not negate the firing of t_1, then a shortest path might alternate between firing t_1 and t_2. This may result in the algorithm constructing an infinite number of zones. A possible solution is to broaden the concept of a stutter step. If a shortest path alternates between a tuple of transitions, the repeated firing of this tuple could be seen as a stutter step in itself, and the sum of the incidence vectors of the individual transitions as the net effect on the marking. Under such a restriction, the space of zones could well be bounded; this is part of ongoing research.

Importance Sampling Efficiency. The importance sampling measure as defined in (5) is inspired by the change of measure proposed in [17], where also the notion of *bounded relative error* comes up. This notion says that as ϵ approaches 0, the ratio of the standard deviation of the estimator to the standard mean remains bounded. This is desirable: since the accuracy of a simulation result is directly linked to this relative error, this means that the time to reach some level of accuracy never crosses a certain threshold value as ϵ becomes smaller. This behaviour is observed in Table 2 of Section 5, so we believe that our method will have bounded relative error, after a slight refinement.

The authors of [17] show that bounded relative error is guaranteed in their setting under the assumption that the state space is finite and that no *high-probability cycles* exist. Essentially, these assumptions imply that the number of paths ω with $\mathbb{P}(\omega) = \Theta(\epsilon^{d(\boldsymbol{x})})$ is finite. If this does not hold, it may be that $\mathbb{P}(\Psi_{\boldsymbol{x}}) \neq \Theta(\epsilon^{d(\boldsymbol{x})})$. A possible remedy would then be to perform a loop-detection algorithm on the initial graph returned by Algorithm 2 in order to detect the high-probability cycles, and remove them. This is also part of ongoing research.

References

1. Ajmone Marsan, M., Balbo, G., Donatelli, S., Franceschinis, G., Conte, G.: Modelling with generalized stochastic Petri nets. John Wiley & Sons, Inc. (1994)
2. Baier, C., D'Argenio, P., Groesser, M.: Partial order reduction for probabilistic branching time. Electronic Notes in Theoretical Computer Science (2006)
3. Baier, C., Katoen, J.P.: Principles of model checking. MIT Press (2008)
4. Blum, A.M., Goyal, A., Heidelberger, P., Lavenberg, S.S., Nakayama, M.K., Shahabuddin, P.: Modeling and analysis of system dependability using the system availability estimator. In: Twenty-Fourth International Symposium on Fault-Tolerant Computing, pp. 137–141. IEEE (1994)
5. Carrasco, J.A.: Failure distance based simulation of repairable fault-tolerant systems. In: Proceedings of the 5th International Conference on Modeling Techniques and Tools for Computer Performance Evaluation, pp. 351–365 (1992)
6. Clark, G., Courtney, T., Daly, D., Deavours, D., Derisavi, S., Doyle, J.M., Sanders, W.H., Webster, P.: The Möbius modeling tool. In: Proceedings of the 9th International Workshop on Petri Nets and Performance Models. IEEE (2001)

7. Devetsikiotis, M., Townsend, J.K.: An algorithmic approach to the optimization of importance sampling parameters in digital communication system simulation. IEEE Transactions on Communications 41(10), 1464–1473 (1993)
8. Glasserman, P., Heidelberger, P., Shahabuddin, P., Zajic, T.: Multilevel splitting for estimating rare event probabilities. Operations Research 47(4), 585–600 (1999)
9. Han, T., Katoen, J.-P.: Counterexamples in probabilistic model checking. In: Grumberg, O., Huth, M. (eds.) TACAS 2007. LNCS, vol. 4424, pp. 72–86. Springer, Heidelberg (2007)
10. Heidelberger, P.: Fast simulation of rare events in queueing and reliability models. In: Donatiello, L., Nelson, R. (eds.) SIGMETRICS 1993 and Performance 1993. LNCS, vol. 729, pp. 165–202. Springer, Heidelberg (1993)
11. Jegourel, C., Legay, A., Sedwards, S.: Cross-entropy optimisation of importance sampling parameters for statistical model checking. In: Madhusudan, P., Seshia, S.A. (eds.) CAV 2012. LNCS, vol. 7358, pp. 327–342. Springer, Heidelberg (2012)
12. Jegourel, C., Legay, A., Sedwards, S.: A platform for high performance statistical model checking – PLASMA. In: Flanagan, C., König, B. (eds.) TACAS 2012. LNCS, vol. 7214, pp. 498–503. Springer, Heidelberg (2012)
13. Júlvez, J.: Basic qualitative properties of Petri nets with multi-guarded transitions. In: American Control Conference, ACC 2009. IEEE (2009)
14. Kelling, C.: A framework for rare event simulation of stochastic Petri nets using "RESTART". In: Proceedings of the 28th Winter Simulation Conference, pp. 317–324. IEEE Computer Society (1996)
15. Kwiatkowska, M., Norman, G., Parker, D.: PRISM: Probabilistic symbolic model checker. In: Field, T., Harrison, P.G., Bradley, J., Harder, U. (eds.) TOOLS 2002. LNCS, vol. 2324, pp. 200–204. Springer, Heidelberg (2002)
16. Law, A., Kelton, W.: Simulation modeling and analysis. McGraw-Hill, New York (1991)
17. L'Ecuyer, P., Tuffin, B.: Approximating zero-variance importance sampling in a reliability setting. Annals of Operations Research 189(1), 277–297 (2011)
18. Miretskiy, D., Scheinhardt, W., Mandjes, M.: On efficiency of multilevel splitting. Communications in Statistics – Simulation and Computation 41(6), 890–904 (2012)
19. Murata, T.: Petri nets: Properties, analysis and applications. Proceedings of the IEEE 77(4), 541–580 (1989)
20. Nicola, V., Shahabuddin, P., Nakayama, M.: Techniques for fast simulation of models of highly dependable systems. IEEE Transactions on Reliability 50(3), 246–264 (2001)
21. Obal, W., Sanders, W.: An environment for importance sampling based on stochastic activity networks. In: Proceedings of the 13th Symposium on Reliable Distributed Systems, pp. 64–73. IEEE (1994)
22. Ridder, A.: Importance sampling simulations of Markovian reliability systems using cross-entropy. Annals of Operations Research 134(1), 119–136 (2005)
23. Rubinstein, R., Kroese, D.: The cross-entropy method: a unified approach to combinatorial optimization, Monte-Carlo simulation and machine learning. Springer (2004)
24. Sanders, W.H., Meyer, J.F.: Stochastic activity networks: Formal definitions and concepts. In: Brinksma, E., Hermanns, H., Katoen, J.-P. (eds.) FMPA 2000. LNCS, vol. 2090, pp. 315–343. Springer, Heidelberg (2001)
25. Shahabuddin, P.: Importance sampling for the simulation of highly reliable Markovian systems. Management Science 40(3), 333–352 (1994)

26. Tuffin, B., Trivedi, K.S.: Implementation of importance splitting techniques in stochastic Petri net package. In: Haverkort, B.R., Bohnenkamp, H.C., Smith, C.U. (eds.) TOOLS 2000. LNCS, vol. 1786, pp. 216–229. Springer, Heidelberg (2000)
27. Villén-Altamirano, M., Villén-Altamirano, J.: RESTART: A method for accelerating rare event simulations. In: Queueing, Performance and Control in ATM, pp. 71–76. Elsevier Science Publishers (1991)
28. Zimmermann, A., Freiheit, J., German, R., Hommel, G.: Petri net modelling and performability evaluation with TimeNET 3.0. In: Haverkort, B.R., Bohnenkamp, H.C., Smith, C.U. (eds.) TOOLS 2000. LNCS, vol. 1786, pp. 188–202. Springer, Heidelberg (2000)

Topology-Based Mobility Models
for Wireless Networks

Ansgar Fehnker[1], Peter Höfner[2,3], Maryam Kamali[4,5], and Vinay Mehta[1]

[1] University of the South Pacific, Fiji
[2] NICTA*, Australia
[3] University of New South Wales, Australia
[4] Turku Centre for Computer Science (TUCS), Finland
[5] Åbo Akademi University, Finland

Abstract. The performance and reliability of wireless network protocols heavily depend on the network and its environment. In wireless networks node mobility can affect the overall performance up to a point where, e.g. route discovery and route establishment fail. As a consequence any formal technique for performance analysis of wireless network protocols should take node mobility into account. In this paper we propose a topology-based mobility model, that abstracts from physical behaviour, and models mobility as probabilistic changes in the topology. We demonstrate how this model can be instantiated to cover the main aspects of the random walk and the random waypoint mobility model. The model is not a stand-alone model, but intended to be used in combination with protocol models. We illustrate this by two application examples: first we show a brief analysis of the Ad-hoc On demand Distance Vector (AODV) routing protocol, and second we combine the mobility model with the Lightweight Medium Access Control (LMAC).

1 Introduction

The performance and reliability of network protocols heavily depend on the network and its environment. In wireless networks node mobility can affect the overall performance up to a point where e.g. route discovery and route establishment fail. As a consequence any formal technique for analysis of wireless network protocols should take node mobility into account.

Traditional network simulators and test-bed approaches usually use a detailed description of the physical behaviour of a node: models include e.g. the location, the velocity and the direction of the mobile nodes. In particular changes in one of these variables are mimicked by the mobility model. It is common for network simulators to use *synthetic models* for protocol analysis [15]. In this class of models, a mobile node randomly chooses a direction and speed to travel from its current location to a new location. As soon as the node reaches the

---

* NICTA is funded by the Australian Government as represented by the Department of Broadband, Communications and the Digital Economy and the Australian Research Council through the ICT Centre of Excellence program.

K. Joshi et al. (Eds.): QEST 2013, LNCS 8054, pp. 389–404, 2013.

new location, it randomly chooses the next direction. Although these models abstract from certain characteristics such as acceleration, they still cover most of the physical attributes of the mobile node. Two well-known synthetic mobility models are the *random walk* (e.g. [1]) and the *random waypoint model* (e.g. [2]).

However, a physical mobility model is often incompatible with models of protocols, in particular protocols in the data link and network layers, due to limitations of the used modeling language and analysis tools. Even if it could be included, it would add a high complexity and make automatic analysis infeasible. From the point of view of the protocol it is often sufficient to model changes on the topology (connectivity matrix) rather than all physical behaviour.

In this paper we propose a topology-based mobility model that abstracts from physical behaviour, and models mobility as probabilistic changes in the topology. The main idea is to identify the position of a node with its current set of neighbours and determine changes in the connectivity matrix by adding or deleting nodes probabilistically to this set. The probabilities are distilled from the random walk or the random waypoint model. The resulting model is not meant to be a stand-alone model, but to be used in combination with protocol models. For this, we provide an Uppaal template for our model, which can easily be added to existing protocol models. The paper illustrates the flexibility of our model by two application examples: the first analyses quantitative aspects of the Ad hoc On-Demand Distance Vector (AODV) protocol [14], a widely used routing protocol, particularly tailored for wireless networks; the second example presents an analysis of the Lightweight Media Access Control (LMAC) [12], a protocol designed for sensor networks to schedule communication, and targeted for distributed self-configuration, collision avoidance and energy efficiency.

The rest of the paper is organised as follows: after a short overview of related work (Sect. 2), we develop the topology-based mobility model in Sect. 3. In Sect. 4 we present a simulator that is used to compute the transition probabilities for two common mobility models. In Sect. 5, we combine the distilled probabilities with our topology-based model to create an Uppaal model. Before concluding in Sect. 7, we illustrate how the model can be used in conjunction with protocol models. More precisely we present a short analysis of AODV and LMAC.

2   Related Work

Mobility models are part of most network simulators such as ns-2. In contrast to this, formal models used for verification or performance analysis usually assume a static topology, or consider a few scenarios with changing topology only. For the purpose of this section, we distinguish two research areas: mobility models for network simulators and models for formal verification methods.

Mobility models for network simulators either replay traces obtained from real world, or they use synthetic models, which abstract from some details and generate mobility scenarios. There are roughly two dozen different synthetic models (see [15,4] for an overview), starting from well-known models such as the *random walk model* (e.g. [1]) and the *random way point model* (e.g. [2]), via *(partially) deterministic models* and *Manhattan models* to *Gauss-Markov* and *gravity mobility models*. All these models are based on the physical behaviour of mobile nodes, i.e. each node has a physical location (in 2D or 3D[1]), a current speed and a direction it is heading to. As these models cover most of the physical behaviour, they are most often very complex (e.g. [13,10]) and include for example mathematics for Brownian motion. Due to this complexity these models cannot be incorporated directly into formal models for model-checking. This paper describes how two of these models, the random waypoint, and the random walk model, can be used to distill transition probabilities for a mobility model, which can easily be combined with formal protocol models.

Including mobility into a model for formal verification is not as common as it is for network simulators. If they are included, then typically in the protocol specification and therefore can rarely be reused for the analysis of different protocols. Moreover, formal verification often abstracts entirely from the underlying mobility model and allows arbitrary topology changes [9,5,8]. Other approaches allow only random, but very limited changes in the topology, often in the form of a scenario that involves deletion or creation of links [6,18,17]. Song and Godskesen propose in [16] a framework for modelling mobility; it models connectivity by distributions and propose a probabilistic mobility function to model mobility, without any specifics. This paper takes a similar approach, but adjacency matrices to model connectivity, and works out and analyses the transition probabilities obtained for two mobility models.

Our contribution is the following: we take the idea that the position of a mobile node can be characterised by a set of neighbours, which determines the topology, and we then define mobility as transitions between these sets. We then analyse the geometry of mobile nodes in a grid and determine which parameters actually influence the transition probabilities. In fact we found that some parameters, such as the step size of the random walk model have no influence on the transition probabilities. Based on this observation we build a topology-based mobility model which can easily be combined with protocol models.

3   Topology-Based Mobility Model

Our model takes up the position of the protocol: for a protocol it only matters whether data packets can be sent to a node, i.e. whether the node is within transmission range. The speed, the direction and other physical attributes are unimportant and irrelevant for the protocol. Hence the topology-based mobility model we introduce abstracts from all physical description of a node, and also largely abstracts from time. It models the node as a set of one-hop neighbours, i.e. nodes that are within transmission range of the node. Movement is modelled as a transition from one set of neighbours to another.

We assume that the node to be modelled moves within a quadratic $N \times N$-grid of stationary nodes. For simplicity we assume that nodes in the grid have a

---

[1] 3D is required when nodes model aerospace vehicles, such as UAVs.

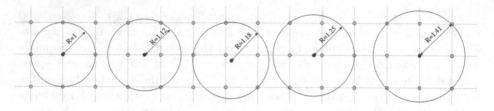

Fig. 1. Transmission ranges 1, $\frac{\sqrt{5}}{2} \approx 1.12$, $\frac{5}{6}\sqrt{2} \approx 1.18$, 1.25 and $\sqrt{2} \approx 1.41$

distance of 1, and that both the stationary and the mobile node have the same transmission range R. Obviously, the model depends on the grid size and the transmission range. We further assume that the transition range R is larger than 1 and strictly smaller than $\sqrt{2}$. If it were smaller than 1 nodes in the grid would be outside of the range of all neighbours, if it were larger than $\sqrt{2}$ nodes could communicate diagonally in the grid.

The network topology of all nodes, including the mobile node, can be represented by an adjacency or *connectivity matrix* A with

$$A_{i,j} = \begin{cases} 1 & \text{if } D(i,j) \leq R \\ 0 & \text{otherwise ,} \end{cases}$$

where $D(i,j)$ is the distance between the nodes i and j using some kind of metric, such as the Euclidean distance. While the connectivity matrix has theoretically 2^{N^2} possible configurations, with N the number of nodes, a network with one mobile node will only reach a small fraction of those. First, the matrix is symmetric. Second, all nodes, except for one, are assumed static, and the connectivity $A_{i,j}$ between two static nodes i and j will be constant. Third, due to the geometry of the plane, even the mobile node can only have a limited number of configurations. For example, neither a completely connected node, nor a completely disconnected node is possible given the transmission range.

The possible topologies depend on the transmission range: the larger the range the larger the number of possible nodes that can be connected to the mobile node. Within the right-open interval $[1, \sqrt{2})$, the set of possible topologies changes at values $\frac{\sqrt{5}}{2}$, $\frac{5}{6}\sqrt{2}$ and 1.25. These values can be computed with basic trigonometry. Fig. 1 illustrates which topologies become possible at those transmission ranges.

By considering the transmission range of the stationary nodes, one can partition the plane into regions in which mobile nodes will have the same set of neighbours. The boundaries of these regions are defined by circles with radius R around the stationary nodes. Fig. 2 depicts three possible regions and a transmission range $R = 1.25$; stationary nodes that are connected to the mobile node (located somewhere in the coloured area) are highlighted. As convention we will number nodes from the top left corner, starting with node 0. This partitioning abstracts from the exact location of the mobile node. Mobility can now be expressed as a change from one region to the next. The topology-based model will capture the changing topology as a Markovian transition function, that assigns to a pair of topologies a transition probability.

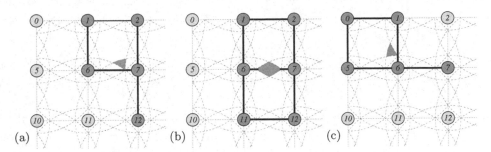

Fig. 2. Three regions and the corresponding set of neighbours for range $R = 1.25$

The number of possible transitions is also limited by the partition, as every region is bounded by a small number of arcs. If a mobile node transits an arc, a static node has to be added to or deleted from its set of neighbours. Consider, for example, the region that corresponds to set $\{1, 2, 6, 7, 12\}$ in Fig. 2(a). If the mobile node crosses the arc to the bottom left, node 11 will be added (Fig. 2(b)). The other two arcs of $\{1, 2, 6, 7, 12\}$ define the only two other transitions that are possible from this set.

We call a mobility model *locally defined* if congruent regions yield the same transition probabilities. Regions are congruent if they can be transformed into each other by rotation, reflection and translation. By extension we call transitions that correspond to congruent arcs in such regions also congruent. The movement of a node in a locally defined mobility model is independent from its exact position in the grid. The changes that can occur depend only on the topology of the current neighbours. For example, the congruent sets $\{1, 2, 6, 7, 12\}$ and $\{0, 1, 5, 6, 7\}$ in Fig. 2(a) and (c), would have the same transition probabilities.

In some cases this principle will uniquely determine the transition probability: the set $\{1, 2, 6, 7, 11, 12\}$ in Fig. 2(b) is bounded by 4 identical arcs. This means that all of them should correspond to a probability of $\frac{1}{4}$. For other regions the partition implies a relation/equation between some probabilities, but does not determine them completely. Considering only transitions in a single cell of the grid yields just a few and very symmetric transitions between possible topologies. Fig. 3 depicts the transitions as transitions between topologies.

One way to assign probabilities is to require that they are proportional to the length of the arc. Alternatively, probabilities may be estimated by simulations of a moving node in the plane. Note, that the resulting probabilistic transition system will be memoryless, i.e. the probability of the next transition depends only on the current region (set of neighbours). In the next section, we will see that the common random waypoint model is not locally defined, i.e. the local topology is not sufficient to determine the transition probabilities.

4   Simulations of Two Mobility Models

In the previous section we proposed a topology-based mobility model, based on transition probabilities; the exact values for the probabilities, however, were

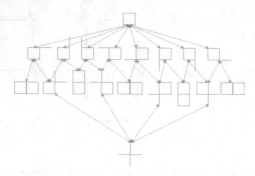

Fig. 3. Possible transitions within a single grid cell for $R = 1.25$

not specified. In this section we use a simulator to compute it for two common mobility models, a random walk model, and a random waypoint model.

4.1   Simulator

The simulator considers a single mobile node in an $N \times N$ grid of stationary nodes. As before, we assume a distance of 1 between the nodes on the grid. The initial position (x_0, y_0) of the mobile node is determined by a uniform distribution over $[0, N-1] \times [0, N-1]$, i.e. $x_0 \sim \mathcal{U}([0, N-1])$ and $y_0 \sim \mathcal{U}([0, N-1])$. Depending on the mobility model chosen, the simulator then selects a finite number of waypoints $(x_1, y_1), \ldots, (x_n, y_n)$, and moves along a straight line from waypoint (x_i, y_i) to the next (x_{i+1}, y_{i+1}).

The *random waypoint* model uses a uniform distribution over the grid to select the next way point, i.e. for all x_i, we have y_i, $x_i \sim \mathcal{U}([0, N-1])$ and $y_i \sim \mathcal{U}([0, N-1])$. The choice of the next waypoint is independent of the previous waypoint. This model is the most common model of mobility for network simulators, even if its merits have been debated [19]. A consequence of the waypoint selection is that the direction of movement is not uniformly distributed; nodes tend to move more towards the centre of the square interval.

As an alternative we are using a simple *random walk* model. Given way point (x_i, y_i) the next way point is computed by $(x_i, y_i) + (x_\Delta, y_\Delta)$ where both x_Δ and x_Δ are drawn from a normal distribution $\mathcal{N}(0, \sigma)$. This also means that the Euclidean distance between waypoints $\|(x_\Delta, y_\Delta)\|$ has an expected value of σ, which defines the average step size in the random walk model. By this definition, the model is unbounded, i.e. the next waypoint may lie outside the grid. If this happens the simulator computes the intersection of the line segment with the grid's boundary and reflects the waypoint at that boundary. In this model the mobile node moves from the first waypoint to the boundary, and from there to the reflected waypoint. For the purposes of this paper the intersections with the boundary do not count as waypoints.

Since the topology-based mobility model introduced in Sect. 3 abstracts from acceleration and speed, these aspects are not included in the simulation either. The simulator checks algebraically for every line segment from (x_i, y_i) to

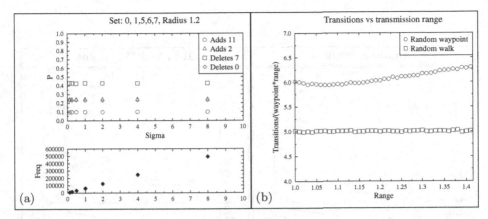

Fig. 4. (a) Transition probabilities and occurrences of set $\{0, 1, 5, 6, 7\}$. (b) The relation between number of transitions, the number of waypoints, and the transmission range.

(x_{i+1}, y_{i+1}) if it intersects with a node's transmission range R (given by a circle with radius R and the node in its centre). The simulator sorts all the events of nodes entering and leaving the transmission range and computes a sequence of sets of neighbours. This sequence is then used to count occurrences of transitions between these sets that are used to compute relative transition probabilities.

4.2 Simulation Results

The simulator is implemented in C++, and used to generate transition probabilities for the topology-based mobility model of Section 3. The simulator allows also a more detailed analysis of these two mobility models, in particular how the choice of parameters (grid size, transmission range, and standard deviation of the normal distribution σ) affects the transition probabilities. In this section we discuss some results for scenarios with a single mobile node on a 5×5 grid.

The simulation of the random walk model demonstrates a few important invariants. One observation is that the transition probabilities do not depend on the size of σ. This fact is illustrated by Fig. 4(a). The top part of this figure shows the probabilities that certain nodes are added or deleted from the set $\{0, 1, 5, 6, 7\}$. While σ ranges from $\frac{1}{8}$ to 8 the probabilities remain constant. The bottom part of the figure depicts the frequency with which the set occurs. Here there is a linear relation between σ and the total number of times that the set is visited. This is explained by the fact that σ is also the average step size, and doubling it means that twice as many transitions should be taken along the path.

Another linear relation exists between the total number of transitions along a path and the transmission range (cf. Fig. 4(b)). This relation is explained by the fact that the length of the boundary of each transmission area is linear to the range. For $\sigma = 1$, and $R = 1$, approximately 5 transitions will occur between any two waypoints. The ratio transition/range is constant for an increasing range. Note, that this number is independent of the grid size, and grows linearly with σ.

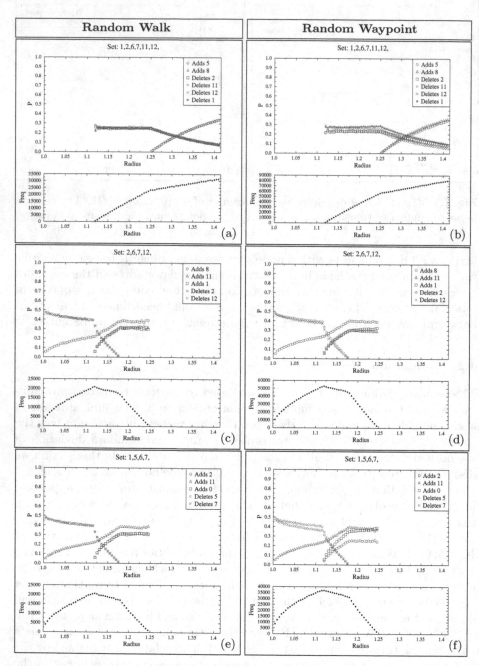

Fig. 5. Selected simulation results of the random walk and the random waypoint model

These invariants do not hold for the random waypoint model. The ratio of transitions to range is not constant, as illustrated in Fig. 4(b). This is because transitions are not evenly distributed but cluster towards the center of the grid. The ratio is also dependent on the size of the grid. In a larger grid the distance between waypoints will be larger, and more transitions occur per waypoint.

For the random walk model we found that the step size σ has no effect on the actual transition probabilities. The effect of the transmission range on the transition probabilities is less trivial. Fig. 5 shows a few illustrative examples. Similar result were obtained for all possible sets of neighbours.

Fig. 5(a) depicts the results for $\{1,2,6,7,11,12\}$, a set of six nodes that form a rectangle. This set cannot occur if the transmission ranges are smaller than $\frac{\sqrt{5}}{2}$ (cf. Sect. 3). For transmission ranges $R \in [\frac{\sqrt{5}}{2}, 1.25]$ the only possible transitions are to delete one of the four vertices located at the corners of the rectangle. In the random walk model the probability for these four transitions is $\frac{1}{4}$. Fig. 5(a) also illustrates that for transmission ranges $R \geq 1.25$, it is possible to add one additional node (either 5 or 8), reaching a set with 7 one-hop neighbours. As the range increases, the probability of this happening increases. At the same time the probability of deleting a vertex decreases.

Fig. 5(b) consider the same set of neighbours as Fig. 5(a), but under the random waypoint model. It demonstrates that this model is not locally defined, as congruent transitions, e.g. deleting vertices, do not have the same probability. The probability also depends of the distance of a node to the centre of the grid.

Fig. 5(c–f) show the transition probabilities for sets of neighbours that occur only if $R \in [1, 1.25]$: if $R < 1$, the transmission range is too small to cover the sets $\{2,6,7,12\}$ and $\{1,5,6,7\}$, resp.; if $R > 1.25$ the transmission range of the mobile will always contain more than four nodes. The observation is that as the transmission range increases, the probability of deleting a node decreases, while the probability of adding nodes increases. The sets $\{2,6,7,12\}$ and $\{1,5,6,7\}$ have the same basic "⊤" shape; one is congruent to the other. Hence, for the random walk model both sets have essentially the same transition probability; but also the frequency with which the sets occur is the same. This confirms that the position or orientation in the grid does not matter.

For the random waypoint model this no longer holds. The transition probabilities of similarly shaped neighbourhoods are not similar, but also determined by the position relative to the centre: the closer the set is to the centre the often it occurs in paths. Note, Fig. 1(d) and (f) use different scales for the frequency.

To conclude this section, we summarise our findings:

Random walk model:

– The transition probabilities are independent of σ and the grid size;
– The number of transitions per waypoint path grows linear with the range;
– The transition probabilities of congruent transitions are the same;
– The probabilities depend only locally on the set of nodes within range.

Random waypoint model: None of the above observations hold.

Table 1. Number of possible topologies, in relation to the range and the grid size[2]

Grid size	Transmission range [1, 1]	(1, 1.12)	(1.12, 1.18)	(1.18, 1.25)	[1.25, 1.25]	(1.25, 1.41)
2×2	9	9	5	5	5	5
3×3	32	41	49	49	37	41
4×4	69	97	133	133	101	117
5×5	120	177	257	257	197	233
6×6	185	281	421	421	325	389
7×7	264	409	625	625	485	585
8×8	357	561	869	869	677	821
9×9	464	737	1153	1153	901	1097
10×10	585	937	1477	1477	1157	1413

5 Uppaal Model

This section describes an Uppaal model that implements the topology-based mobility model described in Sect. 3, and uses the transition probabilities obtained in Sect. 4. The model is not meant to be stand-alone, but meant to be used within other protocol models. It assumes that an adjacency matrix `bool topoloy[N][N]` is used. The constant `N` is the size of the grid plus the mobile node. Depending on whether the random walk or random waypoint model is used, the model includes parameters for grid size and transmission range.

The template provides a list of all possible sets of neighbours. Table 1 shows the numbers of possible sets depending on the size of the grid and the transmission range. The results show that even for relatively large grids the number of possible sets of neighbors of the mobile node is limited. They will increase the potential state space only by three order of magnitude. The reachable space may increase by more when a template for mobility is added, because the protocol might reach more states than it did for static topologies.

The Uppaal template of Fig. 6 implements a lookup table of transition probabilities. After initialisation the template loops through a transition that changes the topology probabilistically. It contains a clock `t`, a guard `t>=minframe` and an invariant `t<=maxframe` to ensure that the change happens once in the interval `[minframe,maxframe]`. The values of `minframe` and `maxframe` determine the frequency of topology changes, and hence simulate the speed of a node.

The lookup is implemented by functions `updatemapindex`, `changeprob` and `changenode`. After every topology change, the function `updatemapindex` maintains the index (`mapindex`); this index into the list of possible sets is used to look up transition probabilities for a smaller set of representative sets of neighbours. Every set of neighbours is congruent to one of these representative sets. This information is used by `changeprob` to look up for a given node i the probability that it will be added or deleted from the current set of neighbours. Function `changenode` implements that change.

[2] Results for the point intervals containing $\frac{\sqrt{5}}{2}$ and $\frac{5}{6}\sqrt{2}$ are omitted.

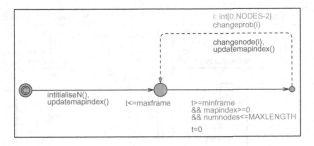

Fig. 6. Uppaal template for the mobility model

6 Application Examples

In this section we illustrate how the topology-based model can be used in combination with protocol models: first we briefly present an analysis of the Ad-hoc On demand Distance Vector (AODV) routing protocol, and second we combine the mobility model with the Lightweight Medium Control (LMAC). A detailed study of these protocols is out of the scope of the paper; we only show the applicability and power of the introduced mobility model.

Since we are interested in quantitative properties of the protocols, we are not using "classical" Uppaal, but SMC-Uppaal, the statistical extension of Uppaal [3]. *Statistical Model Checking (SMC)* [20] combines ideas of model checking and simulation with the aim of supporting quantitative analysis as well as addressing the size barrier that currently prevents useful analysis of large models. SMC trades certainty for approximation, using Monte Carlo style sampling, and hypothesis testing to interpret the results. Parameters setting thresholds on the probability of false negatives and on probabilistic uncertainty can be used to specify the statistical confidence on the result. For this paper, we choose a confidence level of 95%.

6.1 The Ad-Hoc on Demand Distance Vector (AODV) Protocol

AODV is a reactive routing protocol, which means that routes are only established on demand. If a node S needs to send a data packet to node D, but currently does not know a route, it buffers the packet and initiates a route discovery process by broadcasting a route request message in the network. An intermediate node A that receives this message stores a route to S, and re-broadcasts the request. This is repeated until the message reaches D, or alternatively a node with a route to D. In both cases, the node replies to the route request by unicasting a route reply back to the source S, via the previously established route.

An Uppaal model of AODV is proposed in [6]. The analysis performed on this model was done for static topologies and for topologies with very few changes. This limits the scope of the performance analysis. Here, the mobility automaton is added to the model of AODV. Since the mobility automaton is an almost independent component, it can be easily integrated into any Uppaal model that model topologies by adjacency matrices.

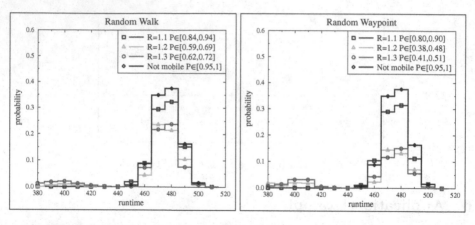

Fig. 7. AODV: probability of packet delivery within a certain time

Our experiments consider scenarios with a single mobile node moving within a 4×4 grid. A data packet destined for a randomly chosen stationary node is injected at a different stationary node. During route discovery the mobile node will receive and forward route requests and replies, as any other node will do.

The experiment determines the probability that the originator of the route request learns a route to the destination within 2000 time units. This time bound is chosen as a conservative upper bound to ensure that the analyser explores paths to a depth where the protocol is guaranteed to have terminated. In (SMC-) Uppaal syntax this property can be expressed as

$$\texttt{Pr[<=2000](<> node(OIP).rt[DIP].nhop!=0)} . \tag{1}$$

The variable `node(OIP).rt` denotes the routing table of the originator `OIP`, and the field `node(OIP).rt[DIP].nhop` represents the next hop on the stored route to the destination `DIP`. In case it is not 0, a route to `DIP` was successfully established. The property was analysed for the random walk and the random waypoint model with three different transmission ranges R: 1.1, 1.2, and 1.3. SMC-Uppaal returns a probability interval for the property (1), as well as a histogram of the probabilities of the runtime needed until the property is satisfied.

The results are presented in Fig. 7. The legend contains, besides the name of the model, the probability interval. For example, the random walk model with $R = 1.1$ satisfies property (1) with a probability $P \in [0.84, 0.94]$. In contrast to that the probability of route establishment in a scenario without a mobile node is $[0.95, 1]$, which indicates that the property is always satisfied. The probability intervals show that all scenarios with a mobile node have a lower probability for route discovery, some dramatically so. The random waypoint model with $R = 1.2$ has a probability interval of $[0.38, 0.48]$, which means that more than half of all route discovery processes fail. It is also notable that the random walk models have better results than the corresponding random waypoint models. Finally, the mobility models with $R = 1.1$ have a significantly higher probability to succeed than the other four models with $R = 1.2$ and $R = 1.3$.

The histograms show another interesting finding. The time it takes for a route reply to be delivered, if it is delivered, can be shorter for the models with the mobile node. Apparently, the mobile node can function as a messenger between originator and destination; not just by forwarding messages, but also by physically creating shortcuts.

6.2   The Lightweight Medium Access Control (LMAC) Protocol

LMAC [11] is a lightweight time division medium access protocol designed for sensor networks to schedule communication, and targeted for distributed self-configuration, collision avoidance and energy efficiency. It assumes that time is divided into frames with a fixed number of time slots. The purpose of LMAC is to assign to every node a time slot different from its one- and two-hop neighbours. If it fails to do so, collisions may occur, i.e. a node receives messages from two neighbours at the same time. However, LMAC contains a mechanism to detect collisions and report them to the nodes involved, such that they choose (probabilistically) a new time slot.

A (non-probabilistic) Uppaal model for LMAC was developed in [7], where it was also used to study static topologies. Based on this model a probabilistic model was developed [11]. This model was then used to study the performance of LMAC for heuristically generated topologies with 10 nodes [3]. The model we use for this paper differs in one aspect from [3]: it uses a smaller frame, with only six time slots, rather than 20. The purpose of LMAC is to assign time slots such that collisions are avoided or resolved, even if the number of time slots is restricted. For a 3×3 grid, it is possible to find a suitable assignment with only five time slots; six time slots should therefore be sufficient to cover a network with 10 nodes (one mobile node), although it might be challenging.

We check the following two properties:

$$\texttt{Pr[<=2000](<> forall (i: int[0,9]) slot\_no[i]>=0)} \qquad (2)$$

$$\texttt{Pr[collisions<=2000](<> time>=2000) .} \qquad (3)$$

The first property holds if, at some time point (before time 2000), all nodes are able to select a time slot. While this does not guarantee the absence of collisions, it does guarantee that all nodes have been able to participate in the protocol. The second property checks whether it is possible to reach 2000 time units, with less than 2000 collisions. This property is true for all runs. It is used merely to obtain a histogram of the number of collisions.

The results are illustrated in the histogram of Fig. 8: for all models with a mobile node, the property (2) is satisfied (the probability interval is $[0.95, 1]$, by a confidence level of 95%). The detailed results show that all runs reach a state in which all nodes have chosen a time slot. For the model without a mobile node, the probability interval is $[0.80, 0.90]$. This means that in at least 10% of all cases LMAC is not able to assign a time slot to all nodes; the histogram shows runs with 80–90, 160–170, and 240–250 collisions. These are runs in which one, or more nodes are engaged in a perpetual collision. Interestingly, this type of perpetual collisions do not occur in models with a mobile node. The mobile

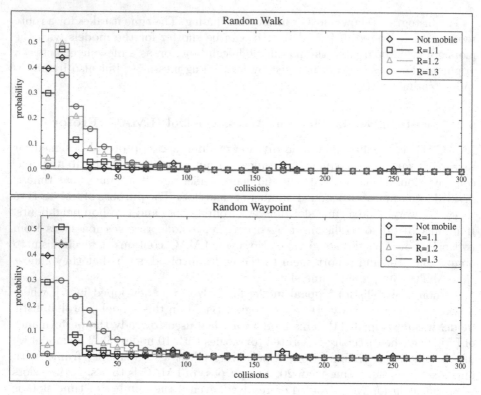

Fig. 8. LMAC: number of collisions within 2000 time units

node functions as an arbiter, which, as it moves around, detects and reports collisions that static nodes could not resolve.

The histograms reveal a few other interesting findings. In the model without mobility about 40% of the runs have no collisions. For both mobility models with transmission range $R = 1.1$ this drops to about 30%. For larger transmission ranges this drops even further to close to 0%, which means that almost all runs have at least some collisions. The differences between range $R = 1.1$, $R = 1.2$ and $R = 1.3$ is explained by the fact that the mobile node for $R = 1.1$ will have at most 5 neighbours, while for $R = 1.3$ it may be 7 neighbours. A larger neighbourhood makes choosing a good time slot more difficult. This is confirmed by another observation, namely that for $R = 1.1$ only a few runs have more than 20 collisions (approx. 12% of the runs, both random walk and random waypoint), while for a range of 1.2 and 1.3 it is in the range from 25% to 45%.

Both application examples show that introducing mobility can change the behaviour of network protocols significantly. As mentioned above, the purpose of these application examples was not to analyse these protocols in detail, but to show that the topology-based mobility models can be used to improve the scope of performance analyses of such protocols.

7   Conclusion

In this paper we have proposed an abstract, reusable, topology-based mobility model for wireless networks. The model abstracts from all physical aspects of a node as well as from time, and hence results in a simple probabilistic model. To choose a right level of abstraction, we have studied possible transitions and configurations of network topologies. To determine realistic transition probabilities regarding existing mobility models, we have performed simulation-based experiments. In particular, we have distilled probabilities for the random walk and the random waypoint model (using different transition ranges). We have then combined the topology-based model with the distilled probabilities and have created a (SMC-)Uppaal model[3]. The generated model is small and can easily be combined with other Uppaal models specifying arbitrary protocols. To illustrate this claim we have combined our model with a model of AODV and LMAC, resp. By this we were able to demonstrate that topology-based mobility models can be used to improve the scope of performance analysis of such protocols.

There are several possible directions for future work. First, we hope that our model is combined with a variety of protocols. Anybody who has some experience with the model checker Uppaal should be able to integrate our model easily. Second, we want to extend our mobility model to more than one mobile node. Having many mobile nodes will most likely increase the state space significantly, but statistical model checking should overcome this drawback. Last, but not least, we plan to use the mobility model to perform a thorough and detailed analysis of AODV and LMAC. In this paper we have only scratched the surface of the analysis; we expect to find unexpected behaviour in both protocols.

References

1. Basu, P., Redi, J., Shurbanov, V.: Coordinated flocking of UAVs for improved connectivity of mobile ground nodes. In: MILCOM 2004, pp. 1628–1634. IEEE (2004)
2. Bettstetter, C., Hartenstein, H., Pérez-Costa, X.: Stochastic properties of the random waypoint mobility model. Wireless Networks 10(5), 555–567 (2004)
3. Bulychev, P., David, A., Larsen, K., Mikučionis, M., Poulsen, D.B., Legay, A., Wang, Z.: UPPAAL-SMC: Statistical model checking for priced timed automata. In: Wiklicky, H., Massink, M. (eds.) Quantitative Aspects of Programming Languages and Systems. EPTCS, vol. 85, pp. 1–16. Open Publishing Association (2012)
4. Camp, T., Boleng, J., Davies, V.: A survey of mobility models for ad hoc network research. In: Wireless Communications & Mobile Computing (WCMC 2002), pp. 483–502 (2002)
5. Fehnker, A., van Glabbeek, R., Höfner, P., McIver, A., Portmann, M., Tan, W.L.: A process algebra for wireless mesh networks. In: Seidl, H. (ed.) ESOP 2012. LNCS, vol. 7211, pp. 295–315. Springer, Heidelberg (2012)
6. Fehnker, A., van Glabbeek, R., Höfner, P., McIver, A., Portmann, M., Tan, W.L.: Automated analysis of AODV using UPPAAL. In: Flanagan, C., König, B. (eds.) TACAS 2012. LNCS, vol. 7214, pp. 173–187. Springer, Heidelberg (2012)

---

[3] The models are available at http://repository.usp.ac.fj/5880

7. Fehnker, A., van Hoesel, L., Mader, A.: Modelling and verification of the LMAC protocol for wireless sensor networks. In: Davies, J., Gibbons, J. (eds.) IFM 2007. LNCS, vol. 4591, pp. 253–272. Springer, Heidelberg (2007)

8. Ghassemi, F., Ahmadi, S., Fokkink, W., Movaghar, A.: Model checking mANETs with arbitrary mobility. In: Arbab, F., Sirjani, M. (eds.) FSEN 2013. LNCS, vol. 8161. Springer, Heidelberg (2013)

9. Godskesen, J.C.: A calculus for mobile ad hoc networks. In: Murphy, A.L., Vitek, J. (eds.) COORDINATION 2007. LNCS, vol. 4467, pp. 132–150. Springer, Heidelberg (2007)

10. Groenevelt, R., Altman, E., Nain, P.: Relaying in mobile ad hoc networks: the Brownian motion mobility model. Wireless Networks 12(5), 561–571 (2006)

11. van Hoesel, L., Havinga, P.: A lightweight medium access protocol (LMAC) for wireless sensor networks: Reducing preamble transmissions and transceiver state switches. In: Networked Sensing Systems, INSS 2004, pp. 205–208. Society of Instrument and Control Engineers (SICE) (2004)

12. van Hoesel, L.: Sensors on speaking terms: schedule-based medium access control protocols for wireless sensor networks. Ph.D. thesis, University of Twente (2007)

13. McGuire, M.: Stationary distributions of random walk mobility models for wireless ad hoc networks. In: Mobile Ad Hoc Networking and Computing (MobiHoc 2005), pp. 90–98. ACM (2005)

14. Perkins, C., Royer, E.: Ad-hoc On-Demand Distance Vector Routing. In: 2nd IEEE Workshop on Mobile Computing Systems and Applications, pp. 90–100 (1999)

15. Roy, R.R.: Handbook of Mobile Ad Hoc Networks for Mobility Models. Springer (2011)

16. Song, L., Godskesen, J.C.: Probabilistic mobility models for mobile and wireless networks. In: Calude, C.S., Sassone, V. (eds.) TCS 2010. IFIP AICT, vol. 323, pp. 86–100. Springer, Heidelberg (2010)

17. Tschirner, S., Xuedong, L., Yi, W.: Model-based validation of qos properties of biomedical sensor networks. In: Embedded Software (EMSOFT 2008), pp. 69–78. ACM (2008)

18. Wibling, O., Parrow, J., Pears, A.N.: Automatized verification of ad hoc routing protocols. In: de Frutos-Escrig, D., Núñez, M. (eds.) FORTE 2004. LNCS, vol. 3235, pp. 343–358. Springer, Heidelberg (2004)

19. Yoon, J., Liu, M., Noble, B.: Random waypoint considered harmful. In: Joint Conference of the IEEE Computer and Communications (INFOCOM 2003). IEEE (2003)

20. Younes, H.: Verification and Planning for Stochastic Processes with Asynchronous Events. Ph.D. thesis, Carnegie Mellon University (2004)

Author Index